D1231984

Quality Engineering Handbook

QUALITY AND RELIABILITY

A Series Edited by

EDWARD G. SCHILLING
Founding Editor
Center for Quality and Applied Statistics
Rochester Institute of Technology
Rochester, New York

1. Designing for Minimal Maintenance Expense: The Practical Application of Reliability and Maintainability, *Marvin A. Moss*
2. Quality Control for Profit: Second Edition, Revised and Expanded, *Ronald H. Lester, Norbert L. Enrick, and Harry E. Mottley, Jr.*
3. QCPAC: Statistical Quality Control on the IBM PC, *Steven M. Zimmerman and Leo M. Conrad*
4. Quality by Experimental Design, *Thomas B. Barker*
5. Applications of Quality Control in the Service Industry, *A. C. Rosander*
6. Integrated Product Testing and Evaluating: A Systems Approach to Improve Reliability and Quality, Revised Edition, *Harold L. Gilmore and Herbert C. Schwartz*
7. Quality Management Handbook, *edited by Loren Walsh, Ralph Wurster, and Raymond J. Kimber*
8. Statistical Process Control: A Guide for Implementation, *Roger W. Berger and Thomas Hart*
9. Quality Circles: Selected Readings, *edited by Roger W. Berger and David L. Shores*
10. Quality and Productivity for Bankers and Financial Managers, *William J. Latzko*
11. Poor-Quality Cost, *H. James Harrington*
12. Human Resources Management, *edited by Jill P. Kern, John J. Riley, and Louis N. Jones*
13. The Good and the Bad News About Quality, *Edward M. Schrock and Henry L. Lefevre*
14. Engineering Design for Producibility and Reliability, *John W. Priest*
15. Statistical Process Control in Automated Manufacturing, *J. Bert Keats and Norma Faris Hubele*
16. Automated Inspection and Quality Assurance, *Stanley L. Robinson and Richard K. Miller*
17. Defect Prevention: Use of Simple Statistical Tools, *Victor E. Kane*
18. Defect Prevention: Use of Simple Statistical Tools, Solutions Manual, *Victor E. Kane*
19. Purchasing and Quality, *Max McRobb*
20. Specification Writing and Management, *Max McRobb*
21. Quality Function Deployment: A Practitioner's Approach, *James L. Bossert*
22. The Quality Promise, *Lester Jay Wollschlaeger*
23. Statistical Process Control in Manufacturing, *edited by J. Bert Keats and Douglas C. Montgomery*
24. Total Manufacturing Assurance, *Douglas C. Brauer and John Cesarone*
25. Deming's 14 Points Applied to Services, *A. C. Rosander*
26. Evaluation and Control of Measurements, *John Mandel*
27. Achieving Excellence in Business: A Practical Guide to the Total Quality Transformation Process, *Kenneth E. Ebel*
28. Statistical Methods for the Process Industries, *William H. McNeese and Robert A. Klein*
29. Quality Engineering Handbook, *edited by Thomas Pyzdek and Roger W. Berger*
30. Managing for World-Class Quality: A Primer for Executives and Managers, *Edwin S. Shecter*
31. A Leader's Journey to Quality, *Dana M. Cound*
32. ISO 9000: Preparing for Registration, *James L. Lamprecht*
33. Statistical Problem Solving, *Wendell E. Carr*
34. Quality Control for Profit: Gaining the Competitive Edge. Third Edition, Revised and Expanded, *Ronald H. Lester, Norbert L. Enrick, and Harry E. Mottley, Jr.*
35. Probability and Its Applications for Engineers, *David H. Evans*
36. An Introduction to Quality Control for the Apparel Industry, *Pradip V. Mehta*
37. Total Engineering Quality Management, *Ronald J. Cottman*
38. Ensuring Software Reliability, *Ann Marie Neufelder*
39. Guidelines for Laboratory Quality Auditing, *Donald C. Singer and Ronald P. Upton*
40. Implementing the ISO 9000 Series, *James L. Lamprecht*

41. Reliability Improvement with Design of Experiments, *Lloyd W. Condra*
42. The Next Phase of Total Quality Management: TQM II and the Focus on Profitability, *Robert E. Stein*
43. Quality by Experimental Design: Second Edition, Revised and Expanded, *Thomas B. Barker*
44. Quality Planning, Control, and Improvement in Research and Development, *edited by George W. Roberts*
45. Understanding ISO 9000 and Implementing the Basics to Quality, *D. H. Stamatis*
46. Applying TQM to Product Design and Development, *Marvin A. Moss*
47. Statistical Applications in Process Control, *edited by J. Bert Keats and Douglas C. Montgomery*
48. How to Achieve ISO 9000 Registration Economically and Efficiently, *Gurmeet Naroola and Robert Mac Connell*
49. QS-9000 Implementation and Registration, *Gurmeet Naroola*
50. The Theory of Constraints: Applications in Quality and Manufacturing: Second Edition, Revised and Expanded, *Robert E. Stein*
51. Guide to Preparing the Corporate Quality Manual, *Bernard Froman*
52. TQM Engineering Handbook, *D. H. Stamatis*
53. Quality Management Handbook: Second Edition, Revised and Expanded, *edited by Raymond J. Kimber, Robert W. Grenier, and John Jourdan Heldt*
54. Multivariate Quality Control: Theory and Applications, *Camil Fuchs and Ron S. Kenett*
55. Reliability Engineering and Risk Analysis: A Practical Guide, *Mohammad Modarres, Mark Kaminskiy, and Vasiliy Krivtsov*
56. Reliability Engineering Handbook, *Bryan Dodson and Dennis Nolan*
57. Quality Engineering Handbook, *Thomas Pyzdek*
58. Product Development and Design for Manufacturing: A Collaborative Approach to Producibility and Reliability: Second Edition, Revised and Expanded, *John W. Priest and José M. Sánchez*
59. Reliability Improvement with Design of Experiments: Second Edition, Revised and Expanded, *Lloyd W. Condra*
60. Quality Engineering Handbook: Second Edition, Revised and Expanded, *Thomas Pyzdek, edited by Paul Keller*

ADDITIONAL VOLUMES IN PREPARATION

To my family: my wife Carol,
my daughters Amy and Angie and my son Andrew.
Writing a book is a lonely task—by being there for me
my family made it bearable.

Quality Engineering Handbook

Second Edition

Thomas Pyzdek

edited by

Paul A. Keller

A CRC title, part of the Taylor & Francis imprint, a member of the Taylor & Francis Group, the academic division of T&F Informa plc.

Published in 2003 by
CRC Press
Taylor & Francis Group
6000 Broken Sound Parkway NW, Suite 300
Boca Raton, FL 33487-2742

Printed in the United States of America on acid-free paper
10 9 8

International Standard Book Number-10: 0-8247-4614-7 (Hardcover)
International Standard Book Number-13: 978-0-8247-4614-8 (Hardcover)
The first edition was edited by Thomas Pyzdek and Roger W. Berger (Marcel Dekker, 1991)

Library of Congress Cataloging-in-Publication Data

Catalog record is available from the Library of Congress

Taylor & Francis Group is the Academic Division of Informa plc.

Visit the Taylor & Francis Web site at
http://www.taylorandfrancis.com

and the CRC Press Web site at
http://www.crcpress.com

♦ ♦ ♦

Preface and Acknowledgements

This work was written to fulfill my long-held desire to provide a single source that covers every topic in the body of knowledge for quality engineering. In teaching courses in the past I found that literally every text book had to be supplemented with materials from other sources (including some massive tomes!). Students were constantly being referred to technical journals, magazine articles and other books. My goal for this book is for it to serve as the single best resource for information about the field of quality engineering, serving as a teaching tool, a desk reference and as a base for certification training, for the student or the quality professional.

It was difficult deciding where to draw the line on including material in the book. My personal library contains literally hundreds of relevant books, not to mention the articles in technical journals and magazines. Virtually every major element in the quality engineering body of knowledge is the topic of entire college courses. It is even possible to obtain college degrees in some quality engineering subject areas, such as statistics. My challenge was to digest this material while providing enough detail to present a coherent picture of the whole. I believe that I have succeeded. Of course, you, the reader, must be the final judge. I welcome your suggestions.

This book is based on one of my previous books: *The Complete Guide to the CQE*. It has been revised and expanded to accommodate the expansion of what is expected of a quality professional. I would like to thank Bryan Dodson and Dennis Nolan for material excerpted from their book: *The Reliability Engineering Handbook*.

Contents

CHAPTER

I

Management and Leadership in Quality Engineering

I.A PROFESSIONAL CONDUCT AND ASQ CODE OF ETHICS

Quality engineers, like other professionals, are expected to conduct themselves in a manner appropriate with their professional standing. In this context, ethics are defined as the rules or standards governing the conduct of a person or the members of a profession. The basic principles of ethical behavior have been very nicely summarized in the Code of Ethics for Members of the American Society for Quality, which are shown in Figure I.1.

Table I.1 lists a number of conflict-of-interest situations likely to be encountered by quality engineers. The appropriate response to these and other situations can be determined by referring to the appropriate item in the ASQ code of ethics. Violations of the code of ethics by a quality professional may be brought before the ASQ Professional Ethics and Qualifications Committee. In extreme cases of ethics violations, ASQ membership privileges may be revoked.

The American Society for Quality Code of Ethics

*To uphold and advance the honor and dignity of the profession, and in keeping with high standards of ethical conduct **I acknowledge that I:***

Fundamental Principles

I. Will be honest and impartial and will serve with devotion my employer, my clients, and the public.
II. Will strive to increase the competence and prestige of the profession.
III. Will use my knowledge and skill for the advancement of human welfare, and in promoting the safety and reliability of products for public use.
IV. Will earnestly endeavor to aid the work of the Society.

Relations with the Public

1.1 Will do whatever I can to promote the reliability and safety of all products that come within my jurisdiction.
1.2 Will endeavor to extend public knowledge of the work of the Society and its members that relates to the public welfare.
1.3 Will be dignified and modest in explaining my work and merit.
1.4 Will preface any public statements that I may issue by clearly indicating on whose behalf they are made.

Relations with Employers and Clients

2.1 Will act in professional matters as a faithful agent or trustee for each employer or client.
2.2 Will inform each client or employer of any business connections, interests, or affiliations which might influence my judgment or impair the equitable character of my services.
2.3 Will indicate to my employer or client the adverse consequences to be expected if my professional judgment is overruled.
2.4 Will not disclose information concerning the business affairs or technical processes of any present or former employer or client without consent.
2.5 Will not accept compensation from more than one party for the same service without the consent of all parties. If employed, I will engage in supplementary employment of consulting practice only with the consent of my employer.

Relations with Peers

3.1 Will take care that credit for the work of others is given to those to whom it is due.
3.2 Will endeavor to aid the professional development and advancement of those in my employ or under my supervision.
3.3 Will not compete unfairly with others; will extend my friendship and confidence to all associates and those with whom I have business relations.

Figure I.1. ASQ code of ethics. From American Society for Quality.

Table I.1. Conflict of interest situations.

- Audits: failing to accurately report audit findings.
- Releasing non-conforming items to a customer without the customer's knowledge.
- Accepting non-conforming supplier materials without proper authorization.
- Ignoring or failing to report unsafe conditions, either in a product or a workplace.
- Plagiarism by yourself or another (e.g., a co-author).
- Revealing proprietary information.
- Failing to reveal a conflict of interest when knowledge of it would affect an important decision.

I.B MANAGEMENT SYSTEMS FOR IMPROVING QUALITY

STRATEGIC PLANNING

Management has no choice but to anticipate the future, to attempt to mold it, and to balance short range and long range goals. The future, however, depends entirely on what is done *now*. The decisions, actions, resource allocation, and work done *now* will create the future. The present and immediate short range require strategic decisions as much as the long range. The long range is based largely on short range plans and decisions. Conversely, unless the long range is considered when making short range plans and decisions, the result will be a lack of direction that will eventually ruin the organization.

The terms "short range" and "long range" are not determined by a span of time. A decision is not short range simply because it can be implemented quickly, nor is it a long range decision to "decide" that we will take some action a decade from now. The range of a decision is the period of time over which the decision will have an effect. If we decide that our business requires that we grow Giant Sequoias, then it is a long range decision indeed.

Despite the inevitability of the future, it cannot be predicted. How can one plan for the unpredictable? The short answer is that one cannot. However, long range planning can still provide valuable benefits. Long range planning's benefits include preventing managers from uncritically extending present trends into the future, from assuming that today's products, services, markets, and technologies will be the products, services, markets and technologies of tomorrow. Above all, long range planning will prevent managers from devoting their energies and resources to the defense of yesterday (Drucker, 1993).

WHAT STRATEGIC PLANNING IS NOT

There are many misunderstandings relating to strategic planning. To avoid confusion, these are listed here.

1. It is not a bundle of techniques. Strategic planning involves answering two simple questions: "What *is* our business?" and "What *should* it be?" The answers cannot be programmed into a computer. Although models and quantitative methods may be used, they do not constitute *planning*. At least one well-known text on planning defines strategic planning as "application of scientific methods to business decisions." This is not correct.
2. Strategic planning is not forecasting. As stated above, the future is unpredictable. Perhaps Drucker said it best: "Forecasting is not a respectable human activity and not worthwhile beyond the shortest of periods." Strategic planning's value is not in its (nonexistent) ability to see into the future; to the contrary, it is necessary precisely because we cannot forecast the future. Strategic planning is an entrepreneurial activity that deliberately seeks to upset the probabilities by innovations and changes to the way people work and live.

3. Strategic planning does not deal with future decisions. *It deals with the futurity of present decisions.* Decisions exist only in the present. Yet our decisions may commit us for a long time, perhaps even permanently.
4. Strategic planning is not an attempt to eliminate or minimize risk. It does, however, seek to assure that the risks taken are the right risks. The end result of successful strategic planning must be capacity to take a greater risk by choosing rationally among risk-taking courses of action.

WHAT STRATEGIC PLANNING IS

Strategic planning is the continuous process of making present entrepreneurial decisions systematically and with the greatest knowledge of their futurity; organizing systematically the efforts needed to carry out these decisions; and measuring the results of these decisions against the expectations through organized, systematic feedback (Drucker, 1993).

Planning starts with the objectives of the business. What do we have to do *now* to achieve our objectives *tomorrow*? Planning must go beyond simply coming up with new things the business can do in the future. The first step is to ask of each present activity, product, process, or market "If we weren't already doing this, would we start?" If the answer is "No," then the organization should stop doing it, ASAP.

After identifying what old baggage to dump, the next step in the planning process is to ask "What *new* and different things do we have to do, and when?" Both the what and the when are important. The essence of planning is to make present decisions with knowledge of their futurity. It is the futurity that determines the time span, not vice versa. Futurity answers the question "What do we have to do today if we want to be in some particular place in the future? What will not get done if we don't commit resources to it today?"

There is no easy way to determine how long "long" is. The question must be answered by each organization.

Strategic planning aims to make changes in the way people work today. The changes are designed to make things different tomorrow than they were yesterday. Plans *must* result in changes in work. If a manager admits that none of her best people are working on the plan today, then she has no plan. A plan shows how to assign scarce resources, and the most scarce resource is always good people. Of course, whenever our most precious resources are being used we must assure that they are used properly. This means accountability must be built into the plan. Deadlines are necessary, as is feedback on progress and measurement of the final result.

FORMAL APPROACHES TO STRATEGIC PLANNING

Unfortunately, most of what has passed as "Strategic Planning" fails to comply with one or more of the fundamental principles just presented. This has resulted in disappointing results and an increasing lack of respect for the practice of strategic planning. This section will present the most popular approaches to strategic planning. We will then discuss the problems with these approaches. Finally, we will present ideas for rescuing strategic planning; after all, as stated earlier, management has no choice but to plan for the future.

There are three main schools of thought on strategy formation:

The design school—Sometimes called the SWOT school, the basis of strategy formation is the comparison of internal Strengths and Weaknesses to external Opportunities and Threats. As shown in Figure I.2, strategy is created at the intersection of an external appraisal of the threats and opportunities facing an organization in its environment, considered in terms of key factors for success, and an internal appraisal of the strengths and weaknesses of the organization itself, distilled into a set of distinctive competencies. Outside opportunities are exploited by inside strengths, while threats are avoided and weaknesses circumvented. Taken into consideration, both in the creation of the strategies and their subsequent evaluation to choose the best, are the values of the leadership as well as the ethics of the society and other aspects of so-called social responsibility. And once a strategy has been chosen, it is implemented.

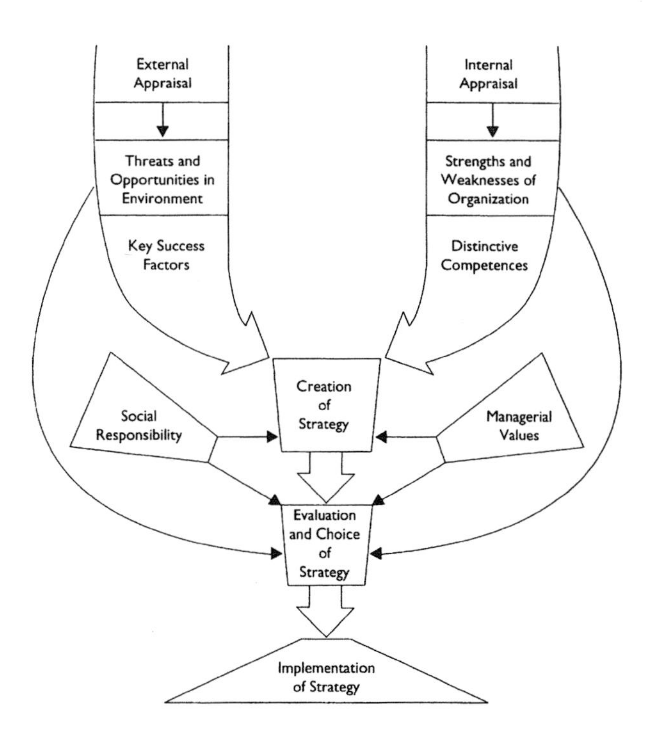

Figure I.2. Design school model of strategy formulation.
From Mintzberg, H. (1994). *The Rise and Fall of Strategic Planning.* Copyright© The Free Press.

A number of premises underlie this model:

1. Strategy formation should be a controlled, conscious process of thought.
2. Responsibility for the process must rest with the CEO, who is *the* strategist. The CEO is seen as the architect whose plans everyone else builds.
3. The model of strategy formation must be kept simple and informal.
4. Strategies should be unique: the best ones result from a process of creative design. Strategies are built upon distinct, core competencies.
5. Strategies must come out of the design process fully developed.
6. The strategies should be made explicit and, if possible, articulated, which means that they have to be kept simple.
7. Finally, once these unique, full-blown, explicit, and simple strategies are fully formulated, they must be implemented.

The planning school—Accepts the premises of the design school, except the assumptions that the process be simple and informal and that the chief executive be the key actor. The basic premises of the planning school are:

1. Strategy formation should be controlled and conscious as well as a formalized and elaborated process, decomposed into distinct steps, each delineated by checklists and supported by techniques.
2. Responsibility for the overall process rests with the chief executive in principle; responsibility for its execution rests with the staff planners in practice.
3. Strategies come out of this process fully developed, typically as generic positions, to be explicated so that they can then be implemented through detailed attention to objectives, budgets, programs, and operating plans of various kinds.

The positioning school—Focuses on the content of strategies (differentiation, diversification, etc.) more than on the process by which they are prescribed to be made (which are generally assumed to be those of the planning school). In other words, the positioning school simply extrapolates the messages of the planning school into the domain of actual strategy content. An example of this approach is the Boston Consulting Group's growth share matrix of cash cows, stars, dogs, and wildcats. This model categorizes companies according to the growth of the market and the company's market share, the company then determines its strategy based on the category it is in.

PLAN FORMATION

The term "strategic plan" was first used by H.I. Ansoff in a 1965 book entitled *Corporate Strategy*. In this seminal work the strategic plan consisted of a product-market strategy, and objectives which, combined with an administrative and design strategy, produced a strategic budget. These relationships are illustrated in Figure I.3.

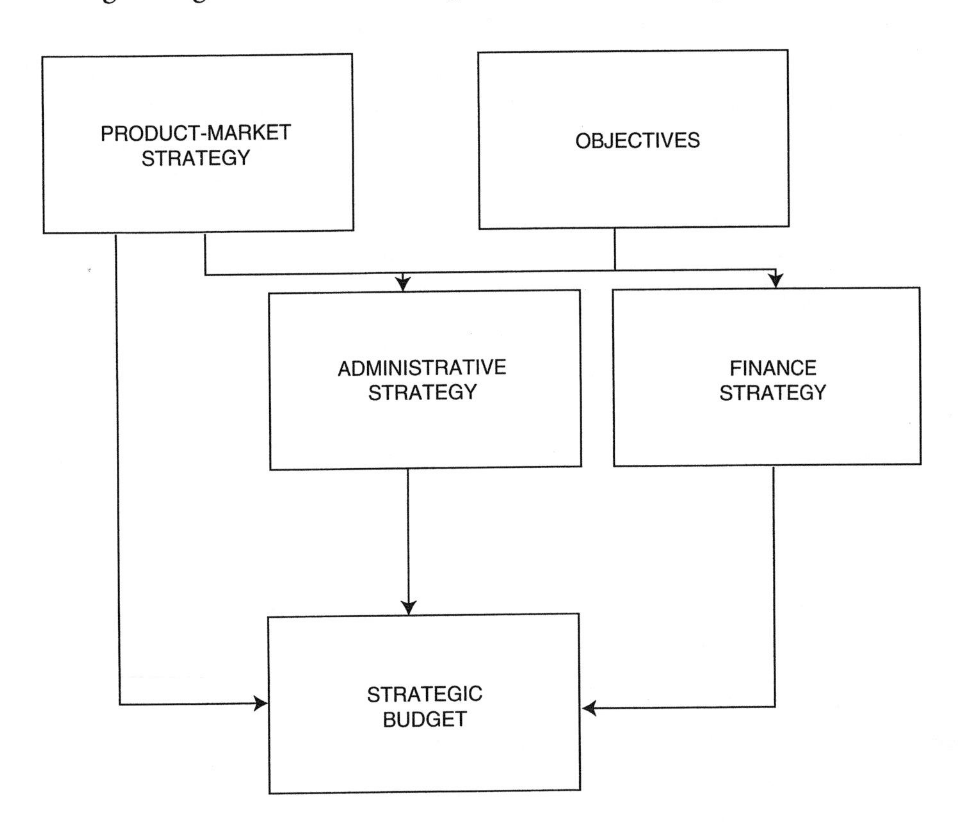

Figure I.3. Ansoff's strategic plan.

The strategic plan is the end result of an elaborate planning process.

RESULTS OF STRATEGIC PLANNING

Strategic planning made its appearance on the business scene in 1965. The results since then have been disappointing. An extensive meta-analysis of empirical studies covering nearly 2,500 organizations concluded that "the overall effect of planning on performance is very weak" (Boyd, 1991). Mintzberg summarizes over twenty years of strategic planning practice and research by stating "At the very least, we have found that planning is not 'the one best way,' that it certainly does not pay in general, and that at best, it may have some suitability in particular contexts, such as larger organizations, those in mass production, etc." (Mintzberg, 1994, p. 97).

In short, the evidence on strategic planning leads to the conclusion that it often doesn't work. Mintzberg postulates that this is so because strategic planning (of which capital budgeting is a part) is based on "The Grand Fallacy:" ***Because analysis is not synthesis, strategic planning is not strategy formation.*** Analysis may precede and support synthesis, by defining the parts that can be combined into wholes. Analysis may follow and elaborate synthesis, by decomposing and formalizing its consequences. But analysis cannot substitute for synthesis. No amount of elaboration will ever enable formal procedures to forecast discontinuities, to inform managers who are detached from their operations, to create novel strategies. Ultimately, the term "strategic planning" has proved to be an oxymoron.

PROPOSED SOLUTIONS TO THE STRATEGIC PLANNING PROBLEM

The first order of business in "fixing" strategic planning is to reintegrate analytic and intuitive thinking. Indeed, before integrating the two, we must first *allow* intuitive thinking, since the conventional approach dismisses it as "too biased." Research indicates that intuitive thinking leads to decisions that are less often precisely correct than analytical thinking, but that have fewer extreme errors. Thus, each approach has advantages and disadvantages. Since normal human beings are capable of performing both types of thinking, why not let them? Since planners tend to be more analytic, and managers more intuitive, the "planning dilemma" may well suggest its own solution. Planners have the time and technique to engage in the analysis. Managers have the intuition and detailed knowledge to suggest projects and to detect possible errors in the analysis. If the two work together, as equal partners, the two modes of thinking can be combined to produce better decisions than either could do without the other.

Mintzberg calls this combined approach to thinking "soft analysis." Soft analysis suggests an approach in which it is more important to pose the right question than to find the precise answer, to incorporate an appreciation for soft data alongside the necessary analysis of hard data. Judgment takes its place alongside formal procedures, and a mutual understanding is allowed to develop between staff planner and line manager. Soft analysis forgets about optimization, accepts the absence of sharply defined goals, downplays elegance in technique. Each issue is approached as a unique, creative challenge. The approach is systematic, but seldom rigorous, distinguishing between "analytic thinking" and "analytic technique."

STRATEGIC PROGRAMMING

One approach to strategic "planning" that incorporates soft analysis as well as hard is called *strategic programming*. Strategic programming provides a framework that delineates the roles of planning, plans, and planners. This framework constitutes an operational definition of planning.

The central premise of strategic programming is that effective organizations engage in formal planning not to create strategies but to program the strategies they already have, that is, to elaborate and operationalize their consequences formally. Thus, strategy is not the result of planning, but its starting point. Programming is a management activity which translates decisions into specific action patterns for implementation. The primary tasks involved in programming are:

1. Scheduling activities in support of decisions
2. Assignment and scheduling of resources in support of decisions (i.e., budgeting)
3. Establishing patterns of work flows in the firm
4. Establishing patterns of authority and responsibility
5. Establishing communication flow networks

Early authors on strategic planning viewed strategic programming as a "narrow context" in which to define planning. However, after decades of experience with planning, strategic programming is coming to be seen as planning's preeminent role. General Electric's chief planner stated in an interview:

> I make a distinction between planning and strategy—they're two different things. Strategy means thinking through a company's basis of competitive advantage . . . Planning, on the other hand, focuses on making the strategy work.

Strategic programming produces action plans. These plans serve two primary purposes: communication of the strategy, and control of behavior. Plans, as communication media, inform people of intended strategy and its consequences. As control devices, they specify what behaviors are expected of particular units and individuals in order to realize strategy, and then being available to feed back into the strategy making process comparisons of these expectations with actual performance.

Strategic programming involves a series of three steps:

1. **Codifying the strategy**—Planning refines the central strategic concepts of management into a relatively few principal thrusts around which the organization can pattern its resource commitments and measure its managers' performance.

Strategies are comprised of visions and decisions made at very high levels of abstraction. It is often necessary, but not always, to concretize these abstractions for communication to others. The planner must keep in mind that, at times, the images are more effective communications devices than precise and measurable plans. Stories still have their place.

2. Elaborating the strategy—The portions of the plan codified in step 1 can be decomposed into a hierarchy that begins with substrategies, proceeds through ad hoc programs of various kinds, and ends with specific action plans—specifying what people must do to realize the strategy as intended. This constitutes action planning: the organization pursues the connections down the strategy hierarchy and across to and down the program hierarchy. They follow a hierarchy of plans. This is the essence of policy deployment. While this sounds a great deal like traditional strategic planning, strategic programming involves two fundamental differences. First, strategy formation is expressly precluded from the model. Second, strategic programming is presented not as an imperative but as a process that is appropriate only under certain conditions (see "When to use strategic programming" below).
3. Converting the elaborated strategy—This step involves going from action planning to action. Objectives are restated, budgets redone, policies and procedures overhauled, and the daily routine is changed to reflect the hierarchy of plans. Unfortunately, very little research has been done to help the manager know how to do this effectively. The best guidance at this time seems to be an abundance of case studies of firms that have successfully done it. The past winners of the Malcolm Baldrige National Quality Award is one source of data.

WHEN TO USE STRATEGIC PROGRAMMING

Strategic programming is not a panacea. Strategic programming makes sense only when the world is expected to hold still or change predictably while intended strategies unfold, so that formulation can logically precede implementation. I.e., it makes sense only after strategic learning has identified predictable patterns. Naturally, strategic programming also assumes that the firm will benefit from the rigor involved and the result of the exercise, a clearly codified and elaborated strategic program. It is entirely possible that, for some firms in some circumstances, the strategic program's rigidity will do more harm than good. Mintzberg lists the following conditions that seem favorable to strategic programming:

- **Stability**—Strategic programming is done to implement strategy. If the conditions on which the strategy depend change radically, then the strategy is no longer viable and the implementation should stop.

- **Industry maturity**—This relates to stability: mature industries tend to be more stable. Research confirms that industries with high rates of growth tend to do less well using formalized planning. Markets in mature industries also seem to be controlled to some extent by the players in that industry.
- **Capital intensity**—Heavy capital investment provides an incentive to engage in planning. The greater the resource commitment to a single program, the more careful the control required. High levels of capital investment also tend to have longer payback periods, which also require greater predictability of returns.
- **Large size**—While planning isn't confined solely to large firms, bigness helps make planning more effective. Larger organizations also tend to have greater market control and higher levels of capital investment, which are additional favorable factors for strategic programming. In addition, formal planning involves considerable expenditure for planning staff, which smaller firms may not be able to justify.
- **Elaborated structure**—The ideal organization for planning is highly structured. Loose organizations do not mesh well with highly ordered plans. Imagine trying to implement a Soviet-style economic plan in the United States without the coercive power of government.
- **Tightly coupled organizations**—An organization is tightly coupled if the activities in one part of the organization are highly dependent on the activities in another part of the organization. A beauty shop is a loosely coupled organization: each beautician's work is largely independent of the work of other beauticians. An operating room is tightly coupled: every activity must be done in a certain order and at a certain time. The same applies to organizations as a whole. Tightly coupled organizations may require the structure imposed by strategic programming.
- **Simple operations**—Tight coupling and simplicity are related ideas. Prior to Frederick Taylor, machine shops were loosely coupled organizations that employed highly skilled machinists who worked more-or-less independently. Taylor decomposed the work into tiny elements that could be done using unskilled workers, but the result was a department of "machine operators" instead of machinists. The machine operator organization was tightly coupled; unless each worker did each operation in the correct order, production would stop. Work now required the coordination of centralized planners. What is generally not recognized is that planning's capacity to coordinate work requires *simplicity* on this scale. Planning's powers of synthesis are weak. In the administrative realm, planning must also follow a simple model. The plan should be capable of being decomposed into simple steps elaborated by a variety of simple checklists and supported by the simple procedures of scheduling and budgeting.
- **External control**—Another factor that encourages planning is an external influence with the power and intent to control an organization from the outside. Planning is a means of providing such control.

In summary, strategic programming works best in stable environments and in simple, tightly coupled organizations with highly elaborated structures. Strategic programming is facilitated when the firm is a large, capital intensive firm in a mature industry. Strategic programming is an effective tool for external control of such an organization.

THE STRATEGIC APPROACH TO QUALITY

Strategic quality requires that quality be defined as more than simply the absence of defects or the reduction of variation. These quality definitions are internally focused. Quality must be defined from the customer's perspective:

- It is not those who offer the product but those whom it serves—the customers, users, and those who influence or represent them—who have the final word on how well a product fulfills needs and expectations.
- Satisfaction is related to competitive offerings.
- Satisfaction, as related to competitive offerings, is formed over the product lifetime, not just at the time of purchase.
- A *composite* of attributes is needed to provide the most satisfaction to those whom the product serves.

In other words, quality is a comparison between your product and those offered by others, a comparison made by people outside of your firm. This means that quality cannot be defined by the internal focus that still dominates the quality profession. The methods described in section I.H of this chapter, formerly the sole domain of the marketing function, become part of the quality profession's toolkit. In addition, focus shifts from price to life-cycle costs, which include expenditures on service and maintenance after the purchase. Customer complaints, customer lifetime value, and a number of other issues suddenly become the concern of senior management and the quality professional.

This new, broader, strategic perspective to quality is embraced by top management because overwhelming evidence indicates, despite the great expense involved, it pays handsome dividends. Quality is clearly a competitive weapon. Once this is established the logical conclusion is "if firms with a clear quality advantage earn better market returns, then it makes sense to beat our competitors by offering higher levels of quality than they do." This is the premise that drives strategic quality management.

This is a radical departure from traditional quality assurance. Traditional quality assurance seeks to detect or prevent *defects*, i.e., departures from internally developed and imposed requirements. The strategic approach is a commitment to a process of continuous improvement relative to competitors or external norms. Such a dramatic change of focus requires the participation of the entire company, it can't simply be delegated to specialists in the quality department. Companywide commitment involves top management because organizational forms must change, e.g., cross-functional management of the quality function, horizontal structures, etc.

The new approach also places new demands on quality professionals. Technical expertise remains desirable, but in addition the quality professional must also understand the company's strategic goals. The activities in the quality department change from a focus on inspection, auditing and procedure writing to education and training, goal setting, and consulting support to internal departments. The role shifts from that of policeman to that of general management. The opportunity is enormous, as is the challenge.

To facilitate the new perspective, quality is now included as an explicit part of the strategy development process. Annual goals are set for quality improvement and made specific and actionable. Goals normally take into account the customer's perspective and are also matched against the expected performance of competitors. Both internal cost of quality measures and external customer-related quality measures are included.

Of course, the strategic approach doesn't ignore the previously learned body of knowledge of quality. SPC, design review, and all of the earlier innovations are incorporated into the approach. These techniques and practices help to improve product quality and reduce total cost. Also, some customer expectations can be translated into internal requirements, which provide operational definitions for employees. Strategic quality management is an extension of the traditional scope of quality control, not a replacement.

Garvin summarized the various quality control "eras" as shown in Table I.2.

Table I.2. The four major quality eras.

STAGE OF THE QUALITY MOVEMENT				
Identifying Characteristics	**Inspection**	**Statistical Quality Control**	**Quality Assurance**	**Strategic Quality Management**
Primary concern	detection	control	coordination	strategic impact
View of quality	a problem to be solved	a problem to be solved	a problem to be solved, but one that is attacked proactively	a competitive opportunity
Emphasis	product uniformity	product uniformity with reduced inspection	the entire production chain	the market and consumer needs

Continued on next page . . .

Table I.2.—*Continued . . .*

STAGE OF THE QUALITY MOVEMENT				
Identifying Characteristics	**Inspection**	**Statistical Quality Control**	**Quality Assurance**	**Strategic Quality Management**
Methods	gauging and measurement	statistical tools and techniques	programs and systems	strategic planning, goal-setting, and mobilizing the organization
Role of quality professionals	inspection, sorting, counting, and grading	troubleshooting and the application of statistical methods	quality measurement, quality planning, and program design	goal-setting, education and training, consultative work with other departments, and program design
Who has responsibility for quality	the inspection department	the manufacturing and engineering departments	all departments, although management is only peripherally involved in designing, planning, and executing quality policies	everyone in the organization, with top management exercising strong leadership
Orientation and approach	"inspects in" quality	"controls in" quality	"builds in" quality	"manages in" quality

DEPLOYMENT OF GOALS

Once strategies and policies have been formulated and goals have been set, the organization must develop a process to administer the policy effectively. Sheridan (1993) describes the following approach to this task:

President—appoints a senior management committee to oversee the policy process.

Senior management committee—develops long-term management policy;appoints and oversees the policy committee; approves annual policy. This committee has the ultimate responsibility to the customer.

Policy committee—conducts policy process. The policy process includes the following sequential steps:

1. Assess customer's needs and reasonable expectations using statistical quality control techniques to collect data.
2. Top executives use this along with data from various areas of expertise to develop the organization's strategy.
3. This is set as the policy in the form of a vision and long-term, mid-term, and short-term management policies.
4. The policy is disseminated throughout the organization along with a set of quantitative indicators and their respective targets.
5. A single executive is assigned to oversee the analysis of a single issue using statistical quality control techniques. Based on the analysis, countermeasures are to be put in place.
6. Effective countermeasures are to be standardized and replicated whenever appropriate. An assessment of the results is made by checking the indicators and targets.
7. Finally, the entire policy process is assessed and improved as needed. The process continues with the assessment of customer needs and reasonable expectations.

Cross-functional committees—assess how well the organization is performing in the areas of quality, delivery, cost, safety, and morale. These committees are executive level and cross-functional.

Coordinating executive—develops plans to attain the goals of their assigned short-term management policy. This high-ranking executive is appointed by the policy committee to oversee one short-term (one to two years) management policy.

Functional managers—conduct projects to develop countermeasures. They work with the coordinating executive through win-win negotiations.

Plans to meet goals—Goals that require a significant outlay of corporate resources should be implemented using formal project planning. This approach is described in the sections that follow below.

Organizational structure—Organizational structures may need to be modified to accomplish the strategic goals of the company. The use of teams for quality improvement is described in I.C and I.D.

Resources—To meet the stated goals, adequate resources must be made available in a timely fashion. Determination of resource requirements for quality activities is discussed in II.A. Determination of resource requirements for quality improvement projects is discussed later in this section.

Measurement feedback/Review of progress—Feedback is necessary to assure that the quality goals are being achieved according to plans. For improvement goals this is provided via the quality information system (see I.B) and project feedback loops (see later this section). For control goals this is provided through the use of SPC (see VI.G).

Rewards based on performance against goals—Many authors still recommend this traditional approach to compensation. However, there is a substantial body of research that indicates that performance-based reward and compensation systems do more harm than good. A better approach is to provide *recognition* based on performance, while treating compensation issues separately. The reader should review the material in I.D.

Training—Quality goals often require that employees acquire new knowledge, skills, and abilities. This is accomplished through training. Strategic plans are used to develop *strategic training plans* which assure that the requisite skills are possessed by employees (see I.E).

PROJECT MANAGEMENT

The dictionary defines the word *project* as follows:

1. A plan or proposal; a scheme. See synonyms at *plan.*
2. An undertaking requiring concerted effort.

Under the synonym *plan* we find:

1. A scheme, program, or method worked out beforehand for the accomplishment of an objective: a plan of attack.
2. A proposed or tentative project or course of action.
3. A systematic arrangement of important parts.

Although truly dramatic improvement in quality often requires transforming the management philosophy and organization culture, the fact is that, sooner or later, projects must be undertaken to make things happen. Projects are the means through which things are systematically changed, projects are the bridge between the planning and the doing.

Frank Gryna makes the following observations about projects (Juran and Gryna, 1988, pp. 22.18–22.19):

- An agreed-upon project is also a legitimate project. This legitimacy puts the project on the official priority list. It helps to secure the needed budgets, facilities, and personnel. It also helps those guiding the project to secure attendance at scheduled meetings, to acquire requested data, to secure permission to conduct experiments, etc.
- The project provides a forum of converting an atmosphere of defensiveness or blame into one of constructive action.

- Participation in a project increases the likelihood that the participant will act on the findings.
- All breakthrough is achieved project by project, and in no other way.

The last item represents both good news and bad news. The bad news is that few projects are truly successful; the good news is that companies can and do become proficient at implementing projects without the need for mystical powers. What is needed is effective project management.

USEFUL PROJECT MANAGEMENT TOOLS AND TECHNIQUES

Project management is a system for planning and implementing change that will produce the desired result most efficiently. There are a number of tools and techniques that have been found useful in project management. Brief descriptions of the major project management methods are provided here. Techniques specific to project management are covered in greater detail elsewhere in this chapter. Many of these tools are used in a wide variety of quality improvement and quality control situations in addition to project management; additional information on each of these more general techniques is found elsewhere in this book; consult the index for details.

Project plan—The project plan shows the "why" and the "how" of a project. A good project plan will include a statement of the goal, a cost/benefit analysis, a feasibility analysis, a listing of the major steps to be taken, a timetable for completion, and a description of the resources required (including human resources) to carry out the project. The plan will also identify objective measures of success that will be used to evaluate the effectiveness of the proposed changes, these are sometimes called the "deliverables" of the project.

Gantt chart—A Gantt chart shows the relationships among the project tasks, along with time constraints. See below for a discussion of Gantt charts.

Milestone charts—A Gantt chart modified to provide additional information on project status. See below for a discussion of Milestone charts.

Pareto analysis—Pareto analysis is a technique that helps one to rank opportunities to determine which of many potential projects should be pursued first. It can also be used sequentially to determine which step to take next. The Pareto principle has been described by Juran as separating the "vital few" from the "trivial many." It is the "why" and the "benefit" of the project plan. See V.C for additional discussion.

Budget—A budget is an itemized summary of estimated or intended expenditures for a given project along with proposals for financing them. Project budgets present management with a systematic plan for the expenditure of the organization's resources, such as money or time, during the course of the project. The resources spent include time of personnel, money, equipment utilization and so on. The budget is the "cost" portion of the project plan.

Process decision program chart (PDPC)—The PDPC technique is used to develop contingency plans. It is modeled after reliability engineering methods such as failure mode, effects, and criticality analysis (FMECA) and fault-tree analysis (FTA). The emphasis of PDPC is the impact of problems on project plans. PDPCs are accompanied by specific actions to be taken should the problems occur to mitigate the impact of the problems. PDPCs are useful in the planning of projects in developing a project plan with a minimum chance of encountering serious problems. Also see V.B.

Quality function deployment (QFD)—Traditionally, QFD is a system for design of a product or service based on customer demands, a system that moves methodically from customer requirements to requirements for the products or services. QFD provides the documentation for the decision-making process. QFD can also be used to show the "whats" and "hows" of a project. Used in this way QFD becomes a powerful project planning tool. Also see I.H.

Matrix chart—A matrix chart is a simplified application of QFD (or, perhaps, QFD is an elaborate application of matrix charts). This chart is constructed to systematically analyze the correlation's between two groups of ideas. When applied to project management the two ideas might be, for example 1) what is to be done? 2) who is to do it? Also see V.B.

Arrow diagrams—Arrow diagrams are simple network representations of project flows. They show which tasks must be completed in the project and the order in which the tasks must be completed. See V.B. Arrow diagrams are a simplification of PERT-type systems (see below).

PLANNING

There are several reasons why one should plan carefully before starting a project (Ruskin and Estes, 1995, p. 44):

1. The plan is a simulation of prospective project work, which allows flaws to be identified in time to be corrected.
2. The plan is a vehicle for discussing each person's role and responsibilities, thereby helping direct and control the work of the project.
3. The plan shows how the parts fit together, which is essential for coordinating related activities.

4. The plan is a point of reference for any changes of scope, thereby helping project managers deal with their customers.
5. The plan helps everyone know when the objectives have been reached and therefore when to stop.

The project plan shows the "why" and the "how" of a project. A good project plan will include the following elements:

- statement of the goal
- cost/benefit analysis
- feasibility analysis
- listing of the major steps to be taken
- timetable for completion
- description of the resources required (including human resources) to carry out the project

The plan will also identify objective measures of success that will be used to evaluate the effectiveness of the proposed changes; these are sometimes called the "deliverables" of the project.

PROJECT DECOMPOSITION

Most projects important enough to have a significant impact on quality are too large to tackle all at once. Instead, large projects must be broken down into smaller projects and, in turn, into specific work elements and tasks. The process of going from project objectives to tasks is called decomposition. Project decomposition begins with the preparation of a preliminary plan. A preliminary project plan will identify, in broad high-level terms, the objectives of the project and constraints in terms of time and resources. The work to be performed should be described and precedence relationships should be sketched out. Preliminary budgets and schedules will be developed. Finally, subplans will be developed for each subproject for the following:

- Control plans
  - Quality control plans
  - Cost control plans
  - Schedule control plans
- Staffing plans
- Material plans
- Reporting plans
- Other plans as deemed necessary

These subplans are developed in parallel for the various subprojects.

INTEGRATED QUALITY INITIATIVES

Also see later this section, cross-functional collaboration.

Projects should be selected consistent with the organization's overall strategy and mission. Because of this global perspective most projects involve the efforts of several different functional areas. Not only do individual quality projects tend to cut across organizational boundaries, different projects are often related to one another. To effectively manage this complexity it is necessary to integrate the planning and execution of projects organization-wide.

(For additional details on teams see I.C.)

Teams are chartered by senior leadership, generally the only group with the necessary authority to designate cross-functional responsibilities and allow access to interdepartmental resources. The team facilitator should ask senior leadership to put the problem statement in writing. The problem statement should be specific enough to help the team identify the scope of the project and the major stakeholders. Problems of gargantuan proportions should be subdivided into smaller projects.

There are six steps in the chartering process:

1. Obtaining a problem statement
2. Identifying the principal stakeholders
3. Creating a macro flowchart of the process
4. Selecting the team members
5. Training the team
6. Selecting the team leader

The project charter should include a statement indicating the mission of the project team.

If the team has not received its mission statement as part of its charter, it should prepare its own team mission statement. The mission statement should be linked to the larger organization's mission via the charter. Scholtes (1996) lists a number of situations which require that the team spend time developing its own mission statement:

- Teams with a mission statement from management which is too broad ("reduce customer complaints")
- Teams with an unclear or undefined mission ("get a team together to work in the shipping department")
- Teams who need to describe the nature and severity of a problem before they can decide how to proceed with their project ("are scratches really a big problem?")
- Teams whose sole mission is to identify and describe problems so that more focused projects can be designed; probably involving different teams
- Teams who have the leeway to select their own project

The mission statements of various project teams should clearly identify the boundaries between them. Where confusion exists, teams should meet to clarify responsibilities.

Several problems with mission statements appear repeatedly:

- Projects have little or no impact on the organization's success, even if successful, no one will really care.
- Missions overlap the missions of other teams. E.g., Team A's mission is to reduce solder rejects, Team B's mission is to reduce wave solder rejects, Team C's mission is to reduce circuit board assembly problems.
- Projects improve processes that are scheduled for extensive redesign or discontinuation. For example, working on improving work flow for a production process that is to be relocated to another factory.
- Studying a huge system ("patient admitting"), rather than a manageable process ("outpatient surgery preadmission").
- Studying symptoms ("touch-up of defective solder joints") rather than root causes ("wave solder defects")
- Project deliverables are undefined. E.g., "Study TQM" rather than "Reduce waiting time in Urgent Care."

SHORT- AND LONG-TERM QUALITY PLANS AND OBJECTIVES

See chapter III for an in-depth discussion of quality planning in general. The discussion here will focus on the role of quality plans in project management.

In general, long-term planning addresses issues of strategy and mission. Projects involve the actual implementation of strategic programs and should be considered short-term and tactical, even projects that take a relatively long time to complete.

Work breakdown structures

Ruskin and Estes (1995) define work breakdown structures (WBS) as a process for defining the final and intermediate products of a project and their relationships. Defining project tasks is typically complex and accomplished by a series of decompositions followed by a series of aggregations. For example, a software project to develop an SPC software application would disaggregate the customer requirements into very specific engineering requirements (e.g., the customer's requirement that the product create x-bar charts would be decomposed into engineering requirements such as subroutines for computing subgroup means and ranges, plotting data points, drawing lines, etc.). Aggregation would involve linking the various modules to produce an x-bar chart displayed on the screen.

The WBS can be represented in a tree-diagram, as shown in Figure I.4.

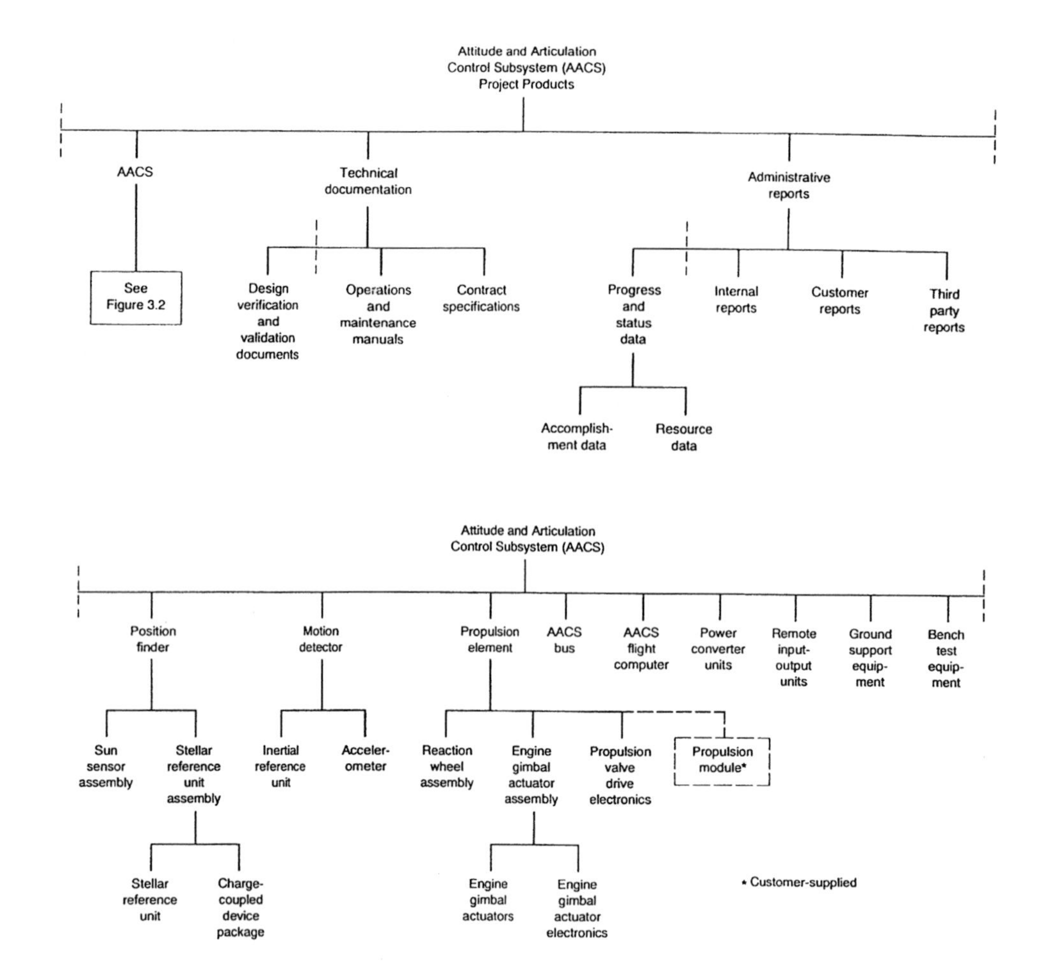

Figure I.4. WBS of a spacecraft system.

From Ruskin, A.M. and Estes, W.E. *What Every Engineer Should Know About Project Management,* Second Edition. Copyright © 1995 by Marcel Dekker, Inc. Reprinted with permission.

Preliminary requirements WBS—is a statement of the overall requirements for the project as expressed by the customer (e.g., the deliverables or "product"), and subsidiary requirements as expressed by management (e.g., billing, reports required).

Detailed plan WBS—breaks down the product into subproducts. Requirements are listed for each subproduct (e.g., tooling, staff). The subproducts are, in turn, broken down into their subproducts, etc., until a reasonable limit is reached. All work begins at the lowest level. Detailed plans for each subsystem include control plans for quality, cost and schedule; staffing plans, materials plans, reporting plans, contingency plans and work authorization plans. In addition, the overall detailed plan covers objectives, constraints, precedence relationships, timetables, budgets, and review and reporting criteria.

Typical subsystem WBS—are created. I.e., the process just described is performed for each subsystem. Subsytems are then built.

Integration WBS—detail how the various subsystems will be assembled into the product deliverables. This usually involves integrating into larger subsystems, then still larger subsystems, etc., to the highest level of integration.

Validation WBS—plans explain how the various systems integrations will be measured and tested to assure that the final requirements will be met.

FEEDBACK LOOPS

The project plan is itself an important feedback tool. It provides details on the tasks that are to be performed, when they are to be performed, and how much resource is to be consumed. The plan should also include explicit provisions for feedback. Typical forms of feedback are:

- **Status reports**—Formal, periodic written reports, often with a standardized format, telling what the project is based on, and where it is supposed to be relative to the plan. Where project performance does not match planned performance, the reports include additional information as to the cause of the problem and what is being done to bring the project into alignment with the plan. Remedial action may, at times, involve revising the plan. When the project is not meeting plan due to obstacles which the project team cannot overcome, the status report will request senior management intervention.
- **Management reviews**—These are meetings, scheduled in advance, where the project leader will have the opportunity to interact with key members of the management team. The chief responsibility for these meetings is management's. The purpose is to brief management on the status of the project, review the project charter and project team mission, discuss those management activities likely to have an impact on the progress of the team, etc. This is the appropriate forum for addressing systems barriers encountered by the team: while the team must work within existing systems, management has the authority to change the systems. At times a minor system change can dramatically enhance the ability of the team to progress.
- **Budget reviews**—While budget reports are included in each status report, a budget *review* is a formal evaluation of actual resource utilization with respect to budgeted utilization. Budget review may also involve revising budgets, either upward or downward, based on developments since the original budget approval. Among those unschooled in the science of statistics there is an unfortunate tendency to react to every random tick in budget variances as if they were due to a special cause of variation. Quality managers should coach finance and management personnel on the principles of variation (see I.G) to preclude tampering with the budgeting process.

- **Customer audits**—The "customer" in this context means the principal stakeholder in the project (see below). This person is the "owner" of the process being modified by the project. The project deliverables are designed to meet the objectives of this customer, and the customer should play an active role in keeping the project on track to the stated goals.
- **Updating plans and timetables**—The purpose of feedback is to provide information to form a basis for modifying future behavior. Since that behavior is documented in the project plans and schedules, these documents must be modified to ensure that the appropriate action is taken. Remember, in the PDCA cycle, *plans* change first.
- **Resource redirection**—The modifications made to the plans and timetables will result in increasing or decreasing resource allocation to the project, or accelerating or decelerating the timetable for resource utilization. The impact of these resource redirections on other projects should be evaluated by management in view of the organization's overall objectives.

PERFORMANCE MEASURES: Timeliness

There are a wide variety of tools and techniques available to help the project manager develop a realistic project timetable, to use the timetable to time the allocation of resources, and to track progress during the implementation of the project plan. We will review two of the most common here: Gantt charts and PERT-type systems.

Gantt charts

Gantt chart—A Gantt chart shows the relationships among the project tasks, along with time constraints. The horizontal axis of a Gantt chart shows the units of time (days, weeks, months, etc.). The vertical axis shows the activities to be completed. Bars show the estimated start time and duration of the various activities. Figure I.5 illustrates a simple Gantt chart.

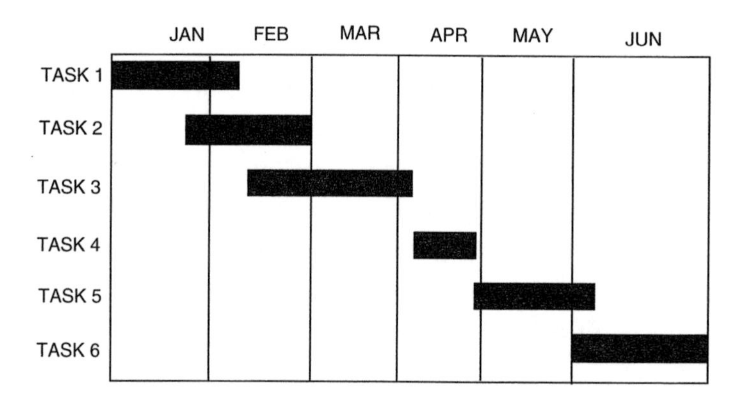

Figure I.5. Gantt chart.

Milestone charts—Gantt charts are often modified in a variety of ways to provide additional information. One common variation is shown in Figure I.6. The *milestone* symbol represents an event rather than an activity; it does not consume time or resources. When Gantt charts are modified in this way they are sometimes called "milestone charts."

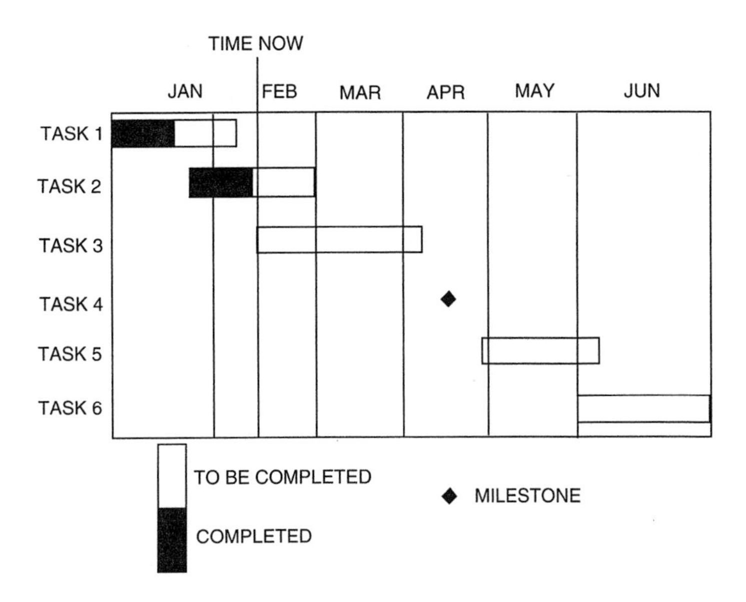

Figure I.6. Enhanced Gantt chart (milestone chart).

Gantt charts and milestone charts can be modified to show additional information, such as who is responsible for a task, why a task is behind schedule, remedial action planned or already taken, etc.

PERT-CPM-type project management systems

While useful, Gantt charts and their derivatives provide limited project schedule analysis capabilities. The successful management of large-scale projects requires more rigorous planning, scheduling and coordinating of numerous interrelated activities. To aid in these tasks, formal procedures based on the use of networks and network techniques were developed beginning in the late 1950s. The most prominent of these procedures have been PERT (**P**rogram **E**valuation and **R**eview **T**echnique) and CPM (**C**ritical **P**ath **M**ethod). The two approaches are usually referred to as PERT-type project management systems. The most important difference between PERT and CPM is that originally the time estimates for the activities were assumed deterministic in CPM and were probabilistic in PERT. Today, PERT and CPM actually comprise one technique and the differences are mainly historical.

Project scheduling by PERT-CPM consists of four basic phases: planning, scheduling, improvement, and controlling. The planning phase involves breaking the project into distinct activities. The time estimates for these activities are then determined and a network (or arrow) diagram is constructed with each activity being represented by an arrow.

PERT-type systems are used to:

- Aid in planning and control of projects
- Determine the feasibility of meeting specified deadlines
- Identify the most likely bottlenecks in a project
- Evaluate the effects of changes in the project requirements or schedule
- Evaluate the effects of deviating from schedule
- Evaluate the effect of diverting resources from the project, or redirecting additional resources to the project.

The ultimate objective of the scheduling phase is to construct a time chart showing the start and finish times for each activity as well as its relationship to other activities in the project. The schedule must identify activities that are "critical" in the sense that they *must* be completed on time to keep the project on schedule.

It is vital not to merely accept the schedule as a given. The information obtained in preparing the schedule can be used to improve the project schedule. Activities that the analysis indicates to be critical are candidates for improvement. Pareto analysis can be used to identify those critical elements that are most likely to lead to significant improvement in overall project completion time. Cost data can be used to supplement the time data, and the combined time/cost information analyzed using Pareto analysis.

The final phase in PERT-CPM project management is project control. This includes the use of the network diagram and time chart for making periodic progress assessments.

Constructing PERT-CPM network charts

All PERT-type systems use a project network chart to graphically portray the interrelationships among the elements of a project. This network representation of the project plan shows all the precedence relationships, i.e., the order in which the tasks must be completed. On the network chart arrows represent *activities*, while boxes or circles represent *events*; in preparing and understanding this technique it is very important to keep these two terms distinct. An arrow goes from one event to another only if the first event is the immediate predecessor of the second. If more than one activity must be completed before an event can occur, then there will be several arrows entering the box corresponding to that event. Sometimes one event must wait for another event, but no activity intervenes between the two events. In this case, the two events are joined with a dotted arrow, representing a *dummy activity*. Dummy activities take no time to complete; they merely show precedence relationships.

These drawing conventions are illustrated in Figure I.7.

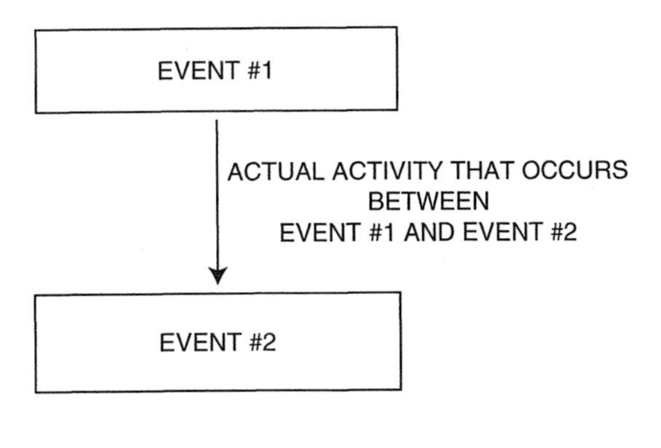

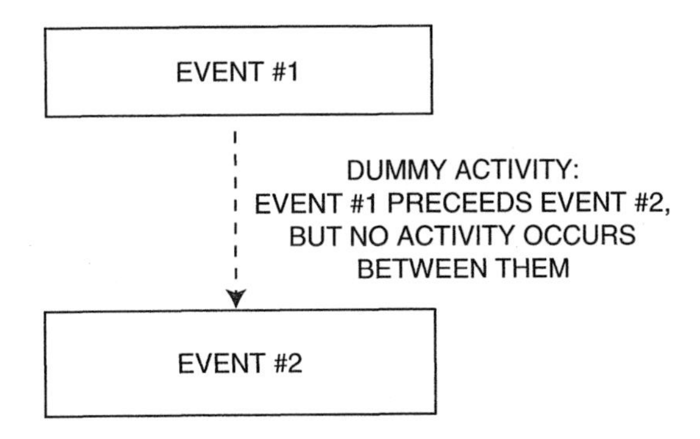

Figure I.7. Network diagram terms and drawing conventions.

The node towards which all activities lead, the final completion of the project, is called the *sink* of the network.

Taha (1976) offers the following rules for constructing the arrow diagram:

Rule 1—*Each activity is represented by one and only one arrow in the network.* No single activity can be represented twice in the network. This does not mean that one activity cannot be broken down into segments.

Rule 2—*No two activities can be identified by the same head-and-tail events.* This situation may arise when two activities can be performed concurrently. The proper way to deal with this situation is to introduce *dummy events* and activities, as shown in Figure I.8.

INCORRECT: VIOLATES RULE #2

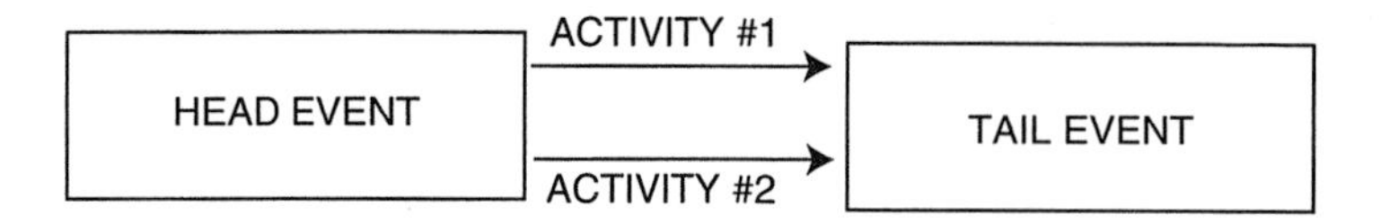

SOLUTION: ADD DUMMY EVENT and ACTIVITY

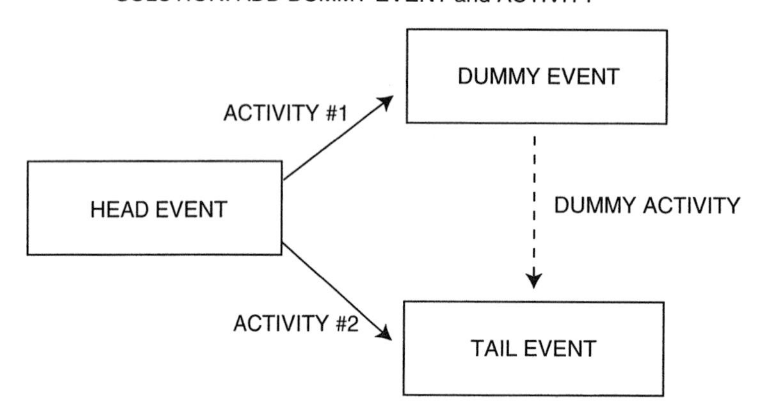

Figure I.8. Parallel activities: network representation.

Rule #2 facilitates the analysis of network diagrams with computer programs for project analysis.

Rule 3—*In order to ensure the correct precedence relationship in the arrow diagram the following questions must be answered as each activity is added to the network:*

a. What activities must be completed immediately before this activity can start?
b. What activities immediately follow this activity?
c. What activities must occur concurrently with this activity?

Example of PERT

The following is based on an example from Hillier and Lieberman (2002). Let's say that we wish to use PERT on a project for constructing a house. The following activities, and their estimated completion times, are presented in Table I.3.

Table I.3. Activities involved in constructing a house.

ACTIVITY	TIME TO COMPLETE (DAYS)
Excavate	2
Foundation	4
Rough wall	10
Rough electrical work	7
Rough exterior plumbing	4
Rough interior plumbing	5
Wall board	5
Flooring	4
Interior painting	5
Interior fixtures	6
Roof	6
Exterior siding	7
Exterior painting	9
Exterior fixtures	2

Now, it is important that certain of these activities be done in a particular order. For example, one cannot put on the roof until the walls are built. This is called a *precedence relationship*, i.e., the walls must *precede* the roof. The network diagram graphically displays the precedence relationships involved in constructing a house. A PERT network for constructing a house is shown in the Figure I.9 (incidentally, the figure is also an arrow diagram).

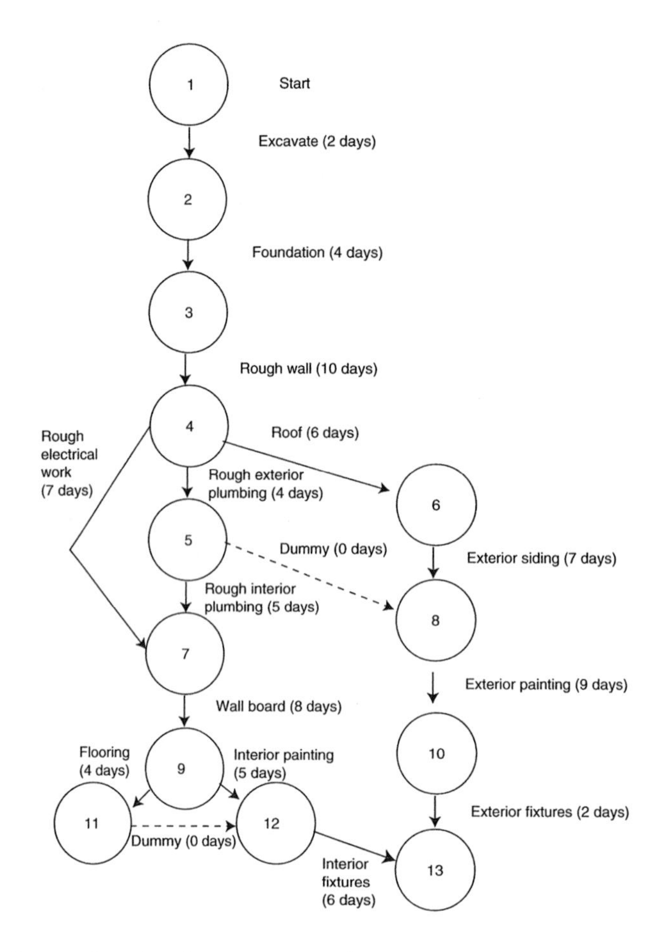

Figure I.9. Project network for constructing a house.
Source: *Based on example from Hillier and Lieberman.*

Finding the critical path

There are two time-values of interest for each event: its *earliest time of completion* and its *latest time of completion.* The earliest time for a given event is the estimated time at which the event will occur if the preceding activities are started as early as possible. The latest time for an event is the estimated time the event can occur without delaying the completion of the project beyond its earliest time. Earliest times of events are found by starting at the initial event and working forward, successively calculating the time at which each event will occur if each immediately preceding event occurs at its earliest time and each intervening activity uses only its estimated time. Table I.4 shows the process of finding the earliest completion time for the house construction example. (Event numbers refer to the network diagram in Figure I.9.) The reader is advised to work through the results in Table I.4, line-by-line, using Figure I.9.

Table I.4. Calculation of earliest completion times for house construction example.

Event	Immediately Preceding Event	Earliest Time + Activity Time	Maximum = Earliest Completion Time
1	—	—	0
2	1	0+2	2
3	2	2+4	6
4	3	6+10	16
5	4	16+4	20
6	4	16+6	22
7	4	16+7	*
	5	20+5	25
8	5	20+0	**
	6	22+7	29
9	7	25+8	33
10	8	29+9	38
11	9	33+4	37
12	9	33+5	38
	11	37+0	
13	10	38+2	
	12	38+6	44

Thus, for example, the earliest time event #8 can be completed is 29 days. Latest times are found by starting at the final event and working backwards, calculating the latest time an event will occur if each immediately following event occurs at its latest time.

---

*The answer in this cell is 23, which is *not* the earliest time which event #7 will occur. Event #7 occurs when *both* events #4 and #5 have been completed.

**The answer in this cell is 20, which is *not* the earliest time which event #8 will occur. Event #8 occurs when *both* events #5 and #6 have been completed.

Table I.5. Calculation of latest completion times for house construction example.

Event	Immediately Following Event	Latest Time – Activity Time	Minimum = Latest Time
13	—	—	44
12	13	44–6	38
11	12	38–0	38
10	13	44–2	42
9	12	38–5	33
	11	38–4	*
8	10	42–9	33
7	9	33–8	25
6	8	33–7	26
5	8	33–0	
	7	25–5	20
4	7	25–7	
	6	26–6	
	5	20–4	16
3	4	16–10	6
2	3	6–4	2
1	2	2–2	0

Slack time for an *event* is the difference between the latest and earliest times for a given event. Thus, assuming everything else remains on schedule, the slack for an event indicates how much delay in reaching the event can be tolerated without delaying the project completion. Slack times for the events in the house construction project are shown in Table I.6.

*The answer in this cell, 34, is not the latest completion time since it is not the minimum time for completing event #9 without delaying the project. Note that if it took 34 days to complete event #9 the project would take 45 days to complete, i.e., it would be delayed by 1 day.

Table I.6. Calculation of slack times for house construction events.

EVENT	SLACK	EVENT	SLACK
1	0–0=0	7	25–25=0
2	2–2=0	8	33–29=4
3	6–6=0	9	33–33=0
4	16–16=0	10	42–38=4
5	20–20=0	11	38–37=1
6	26–22=4	12	38–38=0
Continued . . .	*Continued . . .*	13	44–44=0

The slack time for an *activity x,y* is the difference between

1. The latest time of event y.
2. The earliest time of event x plus the estimated activity time.

Slack time for an *activity* is the difference between the latest and earliest times for a given activity. Thus, assuming everything else remains on schedule, the slack for an activity indicates how much delay in reaching the activity can be tolerated without delaying the project completion. Slack times for the activities in the house construction project are shown in Table I.7.

Table I.7. Calculation of slack times for house construction activities.

ACTIVITY	SLACK
Excavate (1,2)	2–(0+2)=0
Foundation (2,3)	6–(2+4)=0
Rough wall (3,4)	16–(6+10)=0
Rough exterior plumbing (4,5)	20–(16+4)=0
Roof (4,6)	26–(16+6)=4
Rough electrical work (4,7)	25–(16+7)=2
Rough interior plumbing (5,7)	25–(20+5)=0
Exterior siding (6,8)	33–(22+7)=4
Wall board (7,9)	33–(25+8)=0
Exterior painting (8,10)	42–(29+9)=4
Flooring (9,11)	38–(33+4)=1
Interior painting (9,12)	38–(33+5)=0
Exterior fixtures (10,13)	44–(38+2)=4
Interior fixtures (12,13)	44–(38+6)=0

Events and activities with slack times of zero are said to lie on the *critical path* for the project. A critical path for a project is defined as a path through the network such that the activities on this path have *zero slack*. All activities and events having zero slack must lie on a critical path, but no others can. Figure I.10 shows the activities on the critical path for the housing construction project as thick lines.

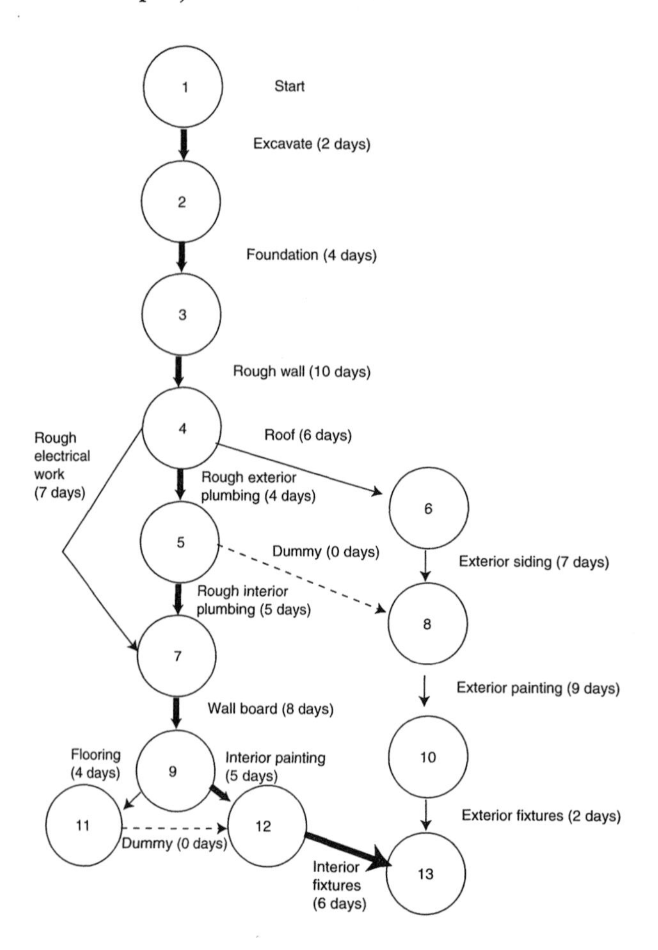

Figure I.10. Critical path for house construction example.

Control and prevention of schedule slippage

Project managers can use the network and the information obtained from the network analysis in a variety of ways to help them manage their projects. One way is, of course, to pay close attention to the activities that lie on the critical path. Any delay in these activities will result in a delay for the project. However, the manager should also consider assembling a team to review the network with an eye towards modifying the project plan to reduce the total time needed to complete the project. The manager should also be aware that the network times are based on *estimates*. In fact, it is likely that the completion times will vary. When this occurs it often happens that a new critical path appears.

Thus, the network should be viewed as a dynamic entity which should be revised as conditions change.

Primary causes of slippage include poor planning and poor management of the project. Outside forces beyond the control of the project manager will often play a role. However, it isn't enough to be able to simply identify "outside forces" as the cause and beg forgiveness. Astute project managers will anticipate as many such possibilities as possible and prepare contingency plans to deal with them. The PDPC technique is useful in this endeavor (see V.B). Schedule slippage should also be addressed rigorously in the schedule control plan, which was mentioned earlier as a primary deliverable from the project planning process. The control plan should make provision for reviews conducted at intervals frequent enough to assure that any unanticipated problems are identified before schedule slippage becomes a problem.

PERFORMANCE MEASURES: **Resources**

Resources are those assets of the firm, including the time of employees, that are used to accomplish the objectives of the project. The project manager should define, negotiate, and secure resource commitments for the personnel, equipment, facilities, and services needed for the project. Resource commitments should be as specific as possible. Generally, resource utilization is specified in the project budget.

The following items should be defined and negotiated:

- What will be furnished?
- By whom?
- When?
- How will it be delivered?
- How much will it cost?
    - Who will pay?
    - When will payment be made?

Resource conflicts

Of course, there are always other opportunities for utilizing resources. On large projects, conflicts over resource allocation are inevitable. It is best if resource conflicts can be resolved between those managers directly involved. However, in some cases, resource conflicts must be addressed by higher levels of management. Senior managers should view resource conflicts as potential indications that the management system for allocating resources must be modified or redesigned. Often, such conflicts create ill will among managers and lead to lack of support, or even active resistance to the project. Too many such conflicts can lead to resentment towards quality improvement efforts in general.

PERFORMANCE MEASURES: Methodology

Cost considerations in project scheduling

Most project schedules can be compressed, if one is willing to pay the additional costs. For the analysis here, costs are defined to include direct elements only. Indirect costs (administration, overhead, etc.) will be considered in the final analysis. Assume that a straight line relationship exists between the cost of performing an activity on a *normal schedule*, and the cost of performing the activity on a *crash schedule*. Also assume that there is a crash time beyond which no further time savings is possible, regardless of cost. Figure I.11 illustrates these concepts.

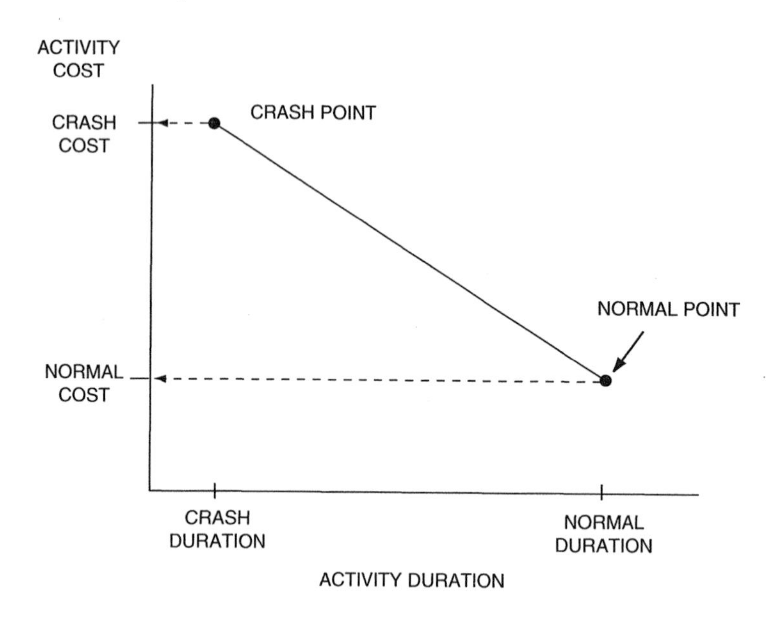

Figure I.11. Cost-time relationship for an activity.

For a given activity the cost-per-unit-of-time saved is found as

$$\frac{\text{crash cost} - \text{normal cost}}{\text{normal time} - \text{crash time}} \tag{I.1}$$

When deciding which activity on the critical path to improve, one should begin with the activity that has the smallest cost-per-unit-of-time saved. The project manager should be aware that once an activity time has been reduced there may be a new critical path. If so, the analysis should proceed using the updated information, i.e., activities on the new critical path should be analyzed.

The data for the house construction example is shown below, with additional data for costs and crash schedule times for each activity.

Table I.8. Schedule costs for activities involved in constructing a house.

	Normal Schedule		Crash Schedule		
ACTIVITY	Time (days)	Cost	Time (days)	Cost	Slope
Excavate	2	**1000**	1	**2000**	**1000**
Foundation	4	**1600**	3	**2400**	**800***
Rough wall	**10**	7500	6	**14000**	**1625**
Rough electrical work	7	7000	4	14000	2333
Rough exterior plumbing	4	**4400**	3	**6000**	**1600**
Rough interior plumbing	5	**3750**	3	**7500**	**1875**
Wall board	5	**3500**	3	**7000**	**1750**
Flooring	4	3200	2	5600	1200
Interior painting	5	**3000**	3	**5500**	**1250**
Interior fixtures	**6**	**4800**	2	**11000**	**1550**
Roof	6	4900	2	12000	1775
Exterior siding	7	5600	3	12000	1600
Exterior painting	9	4500	5	9000	1125
Exterior fixtures	2	1800	1	3200	1400

Activities shown in bold are on the critical path; only critical path activities are being considered since only they can produce an improvement in overall project duration. Thus, the first activity to consider improving would be foundation work, which costs $800 per day saved on the schedule (identified with an asterisk [*] in Table I.8). If additional resources could be directed towards this activity it would produce the best "bang for the buck" in terms of reducing the total time of the project. Next, assuming the critical path doesn't change, would be excavation, then exterior painting, etc.

As activities are addressed one-by-one, the time it takes to complete the project will decline, while the direct costs of completing the project will increase. Figure I.12 illustrates the cost-duration relationship graphically.

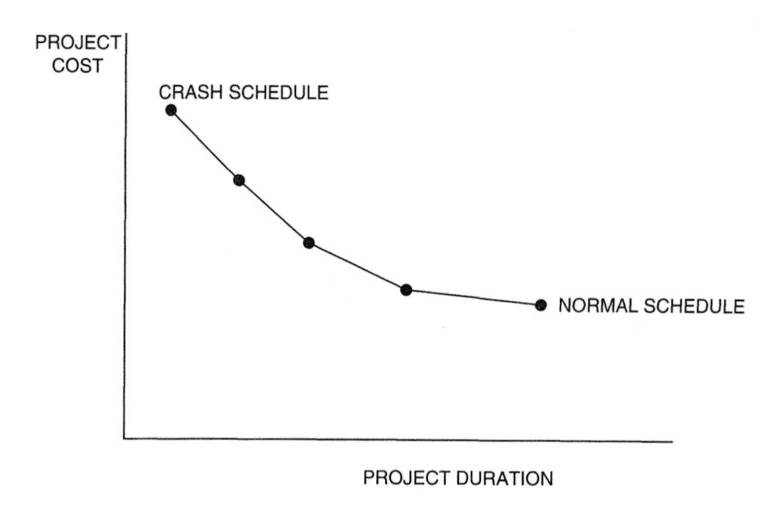

Figure I.12. Direct costs as a function of project duration.

Conversely, *indirect costs* such as overhead, etc., are expected to *increase* as projects take longer to complete. When the indirect costs are added to the direct costs, total costs will generally follow a pattern similar to that shown in Figure I.13 results.

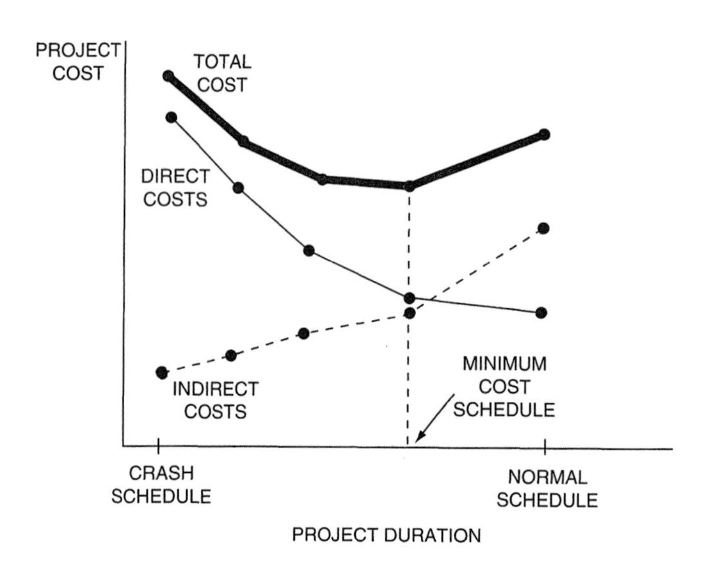

Figure I.13. Total costs as a function of project duration.

To optimize resource utilization, the project manager will seek to develop a project plan that produces the minimum cost schedule. Of course, the organization will likely have multiple projects being conducted simultaneously, which places additional constraints on resource allocation.

Other performance measurement methodology

Project information should be collected on an ongoing basis as the project progresses. Information obtained should be communicated in a timely fashion to interested parties and decision-makers. The people who receive the information can often help the project manager to maintain or recover the schedule. There are two types of communication involved: feedback and feedforward. Feedback is historical in nature and includes such things as performance to schedule, cost variances (relative to the project budget), and quality variances (relative to the quality plan). The reader will recall that initial project planning called for special control plans in each of these three areas. Feedforward is oriented towards the future and is primarily concerned with heading off future variances in these three areas. Information reporting formats commonly fall into one of the following categories:

- formal, written reports
- informal reports and correspondence
- presentations
- meetings
- guided tours of the project, when feasible
- conversations

The choice of format for the communication should consider the nature of the audience and their needs and the time and resources available. Audiences can be characterized along five dimensions (Ruskin, 1995):

1. Audience diversity
2. Audience sophistication
3. Audience familiarity with the subject matter
4. Audience size and geographic location
5. Audience need to know

The report or presentation should be planned to avoid wasting the time of the audience members, or the time of those preparing the report or presentation. Objectives should be stated and the steps necessary to meet the objectives should be clearly stated in the plan. It may help to consider the communication as a "lesson" and the plan as a "lesson plan." Provision should be made for assuring that the objectives were, in fact, met.

Project communication is a process and, like all processes, it can be improved. The tools of CQI are designed for just this purpose. Measurements can be analyzed using the quality control tools described in section V.C and chapter VI and used to improve the process of project management. The PDCA cycle also applies to project management. See below for additional details on reporting and feedback methodology.

RELEVANT STAKEHOLDERS

Large quality improvement projects impact large numbers of people within the organization. Those impacted are known as "stakeholders" in the project. As far as is practicable, the interests of stakeholders should be aligned with the objectives of the project. If this is not the case, when stakeholders act according to their own interests they will be acting to sabotage the project, intentionally or unintentionally.

Identifying project stakeholders begins with obtaining a project charter. The charter and the chartering process was discussed earlier, in "integrated quality initiatives". Once the project charter has been finalized, the project team should prepare a list of potential stakeholders and their roles. If the project will have significant impact on hourly employees, they should be involved as well. If the workers are unionized, the union should be informed. Sell all stakeholders on the merits of the project. People resist changes unless they see the value in them and the urgency to take action. Stakeholders must be identified and their needs analyzed so that an action plan can be created to meet the needs and gain commitment. To avoid problems, the project team must constantly communicate with the stakeholders.

Stakeholder focus groups is a method that allows group members to evaluate the potential impact of a plan by identifying the stakeholders affected by or having influence over the project plan. The focus group approach is a highly structured method in which the project team first identify the stakeholders and their assumptions, then brings those identified together to elicit their responses to the proposed project (see I.H for a discussion of the focus group technique). The team then rates these assumptions for importance to the stakeholders and importance to the plan. A stakeholder satisfaction plan may be developed to assure the support of key individuals and groups.

As soon as possible the project manager should arrange a short, informal meeting with all of these individuals identified as being impacted, including the executive who sits on the quality council (but no other council members). The project manager and process owner are letting the stakeholders know that a project is about to be undertaken in "their" area, with the permission and direction of the senior executives. This meeting also represents an informal invitation for the middle managers to challenge the decision to conduct the project. It is important to allow the managers about a week to attempt to reverse the leadership's decision to pursue the project. If concrete information suggests that tampering or sabotage is occurring, the project manager or process owner should immediately bring it to the attention of the senior executives who approved the project charter. The senior leadership should resolve the issue promptly.

If a week or so passes without clear opposition to the project, the project manager should proceed with the implementation of the project plan. Of course, the lines of communication should remain open throughout the implementation of the project plan.

BENCHMARKING

Benchmarking is a topic of general interest in quality management. Thus, our discussion here goes beyond the use of benchmarking in project management alone.

Benchmarking is a popular method for developing requirements and setting goals. In more conventional terms, benchmarking can be defined as measuring your performance against that of best-in-class companies, determining how the best-in-class achieve those performance levels, and using the information as the basis for your own company's targets, strategies, and implementation. Benchmarking involves research into the best practices at the industry, firm, or process level. Benchmarking goes beyond a determination of the "industry standard," it breaks the firm's activities down to process operations and looks for the best-in-class for a particular operation. For example, Xerox corporation studied the retailer LL Bean to help them improve their parts distribution process.

Benchmarking goes beyond the mere setting of goals. It focuses on practices that produce superior performance. Benchmarking involves setting up partnerships that allow both parties to learn from one another. Competitors can also engage in benchmarking, providing they avoid proprietary issues.

Benchmarking projects are like any other major project. Benchmarking must have a structured methodology to ensure successful completion of thorough and accurate investigations. However, it must be flexible to incorporate new and innovative ways of assembling difficult-to-obtain information. It is a discovery process and a learning experience. It forces the organization to take an external view, to look beyond itself.

The benchmarking process

Camp (1989) lists the following steps for the benchmarking process:

1. Planning
    - 1.1. Identify what is to be benchmarked
    - 1.2. Identify comparative companies
    - 1.3. Determine data collection method and collect data
2. Analysis
    - 2.1. Determine current performance "gap"
    - 2.2. Project future performance levels
3. Integration
    - 3.1. Communicate benchmark findings and gain acceptance
    - 3.2. Establish functional goals
4. Action
    - 4.1. Develop action plans
    - 4.2. Implement specific actions and monitor progress
    - 4.3. Recalibrate benchmarks

5. Maturity
   5.1. Leadership position attained
   5.2. Practices fully integrated into process

The first step in benchmarking is determining what to benchmark. To focus the benchmarking initiative on critical issues, begin by identifying the process outputs most important to the customers of that process (i. e., the key quality characteristics). This step applies to every organizational function, since each one has outputs and customers. The QFD/customer needs assessment is a natural precursor to benchmarking activities (see I.H).

Getting started with benchmarking

The essence of benchmarking is the acquisition of information. The process begins with the identification of the process that is to be benchmarked. The process chosen should be one that will have a major impact on the success of the business. The rules used for identifying candidates for business process reengineering can also be used here (see V.A).

Once the process has been identified, contact a business library and request a search for the information relating to your area of interest. The library will identify material from a variety of external sources, such as magazines, journals, special reports, etc. You should also conduct research using the Internet and other electronic networking resources. However, be prepared to pare down what will probably be an extremely large list of candidates (e.g., an Internet search on the word "benchmarking" produced 20,000 hits). Don't forget your organization's internal resources. If your company has an "Intranet" use it to conduct an internal search. Set up a meeting with people in key departments, such as R&D. Tap the expertise of those in your company who routinely work with customers, competitors, suppliers, and other "outside" organizations. Often your company's board of directors will have an extensive network of contacts.

The search is, of course, not random. You are looking for the best of the best, not the average firm. There are many possible sources for identifying the elites. One approach is to build a compendium of business awards and citations of merit that organizations have received in business process improvement. Sources to consider are *Industry Week*'s Best Plant's Award, National Institute of Standards and Technology's Malcolm Baldrige Award, *USA Today* and the Rochester Institute of Technology's Quality Cup Award, European Foundation for Quality Management Award, Occupational Safety and Health Administration (OSHA.), Federal Quality Institute, Deming Prize, Competitiveness Forum, *Fortune* magazine, United States Navy's Best Manufacturing Practices, to name just a few. You may wish to subscribe to an "exchange service" that collects benchmarking information and makes it available for a fee. Once enrolled, you will have access to the names of other subscribers—a great source for contacts.

Don't overlook your own suppliers as a source for information. If your company has a program for recognizing top suppliers, contact these suppliers and see if they are willing to share their "secrets" with you. Suppliers are predisposed to cooperate with their customers; it's an automatic door-opener. Also contact your customers. Customers have a vested interest in helping you do a better job. If your quality, cost, and delivery performance improves, your customers will benefit. Customers may be willing to share some of their insights as to how their other suppliers compare with you. Again, it isn't necessary that you get information about direct competitors. Which of your customer's suppliers are best at billing? Order fulfillment? Customer service? Keep your focus at the process level and there will seldom be any issues of confidentiality. An advantage to identifying potential benchmarking partners through your customers is that you will have a referral that will make it easier for you to start the partnership.

Another source for detailed information on companies is academic research. Companies often allow universities access to detailed information for research purposes. While the published research usually omits reference to the specific companies involved, it often provides comparisons and detailed analysis of what separates the best from the others. Such information, provided by experts whose work is subject to rigorous peer review, will often save you thousands of hours of work.

After a list of potential candidates is compiled, the next step is to choose the best three to five targets. A candidate that looked promising early in the process might be eliminated later based on the following criteria (Vaziri, 1992):

- Not the best performer
- Unwilling to share information and practices (i.e., doesn't view the benchmarking process as a mutually beneficial learning opportunity)
- Low availability and questionable reliability of information on the candidate

As the benchmarking process evolves, the characteristics of the most desirable candidates will be continually refined. This occurs as a result of a clearer understanding of your organization's key quality characteristics and critical success factors and an improved knowledge of the marketplace and other players. This knowledge and the resulting actions tremendously strengthen an organization.

Why benchmarking efforts fail

The causes of failed benchmarking projects are the same as those for other failed projects (DeToro, 1995):

☐ **Lack of sponsorship**—A team should submit to management a one- to four-page benchmarking project proposal that describes the project, its objectives, and potential costs. If the team can't gain approval for the project or get a sponsor, it makes little sense to proceed with a project that's not understood or appreciated or that is unlikely to lead to corrective action when completed.

- ☐ **Wrong people on team**—Who are the right people for a benchmarking team? Individuals involved in benchmarking should be the same ones who own or work in the process. It's useless for a team to address problems in business areas that are unfamiliar or where the team has no control or influence.
- ☐ **Teams don't understand their work completely**—If the benchmarking team didn't map, flowchart, or document its work process, and if it didn't benchmark with organizations that also documented their processes, there can't be an effective transfer of techniques. The intent in every benchmarking project is for a team to understand how its process works and compare it to another company's process at a detailed level. The exchange of process steps is essential for improved performance.
- ☐ **Teams take on too much**—The task a team undertakes is often so broad that it becomes unmanageable. This broad area must be broken into smaller, more manageable projects that can be approached logically. A suggested approach is to create a functional flowchart of an entire area, such as production or marketing, and identify its processes. Criteria can then be used to select a process to be benchmarked that would best contribute to the organization's objectives.
- ☐ **Lack of long-term management commitment**—Since managers aren't as familiar with specific work issues as their employees, they tend to under-estimate the time, cost, and effort required to successfully complete a benchmarking project. Managers should be informed that while it's impossible to know the exact time it will take for a typical benchmarking project, a rule of thumb is that a team of four or five individuals requires a third of their time for five months to complete a project.
- ☐ **Focus on metrics rather than processes**—Some firms focus their bench-marking efforts on performance targets (metrics) rather than processes. Knowing that a competitor has a higher return on assets doesn't mean that its performance alone should become the new target (unless an understanding exists about how the competitor differs in the use of its assets and an evaluation of its process reveals that it can be emulated or surpassed).
- ☐ **Not positioning benchmarking within a larger strategy**— Benchmarking is one of many total quality management tools—such as problem solving, process improvement, and process reengineering—used to shorten cycle time, reduce costs, and minimize variation. Benchmarking is compatible with and complementary to these tools, and they should be used together for maximum value.
- ☐ **Misunderstanding the organization's mission, goals, and objectives**—All benchmarking activity should be launched by management as part of an overall strategy to fulfill the organization's mission and vision by first attaining the short-term objectives and then the long-term goals.

- ☐ **Assuming every project requires a site visit**—Sufficient information is often available from the public domain, making a site visit unnecessary. This speeds the benchmarking process and lowers the cost considerably.
- ☐ **Failure to monitor progress**—Once benchmarking has been completed for a specific area or process benchmarks have been established and process changes implemented, managers should review progress in implementation and results.

The issues described here are discussed in other parts of this chapter and in other parts of this book. The best way of dealing with them is to prevent their occurrence by carefully planning and managing the project from the outset. This list can be used as a checklist to evaluate project plans; if the plans don't clearly preclude these problems, then the plans are not complete.

The benefits of benchmarking

The benefits of competitive benchmarking include:

- Creating a culture that values continuous improvement to achieve excellence
- Enhancing creativity by devaluing the not-invented-here syndrome
- Increasing sensitivity to changes in the external environment
- Shifting the corporate mind-set from relative complacency to a strong sense of urgency for ongoing improvement
- Focusing resources through performance targets set with employee input
- Prioritizing the areas that need improvement
- Sharing the best practices between benchmarking partners

Some dangers of benchmarking

Benchmarking is based on learning from others, rather than developing new and improved approaches. Since the process being studied is there for all to see, a firm will find that benchmarking cannot give them a sustained competitive advantage. Although helpful, benchmarking should never be the *primary* strategy for improvement.

Competitive analysis is an approach to goal setting used by many firms. This approach is essentially benchmarking confined to one's own industry. Although common, competitive analysis virtually guarantees second-rate quality because the firm will always be following their competition. If the entire industry employs the approach it will lead to stagnation for the entire industry, setting them up for eventual replacement by outside innovators.

MANAGEMENT SUPPORT AND ORGANIZATIONAL ROADBLOCKS

Internal roadblocks

Most organizations still have a hierarchical, command-and-control organizational structure, sometimes called "smoke stacks" or "silos." The functional specialists in charge of each smoke stack tend to focus on optimizing their own functional area, often to the detriment of the organization as a whole. In addition, the hierarchy gives these managers a monopoly on the authority to act on matters related to their functional specialty. The combined effect is both a desire to resist change and the authority to resist change, which often creates insurmountable roadblocks to quality improvement projects.

It is important to realize that organizational rules are, by their nature, a barrier to change. The formal rules take the form of written standard operating procedures (SOPs). The very purpose of SOPs is to standardize behavior. The quality profession has (in this author's opinion) historically overemphasized formal documentation, and it continues to do so by advocating such approaches as ISO 9000 and ISO 14000. Formal rules are often responses to past problems and they often continue to exist long after the reason for their existence has passed. In an organization that is serious about its written rules even senior leaders find themselves helpless to act without submitting to a burdensome rule-changing process. The true power in such an organization is the bureaucracy that controls the procedures. If the organization falls into the trap of creating written rules for too many things, it can find itself moribund in a fast-changing external environment. This is a recipe for disaster.

Restrictive rules need not take the form of management limitations on itself, procedures that define hourly work in great detail also produce barriers, e.g., union work rules. Projects almost always require that work be done differently and such procedures prohibit such change. Organizations that tend to be excessive in SOPs also tend to be heavy on work rules. The combination is often deadly to quality improvement efforts.

Organization structures preserve the status quo in other ways besides formal, written restrictions in the form of procedures and rules. Another effective method of limiting change is to require permission from various departments, committees, councils, boards, experts, etc. Even though the organization may not have a formal requirement, that "permission" be obtained, the effect may be the same, e.g., "You should run that past accounting" or "Ms. Reimer and Mr. Evans should be informed about this project." When permission for vehicles for change (e.g., project budgets, plan approvals) is required from a group that meets infrequently it creates problems for project planners. Plans may be rushed so they can be presented at the next meeting, lest the project be delayed for months. Plans that need modifications may be put on hold until the next meeting, months away. Or, projects may miss the deadline and be put off indefinitely.

External roadblocks

Modern organizations do not exist as islands. Powerful external forces take an active interest in what happens within the organization. Government bodies have created a labyrinth of rules and regulations that the organization must negotiate to utilize its human resources without incurring penalties or sanction. The restrictions placed on modern businesses by outside regulators is challenging to say the least. When research involves people, ethical and legal concerns sometimes require that external approvals be obtained. The approvals are contingent on such issues as informed consent, safety, cost and so on.

Many industries have "dedicated" agencies to deal with. For example, the pharmaceutical industry must deal with the Food and Drug Administration (FDA). These agencies must often be consulted before undertaking projects. For example, a new treatment protocol involving a new process for treatment of pregnant women prior to labor may involve using a drug in a new way (e.g., administered on an outpatient basis instead of on an inpatient basis).

Many professionals face liability risks that are part of every decision. Often these fears create a "play it safe" mentality that acts as a barrier to change. The fear is even greater when the project involves new and untried practices and technology.

Individual barriers to change

Perhaps the most significant change, and therefore the most difficult, is to change ourselves. It seems to be a part of human nature to resist changing oneself. By and large, we worked hard to get where we are, and our first impulse is to resist anything that threatens our current position. Forsha (1992) provides the process for personal change shown in Figure I.14.

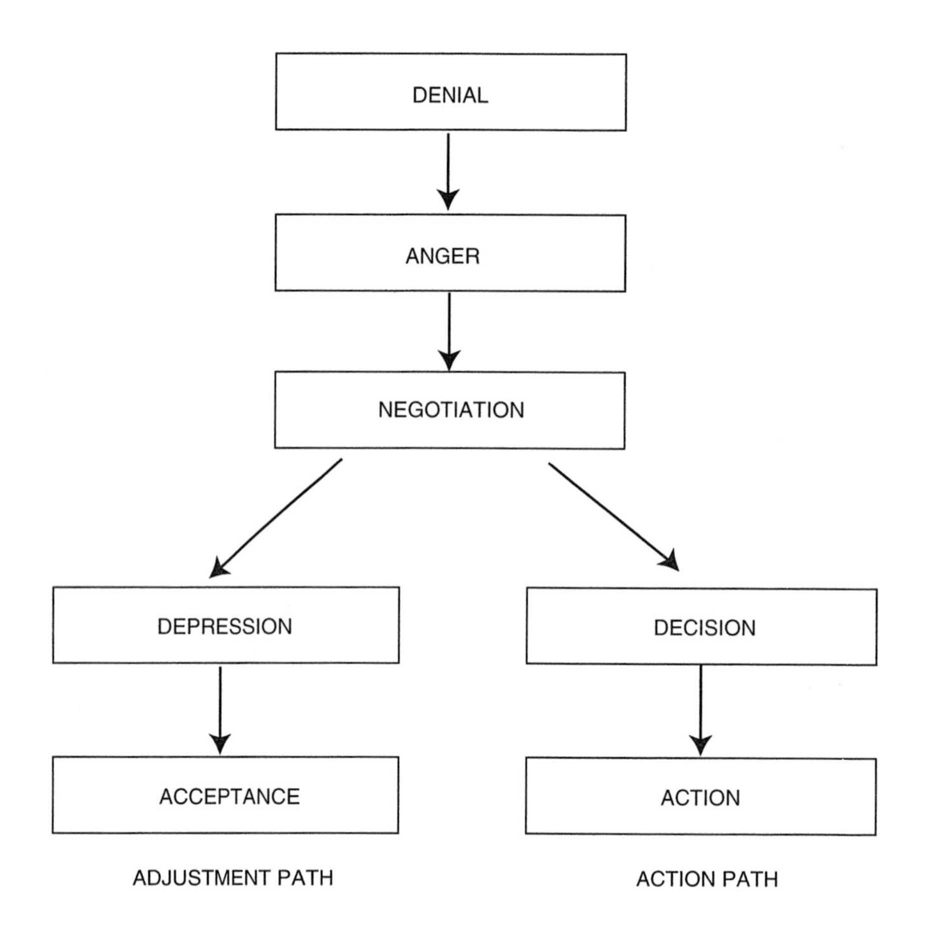

Figure I.14. The process of personal change.
From Forsha. Used by permission of the publisher.

The adjustment path results in preservation of the status quo. The action path results in change. The well-known PDCA cycle can be used once a commitment to action has been made by the individual. The goal of such change is continuous *self*-improvement.

Within an organizational context, the individual's reference group plays a part in personal resistance to change. A reference group is the aggregation of people a person thinks of when they use the word "we." If "we" refers to the company, then the company is the individual's reference group and he or she feels connected to the company's success or failure. However, "we" might refer to the individual's profession or trade group, e.g., "We doctors," "We engineers," "We union members." In this case the leaders shown on the formal organization chart will have little influence on the individual's attitude towards the success or failure of the project. When a project involves external reference groups with competing agendas, the task of building buy-in and consensus is daunting indeed.

Ineffective management support strategies

Strategy #1: command people to act as you wish—With this approach the senior leadership simply commands people to act as the leaders wish. The implication is that those who do not comply will be subjected to disciplinary action. People in less senior levels of an organization often have an inflated view of the value of raw power. The truth is that even senior leaders have limited power to rule by decree. Human beings by their nature tend to act according to their own best judgment. Thankfully, commanding that they do otherwise usually has little effect. The result of invoking authority is that the decision-maker must constantly try to divine what the leader wants them to do in a particular situation. This leads to stagnation and confusion as everyone waits on the leader. Another problem with commanding as a form of "leadership" is the simple communication problem. Under the best of circumstances people will often simply misinterpret the leadership's commands.

Strategy #2: change the rules by decree—When rules are changed by decree the result is again confusion. What are the rules today? What will they be tomorrow? This leads again to stagnation because people don't have the ability to plan for the future. Although rules make it difficult to change, they also provide stability and structure that may serve some useful purpose. Arbitrarily changing the rules based on force (which is what "authority" comes down to) instead of a set of guiding principles does more harm than good.

Strategy #3: authorize circumventing of the rules—Here the rules are allowed to stand, but exceptions are made for the leader's "pet projects." The result is general disrespect for and disregard of the rules, and resentment of the people who are allowed to violate rules that bind everyone else. An improvement is to develop a formal method for circumventing the rules, e.g., deviation request procedures. While this is less arbitrary, it adds another layer of complexity and still doesn't change the rules that are making change difficult in the first place.

Strategy #4: redirect resources to the project—Leaders may also use their command authority to redirect resources to the project. A better way is to develop a fair and easily understood system to assure that projects of strategic importance are adequately funded as a matter of policy. In our earlier discussion of project scheduling we discussed "crash scheduling" as a means of completing projects in a shorter time frame. However, the assumption was that the basis for the allocation was cost or some other objective measure of the organization's best interest. Here we are talking about political clout as the basis of the allocation.

Effective management support strategies

Strategy #1: transform the formal organization and the organization's culture—By far the best solution to the problems posed by organizational roadblock is to transform the organization to one where these roadblocks no longer exist. As discussed earlier, this process can't be implemented by decree. As the leader helps project teams succeed, he will learn about the need for transformation. Using his persuasive powers the leader-champion can undertake the exciting challenge of creating a culture that embraces change instead of fighting it.

Strategy #2: mentoring—In Greek mythology, Mentor was an elderly man, the trusted counselor of Odysseus, and the guardian and teacher of his son Telemachus. Today the term, "mentor" is still used to describe a wise and trusted counselor or teacher. When this person occupies an important position in the organization's hierarchy, he or she can be a powerful force for eliminating roadblocks. Modern organizations are complex and confusing. It is often difficult to determine just where one must go to solve a problem or obtain a needed resource. The mentor can help guide the project manager through this maze by clarifying lines of authority. At the same time, the mentor's senior position enables him to see the implications of complexity and to work to eliminate unnecessary rules and procedures.

Strategy #3: identify informal leaders and enlist their support—Because of their experience, mentors often know that the person whose support the project really needs is not the one occupying the relevant box on the organization chart. The mentor can direct the project leader to the person whose opinion really has influence. For example, a project may need the approval of, say, the vice-president of engineering. The engineering VP may be balking because his senior metallurgist hasn't endorsed the project.

Strategy #4: find legitimate ways around people, procedures, resource constraints and other roadblocks—It may be possible to get approvals or resources through means not known to the project manager. Perhaps a minor change in the project plan can bypass a cumbersome procedure entirely. For example, adding an engineer to the team might automatically place the authority to approve process experiments within the team rather than in the hands of the engineering department.

SHORT-TERM (TACTICAL) PLANS

Conceptually, project plans are subsets of bigger plans, all of which are designed to carry out the organization's mission. The project plan must be broken down further. The objective is to reach a level where projects are "tiny." A tiny project is reached when it is possible to easily answer two questions:

1. Is the project complete?
2. Is the project done correctly?

For example, a software development team concluded that a tiny computer module had the following characteristics: 1) it implemented only one concept; 2) it could be described in 6 lines or less of easily understood pseudo-code (English-like descriptions of what the program would do); and 3) the programming would fit on a single sheet of paper. By looking at the completed programming for the module, the team felt that it could answer the two questions.

On quality improvement projects, tactical plans are created by developing work breakdown structures. The process of creating work breakdown structures was discussed above. Tactical planning takes place at the bottom-most level of the work breakdown structures. If the project team doesn't agree that the bottom level is tiny, then additional work breakdown must take place.

Creating WBS employs the tree diagram technique. Tree diagrams are described in section V.B. Tree diagrams are used to break down or stratify ideas in progressively greater detail. The objective is to partition a big idea or problem into its smaller components. By doing this, you will make the idea easier to understand, or the problem easier to solve. The basic idea behind this is that, at some level, a problem's solution becomes relatively easy to find. This is the tiny level. Work takes place on the smallest elements in the tree diagram.

Tactical plans are still project plans, albeit for tiny projects. As such, they should include all of the elements of any well-designed project plan.

Contingency plans should be prepared to deal with unexpected but potentially damaging events. The process decision program chart (PDPC) is a useful tool for identifying possible events that might be encountered during the project. The emphasis of PDPC is the impact of the "failures" (problems) on project schedules. Also, PDPC seeks to describe specific actions to be taken to prevent the problems from occurring in the first place, and to mitigate the impact of the problems if they do occur. An enhancement to classical PDPC is to assign subjective probabilities to the various problems and to use these to help assign priorities. The amount of detail that should go into contingency plans is a judgment call. The project manager should consider both the seriousness of the potential problem and the likelihood of its occurring. See V.B for additional information on PDPC.

CROSS-FUNCTIONAL COLLABORATION

Quality improvement projects are process-oriented and most processes that have significant impact on quality cut across several different departments. Modern organizations, however, are hierarchical, i.e., they are defined by superior/subordinate relationships. These organizations tend to focus on specialized functions (e.g., accounting, engineering). But adding value for the customer requires that several different functions work together. The ideal solution is the transformation of the organization into a structure designed to produce value without the need for a hierarchical structure. However, until that is accomplished, QI project managers will need to deal with the conflicts inherent in doing cross-functional projects in a hierarchical environment.

Project managers "borrow" people from many departments for their projects, which creates matrix organizational structures. The essential feature of a matrix organization is that some people have two or more bosses or project customers. These people effectively report to multiple bosses, e.g., the project manager and their own boss. Ruskin and Estes refer to people with more than one boss as *multibossed individuals*, and their bosses and customers as *multiple bosses*. Somewhere in the organization is a *common boss*, who resolves conflicts between multiple bosses when they are unable to do so on their own. Of course, multiple bosses can prevent conflicts by cooperation and collaboration before problems arise.

Often multibossed individuals are involved with several projects, further complicating the situation. Collaborative planning between the multi-bosses is necessary to determine how the time of multi-bossed individuals, and other resources, will be shared. Figure I.15 illustrates the simplest multi-bossing structure where the multibossed individual has just two multiple bosses, and the common boss is directly above the multiple bosses on the organizational hierarchy. For additional discussion of more complex matrix structures see Ruskin and Estes, 1995, pp. 169–182.

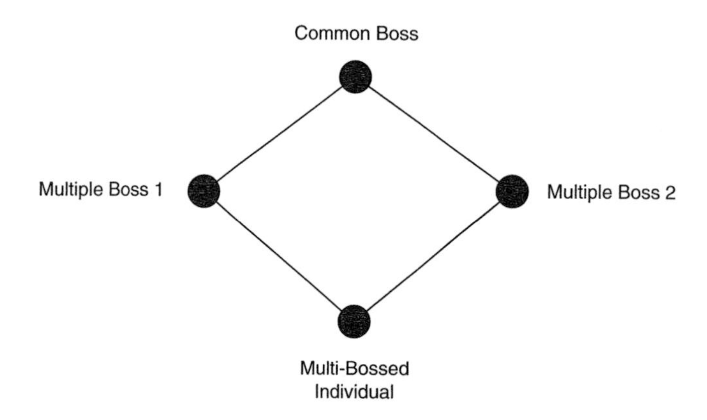

Figure I.15. Multi-boss collaborative planning.

The matrix structure raises a number of questions regarding project planning and coordination. What should be planned? Who should organize planning activities? Who should participate in the planning? These issues were addressed above, especially in the section "Relevant Stakeholders".

Good communication is helpful in preventing problems. Perhaps the most important communication is frequent, informal updates of all interested parties by the project manager. More formal status reports should also be specified in the project plan and sent to people with an interest in the project. The project manager should determine who gets what information, which is often tricky due to the multi-boss status of the project manager. Some managers may not want "too much" information about their department's "business" shared with their peers from other areas. Other managers may be offended if they receive less information than everyone else. The project manager's best diplomatic skills may be required to find the right balance.

Status reports invariably indicate that the project plan is less than perfect. The process by which the plans will be adjusted should be understood in advance. The process should specify who will be permitted to make adjustments, when the adjustments will be allowed and how much authority the bosses and project manager have in making adjustments.

Negotiated agreements should be documented, while generating the minimum possible amount of additional red tape and paperwork. The documentation will save the project manager a great deal of time in resolving disputes down the road regarding who agreed to what.

CONTINUOUS REVIEW AND ENHANCEMENT OF QUALITY PROCESS

The project management system can be improved, just as any system can be improved. The quality management professional's arsenal of tools and principles offer the means. Address such issues as cycle time, supplier management, customer service, etc., just as you would for any other critical management system. Continuous improvement principles, tools, and techniques are described in detail in chapters V and VI.

Projects have customers, usually internal. The techniques described in section H can be used to evaluate the satisfaction of the customers whose needs are being addressed by the project.

The records of the project provide the raw data for process improvement. These records, combined with customer satisfaction analysis, tell management where the project planning and implementation process can be improved. The project manager should include recommendations for process improvement in the final project report. Organizational management, in particular common bosses, should aggregate the recommendations from several projects to identify systemic problems. Project schedules and budgets should be compared to actual results to evaluate the accuracy of these forecasts.

The results can be analyzed using the techniques described in VI.G. Where special causes of variation in the results are present, they should be identified and corrected quickly. Common causes of variation indicate systemic problems.

DOCUMENTATION AND PROCEDURES

Project records provide information that is useful both while the project is underway, and afterwards. Project records serve three basic purposes:

- cost accounting requirements
- legal requirements
- learning

Project records should be organized and maintained as if they were part of a single database, even if it isn't possible to keep all of the records in a single location. There should be one "official" copy of the documentation, and a person designated to act as the caretaker of this information while the project is active. Upon completion of the project, the documentation should be sent to the organization's archives. Large quality improvement projects are expensive, time-consuming undertakings. The process involved is complex and often confusing. However, a great deal can be learned from studying the "project process." To do this requires that there be data. Archives of a number of projects can be used to identify common problems and patterns between the various projects. For example, project schedules may be consistently too optimistic or too pessimistic.

The following records should be kept:

- statement of work
- plans and schedules for projects and subprojects
- correspondence (written and electronic)
- written agreements
- meeting minutes
- action items and responsibilities
- budgets and financial reports
- cost-benefit analyses
- status reports
- presentation materials
- documentation of changes made to plans and budgets
- procedures followed or developed
- notes of significant lessons learned

It is good practice for the project team to have a final meeting to perform a "post mortem" of the project. The meeting should be conducted soon after the project's completion, while memories are still fresh. The meeting will cover the lessons learned from conducting the project, and recommendations for improving the process. The minutes from these meetings should be used to educate project managers.

The author believes that former guidelines for record retention are now outdated. In the past, record storage involved warehousing costs, insurance, aging of the paper, protecting the records from damage, etc. Furthermore, using the records required that one sift through endless boxes of disorganized material thrown haphazardly in a box. Today it is possible to reduce mountains of paper to electronic form. Low-cost software can automatically catalog the information and provide the ability to search the entire database quickly and easily. There seems to be little reason not to store complete project information indefinitely.

Formal versus informal reports

When people speak of "reports," they usually mean formal written reports or, perhaps, formal verbal presentations. These forms of communication have certain advantages. They are relatively self-contained and complete and thus useful to personnel not intimately involved with the project. Their form lends itself to long-term retention. They usually include additional background materials. Finally, formal written reports are usually designed to address the concerns of all parties. However, formal written reports also have some drawbacks. Their preparation takes considerable time and effort, which makes them costly. Also, by trying to address everyone's concern the formal written report usually contains a lot of information that is of little interest to the majority of the recipients. Of course, this latter drawback can be mitigated by creating a good table of contents and index and by carefully organizing the material.

Informal reports and correspondence are used to keep everyone up to date on the project. Unlike formal reports, this form of communication generally addresses a specific issue involving both the sender and the receiver. Because the parties to the communication usually bring a great deal of background information to the communication, informal reports tend to do poorly as stand-alone documents. However, informal reports can be prepared quickly and inexpensively and they tend to be short.

ELEMENTS OF A QUALITY INFORMATION SYSTEM

A Quality Information System (QIS) is the system used in collecting, analyzing, reporting, or storing information for the Total Quality System. This section covers the fundamental principles of QIS.

QIS contains both manual and computerized elements and it requires inputs from many different functions within the company. In many companies, the QIS is one part of an even more comprehensive system, known simply as the Information System or IS. IS is an effort to bring the data collection activities of the entire organization into a single integrated system. IS, when it works properly, has a number of advantages. For one thing, it eliminates the duplication of effort required if IS does not exist. Also, IS creates a globally optimized system, i.e., the whole system is designed to be of maximum benefit to the company. Without IS, it is likely that the information system will be a collection of optimized subsystems which are "best" for the individual departments they serve, but are probably not optimal for the company as a whole.

IS also has its down side. Even the concept itself can be questioned. "Dictatorship of data processing" is one common complaint. In designing the one grand system, data processing has the ultimate say in whose needs are served and when they will be served. Second, IS systems are often so large and complex that changing them is nearly impossible. IS is concerned almost completely with computerized systems, thereby making it incomplete since much vital information never makes it into a computerized database.

SCOPE AND OBJECTIVES OF A QUALITY INFORMATION SYSTEM

As mentioned earlier, QIS requires inputs from a wide variety of different functions within the company. This means that planning the system will involve a large number of people from these different groups. The quality department should not develop QIS in a vacuum and expect it to be accepted. QIS is not a quality department system, it is a *companywide system*. However, the quality department usually bears the major responsibility for operating the system.

While QIS means "computerized" to most people, it is important to note that most successful QIS applications begin as successful *manual* applications. Usually a system element important enough to be made a part of the computerized QIS is too important to wait until it can be programmed, thus some form of a manual system is established, at least as a stopgap measure. Planning the manual system is no trivial task; data collection needs are established; personnel are selected and trained; reports are designed and their distribution lists established, etc. Because of this vested work, manual systems provide a great deal of information that can be used in planning the more elaborate computer-based system.

The flow chart is a tool that has been used in planning computer systems since the beginning of the computer age. Flow charts can be applied at all stages in the process, from system design to writing the actual computer program code. The standard symbols used in creating information system flow charts are established by the American National Standard Institute (ANSI) and are shown in Figure I.16.

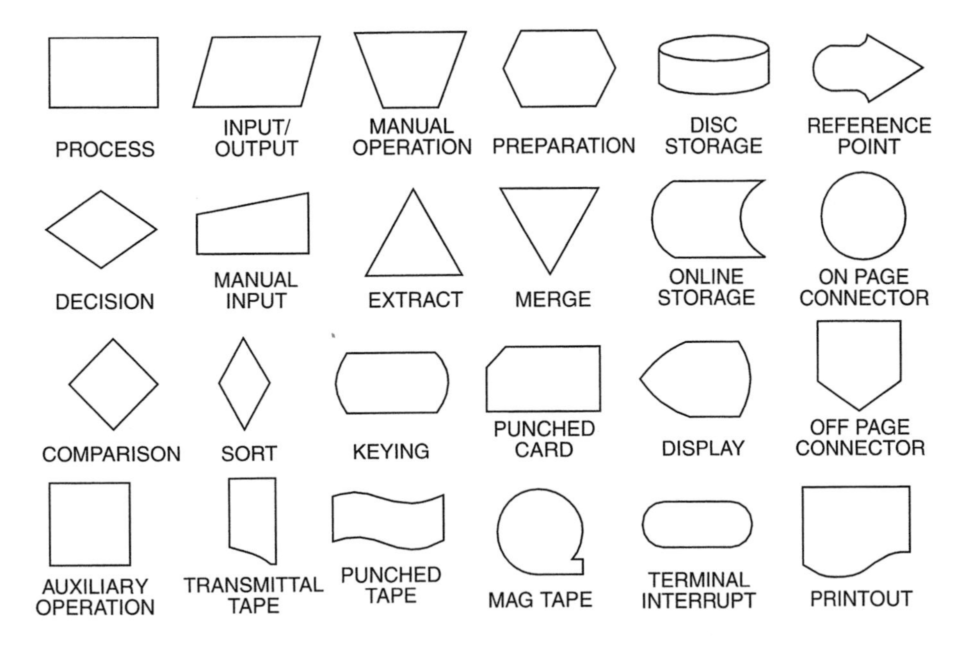

Figure I.16. Information systems flow chart symbols.

TECHNIQUES FOR ASSURING DATA ACCURACY AND INTEGRITY

Good QIS planning always starts with defining the benefits of the system. This drives the rest of the process. Benefits should be expressed in dollars when possible. However, the most important benefits are sometimes the most difficult to quantify, such as safety. At other times, the QIS elements are required by law; for example, FDA traceability requirements.

After deciding that the benefits justify the creation of the QIS element, the next step is to define the outputs required. This should be done in as much detail as possible, as it will be useful in later stages of the planning process. Output requirements should be classified into "must have" and "would like to have" categories. Often the cost of a system is increased dramatically because of the effort required to add some optional item. Data processing professionals refer to these options as "bells and whistles." The form of the output—on line data, periodic reports, graphs, etc.—can have a significant impact on the total cost of the system.

How often the output is needed is also important. Continuous output in the form of a real-time system is at one extreme, while annual reports are at another.

Figure I.17 illustrates the elements of a basic information system.

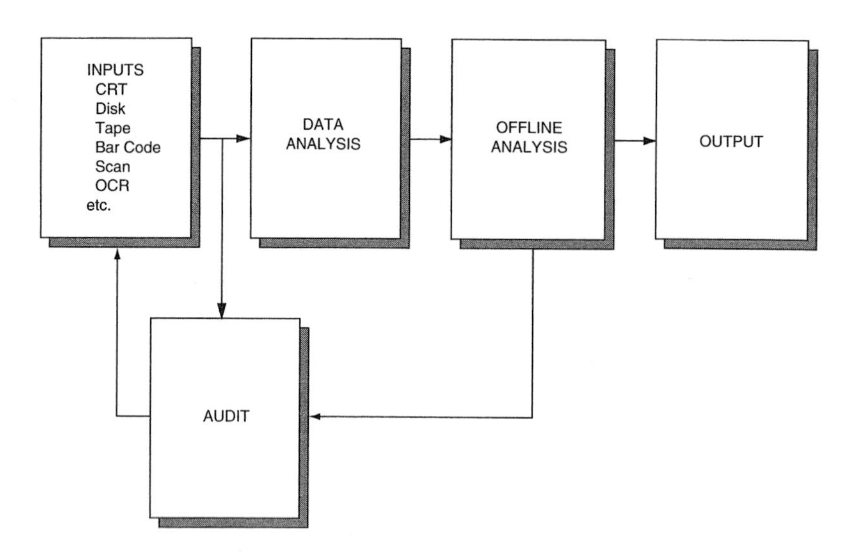

Figure I.17. Elements of an information system.

Inputs

Inputs, including personnel and equipment requirements, can be determined at this point. The resources required to implement the system can be substantial.

The computer can receive information in a variety of different ways. The methods can be divided into two different categories: on-line systems and batch systems. On-line systems enter data directly into the computer database (in "real time"), while batch systems go through an intermediate stage, such as keyboard data entry, and the database is updated periodically. Which system to use depends on the use to be made of the data. An inventory tracking system should be an on-line system since people will be using it to locate parts in real time. A quality cost reporting system could be a batch system since it only needs to be up to date at the time the monthly report is run. In general, batch systems are less expensive.

On-line systems involve entering data directly from the data collection point into the database. These systems may involve typing the information in at an on-line terminal, reading data in via a bar code system, voice data entry, scanned input, or input directly from automatic sensors. The technology is constantly changing, and it is beyond the scope of this book to cover the subject completely. In selecting the appropriate technology the system analyst must consider the environment in which the equipment will be used, costs, the amount of data to be collected, and the level of training and number of trained personnel required. Weighing the costs and benefits of the various technologies can be challenging.

Batch systems are generally more paper-oriented. Manual systems being converted to computers often go through a batch processing phase on the way to becoming a real-time system. Because most batch systems work from a physical document, the training is directed at getting the correct information on the document, and then at getting the information from the document into the computer. The existence of a physical document provides a built-in audit trail and greatly simplifies the task of verifying the accuracy of the information. Optical character readers can convert information from typed, printed, or even handwritten documents directly into an electronic database. Computers can also tie in to other computers over the phone lines or through direct links, and radio transmission of computer signals is also possible. As with real-time data entry, the pace of change requires that you research the technology at the time of your project.

More important than the technology is the *accuracy* of the input. The computer trade coined the phrase "garbage in-garbage out," or GIGO, to describe the common problem of electronically analyzing inaccurate input data. Accuracy can be improved in a variety of ways. Choosing the appropriate input technology is one way. Another is to have computer software check inputs for validity, rejecting obvious errors as soon as they are entered. However, no matter how much input checking takes place, it is usually necessary to have someone audit the input data to assure accuracy.

Data processing

When confident that the input data is accurate, you are ready to have the computer analyze the data. Whether to do the analysis on a micro computer, a mini computer, a mainframe computer, or a super computer is a question of balancing the need for speed and accuracy with that of cost. You must also consider the need for other information stored elsewhere in the system, such as the need to access the part number database. Again, the pace of change makes it essential that you review this question at the time you are designing your system.

Off-line analysis

A truism of quality is that you can never get all of the defects out of the system. A parallel truism of data processing is "all software has bugs." The result of these two in combination is the production of computer reports that contain ridiculous errors. Thus, before distribution of any computer report you should look it over for glaring mistakes. Is the scrap total larger than the sales of the company for the year? Are people who left the company last year being listed as responsible for current activities? Are known problem parts not listed in the top problems? This type of review can save you much embarrassment and it can also protect the integrity of the information system you worked so hard to create.

Reporting

Reports should be sent only to those who need them. And people should be sent only what they need. These simple guidelines are often ignored, creating an information overload for management. As a rule, the level of detail for a report is greatest at the lower levels of the organization and least at the top. A storeroom clerk may need a complete printout of all the items in his storeroom, breaking out the number of parts in each location. The company president only needs to know the total investment in inventory and the trend. It is recommended that charts be used to condense and summarize large quantities of tabular data when possible.

The distribution of reports should be carefully monitored. From time to time, report recipients should be asked if they still need the report. Also, the report recipients are your *customers* and their input regarding changes to the reports should be actively solicited and fully integrated into the report design and distribution systems.

Rapid pace of technology

It is virtually certain that anything said here regarding technology will be obsolete tomorrow. Few fields have advanced faster than microelectronics. Thus, the engineer must constantly update his or her knowledge of available solutions to information system problems. What is presented here is a partial list of recent innovations that promise to have an impact on quality engineering.

Expert systems

Although the field of artificial intelligence (AI) has been unable to approximate normal human thought, it has been much more successful in modeling the decision-making processes of experts. Even more impressive is the fact that, once the experts' process has been modeled, the computer does *better* than the human expert! Expert models currently exist for doctors and consultants in various fields. The engineer in need of expert assistance should avail himself of the "expert in a box" if at all possible.

Decision aids

Human beings are notoriously poor at using available information in certain types of decision-making. For example, customers are notoriously unreliable when asked to rank-order their preference for product features; repeated rankings are uncorrelated, and buying behavior fails to match stated preferences. However, using software, customers can be asked to make simple judgments which the computer can then analyze to determine conjoint preference weights which are much better predictors. The same approach can be used to determine subjective priorities for managers, something akin to doing a Pareto analysis without hard data.

Training

Mutimedia offers an incredibly effective method of delivering learning materials to students at their desks. At its best, the medium makes it possible for teams of experts to collaborate in the design of training programs that are better than anything instructors could deliver in a lecture format. In a well-designed instructional package, the student is totally involved in the learning experience. The combination of voice, sound, graphics, animation, video and text gives the instructional designer a wider range of options for presenting the material than any other medium. The cost per student is low, and dropping rapidly. In addition, scheduling considerations are much easier to manage. Of course, it is still possible to design ineffective multimedia training programs.

Archiving

Warehousing of vast quantities of paper documents is rapidly becoming a thing of the past. With the advent of low-cost automated scanners, optical character recognition (OCR) software, and truly massive storage capacities costing pennies per megabyte, the long-promised "paperless office" may finally come to pass. Digital archiving has other advantages:: the data can be electronically searched; the information retrieved can be easily included in other documents; data analysis can be performed without the data being re-entered into the computer; fewer errors can produce significant space savings and lower cost.

Data input

Data can now be entered directly into the computer in a wide variety of ways, including bar codes, scanned, direct gage input and voice data entry. Local area networks (LANs) and wide area networks (WANs) make computer-to-computer data exchange easy. Software exists which can scan data from completed forms directly into databases or analytical software packages such as SPSS.

Presentation aids

Modern presentation software makes it possible for everyone to deliver top-notch presentations complete with full-color graphics, sound, animation, video, and so on. These presentations involve all of the senses of the audience, thus making the presentation more interesting and increasing the retention rate. However, most engineers are untrained in commercial art and they make poor graphic artists. Multimedia presentations offer infinite opportunities to produce garish and ugly designs. Fortunately, most software packages also include "boilerplate" presentations with pre-set color schemes and graphic designs. Unless the engineer has some training and experience in such matters, it is best to use the standard designs that came with the computer, or seek professional help.

Team aids

Programs exist that allow people to participate in "electronic brainstorming." Members sit in a virtual electronic conference room and each member has a computer. The computer screen shows all of the ideas or comments that have been typed by the group. However, the person who made the comment is not shown, thus assuring anonymity. Research shows that this approach produces more ideas than conventional brainstorming.

Systems modeling and simulation

Modern simulation packages are a far cry from the user-hostile packages of yesterday. People with minimal training and non-technical backgrounds can build models by dragging icons and graphic elements around a computer screen. After doing a bit of data collection, the model can be run to see the entire system in operation; the graphic representations of people, equipment, supplies, etc., actually move around the screen. Many system problems become obvious by simply watching the animation. Other problems can be discovered by reviewing the statistics collected during the simulation. Proposed system changes can be evaluated electronically before committing resources. E.g., what would happen if we hired additional personnel? Would we be better off upgrading the equipment?

Networks and commercial computer information services

It is now easy to find people interested in precisely what you are interested in. Public-access computer networks such as the Internet or CompuServe allow people around the world to communicate with one another instantly. Participants can keep up on the latest news or learn about the solutions discovered by others doing the same thing they are doing. For example, there are user groups for TQM and ISO 9000 (and many others).

Research

The resources of hundreds of libraries can be accessed from any desktop computer with a modem. Literally millions of articles and technical papers can be searched in minutes. Some university libraries offer free services for searchers. By using interlibrary loans, books can be obtained at even the most remote locations. Some services will, for a fee, find the paper you want and fax it to you.

I.C LEADERSHIP PRINCIPLES AND TECHNIQUES

MOTIVATION THEORIES AND PRINCIPLES

The science of psychology, while still in its infancy, has much to offer anyone interested in motivating people to do a better job.

MASLOW'S HIERARCHY OF NEEDS

Professor A.S. Maslow of Brandeis University has developed a theory of human motivation elaborated on by Douglas McGregor. The theory describes a "hierarchy of needs." Figure I.18 illustrates this concept.

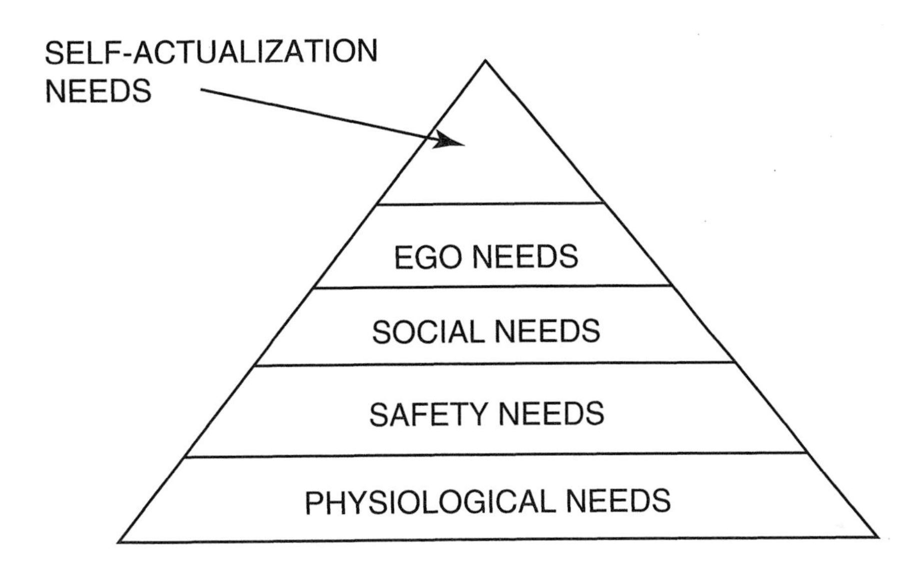

Figure I.18. Maslow's hierarchy of needs.

Maslow postulated that the lower needs must be satisfied before one can be motivated at higher levels. Furthermore, as an individual moves up the hierarchy the motivational strategy must be modified because *a satisfied need is no longer a motivator*; for example, how much would you pay for a breath of air right now? Of course, the answer is nothing because there is a plentiful supply of free air. However, if air were in short supply, you would be willing to pay plenty.

The hierarchy begins with physiological needs. At this level a person is seeking the simple physical necessities of life, such as food, shelter, and clothing. A person whose basic physiological needs are unmet will not be motivated with appeals to personal pride. If you wish to motivate personnel at this level, provide monetary rewards such as bonuses for good quality. Other motivational strategies include opportunities for additional work, promotions, or simple pay increases. As firms continue doing more business in underdeveloped regions of the world, this category of worker will become more commonplace.

Once the simple physiological needs have been met, motivation tends to be based on safety. At this stage issues such as job security become important. Quality motivation of workers in this stage was once difficult. However, since the loss of millions of jobs to foreign competitors who offer better quality goods, it is easy for people to see the relationship between quality, sales, and jobs.

Social needs involves the need to consider oneself as an accepted member of a group. People who are at this level of the hierarchy will respond to group situations and will work well on quality circles, employee involvement groups, or quality improvement teams.

The next level, ego needs, involve a need for self-respect and the respect of others. People at this level are motivated by their own craftsmanship, as well as by recognition of their achievements by others.

The highest level is that of self-actualization. People at this level are self-motivated. This type of person is characterized by creative self-expression. All you need do to "motivate" this group is to provide an opportunity for him or her to make a contribution.

HERZBERG'S HYGIENE THEORY

Frederick Herzberg is generally given credit for a theory of motivation known as the hygiene theory. The basic underlying assumption of the hygiene theory is that job satisfaction and job dissatisfaction are not opposites. Satisfaction can be increased by paying attention to "satisfiers," and dissatisfaction can be reduced by dealing with "dissatisfiers." The theory is illustrated in Figure I.19.

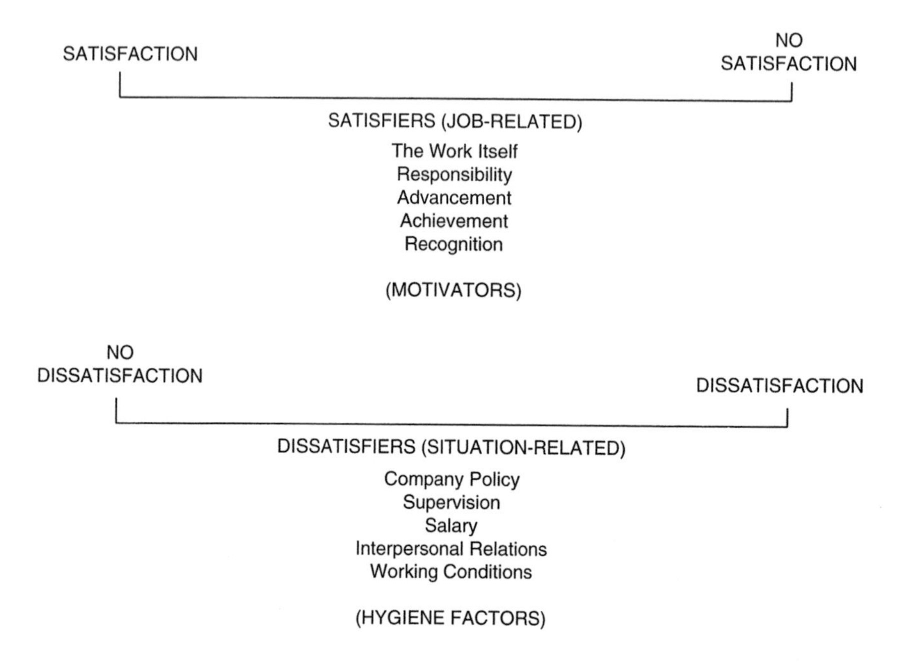

Figure I.19. Herzberg's hygiene theory.

THEORIES X, Y, AND Z

All people seem to seek a coherent set of beliefs that explain the world they see. The belief systems of managers were classified by McGregor into two categories, which he called Theory X and Theory Y.

Under Theory X, workers have no interest in work in general, including the quality of their work. Because civilization has mitigated the challenges of nature, modern man has become lazy and soft. The job of managers is to deal with this by using "carrots and sticks." The carrot is monetary incentive, such as piece rate pay. The stick is docked pay for poor quality or missed production targets. Only money can motivate the lazy, disinterested worker.

Theory Y advocates believe that workers are internally motivated. They take satisfaction in their work, and would like to perform at their best. Symptoms of indifference are a result of the modern workplace, which restricts what a worker can do and separates him from the final results of his efforts. It is management's job to change the workplace so that the worker can, once again, recapture his pride of workmanship.

Theories X and Y have been around for decades. Much later, in the 1980s, Theory Z came into vogue. Z organizations have consistent cultures where relationships are holistic, egalitarian, and based on trust. Since the goals of the organization are obvious to everyone, and integrated into each person's belief system, self-direction is predominant. In the Z organization, Theories X and Y become irrelevant. Workers don't need the direction of Theory X management, nor does management need to work on the removal of barriers since there are none.

MANAGEMENT STYLES

In this section we will discuss the question of "management style" from several different perspectives. To provide a context, we will first examine Deming's view of the proper role of a manager of people. Next we evaluate management styles as they relate to managers-as-decision-makers. Management decisions are classified as being judgmental, data-based, or scientific (Weaver 1995, pp. 223–231). We then look at management style from a psychological perspective and examine the premises upon which participatory and autocratic management styles are based. Management-by-wandering-around (MBWA) is briefly described. We conclude with a discussion of "Fourth-generation management," which is an attempt to incorporate Deming's guidelines into management.

Before discussing specific management styles per se, it is fair to ask "what is the proper role of a manager?" The following material is from Dr. W. Edwards Deming.*

*Deming, W. Edwards, (2000). *The New Economics for Industry, Government, Education, 2nd ed.*, Cambridge, MA: MIT Press.

Role of a manager of people

This is the new role of a manager of people after transformation:

1. A manager understands and conveys to his people the meaning of a system. He explains the aims of the system. He teaches his people to understand how the work of the group supports these aims.
2. He helps his people to see themselves as components in a system, to work in cooperation with preceding stages and with following stages toward optimization of the efforts of all stages toward achievement of the aim.
3. A manager of people understands that people are different from each other. He tries to create for everybody interest and challenge, and joy in work. He tries to optimize the family background, education, skills, hopes, and abilities of everyone.

   This is not ranking people. It is, instead, recognition of differences between people, and an attempt to put everybody in position for development.
4. He is an unceasing learner. He encourages his people to study. He provides, when possible and feasible, seminars and courses for advancement of learning. He encourages continued education in college or university for people who are so inclined.
5. He is coach and counsel, not a judge.
6. He understands a stable system. He understands the interaction between people and the circumstances that they work in. He understands that the performance of anyone that can learn a skill will come to a stable state—upon which further lessons will not bring improvement of performance. A manager of people knows that in this stable state it is distracting to tell the worker about a mistake.
7. He has three sources of power:
   1. Authority of office
   2. Knowledge
   3. Personality and persuasive power; tact

   A successful manager of people develops Nos. 2 and 3; he does not rely on No. 1. He has nevertheless the obligation to use No. 1, as this source of power enables him to change the process—equipment, materials, methods—to bring improvement, such as to reduce variation in output. (Dr. Robert Klekamp.)

   He is in authority, but lacking knowledge or personality (No. 2 or 3), must depend on his formal power (No. 1). He unconsciously fills a void in his qualifications by making it clear to everybody that he is in position of authority. His will be done.
8. He will study results with the aim to improve his performance as a manager of people.
9. He will try to discover who if anybody is outside the system, in need of special help. This can be accomplished with simple calculations, if there be individual figures on production or on failures. Special help may be only simple rearrangement of work.

It might be more complicated. He in need of special help is not in the bottom 5% of the distribution of others: he is clean outside that distribution. (See Figure I.20).

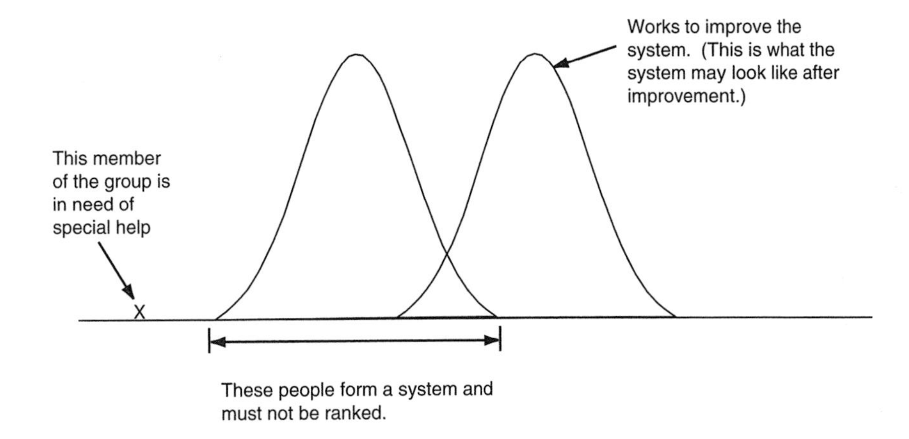

Figure I.20. Figures on production or on failures, if they exist, can be plotted. Study of the figures will show the system, and outliers if any.

10. He creates trust. He creates an environment that encourages freedom and innovation.
11. He does not expect perfection.
12. He listens and learns without passing judgment on him that he listens to.
13. He will hold an informal, unhurried conversation with every one of his people at least once a year, not for judgment, merely to listen. The purpose would be development of understanding of his people, their aims, hopes, and fears. The meeting will be spontaneous, not planned ahead.
14. He understands the benefits of cooperation and the losses from com-petition between people and between groups.*

Deming's perspective is important in that it represents a radical departure from the traditional view of the manager's role. As of this writing, it represents a set of normative guidelines that few organizations incorporate completely.

Judgmental management style

If the traditional organizational hierarchy is viewed as a "chain-of-command," then ultimate authority can be viewed as residing in the topmost position on the organization chart. The individual occupying this position "delegates" authority to subordinates who may, in turn, delegate authority to positions further down in the hierarchy. In this system,

*Alfie Kohn, *No Contest: The Case Against Competition*. (Houghton Mifflin, 1986).

managers are expected to use their authority to get the work done via command-and-control. Action is based on the manager's judgment. This system effectively requires that managers possess complete knowledge of the work being done by their subordinates—how else could the manager "command-and-control" the work? Of course, this omniscience does not exist. Thus, the manager who attempts to follow this metaphor too closely finds herself making decisions based on guesswork to a greater or lesser degree. The result is mistakes, for which the manager is held responsible. Managers who make too many mistakes may be fired, demoted, or disciplined. A natural response to this threat is fear, which may result in the manager blaming his subordinates for their "failures." Because of the authoritarian mindset, problems are ascribed to individuals, not systems. This produces the classic approach to performance appraisal, including ranking of employees, merit pay, etc. Another outcome of this approach is to act only when it is absolutely necessary. Since actions are based on judgments, judgments can lead to mistakes, and mistakes are punished; the manager who can minimize action will minimize the chance that a mistake will occur that can be blamed on her. Of course, this tendency is partially offset by the threat of being blamed for not meeting goals set by higher authorities.

Data-based management style

One reaction to the obvious shortcomings of the judgmental management has been to try to improve the judgments by relying on "facts." Managers solicit feedback from employees and review data in reports before making a decision. Ostensibly, this "data-based approach" changes the basis for action from the manager's judgment to data. Results are marginally better than with the purely judgmental approach. However, data are always incomplete and the element of judgment can never be completely removed. To the extent that the manager abdicates her responsibility for making a judgment, the quality of the decision will suffer. Another problem is the time involved in collecting data. The time (and expense) required increases exponentially to the extent that the manager wishes to remove all judgment from the decision and insists on "complete" data.

Combination data-based/judgment management style

Most experts in management advocate making management decisions based on a combination of the manager's judgment and reasonable amounts of data analysis. Managers, working with all parties impacted, formulate a coherent model of the system. The model (or theory) is used to predict the outcome that would result from operating the system in a certain manner. The system is operated and data is collected on the results obtained. The results are compared to the results predicted by the model, and the theory and systems are updated accordingly. This is the classic Shewhart Plan-Do-Check-Act (PDCA) cycle, or Deming Plan-Do-Study-Act (PDSA) cycle. It closely resembles the scientific method, the hypothesize-experiment-test hypothesis.

With this management style *systems* are evaluated rather than *people.* The change in focus is fundamental and profound. Here judgment is a source of generating hypotheses about systems or problems, and data is used to evaluate the quality of the hypotheses. People are asked to work to stabilize, then improve, the systems and the organization as a whole.

Participatory management style

The premise of the participatory management style is the belief that the worker can make a contribution to the design of their own work. The belief system that leads managers to this conclusion was originally put forth as a management theory by McGregor, who called it Theory Y. Theory Y advocates believe that workers are internally motivated. They take satisfaction in their work, and would like to perform at their best. Symptoms of indifference are a result of the modern workplace, which restricts what a worker can do and separates him from the final results of his efforts. It is management's job to change the workplace so that the worker can, once again, recapture his pride of workmanship. Elements of Theory Y are evident in Deming's discussion of the role of a manager of people, presented earlier.

Managers who practice the participatory style of management tend to engage in certain types of behavior. To engage the workers they establish and communicate the purpose and direction of the organization. This is used to help develop a shared vision of what the organization should be, which is used to develop a set of shared plans for achieving the vision. The manager's role is that of a leader. By her actions and words she shows the way to her employees. She is also a coach, evaluating the results of her people's efforts and helping them use the results to improve their processes. She works with the leaders above her in the organization to improve the organization's systems and the organization as a whole.

Autocratic management style

The premise of the autocratic management style is the belief that in most cases the worker *cannot* make a contribution to their own work, and that even if they could, they wouldn't. McGregor called the belief system that leads to this mindset Theory X. Under Theory X workers have no interest in work in general, including the quality of their work. Because civilization has mitigated the challenges of nature, modern man has become lazy and soft. The job of managers is to deal with this by using "carrots and sticks." The "carrot" is usually a monetary incentive, such as piece-rate pay schemes. The "stick" is docked pay for poor quality or missed production targets. Only money and threats can motivate the lazy, disinterested worker.

The natural management style that a manager with this belief system would favor is the autocratic management style. Autocratic managers attempt to control work to the maximum extent possible. A major threat to control is complexity; complex jobs are more difficult to learn and workers who master such jobs are scarce and possess a certain amount of control over how the job is done. Thus, autocratic managers attempt to simplify work to gain maximum control. Planning of work, including quality planning, is centralized. A strict top-down, chain-of-command approach to management is practiced. Procedures are maintained in exquisite detail and enforced by frequent audits. Product and process requirements are recorded in equally fine detail and in-process and final inspection are used to control quality.

Management By Wandering Around

Peters and Austin (1985, p. 8) call MBWA "The technology of the obvious." MBWA addresses a major problem with modern managers: lack of direct contact with reality. Many, perhaps most, managers don't have enough direct contact with their employees, their suppliers, or, especially, their customers. They maintain superficial contact with the world through meetings, presentations, reports, phone calls, email, and a hundred other ways that don't engage all of their senses. This is not enough. Without more intense contact the manager simply can't fully internalize the other person's experience. They need to give reality a chance to make them *really* experience the world. The difference between reality and many managers' perception of reality is as great as the difference between an icy blast of arctic air piercing thin indoor clothing versus watching a weather report of a blizzard from a sunny beach in the Bahamas.

MBWA is another, more personal way, to collect data. Statistical purists disdain and often dismiss data obtained from opportunistic encounters or unstructured observations. But the information obtained from listening to an employee or a customer pour their heart out is no less "scientifically valid" than a computer printout of customer survey results. And MBWA data is of a different type. Science has yet to develop reliable instruments for capturing the information contained in angry or excited voice pitch, facial expressions, the heavy sigh—but humans have no trouble understanding the meaning these convey in the context of a face-to-face encounter. It may be that nature has hardwired us to receive and understand these signals through eons of evolution.

The techniques employed by managers who practice MBWA are as varied as the people themselves. The important thing is to get yourself into direct contact with the customer, employee, or supplier, up close and personal. This may involve visiting a customer at his place of business, or bringing them to yours, manning the order desk or complaint line every month, spontaneously sitting down with employees in the cafeteria, either one-on-one or in groups, inviting the supplier's truck driver to your office for coffee. Use your imagination. One tip: be sure to schedule regular MBWA time. If it's not on your calendar, you probably won't do it.

Fourth generation management

In his book *Fourth Generation Management,* Brian Joiner develops four categories of management styles, which he calls "generations" (Joiner 1994, pp. 8–9):

- **1st Generation—Management by doing**—We simply do the task ourselves. Assuming we possess the necessary knowledge, skills, abilities, and technology, this is an effective way of assuring that tasks are done to our personal requirements. Its main problem is, of course, limited capacity. As individuals we lack the needed prerequisites to do all but a limited range of tasks, as well as the time to do more than a few things. Of course, there will always be some tasks performed using this approach.
- **2nd Generation—Management by directing (micromanagement)**—People found that they could expand their capacity by telling others exactly what to do and how to do it: a master craftsman giving detailed directions to apprentices. This approach allows an expert to leverage his or her time by getting others to do some of the work, and it maintains strict compliance to the expert's standards. Although the capacity of this approach is better than 1st generation management, micro-management still has limited capacity.
- **3rd Generation—Management by results**—People get sick and tired of you telling them every detail of how to do their jobs and say, "Just tell me what you want by when, and leave it up to me to figure out how to do it." So you say, "OK. Reduce inventories by 20% this year. I'll reward you or punish you based on how well you do it. Good luck." This is the current approach to management practiced by most modern organizations, with all of the problems of suboptimization discussed earlier. Suboptimizing a single department, division or other unit is the most commonly observed problem with this approach.
- **4th Generation—Systems approach.** The core elements of 4th generation management are shown in Figure I.21.

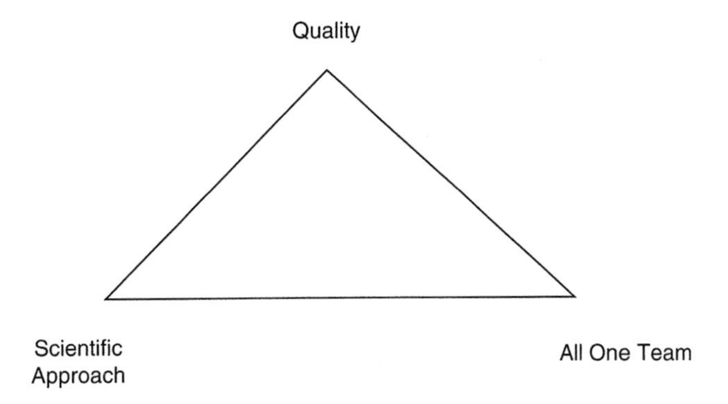

Figure I.21. The Joiner Triangle.

From Joiner, B.L. (1994). *Fourth Generation Management: The New Business Consciousness,* page 11.

The quality corner of the triangle represents an obsession with customer-perceived quality. The organization seeks to delight its customers, not to merely satisfy them. The scientific-approach corner indicates learning to manage the organization as a system, developing process thinking, basing decisions on data, and understanding data. "All one team" means believing in people; treating everyone with dignity, trust, and respect; working towards win-win for customers, employees, shareholders, suppliers, and the communities in which we live.

CONFLICT RESOLUTION

This section will discuss general principles of conflict resolution. For a discussion of conflict resolution in the context of teams, see section I.D later in this chapter.

Conflict is a natural result of people working together. Legitimate differences of opinion will arise in the course of normal work. It is unrealistic to expect a conflict-free working environment. In fact it is undesirable. Conflict is an indication that people see things differently. Identifying the reasons that underlie this situation is a crucial element in the creative problem-solving process. If conflicts are avoided at all costs, the organization will suffer from a lack of creativity.

When conflict is improperly handled, however, it can result in hurt feelings, lack of cooperation, destructive competition between parties, and a variety of other behaviors that are not in the best interest of the organization. Here are some general guidelines for resolving conflicts: for a more detailed discussion, see Johnson (1993a, chapter 16).

1. *Avoid taking sides.* Your position as mediator requires impartiality. Your job is to help the parties resolve their dispute, not to pick a winner. Stress win-win.
2. *Declare a cease-fire.* When people are in conflict, emotions can run high, making it difficult to reason. The first task of the mediator is to defuse the emotional situation with an appeal to reason. Get the parties to agree to a "cease-fire." Allow a period of time to pass so people can cool off. Even several minutes will help, although a full day is better. Until the ability to reason has been restored, it will not be possible to resolve the conflict.
3. *Meet in private.* People tend to have greater difficulty resolving conflicts in public. They may not want to be seen as "weak" or as having "backed down." Arrange a private meeting where there are no audiences. Set ground rules in advance, e.g., no personal attacks, no discussion of the past—only the future. Emphasize cooperation and forbid confrontation.
4. *Fact finding.* Examine the basis of the conflict. What are the *facts.* What goals and values are shared by the parties to the conflict? What benefits will be gained by both parties if the conflict is resolved?
5. *Summarize.* Stop frequently to restate each party's point of view. Use neutral wording. Be as accurate and factual as possible.

6. *Guide toward resolution.* Have the parties describe to each other how they wish to see the conflict resolved. If they won't do this, have them describe it to you and act as messenger. Create an atmosphere of urgency to overcome the natural tendency to "deal with" conflict by avoidance. Facilitate consensus.
7. *Arbitration.* If consensus can't be reached, provide arbitration as a third-party. This option is a last resort and should be avoided if at all possible.

TEAM FORMATION AND EVOLUTION

Perhaps the most important thing management can do for a group is to give it time to become effective. This requires, among other things, that management work to maintain consistent group membership. Group members must not be moved out of the group without very good reason. Nor should there be a constant stream of new people temporarily assigned to the group. If a group is to progress through the four stages described later in this chapter, to the crucial performing stage, it will require a great deal of discipline from both the group and management. Another area where management must help is creating an atmosphere within the company where groups can be effective. The methods for accomplishing this are discussed throughout this book.

The structure of modern organizations is based on the principle of division of labor. Most organizations today consist of a number of departments, each devoted to their own specialty. A fundamental problem is that the separate functional departments tend to optimize their own operations, often to the detriment of the organization as a whole.

Traditional organizations, in effect, create barriers between departments. Departmental managers are often forced to compete for shares of limited budgets; in other words, they are playing a "zero sum game" where another manager's gain is viewed as their department's loss. Behavioral research has shown that people engaged in zero sum games, i.e., they think in terms of win-lose. This leads to self-destructive, cut-throat behavior. Overcoming this tendency requires improved communication and cooperation between departments.

Interdepartmental teams are groups of people with the skills needed to deliver the value desired. Processes are designed by the team to create the value in an effective and efficient manner. Management must see to it that the needed skills exist in the organization. It is also management's job to see that they remove barriers to cooperation.

There are two ways to make quality improvements: improve performance given the current system, or improve the system itself. Much of the time improving performance given the current system can be accomplished by individuals working alone. For example, an operator might make certain adjustments to the machine. Studies indicate that this sort of action will be responsible for about 5%–15% of the improvements. The remaining 85%–95% of all improvements will require changing the system itself. This is seldom accomplished by individuals working alone. It requires group action. Thus, the vast majority of quality improvement activity will take place in a group setting. As with nearly everything, the group process can be made more effective by acquiring a better understanding of the way it works.

PROCESS IMPROVEMENT TEAMS

Process improvement philosophy and techniques were discussed at length in V.A. Management of cross-functional projects was discussed earlier. In this section we will focus on the team aspect of process improvement activity.

Process improvement teams focus on improving one or more important characteristics of a process, e.g., quality, cost, cycle time, etc. The focus is on an entire process, rather than on a particular aspect of the process. A process is an integrated chain of activities that add value. A process can be identified by its beginning and ending states, e.g., manufacturing's beginning state is procurement, its ending state is shipment. Methods of analyzing and characterizing processes are discussed throughout this book. Usually several departments are involved in any given value-added process.

Process improvement teams work on both incremental improvement (KAIZEN) and radical change (reengineering). The team is composed of members who work with the process on a routine basis. Team members typically report to different bosses, and their positions can be on different levels of the organization's hierarchy.

Process improvement projects must be approved by the process owner, usually a senior leader in the organization. Process improvement teams must be chartered and authorized to pursue process improvement. All of this falls in the area of project management. Project management is discussed in I.B.

WORK GROUPS

Work groups focus on improvement within a particular work area. The work area is usually contained within a single department or unit. The process owner is usually the department manager. Team members are usually at the same level within the organization's hierarchy and they usually report to one boss.

Work group members are trained in the use of quality control techniques and supported by management. The idea is that all workers have an important contribution to make to the quality effort and the work group is one mechanism for allowing them the opportunity to make their contribution.

Quality circles

An example of a work group is the quality circle. Quality circles originated in Japan and Japanese companies continue to use quality circles on a massive scale. Quality circles were tried on a massive scale in America, with only limited success. However, the quality circle is the historical forerunner of the modern quality improvement work team; a study of them reveals a great deal about the success factors needed for successful use of other types of work groups.

Quality circles (circles) are local groups of employees who work to continuously improve those processes under their direct control. Here are some necessary steps that must be completed before circles can succeed:

- Management from the top level to the supervisory level must have a clear idea of their organization's purpose. Everyone in the organization must be committed to helping the organization achieve its purpose.
- Senior leadership must have an effective organization for dealing with companywide issues such as quality, cost, cycle time, etc. (e.g., the cross functional form discussed earlier).
- Attention must be focused on processes rather than on internal politics and reporting relationships.
- Personnel involved must be trained in cooperation skills (e.g., team work, group dynamics, communication and presentation skills). This applies to area supervisors and managers, not just circle members.
- Personnel involved must be trained in problem-solving skills (e.g., the traditional QC tools, the 7M tools, brainstorming, etc.)
- Circle participation must be encouraged by local management.

However, this author believes that circles have an important place and that they can succeed anywhere providing the proper corporate environment exists. This environment did not exist in Western business organizations in the 1970s, and for the most part they still do not exist. Merely grafting quality circles onto a traditional command-and-control hierarchy won't work. There were many reasons why quality circles failed in America; they are the same reasons why work groups fail to this day.

1. The quality circle in an American firm was isolated, not part of a companywide quality control effort. As a result, circles were usually unable to deal successfully with problems involving other areas of the company. There were no resources in other areas to draw upon.
2. Key management personnel moved about too frequently and circles were not provided with consistent leadership and management support.
3. Employees transferred in and out of circle work areas too frequently. Without stability in the membership, circles never developed into effective groups. Building effective teams takes time.

OTHER SELF-MANAGED TEAMS

In addition to process-improvement teams and work groups, there are many other types of teams and groups involved to some extent in quality improvement. Self-managed teams are a way to reintegrate work and flatten the management hierarchy. If properly implemented and managed, the result can be improved quality and productivity. If poorly implemented and managed, the result can be added problems.

Self-managed teams are often given some of the responsibilities that, in traditional organizations, are reserved to management. This includes the authority to plan and schedule work, hiring, performance assessment, etc. This is essentially a reversal of over 90 years of scientific management. While difficult to implement successfully, the result is a leaner, more efficient organization, higher employee morale, and better quality. Several preconditions are necessary to assure success:

1. **Communicate and listen**—Encourage two-way, honest, open, frequent communication. The more informed employees are, the more secure and motivated they will be.
2. **Train employees**—An empowering culture is built on the bedrock of continuing education in every form imaginable. If an employee doesn't know what to do, how to do it right, or most important, why it is done a certain way and what difference it makes, don't expect him to feel or act empowered.
3. **Team employees**—No one has found a technological alternative to cooperation when it comes to building a positive work climate. Teams make it possible for people to participate in decision-making and implementation that directly affects them.
4. **Trust employees**—Support team decisions even if they aren't the outcomes you had in mind. Trust teams with information and allow them to fail.
5. **Feedback**—Find people doing things right. Recognize efforts as well as results by finding ways to frequently and creatively say thank you. Share the glory in every way possible. Give frequent specific performance feedback (good news as well as bad).

I.D FACILITATION PRINCIPLES AND TECHNIQUES

WHEN TO USE AN OUTSIDE FACILITATOR

It is not always necessary to have an outside party facilitate a group or team. While facilitators can often be of benefit, they may also add cost and the use of facilitators should, therefore, be carefully considered. The following guidelines can be used to determine if outside facilitation is needed (Schuman, 1996):

1. **Distrust or bias**—In situations where distrust or bias is apparent or suspected, groups should make use of an unbiased outsider to facilitate (and perhaps convene) the group.

2. **Intimidation**—The presence of an outside facilitator can encourage the participation of individuals who might otherwise feel intimidated.
3. **Rivalry**—Rivalries between individuals and organizations can be mitigated by the presence of an outside facilitator.
4. **Problem definition**—If the problem is poorly defined, or is defined differently by multiple parties, an unbiased listener and analyst can help construct an integrated, shared understanding of the problem.
5. **Human limits**—Bringing in a facilitator to lead the group process lets members focus on the problem at hand, which can lead to better results.
6. **Complexity or novelty**—In a complex or novel situation, a process expert can help the group do a better job of working together intellectually to solve the problem.
7. **Timelines**—If a timely decision is required, as in a crisis situation, the use of a facilitator can speed the group's work.
8. **Cost**—A facilitator can help the group reduce the cost of meeting—a significant barrier to collaboration.

SELECTING A FACILITATOR

Facilitators should possess four basic capabilities (Schuman, 1996):

1. He or she should be able to anticipate the complete problem-solving and decision-making process.
2. He or she should use procedures that support both the group's social and cognitive processes.
3. He or she should remain neutral regarding content issues and values.
4. He or she should respect the group's need to understand and learn from the problem-solving process.

Facilitation works best when the facilitator:

- Takes a strategic and comprehensive view of the problem-solving and decision-making processes and selects, from a broad array, the specific methods that match the group's needs and the tasks at hand.
- Supports the group's social and cognitive processes, freeing the group members to focus their attention on substantive issues.
- Is trusted by all group members as a neutral party who has no biases or vested interest in the outcome.
- Helps the group understand the techniques being used and helps the group improve its own problem-solving processes.

PRINCIPLES OF TEAM LEADERSHIP AND FACILITATION

Human beings are social by nature. People tend to seek out the company of other people. This is a great strength of our species, one that enabled us to rise above and dominate beasts much larger and stronger than ourselves. It is this ability that allowed men to control herds of livestock, to hunt swift antelope, and to protect themselves against predators. However, as natural as it is to belong to a group, there are certain behaviors that can make the group function more (or less) effectively than their members acting as individuals.

We will define a group as a collection of individuals who share one or more common characteristics. The characteristic shared may be simple geography, i.e., the individuals are gathered together in the same place at the same time. Perhaps the group shares a common ancestry, like a family. Modern society consists of many different types of groups. The first group we join is, of course, our family. We also belong to groups of friends, sporting teams, churches, PTAs, and so on. The groups differ in many ways. They have different purposes, different time frames, and involve varying numbers of people. However, all effective groups share certain common features. In their work, *Joining Together*, Johnson and Johnson list the following characteristics of an effective group:

- Group goals must be clearly understood, be relevant to the needs of group members, and evoke from every member a high level of commitment to their accomplishment.
- Group members must communicate their ideas and feelings accurately and clearly. Effective, two-way communication is the basis of all group functioning and interaction among group members.
- Participation and leadership must be distributed among members. All should participate, and all should be listened to. As leadership needs arise, members should all feel responsibility for meeting them. The equalization of participation and leadership makes certain that all members will be involved in the group's work, committed to implementing the group's decisions, and satisfied with their membership. It also assures that the resources of every member will be fully utilized, and increases the cohesiveness of the group.
- Appropriate decision-making procedures must be used flexibly if they are to be matched with the needs of the situation. There must be a balance between the availability of time and resources (such as member's skills) and the method of decision-making used for making the decision. The most effective way of making a decision is usually by consensus (see below). Consensus promotes distributed participation, the equalization of power, productive controversy, cohesion, involvement, and commitment.

- Power and influence need to be approximately equal throughout the group. They should be based on expertise, ability, and access to information, not on authority. Coalitions that help fulfill personal goals should be formed among group members on the basis of mutual influence and interdependence.
- Conflicts arising from opposing ideas and opinions (controversy) are to be *encouraged.* Controversies promote involvement in the group's work, quality, creativity in decision-making, and commitment to implementing the group's decisions. Minority opinions should be accepted and used. Conflicts prompted by incompatible needs or goals, by the scarcity of a resource (money, power), and by competitiveness must be negotiated in a manner that is mutually satisfying and does not weaken the cooperative interdependence of group members.
- Group cohesion needs to be high. Cohesion is based on members liking each other, each member's desire to continue as part of the group, the satisfaction of members with their group membership, and the level of acceptance, support, and trust among the members. Group norms supporting psychological safety, individuality, creativeness, conflicts of ideas, growth, and change need to be encouraged.
- Problem-solving adequacy should be high. Problems must be resolved with minimal energy and in a way that eliminates them permanently. Procedures should exist for sensing the existence of problems, inventing and implementing solutions, and evaluating the effectiveness of the solutions. When problems are dealt with adequately, the problem-solving ability of the group is increased, innovation is encouraged, and group effectiveness is improved.
- The interpersonal effectiveness of members needs to be high. Interpersonal effectiveness is a measure of how well the consequences of your behavior match intentions.

These attributes of effective groups apply regardless of the activity in which the group is engaged. It really doesn't matter if the group is involved in a study of air defense, or planning a prom dance. The common element is that there is a group of human beings engaged in pursuit of group goals.

FACILITATING THE GROUP TASK PROCESS

Team activities can be divided into two subjects: task-related and maintenance-related. Task activities involve the reason the team was formed, its charter, and its explicit goals.

The facilitator should be selected before the team is formed and he or she should assist in identifying potential team members and leaders, and in developing the team's charter. The subject of team formation and chartering is discussed in detail in I.C.

The facilitator also plays an important role in helping the team develop specific goals based on their charter. Goal-setting is an art and it is not unusual to find that team goals bear little relationship to what management actually had in mind when the team was formed. Common problems are goals that are too ambitious, goals that are too limited and goals that assume a cause and effect relationship without proof. An example of the latter would be a team chartered to reduce scrap assuming that Part X had the highest scrap loss (perhaps based on a week's worth of data) and setting as its goal the reduction of scrap for that part. The facilitator can provide a channel of communication between the team and management.

Facilitators can assist the team leader in creating a realistic schedule for the team to accomplish its goals. The issue of scheduling projects is covered in I.B.

Facilitators should assure that adequate records are kept on the team's projects. Records should provide information on the current status of the project. Records should be designed to make it easy to prepare periodic status reports for management. The facilitator should arrange for clerical support with such tasks as designing forms, scheduling meetings, obtaining meeting rooms, securing audio visual equipment and office supplies, etc.

Other activities where the facilitator's assistance is needed includes:

Meeting management—Schedule the meeting well ahead of time. Be sure that key people are invited and that they plan to attend. Prepare an agenda and stick to it! Start on time. State the purpose of the meeting clearly at the outset. Take minutes. Summarize from time-to-time. Actively solicit input from those less talkative. Curtail the overly talkative members. Manage conflicts. Make assignments and responsibilities explicit and specific. End on time.

Communication—The idea that "the quality department" can "assure" or "control" quality is now recognized as an impossibility. To achieve quality the facilitator must enlist the support and cooperation of a large number of people outside of the team. The facilitator can relay written and verbal communication between the team and others in the organization. Verbal communication is valuable even in the era of instantaneous electronic communication. A five minute phone call can provide an opportunity to ask questions and receive answers that would take a week exchanging email and faxes. Also, the team meeting is just one communication forum, the facilitator can assist team members in communicating with one another between meetings by arranging one-on-one meetings, acting as a go-between, etc.

FACILITATING THE GROUP MAINTENANCE PROCESS

Study the group process. The facilitator is in a unique position to stand back and observe the group at work. Are some members dominating the group? Do facial expressions and body language suggest unspoken disagreement with the team's direction? Are quiet members being excluded from the discussion?

When these problems are observed, the facilitator should provide feedback and guidance to the team. Ask the quiet members for their ideas and input. Ask if anyone has a problem with the team's direction. Play devil's advocate to draw out those with unspoken concerns.

TEAM DYNAMICS MANAGEMENT, INCLUDING CONFLICT RESOLUTION

Conflict management is a duty shared by the facilitator and the team leader (see section I.C for additional information on conflict management). The facilitator can assist the leader by assuring that creative conflict is not repressed, but encouraged. Explore the underlying reasons for the conflict. If "personality disputes" are involved that threaten to disrupt the team meeting, arrange one-on-one meetings between the parties and attend the meetings to help mediate.

The first step in establishing an effective group is to create a consensus decision rule for the group, namely:

> No judgment may be incorporated into the group decision until it meets at least tacit approval of every member of the group.

This minimum condition for group movement can be facilitated by adopting the following behaviors:

- *Avoid arguing for your own position.* Present it as lucidly and logically as possible, but be sensitive to and consider seriously the reactions of the group in any subsequent presentations of the same point.
- *Avoid "win-lose" stalemates in the discussion of opinions.* Discard the notion that someone must win and someone must lose in the discussion; when impasses occur, look for the next most acceptable alternative for all the parties involved.
- *Avoid changing your mind only to avoid conflict and to reach agreement and harmony.* Withstand pressures to yield which have no objective or logically sound foundation. Strive for enlightened flexibility; but avoid outright capitulation.
- *Avoid conflict-reducing techniques such as the majority vote, averaging, bargaining, coin-flipping, trading out, and the like.* Treat differences of opinion as indicative of an incomplete sharing of relevant information on someone's part, either about task issues, emotional data, or gut level intuitions.
- *View differences of opinion as both natural and helpful rather than as a hindrance in decision-making.* Generally, the more ideas expressed, the greater the likelihood of conflict will be; but the richer the array of resources will be as well.
- *View initial agreement as suspect.* Explore the reasons underlying apparent agreements; make sure people have arrived at the same conclusions for either the same basic reasons or for complementary reasons before incorporating such opinions into the group decision.

- *Avoid subtle forms of influence and decision modification.* E.g., when a dissenting member finally agrees, don't feel that he must be rewarded by having his own way on some subsequent point.
- *Be willing to entertain the possibility that your group can achieve all the fore-going and actually excel at its task.* Avoid doomsaying and negative predictions for group potential.

Collectively, the above steps are sometimes known as the "consensus technique." In tests it was found that 75% of the groups who were instructed in this approach significantly outperformed their best individual resources.

STAGES IN GROUP DEVELOPMENT

Groups of many different types tend to evolve in similar ways. It often helps to know that the process of building an effective group is proceeding normally. Bruce W. Tuckman identified four stages in the development of a group: forming, storming, norming, and performing.

During the *forming* stage a group tends to emphasize procedural matters. Group interaction is very tentative and polite. The leader dominates the decision-making process and plays a very important role in moving the group forward.

The *storming* stage follows forming. Conflict between members, and between members and the leader, are characteristic of this stage. Members question authority as it relates to the group objectives, structure, or procedures. It is common for the group to resist the attempts of its leader to move them toward independence. Members are trying to define their role in the group.

It is important that the leader deal with the conflict constructively. There are several ways in which this may be done:

- Do not tighten control or try to force members to conform to the procedures or rules established during the forming stage. If disputes over procedures arise, guide the group toward new procedures based on a group consensus.
- Probe for the true reasons behind the conflict and negotiate a more acceptable solution.
- Serve as a mediator between group members.
- Directly confront counterproductive behavior.
- Continue moving the group toward independence from its leader.

During the *norming* stage the group begins taking responsibility, or ownership, of its goals, procedures, and behavior. The focus is on working together efficiently. Group norms are enforced on the group by the group itself.

The final stage is *performing.* Members have developed a sense of pride in the group, its accomplishments, and their role in the group. Members are confident in their ability to contribute to the group and feel free to ask for or give assistance.

COMMON PROBLEMS

Table I.9 lists some common problems with teams, along with recommended remedial action (Scholtes, 1996).

Table I.9. Common team problems and remedial action.

PROBLEM	ACTION
Floundering	• Review the plan • Develop a plan for movement
The expert	• Talk to offending party in private • Let the data do the talking • Insist on consensus decisions
Dominating participants	• Structure participation • Balance participation • Act as gate-keeper
Reluctant participants	• Structure participation • Balance participation • Act as gate-keeper
Using opinions instead of facts	• Insist on data • Use scientific method
Rushing things	• Provide constructive feedback • Insist on data • Use scientific method
Attribution (i.e., attributing motives to people with whom we disagree)	• Don't guess at motives • Use scientific method • Provide constructive feedback
Ignoring some comments	• Listen actively • Train team in listening techniques • Speak to offending party in private
Wanderlust	• Follow a written agenda • Restate the topic being discussed
Feuds	• Talk to offending parties in private • Develop or restate groundrules

MEMBER ROLES AND RESPONSIBILITIES

Productive group roles

There are two basic types of roles assumed by members of a group: task roles and group maintenance roles. Group task roles are those functions concerned with facilitating and coordinating the group's efforts to select, define, and solve a particular problem. The group task roles shown in Table I.10 are generally recognized:

Table I.10 Group task roles.

ROLE I.D.	DESCRIPTION
Initiator	Proposes new ideas, tasks, or goals; suggests procedures or ideas for solving a problem or for organizing the group.
Information seeker	Asks for relevant facts related to the problem being discussed.
Opinion seeker	Seeks clarification of values related to problem or suggestion.
Information giver	Provides useful information about subject under discussion.
Opinion giver	Offers his/her opinion of suggestions made. Emphasis is on values rather than facts.
Elaborator	Gives examples.
Coordinator	Shows relationship among suggestions; points out issues and alternatives.
Orientor	Relates direction of group to agreed-upon goals.
Evaluator	Questions logic behind ideas, usefulness of ideas, or suggestions.
Energizer	Attempts to keep the group moving toward an action.
Procedure technician	Keeps group from becoming distracted by performing such tasks as distributing materials, checking seating, etc.
Recorder	Serves as the group memory.

Another type of role played in small groups are the group maintenance roles. Group maintenance roles are aimed at building group cohesiveness and group-centered behavior. They include those behaviors shown in Table I.11.

Table I.11. Group maintenance roles.

ROLE I.D.	DESCRIPTION
Encourager	Offers praise to other members; accepts the contributions of others.
Harmonizer	Reduces tension by providing humor or by promoting reconciliation; gets people to explore their differences in a manner that benefits the entire group.
Compromiser	This role may be assumed when a group member's idea is challenged; admits errors, offers to modify his/her position.
Gate-keeper	Encourages participation, suggests procedures for keeping communication channels open.
Standard setter	Expresses standards for group to achieve, evaluates group progress in terms of these standards.
Observer/commentator	Records aspects of group process; helps group evaluate its functioning.
Follower	Passively accepts ideas of others; serves as audience in group discussions.

The development of task and maintenance roles is a vital part of the team- building process. Team building is defined as the process by which a group learns to function as a unit, rather than as a collection of individuals.

Counter-productive group roles

In addition to developing productive group-oriented behavior, it is also important to recognize and deal with individual roles which may block the building of a cohesive and effective team. These roles are shown in Table I.12.

Table I.12. Counterproductive group roles.

ROLE I.D.	DESCRIPTION
Aggressor	Expresses disapproval by attacking the values, ideas, or feelings of other. Shows jealousy or envy.
Blocker	Prevents progress by persisting on issues that have been resolved; resists attempts at consensus; opposes without reason.
Recognition-seeker	Calls attention to himself/herself by boasting, relating personal achievements, etc.
Confessor	Uses group setting as a forum to air personal ideologies that have little to do with group values or goals.
Playboy	Displays lack of commitment to group's work by cynicism, horseplay, etc.
Dominator	Asserts authority by interrupting others, using flattery to manipulate, claiming superior status.
Help-seeker	Attempts to evoke sympathy and/or assistance from other members through "poor me" attitude.
Special-interest pleader	Asserts the interests of a particular group. This group's interest matches his/her self-interest.

The leader's role includes that of process observer. In this capacity, the leader monitors the atmosphere during group meetings and the behavior of individuals. The purpose is to identify counterproductive behavior. Of course, once identified, the leader must tactfully and diplomatically provide feedback to the group and its members. The success of SPC is, to a great extent, dependent on the performance of groups.

Management's role

Perhaps the most important thing management can do for a group is to give it time to become effective. This requires, among other things, that management work to maintain consistent group membership. Group members must not be moved out of the group without very good reason. Nor should there be a constant stream of new people temporarily assigned to the group. If a group is to progress through the four stages described earlier in this chapter, to the crucial performing stage, it will require a great deal of discipline from both the group and management.

Another area where management must help is creating an atmosphere within the company where groups can be effective.

PERFORMANCE EVALUATION

Evaluating team performance involves the same principles as evaluating performance in general. Before one can determine how well the team's task has been done, a baseline must be established and goals must be identified. Setting goals using benchmarking and other means is discussed elsewhere in this book (see sections I.B and II.A). Records of progress should be kept as the team pursues its goals.

Performance measures generally focus on group tasks, rather than on internal group issues. Typically, financial performance measures show a payback ratio of between 2:1 and 8:1 on team projects. Some examples of tangible performance measures are:

- productivity
- quality
- cycle time
- grievances
- medical usage (e.g., sick days)
- absenteeism
- service
- turnover
- dismissals
- counseling usage

Many intangibles can also be measured. Some examples of intangibles effected by teams are:

- employee attitudes
- customer attitudes
- customer compliments
- customer complaints

The performance of the team process should also be measured. Project failure rates should be carefully monitored. A *p* chart can be used to evaluate the causes of variation in the proportion of team projects that succeed. Failure analysis should be rigorously conducted. The continuous improvement methods discussed in chapter V should be utilized.

Aubrey and Felkins (1988) list the effectiveness measures shown below:

- leaders trained
- number of potential volunteers
- number of actual volunteers
- percent volunteering
- projects started
- projects dropped
- projects completed/approved
- projects completed/rejected
- improved productivity
- improved work environment
- number of teams
- inactive teams
- improved work quality
- improved service
- net annual savings

RECOGNITION AND REWARD

Recognition is a form of employee motivation in which the company identifies and thanks employees who have made positive contributions to the company's success. In an ideal company, motivation flows from the employees' pride of workmanship. When employees are enabled by management to do their jobs and produce a product or service of excellent quality, they will be motivated.

The reason recognition systems are important is not that they improve work by providing incentives for achievement. Rather, they make a statement about what is important to the company. Analyzing a company's employee recognition system provides a powerful insight into the company's values in action. These are the values that are actually driving employee behavior. They are not necessarily the same as management's stated values. For example, a company that claims to value customer satisfaction but recognizes only sales achievements probably does not have customer satisfaction as one of its values in action.

Public recognition is often better for two reasons:

1. Some (but not all) people enjoy being recognized in front of their colleagues.
2. Public recognition communicates a message to all employees about the priorities and function of the organization.

The form of recognition can range from a pat on the back to a small gift to a substantial amount of cash. When substantial cash awards become an established pattern, however, it signals two potential problems:

1. It suggests that several top priorities are competing for the employee's attention, so that a large cash award is required to control the employee's choice.
2. Regular, large cash awards tend to be viewed by the recipients as part of the compensation structure, rather than as a mechanism for recognizing support of key corporate values.

Carder and Clark (1992) list the following guidelines and observations regarding recognition:

Recognition is not a method by which management can manipulate employees. If workers are not performing certain kinds of tasks, establishing a recognition program to raise the priority of those tasks might be inappropriate. Recognition should not be used to get workers to do something they are not currently doing because of conflicting messages from management. A more effective approach is for management to first examine the current system of priorities. Only by working on the system can management help resolve the conflict.

Recognition is not compensation. In this case, the award must represent a significant portion of the employee's regular compensation to have significant impact. Recognition and compensation differ in a variety of ways:

- Compensation levels should be based on long-term considerations such as the employee's tenure of service, education, skills, and level of responsibility. Recognition is based on the specific accomplishments of individuals or groups.
- Recognition is flexible. It is virtually impossible to reduce pay levels once they are set, and it is difficult and expensive to change compensation plans.
- Recognition is more immediate. It can be given in timely fashion and therefore relate to specific accomplishments.
- Recognition is personal. It represents a direct and personal contact between employee and manager.

Recognition should be personal. Recognition should not be carried out in such a manner that implies that people of more importance (managers) are giving something to people of less importance (workers).

Positive reinforcement is not always a good model for recognition. Just because the manager is using a certain behavioral criterion for providing recognition, it doesn't mean that the recipient will perceive the same relationship between behavior and recognition.

Employees should not believe that recognition is based primarily on luck. An early sign of this is cynicism. Employees will tell you that management says one thing but does another.

Recognition meets a basic human need. Recognition, especially public recognition, meets the needs for belonging and self-esteem. In this way, recognition can play an important function in the workplace. According to Maslow's theory, until these needs for belonging and self-esteem are satisfied, self-actualizing needs such as pride in work, feelings of accomplishment, personal growth, and learning new skills will not come into play.

Recognition programs should not create winners and losers. Recognition programs should not recognize one group of individuals time after time while never recognizing another group. This creates a static ranking system, with all of the problems discussed earlier.

Recognition should be given for efforts, not just for goal attainment. According to Imai, a manager who understands that a wide variety of behaviors are essential to the company will be interested in criteria of discipline, time management, skill development, participation, morale, and communication, as well as direct revenue production. To be able to effectively use recognition to achieve business goals, managers must develop the ability to measure and recognize such process accomplishments.

Employee involvement is essential in planning and executing a recognition program. It is essential to engage in extensive planning before instituting a recognition program or before changing a bad one. The perceptions and expectations of employees must be surveyed.

Principles of effective reward systems

Kohn (1999) believes that nearly all existing reward systems share the following characteristics:

1. they punish the recipients
2. they rupture relationships
3. they ignore reasons for behavior
4. they discourage risk-taking

Most existing reward systems (including many compensation systems) are an attempt by management to manipulate the behaviors of employees. Kohn convincingly demonstrates, through solid academic research into the effects of rewards, that people who receive the rewards as well as those who hand them out suffer a *loss* of incentive—hardly the goal of the exercise!

Rather than provide cookbook solutions to the problem of rewards and incentives, Kohn offers some simple guidelines to consider when designing reward systems.

1. *Abolish incentive pay* (something Deming advocated as well). Hertzberg's hygiene theory tells us that money is not a motivator, but it can be a demotivator. Pay people generously and equitably, then do everything in your power to put money out of the employee's mind.
2. *Reevaluate evaluation.* See Pyzdek (2000) for information on performance appraisals and alternatives.
3. *Create conditions for authentic motivation.* Money is no substitute for the real thing—interesting work. Here are some principles to use to make work more interesting:

a. *Design interesting jobs.* Give teams projects that are intrinsically motivating, e.g., meaningful, challenging, achievable.
b. *Collaborative.* Help employees work together, then provide the support needed to make it possible for the teams to accomplish their goals.
c. *Freedom.* Trust people to make the right choices. Encourage them when they make mistakes.

I.E TRAINING

Although the terms are loosely synonymous, a distinction can be made between *training* and *education.* The dictionary definitions of the two terms are the following:

Educate

1. To develop the innate capacities of, especially by schooling or instruction.
2. To stimulate or develop the mental or moral growth of.

Train

1. To coach in or accustom to a mode of behavior or performance.
2. To make proficient with specialized instruction and practice.

Education is seen to be more general and more focused on mental processes. In other words, education teaches people how to *think.* It enhances the mind's ability to deal with reality. Education is more theoretical and conceptual in nature. The *essentials* of past knowledge are taught for the purpose of teaching students how to appraise new events and situations. Education equips students to acquire new knowledge. Generally, people bring a certain *educational background* to their job with them. Employers typically fund ongoing education of employees, but they do not usually provide it for the employees. Education is more likely the responsibility of the individual employee. This may be because education is more generally applicable and more "portable" than is training. Thus, the employee's education is readily transferable to new employers.

Training focuses more on *doing.* Training is more job-focused than education. The emphasis is on maintaining or improving skills needed to perform the current job, or on acquiring new job skills. Training is usually paid for, and often provided by the employer. Often, training is employer-specific and the skills acquired may not always transfer readily to another employer.

Quality improvement requires change and change starts with people. People change when they understand why they must change and how they must change to meet their own personal goals. People join organizations because doing so helps them meet certain of their own goals. Conversely, organizations hire people to help achieve the organization's goals. When organizations set new goals they are, in effect, asking their employees to think differently, perform new tasks, and to engage in new behaviors. Organizations must be prepared to help employees acquire the knowledge, skills, and abilities (KSAs) required by these new expectations. Training and education are the means by which these new KSAs are acquired.

TRAINING NEEDS ANALYSIS

The first step in the development of the strategic training plan is a training needs assessment. The training needs assessment provides the background necessary for designing the training program and preparing the training plan. The assessment proceeds by performing a task-by-task audit to determine what the organization is doing, and comparing it to what the organization should be doing. The assessment process focuses on three major areas:

Process audit—As stated numerous times, all work is a process. Processes are designed to add values to inputs and deliver values to customers as outputs. Are they operating as designed? Are they operated consistently? Are they measured at key control points? If so, do the measurements show statistical control? The answers to these questions, along with detailed observations of how the process is operated, are input to the development of the training plan.

Assessment of knowledge, skills and abilities—In all probability, there will be deficiencies (opportunities for improvement) observed during the process audits. Some of these deficiencies will involve employee knowledge, skills, or abilities (KSAs). The first principle of self-control is that employees must know what they are doing.

Management's job doesn't end by simply giving an employee responsibility for a particular process or task; they must also provide the employee with the opportunity to acquire the KSAs necessary to successfully perform their new duties. This means that if the employee is asked to assume new duties as a member of a quality improvement team, they are given training in team skills; if they are to keep a control chart, they receive training in the maintenance and interpretation of the charts, etc. Since employees are expected to contribute to the implementation of the organization's strategic plan, they should be told what the plan is, and how their job contributes to the plan.

Assessment of employee attitudes—Attitudes are emotions that reflect a response to something taking place within an organization. A person's attitude is, in essence, a judgment about the wisdom of a particular course of events. Without a positive attitude an employee's KSA will not be as effective in helping the organization. Negative employee attitudes about the direction being taken by the organization indicate that the employee either questions the wisdom of the proposed changes, or doubts the sincerity of the leadership. Regardless, it represents a problem that must be addressed by the training plan.

The assessments above can be conducted using the audit techniques described in section II.D, or the survey techniques described in section I.H. Assessments can be conducted by either internal or external personnel. In general, employees are more likely to be open and honest when confidentiality is assured, which is more likely when assessments are conducted by outside parties. However, internal assessments can reveal valuable information if proper safeguards are observed to assure the employee's privacy.

It is important that follow-up assessments be made to determine if the training conducted closed the gap between the "is" and the "should be." The follow up will also provide a basis for additional training. Reassessment should be conducted first to assure that the desired KSAs were acquired by the target group of employees; then the process should be reassessed to determine if the new KSAs improved the process as predicted. It's common to discover that we made a mistake in assuming that the root cause of the process "is/should-be" gap is a KSA deficiency. If the reassessments indicate that the KSA gap was closed but the process gap persists, another approach must be taken to close the process gap.

MATERIAL DEVELOPMENT

This section covers some of the technical aspects of training material development and presentation. The reader should understand that training and education are, in essence, *communication.*

Principles of effective communication

Zaremba (1989) points out that the foundation of effective communication is the principle of "receiver orientation." This means that communication does not occur until the receiver makes sense of the message that has been received. Simply generating a message does not constitute communication. Furthermore, communication can occur even when there was no intent to send a message. An example of this is when leaders say one thing but do another.

The goal of communication is to accurately transmit a message to a receiver. Certain elements can improve the likelihood that this will occur.

- Create information networks for transmitting messages efficiently.
- Cultivate an organizational climate that encourages communication.
- Recognize the effect and importance of nonverbal factors in communication.
- Understand the value of symbolism.
- Provide formal training in making presentations, one-on-one interpersonal interaction, and facilitating group discussions.
- Assure that meetings and conferences are managed in accordance with the above principles.

Johnson (1993a) lists four specific parts to successful communications:

1. **Transmission**—There must be a message that is put forth by speaking, writing, drawing, or some other form of transmission
2. **Reception**—The information must be received by listening, reading, seeing, feeling, etc.
3. **Interpretation**—The message must be successfully interpreted. This means the receiver must understand exactly what the sender intended.
4. **Correct use**—The information must be used in the way it was intended. This could be active use to initiate or change policy, storage for later use when appropriate, or as a basis for returned communications at the present time.

Johnson points out that communication has two characteristics which impact on the way the transmitted information is received: *fact*—what actually happened or is expected to happen, and *feeling*—how the transmitter feels about what happened or is expected to happen. Both feeling and fact have a place in communications, often together. The problem is that in too many instances these characteristics are not considered, which causes communication of feelings when facts are appropriate or vice versa.

Criteria for effective communication

Zaremba (1989, pp. 14–15) lists five essential criteria for providing information:

1. **Timely**—Messages must be received in time for the receiver to take the action desired by the sender. Messages that force the recipient to reorganize urgent priorities are disruptive. Messages received too late to act, or even after a deadline has passed, are annoying.
2. **Lucid**—Messages that are not clear can lead to inappropriate behavior, misunderstandings, and mistakes. At best, an unclear message will result in wasted time when the receiver contacts the source to clarify the message.
3. **Accurate**—Clear messages with inaccurate information are also a problem.
4. **Pertinent**—A common problem in today's business world is "information overload" (IO). The problem would be bad enough if only worth-while information were sent; it is made worse by adding useless and irrelevant information to the load. The business equivalent of "junk mail" contributes to IO. The problem is most acute in organizations with a culture of fear. Employees will "CC" anyone with a remote chance of being interested in a topic just to avoid offending a powerful manager. Another source of IO is broadcast email messages. Access to broadcasting capabilities on an email network should be carefully monitored for abuse, lest the "signal" of useful information get lost in the "noise" of irrelevant messages.
5. **Credible**—For communication to occur, employees must believe the messages they are sent. This is easier to assure if the messages meet the above criteria. However, it is also important that the explicit message be consistent with the message conveyed by actions and other behaviors with implicit meaning.

Johnson (1993a, p. 217) lists "six 'Cs' " of effective communication: clear, concise, correct, complete, courteous, and convincing.

Of course, the receiver also shares responsibility with the sender. Listening effectively requires effort and skill. Good listeners stop talking and concentrate on the message being sent. They focus their attention on the speaker and avoid distractions, and they look at the speaker to detect nonverbal cues related to the message. Listeners will ask for clarification when needed. Listeners put their emotions on hold and refrain from judgment until the entire message has been communicated and understood.

Types of instruction

There are three basic types of instructional formats:

1. **Learner-centered instruction**—Learner-centered instruction begins with the needs of the learner and tailors the course accordingly. This approach creates challenges for instructors and administrators when learners bring a wide variety of abilities, backgrounds, and needs to the class. Barring strong reasons to the contrary, this approach should be the first choice for most training programs.
2. **Instructor-centered**—The instructor determines the direction and pace of the learning. Instructor-centered training takes a "one size fits all" approach to learning. Since student-to-student variability exists, this approach fails a high percentage of students.
3. **Material-centered**—This includes most self-paced instruction such as programmed learning, interactive CD ROM, videotape, etc. Students trained using this approach have no instructor with whom they interact.

Lesson plans

Lesson plans are "roadmaps" of the training material to be presented during a particular training session. Johnson (1993b) provides the following guidelines for developing lesson plans:

1. Lesson plans are crucial for training success.
2. Every subject must be adequately researched.
3. Develop twice as much material as you think you will need.
4. Ensure trainees know "what's in it for me" early on so they are motivated to learn.
5. Practice, practice, practice prior to presentation.
6. Use examples that pertain to the operation where possible.
7. Stimulate questions.
8. Use training aids.
9. Adapt the training to the group's level.
10. Never make excuses (e.g., "I've never taught this material before.").
11. Remain positive.
12. Follow your lesson plan.

There are four main types of lesson plans:

- **Topical outline**—A lesson plan set up in a standard outline format, i.e., major sections are identified using Roman numerals, topics by upper case letters, etc. Most useful when the instructor is intimately familiar with the subject matter and has presented it previously. These lesson plans are easy to construct and can be used as agendas to help keep the class oriented.
- **Manuscript**—Lessons are described in prose in detail. Most useful when teaching complex subjects for the first time. While this format provides a lot of information, the sheer volume of information may make it difficult to locate specific information when required. Careful organization (e.g., tab indexing) is essential. Development time can be considerable.
- **Sentence**—An abbreviated version of the manuscript type of lesson plan. Used for less complex subjects. While providing less information than the manuscript format, this approach is easier to prepare and it is lessdifficult to find the information.
- **Key word**—A "memory jogger" for the experienced instructor to use to pace the presentation and assure that all subjects are covered. This type of lesson plan normally does not provide information on sequence or time-per-topic. This "quick-and-dirty" approach provides little detail and should not be used unless the instructor is intimately familiar with the material. An advantage to this approach is that it tends to produce less formal presentations.

Example of a lesson plan

The lesson plan in Figure I.22 was actually used by the author to conduct in-house training for a client, an integrated health-care provider. The lessons in the lesson plan were selected from a longer seminar that was provided to another group of employees. The lesson numbers were left as they were in the longer seminar to allow this group of trainees to easily cross-reference their learning with others who had taken the complete course. (Only a portion of the lesson plan is shown here.)

HEALTHCARE QUALITY IMPROVEMENT SEMINAR

This document describes the seminar to be conducted for Healthcare of Arizona by Thomas Pyzdek, an internationally known author, lecturer, and consultant on quality. The seminar will be conducted on Friday, May 5 and Monday, May 15, 1995. The seminar objective is to provide participants with the knowledge and skills necessary to systematically improve quality and productivity in their area of responsibility. Participants will learn the principles of variation and understand the importance of these principles to all decision-making.

Figure I.22—*Continued on next page . . .*

Figure I.22—*Continued . . .*

DAY #1

Lesson 4. DATA COLLECTION

Approximate start: Day 1, 8am.

Approximate duration: 1:00 hour total

Objectives

- understand the importance of data collection
- learn to locate existing data
- know how to validate data
- know how to construct data collection sheets
- understand the different types of data and their uses
- know how data-collection activities typically occur in quality improvement

Key topics

- why collect data
- sources of data
- obtaining new data
- variables, attributes, and ordinal data
- data collection tools
- layout of data sheets
- location diagrams
- check sheets

Discussion

The process generates data that can be used to facilitate continuous improvement. Knowing what data to look for and where to look for the data makes valuable information accessible to the quality team and utilizes existing data to maximum advantage.

Lesson plan

0:30 Tom Pyzdek lecture

0:30 IS department presentation

Lesson 7. FUNDAMENTAL CONCEPTS OF STATISTICAL CONTROL

Approximate start: Day 1, 9:00am.

Approximate duration: 1:30

Objectives

- understand the difference between variation from special causes and variation from common causes
- understand the effect of tampering with a controlled process
- understand the concept of distributions
- know how to represent variation in a process
- understand the difference between problem prevention and problem management

Figure I.22—*Continued on next page . . .*

Figure I.22—*Continued . . .*

- understand why meeting requirements isn't good enough to remain competitive
- understand the meaning of the term "requirements"
- understand the value of managing processes rather than managing outcomes directly

Key topics

- history of SPC
- special and common causes
- distributions; normal distribution
- central-limit theorem
- measures of central tendency; measures of dispersion
- prevention-versus-detection strategies
- process control and continuous improvement versus meeting requirements
- Taguchi's loss function concept as a rationale for continuous improvement

Discussion

This lesson presents the fundamental principles of variation and relates these principles to the philosophy of continuous improvement. Several populations are represented using colored chips to illustrate the distribution concepts of central tendency, shape, and spread. After presentation of the material, participants will be given an in-class exercise and a quiz.

Lesson plan

0:30 Tom Pyzdek lecture

0:10 Quiz

0:50 Group discussion, topic: discuss the fundamental principles of variation as they relate to the conclusions shown on form entitled "Nursing services patient education quality improvement monitor"

Figure I.22. Sample lesson plan.

The lesson plan shown in Figure I.22 is a composite of the common types of lesson plans discussed above.

Presentation skills

Unlike a meeting where all in attendance are expected to contribute and participate, a presentation involves a speaker who is trying to communicate with an audience. There are two common reasons why a quality manager might want to address an audience:

1. To inform or educate them. This is the purpose of business presentations, training classes, technical reports, and so on. The speaker discusses, explains, describes events or ideas or teaches the audience how to do something. This type of presentation often involves audio-visual aids such as charts, graphs, recordings, video presentations, or computer presentations. Success is defined by an audience leaving with more knowledge than they had when they arrived.
2. To convince or persuade them. The audience may have an understanding of a particular topic but lack the desire to change. The speaker's goal is to get the audience to take a certain type of action. In these presentations the speaker may seek to generate an emotional response rather than communicating factual data. Success is accomplished when the action is taken.

An old saw on speaking states that there are three steps involved in a presentation:

1. Tell them what you're going to tell them
2. Tell them
3. Tell them what you told them.

The effective speaker understands that he or she must provide the audience with a compelling reason for attending the presentation. Thus, the focus of preparation should be the audience, not the speaker. The beginning of the presentation should summarize the content of the presentation and explain why the audience should be interested. The body of the presentation should present the information in a way that is clear, entertaining, and concise. The ending should review the major points of the presentation and solicit the desired audience response (e.g., ask the audience to provide resources for a project).

Preparing the presentation

Many speakers find it difficult to organize their presentations. Here are a few guidelines that the author has found helpful.

1. Prepare a list of every topic you want to cover. Don't be selective or critical, write down everything that comes to mind. When you have finished, take some time off, then do it again.
2. Cull the list to those select few ideas that are most important. You should try to keep the list of major ideas down to three or less. Where possible, group the remaining points as subtopics under the key ideas. Eliminate the rest.

3. Number your points. The numbers help the listener understand and remember the points. Numbers set the point apart and help with retention. For example, "there are three reasons why we should proceed: first, lower cost; second, higher quality; and third, improved customer satisfaction." However, keep the number of items small; a speaker who announces "I have fifteen items to discuss" will frighten the audience.
4. Organize the presentation's ideas. Some speech experts recommend the "buildup" approach: good ideas first, better ideas next, best points last.
5. Analyze each major point. Tell the audience why the point is important to them. Make the presentation relevant and entertaining.

Visual aids

A visual aid in a speech is a pictorial used by a speaker to convey an idea. Well-designed visual aids add power to a presentation by showing the idea more clearly and easily than words alone. Whereas only 10% of presented material is retained from a verbal presentation after 3 days, 65% is retained when the verbal presentation is accompanied by a visual aid. ASQ reports that poor-quality visuals generate more negative comment from conference attendees than any other item. The visual aid must be easy for everyone to see. Small type which cannot be read from the back row of the room defeats the purpose of the visual aid. There should be good contrast between the text and the background color. Visuals should have text that is large enough to see easily from the worst seat in the house. The speaker must also reevaluate the visuals when the room size changes. A presentation that is perfectly acceptable to a group of 30 may be completely inadequate for a group of 300.

Here are a few rules recommended by ASQ for effective visual aids:

- NEVER read the slide!
- Each visual should address only one idea.
- Only the most important points should be the subject of a visual.
- The visual should be in landscape format.
- The maximum viewing distance should be less than 8 times the height of the projected image.
- The original artwork should be readable from a distance 8 times the height of the original; e.g., a 8" x 10" visual should be readable from 64" away.
- Use no more than 5 lines on a single visual with 7 or fewer words per line. Ideally, 20 words maximum per visual.
- Bar charts should have no more than 5 vertical columns.
- Tables should have no more than 4-6 columns.

- Graphs should have only 2-3 curves.
- Lines on charts or in tables should be heavy.
- Avoid sub-bullet points whenever possible. Use "build slides" instead.
- Show each slide for 1.5 minutes maximum, e.g., 10 slides for a 15-minute presentation.
- Put a black slide on the screen when you want to divert attention back to yourself.
- Spacing between words should be approximately the width of the letter "n."
- Spacing between lines should be 75% of the height of the letters, or double-spaced.
- Use sans-serif fonts (e.g., H) instead of serif fonts (e.g., H).
- Use both upper-case and lower-case letters.
- Avoid italics. Use bold lettering for emphasis.
- Maintain consistency in type size and fonts. The minimum letter height should be at least 1/25 the height of artwork (approximately 24-point type). Labels on graphics should be at least 1/50 the size of the artwork.
- Never use handwritten or typed slides.
- Maintain at least a 1" margin on all sides.
- Recommended background colors (white lettering): deep blue, deep green, deep maroon or black. Use yellow lettering for highlighting text.

When using visual aids, it is sometimes necessary to darken the room. However, the speaker should never be in the dark. The visual presentation supports the speaker; it should not be allowed to replace him. The speaker must always be the most important object in the room. If the lights must be lowered, arrange to have a small light on yourself, such as a podium light.

When using visual aids a right-handed speaker usually stands to the left of the visual and directs the attention of the audience to the material. If using a pointer the speaker may stand to either side. Never stand in front of the material you are presenting to the audience. Direct the eye of the viewer to the particular portion of the visual that you are emphasizing, don't just wave at the visual in a random manner. Layout the visual so that the viewer's eye moves in a natural flow from topic to topic.

Speak to the audience, not to the screen. Always face the audience. A microphone may make your speech audible when you face away from the audience, but it is still bad form to turn your back on the audience.

Color plays an important role in the design of effective visuals, if used properly. However, the improper use of color can make visuals ugly. Most computer software for preparing presentations comes with preset color schemes. Unless you have some skill and training in designing visuals, it is recommended that you use one of the schemes or contact a graphic artist.

Line and bar graphs are an effective way to convey numerical data at a glance. People understand more quickly things they see than things they hear. The eye is more effective in gathering and storing information than the ear. Auditory stimuli is presented and processed sequentially, one word at a time. Visual information is presented simultaneously. This is part of the reason why visuals are so effective at displaying patterns.

Business and industry are number-driven entities. Much of the information presented in meetings is numerical. Many, perhaps most, decisions rely on numbers. The effective use of graphs makes numbers easier to assimilate and understand. In most cases, one of three types of graph can be used: line graph, bar graph, and scatter plot. There are endless refinements on these basic three, e.g., multiple lines, grouped or stacked bars, stratified scatter plots. Line graphs are most often used to display time-series data (care must be taken to standardize time-series data, e.g., using constant dollars to correct for inflation). Bar charts are used most often to compare different items or classifications to one another. Scatter plots examine the association between two variables.

Regardless of the type of graph being used, it is important that the graphic have integrity, i.e., it must accurately portray the data. In his book, *The Visual Display of Quantitative Information*, Tufte lists six principles that enhance graphical integrity:

1. The representation of numbers, as physically measured on the surface of the graphic itself, should be directly proportional to the numerical quantities represented.
2. Clear, detailed and thorough labeling should be used to defeat graphical distortion and ambiguity. Write out explanations of the data on the graphic itself. Label important events in the data.
3. Show data variation, not design variation.
4. In time-series displays of money, deflated and standardized units are nearly always better than nominal money units.
5. The number of information-carrying (variable) dimensions depicted should not exceed the number of dimensions in the data.
6. Graphics must not quote data out of context.

Post-training evaluation and reinforcement

Training is said to have "worked" if it accomplishes its objectives. Since the training objectives are (or should be) derived from the strategic plan, the ultimate test is whether or not the organization has accomplished it strategic objectives. However, training is only one of dozens of factors that determine if an organization accomplishes its strategic objectives, and one that is often far removed in time from the final result. To assess training effectiveness we need more direct measures of success, and we need to measure near the time the training has been completed.

Except in academic settings, imparting knowledge or wisdom is seldom the ultimate goal of training. Instead, it is assumed that the knowledge or wisdom will result in improved judgments, lower costs, better quality, higher levels of customer satisfaction, etc. In other words, the training will produce *observable* results. These results were the focus of the training plan development and training needs analysis described earlier in this chapter. Training evaluation requires that they be converted to training measurables or objectives.

Regardless of the format of the presentation, the basic unit of training is the *lesson*. A lesson is a discrete "chunk" of information to be conveyed to a learner. The training objectives form the basis of each lesson, and the lessons provide guidance for development of measurements of success.

Lesson plans provide the basis for measurement at the lowest level. The objectives in the lesson plan are specific and the lesson is designed to accomplish these specific objectives. The assumption is that by accomplishing the set of objectives for each lesson, the objectives of the seminar or other training activity will be met. A further assumption is that by meeting the objectives of all of the training activities, the objectives of the training plan will be met. Finally, it is assumed that by meeting the objectives of the training plan, the objectives of the strategic plan (or strategic quality plan) will be met, or at least will not be compromised due to training inadequacies. All of these assumptions should be subjected to evaluation.

Evaluation

The evaluation process involves four elements (Kirkpatrick, 1996):

1. **Reaction**—How well did the conferees like the program? This is essentially customer satisfaction measurement. Reaction is usually measured using comment sheets, surveys, focus groups and other customer communication techniques. See I.H for additional information.
2. **Learning**—What principles, facts, and techniques were learned? What attitudes were changed? It is entirely possible that conferees react favorably to training, even if learning does not occur. The learning of each conferee should be quantified using pre- and post-tests to identify learning imparted by the training. Results should be analyzed using proper statistical methods (see V.C). In exceptional cases, e.g., evaluating a consulting company for a large training contract, a formal designed experiment may be justified.

3. **Behavior**—What changes in behavior on-the-job occurred? If the conferee leaves the SPC seminar and immediately begins to effectively apply control charts where none were used before, then the training had the desired effect on behavior. However, if the conferee's tests indicate that she gained competence in the subject matter from the training, but no change in behavior took place, the training investment was wasted. Note that behavior change is dependent on a great number of factors besides the training, e.g., management must create systems where the newly learned behaviors are encouraged.
4. **Results**—What were the tangible results of the program in terms of reduced cost, improved quality, improved quantity, etc.? This is the real payback on the training investment. The metrics used for measuring results are typically built into the action plans, project plans, budgets, etc. Again, as with behavior change, there are many factors other than training that produce the desired results.

Phillips adds a fifth item to the above list (Phillips, 1996, p. 316):

5. **Return on investment**—Did the monetary value of the results exceed the cost for the program?

Phillips considers these five items to be different *levels* of evaluation. Each evaluation level has a different value, as shown in Figure I.23.

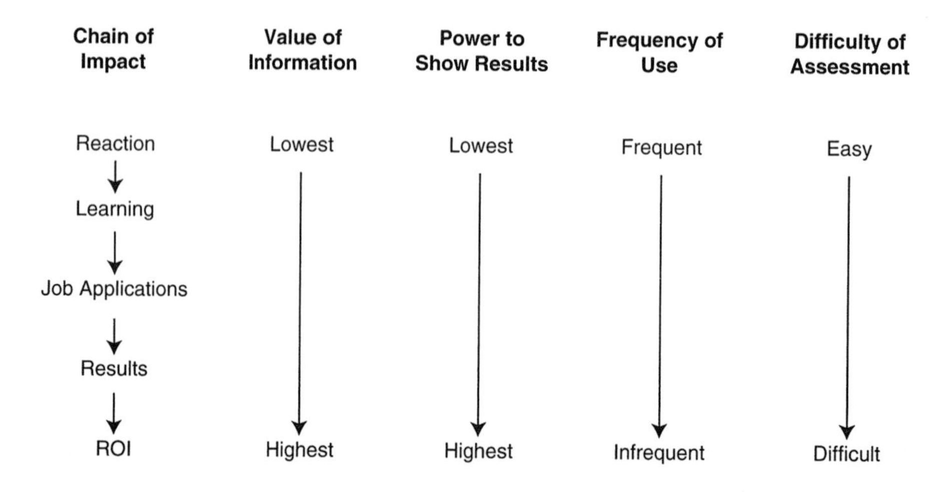

Figure I.23. Characteristics of evaluation levels.
From Phillips, J.J. (1996). "Measuring the Results of Training," in Craig, R.L., editor in chief. *The ASTD Training & Development Handbook: A Guide to Human Resources Development.* New York: McGraw-Hill, p. 317.

Due to the difficulty and cost involved, it is impractical and uneconomical to insist that every program be evaluated at all five levels. Sampling can be used to obtain evaluations at the higher levels. As an example, one large electric utility set the following sampling targets.

Table I.13. Targets for percentages of programs to be evaluated.
From Phillips, J.J. (1996). "Measuring the Results of Training," in Craig, R.L., editor in chief. *The ASTD Training & Development Handbook: A Guide to Human Resources Development.* New York: McGraw-Hill, p. 317.

LEVEL	PERCENTAGE
Participant's satisfaction	100
Learning	70
On-the-job-applications (behavior)	50
Results	10
ROI	5

Where sampling is used, programs should be selected using a randomization procedure such as random numbers tables. ROI calculations are not difficult and they are described in V.A.8. However, to make the results credible, finance and accounting personnel should be involved in calculating financial ratios of this type.

Reinforcement

When the subject of reinforcement is raised, monetary remuneration usually comes to mind first. Skill-based pay is gaining favor in some quarters for a variety of reasons:

- encourage employees to acquire additional skills
- reward people for training and education
- as a reaction to the negative aspects of performance-based pay

While skill-based pay may have merit, cash awards and other such "rewards" are of dubious value and should probably not be used (see I.D for a discussion of this topic).

Rather than assuming that employees will only engage in training if they receive an immediate tangible reward, research and experience indicate that most employees find value in training that helps them better achieve their personal, job, and career goals. Thus, reinforcement is accomplished by providing the trainee with the opportunity to use the skills they learned. Proficiency is gained with practice soon after the learning has taken place. Management should provide an environment where the new skills can be honed without pressure and distraction. The "just-in-time"(JIT) principle applies here. Don't provide training for skills that won't be used in the near future.

People who have just learned something new, be it a job skill or a new philosophy such as quality-focus, often have questions arise as they attempt to integrate their new knowledge into their daily thoughts and routine. User groups are very helpful. A user group consists of a number of people who have received similar training. User groups meet from time to time to discuss their understanding of the material with others. The focus of the group varies from "How are you using this?" to "I don't understand this" to "Here's what I am doing!" At times, well-versed speakers will be invited to clarify particular aspects of the subject. Presentations of successful applications may be made. Management can encourage user groups by providing facilities, helping with administrative details, and, especially, by attending their meetings on occasion.

Electronic forums are gaining in popularity. Trainers will often make themselves available to answer questions via email. Forum subscribers will send their questions or comments to a "list server." The list server then automatically broadcasts the question or comment to other subscribers on the list. Every subscriber receives every user's message, along with the responses to the message. This often produces "threads." A thread is an exchange of information on a particular topic. E.g., subscriber A has a question about using control charts on financial data, subscriber B responds, then C responds to B and so on. These threads look remarkably like a free-wheeling face-to-face discussion. The result is that a great deal of learning takes place in a format that everyone finds to be more natural (and more interesting) than a traditional classroom environment.

Refreshers

Learning that isn't used right away tends to fade. Even when just-in-time training (JITT) is used, it is unlikely that every skill will be put to immediate and routine use. It is wise to plan for periodic refresher courses to hone the skills acquired during prior training. A refresher course is usually shorter, faster, and more intense than new learning. Instructors need not have the same subject matter mastery as those teaching new material. In fact, it may not even be necessary to have an instructor available. Media such as video and audio tape programs, CD ROM, slide presentations, etc., may be sufficient. These self-study media offer a number of cost and scheduling advantages.

If in-house instructors are used, they may be available to answer occasional questions from previous trainees. Of course, when the instructor is not a full-time trainer, this must be strictly limited. There are a number of ways to reduce the demands on the trainer's time while still answering most of the questions from participants. If the need is not urgent, the question may be asked using mail or an online forum. House newsletters or bulletin can be used to provide answers to frequently asked questions. More companies now have "intranets" where such information is made available. The trainee may be able to find an Internet news group devoted to the subject of their concern. There are thousands of news groups covering a huge variety of subjects, many relating to quality and training.

I.F COST OF QUALITY

The history of quality costs dates back to the first edition of *Juran's QC Handbook* in 1951. Today, quality cost accounting systems are part of every modern organization's quality improvement strategy. Indeed, quality cost accounting and reporting are part of many quality standards. Quality cost systems help management plan for quality improvement by identifying opportunities for greatest return on investment. However, the quality manager should keep in mind that quality costs address only half of the quality equation. The quality equation states that quality consists of doing the right things and not doing the wrong things. "Doing the right things" means including product and service features that satisfy or delight the customer. "Not doing the wrong things" means avoiding defects and other behaviors that cause customer *dissatisfaction.* Quality costs address only the latter aspect of quality. It is conceivable that a firm could drive quality costs to zero and still go out of business.

A problem exists with the very name "cost of quality." By using this terminology, we automatically create the impression that quality is a cost. However, our modern understanding makes it clear that quality is *not* a cost. Quality represents a driver that produces higher *profits* through *lower* costs and the ability to command a premium price in the marketplace. This author concurs with such quality experts as H.J. Harrington and Frank M. Gryna that a better term would be "cost of poor quality." However, we will bow to tradition and use the familiar term "cost of quality" throughout this book.

The fundamental principle of the cost of quality is that any cost that would not have been expended if quality were perfect is a cost of quality. This includes such obvious costs as scrap and rework, but it also includes many costs that are far less obvious, such as the cost of reordering to replace defective material. Service businesses also incur quality costs; for example, a hotel incurs a quality cost when room service delivers a missing item to a guest. Specifically, quality costs are a measure of the costs specifically associated with the achievement or nonachievement of product or service quality—including all product or service requirements established by the company and its contracts with customers and society. Requirements include marketing specifications, end-product and process specifications, purchase orders, engineering drawings, company procedures, operating instructions, professional or industry standards, government regulations, and any other document or customer needs that can affect the definition of product or service. More specifically, quality costs are the total of the cost incurred by a) investing in the *prevention* of nonconformances to requirements; b) *appraising* a product or service for conformance to requirements; and c) *failure* to meet requirements (Figure I.24).

PREVENTION COSTS

The costs of all activities specifically designed to prevent poor quality in products or services. Examples are the costs of new product review, quality planning, supplier capability surveys, process capability evaluations, quality improvement team meetings, quality improvement projects, quality education and training.

APPRAISAL COSTS

The costs associated with measuring, evaluating or auditing products or services to assure conformance to quality standards and performance requirements. These include the costs of incoming and source inspection/test of purchased material, in process and final inspection/test, product, process, or service audits, calibration of measuring and test equipment, and the costs of associated supplies and materials.

FAILURE COSTS

The costs resulting from products or services not conforming to requirements or customer/user needs. Failure costs are divided into internal and external failure cost categories.

INTERNAL FAILURE COSTS

Failure costs occurring prior to delivery or shipment of the product, or the furnishing of a service, to the customer. Examples are the costs of scrap, rework, reinspection, retesting, material review, and down grading.

EXTERNAL FAILURE COSTS

Failure costs occurring after delivery or shipment of the product, and during or after furnishing of a service, to the customer. Examples are the costs of processing customer complaints, customer returns, warranty claims, and product recalls.

TOTAL QUALITY COSTS

The sum of the above costs. It represents the difference between the actual cost of a product or service, and what the reduced cost would be if there was no possibility of substandard service, failure of products, or defects in their manufacture.

Figure I.24. Quality costs—general description.

From Campanella (1999). Used by permission of the publisher.

For most organizations, quality costs are hidden costs. Unless specific quality cost identification efforts have been undertaken, few accounting systems include provision for identifying quality costs. Because of this, unmeasured quality costs tend to increase. Poor quality impacts companies in two ways: higher cost and lower customer satisfaction. The lower satisfaction creates price pressure and lost sales, which results in lower revenues. The combination of higher cost and lower revenues eventually brings on a crisis that may threaten the very existence of the company. Rigorous cost of quality measurement is one technique for preventing such a crisis from occurring. Figure I.25 illustrates the hidden cost concept.

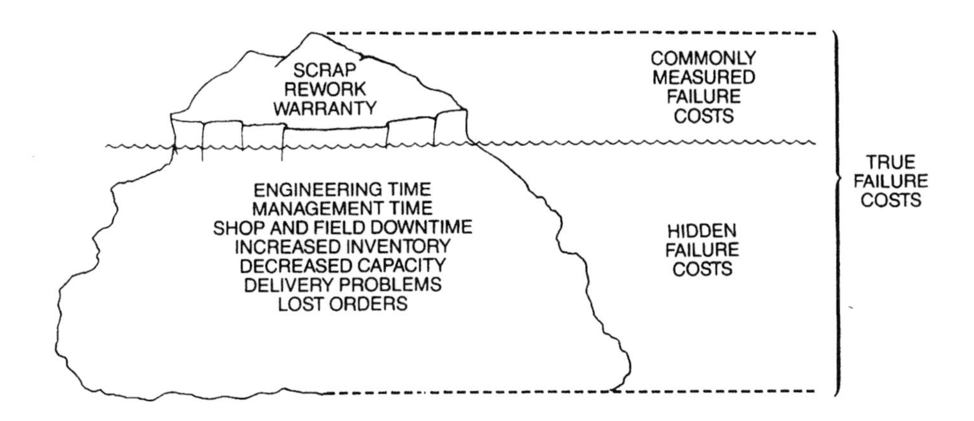

Figure I.25. Hidden cost of quality and the multiplier effect.

From Campanella (1999). Used by permission of the publisher.

GOAL OF QUALITY COST SYSTEM

The goal of any quality cost system is to reduce quality costs to the lowest practical level. This level is determined by the total of the costs of failure and the cost of appraisal and prevention. Juran and Gryna (1988) present these costs graphically as shown in Figure I.26. In the figure it can be seen that the cost of failure declines as conformance quality levels improve toward perfection, while the cost of appraisal plus prevention increases. There is some "optimum" target quality level where the sum of prevention, appraisal, and failure costs is at a minimum. Efforts to improve quality to better than the optimum level will result in increasing the total quality costs.

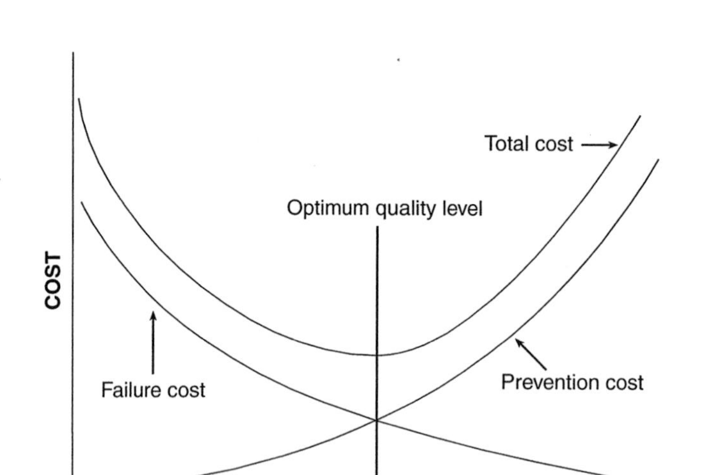

Figure I.26. Classical model of optimum quality costs.
From *Juran's Quality Control Handbook, 4th edition.* J.M. Juran, editor.
Copyright © 1988, McGraw-Hill.

Juran acknowledged that in many cases the classical model of optimum quality costs is flawed. It is common to find that quality levels can be economically improved to literal perfection. For example, millions of stampings may be produced virtually error-free from a well-designed and built stamping die. The classical model created a mindset that resisted the idea that perfection was a possibility. No obstacle is as difficult to surmount as a mindset. The new model of optimum quality cost incorporates the possibility of zero defects and is shown in Figure I.27.

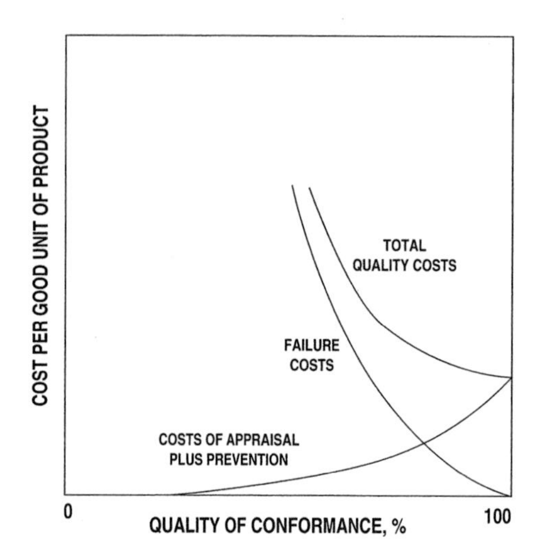

Figure I.27. New model of optimum quality costs.
From *Juran's Quality Control Handbook, 4th edition.* J.M. Juran, editor.
Copyright © 1988, McGraw-Hill.

Quality costs are lowered by identifying the root causes of quality problems and taking action to eliminate these causes. The tools and techniques described in V.B and V.C are useful in this endeavor. KAIZEN, reengineering, and other continuous improvement approaches are commonly used.

STRATEGY FOR REDUCING QUALITY COSTS

As a general rule, quality costs increase as the detection point moves further up the production and distribution chain. The lowest cost is generally obtained when nonconformances are prevented in the first place. If nonconformances occur, it is generally least expensive to detect them as soon as possible after their occurrence. Beyond that point there is loss incurred from additional work that may be lost. The most expensive quality costs are from nonconformances detected by customers. In addition to the replacement or repair loss, a company loses customer goodwill and their reputation is damaged when the customer relates his experience to others. In extreme cases, litigation may result, adding even more cost and loss of goodwill.

Another advantage of early detection is that it provides more meaningful feedback to help identify root causes. The time lag between production and field failure makes it very difficult to trace the occurrence back to the process state that produced it. While field failure tracking is useful in *prospectively* evaluating a "fix," it is usually of little value in *retrospectively* evaluating a problem.

ACCOUNTING SUPPORT

We have said it before, but it bears repeating, that the support of the accounting department is vital whenever financial and accounting matters are involved. In fact, the accounting department bears *primary* responsibility for accounting matters, including cost of quality systems. The quality department's role in development and maintenance of the cost of quality system is to provide guidance and support to the accounting department.

The cost of quality system must be integrated into the larger cost accounting system. It is, in fact, merely a subsystem. Terminology, format, etc., should be consistent between the cost of quality system and the larger system. This will speed the learning process and reduce confusion. Ideally, the cost of quality will be so fully integrated into the cost accounting system that it will not be viewed as a separate accounting system at all, it will be a routine part of cost reporting and reduction. The ideal cost of quality accounting system will simply aggregate quality costs to enhance their visibility to management and facilitate efforts to reduce them. For most companies, this task falls under the jurisdiction of the controller's office.

Quality cost measurement need not be accurate to the penny to be effective. The purpose of measuring such costs is to provide broad guidelines for management decision-making and action. The very nature of cost of quality makes such accuracy impossible. In some instances it will only be possible to obtain periodic rough estimates of such costs as lost customer goodwill, cost of damage to the company's reputation, etc. These estimates can be obtained using special audits, statistical sampling, and other market studies. These activities can be jointly conducted by teams of marketing, accounting, and quality personnel. Since these costs are often huge, these estimates must be obtained. However, they need not be obtained every month. Annual studies are usually sufficient to indicate trends in these measures.

MANAGEMENT OF QUALITY COSTS

In our discussion of the cost of quality subsystem, we emphasized the importance of not creating a unique accounting system. The same holds true when discussing management of quality costs. Quality cost management should be part of the charter of the senior level cross-functional cost management team. It is one part of the broader business effort to control costs. However, in all likelihood, the business will find that quality cost reduction has greater potential to contribute to the bottom line than the reduction of other costs. This is so because, unlike other costs, quality costs are *waste costs* (Pyzdek, 1976). As such, quality costs contribute no value to the product or service purchased by the customer. Indeed, quality costs are often indicators of *negative customer value.* The customer who brings his car in for a covered warranty expense suffers uncompensated inconvenience, the cost of which is not captured by most quality cost systems (although, as discussed above, we recommend that such costs be estimated from time-to-time). All other costs incurred by the firm purchase at least some value.

Effective cost of quality programs consist of taking the following steps (Campanella, 1999):

- Establish a quality cost measurement system
- Develop a suitable long-range trend analysis
- Establish annual improvement goals for total quality costs
- Develop short-range trend analyses with individual targets which, when combined, meet the annual improvement goal
- Monitor progress towards the goals and take action when progress falls short of targets

The tools and techniques described in chapter I are useful for managing cost of quality reduction projects.

Quality cost management helps firms establish priorities for corrective action. Without such guidance, it is likely that firms will misallocate their resources, thereby getting less than optimal return on investment. If such experiences are repeated frequently, the organization may even question or abandon their quality cost reduction efforts. The most often-used tool in setting priorities is Pareto analysis (see V.C). Typically at the outset of the quality cost reduction effort, Pareto analysis is used to evaluate failure costs to identify those "vital few" areas in most need of attention. Documented failure costs, especially external failure costs, almost certainly understate the true cost and they are highly visible to the customer. Pareto analysis is combined with other quality tools, such as control charts and cause-and-effect diagrams, to identify the root causes of quality problems. Of course, the analyst must constantly keep in mind the fact that most costs are hidden. Pareto analysis cannot be effectively performed until the hidden costs have been identified. Analyzing only those data easiest to obtain is an example of the GIGO (garbage-in, garbage-out) approach to analysis.

After the most significant failure costs have been identified and brought under control, appraisal costs are analyzed. Are we spending too much on appraisal in view of the lower levels of failure costs? Here quality cost analysis must be supplemented with risk analysis to assure that failure and appraisal cost levels are in balance. Appraisal cost analysis is also used to justify expenditure in prevention costs.

Prevention costs of quality are investments in the discovery, incorporation, and maintenance of defect prevention disciplines for all operations affecting the quality of product or service (Campanella, 1999). As such, prevention needs to be applied correctly and *not* evenly across the board. Much improvement has been demonstrated through reallocation of prevention effort from areas having little effect to areas where it really pays off; once again, the Pareto principle in action.

COST OF QUALITY EXAMPLES

I. **Prevention costs**—Costs incurred to prevent the occurrence of non-conformances in the future, such as*
   - A. Marketing/customer/user
     1. Marketing research
     2. Customer/user perception surveys/clinics
     3. Contract/document review
   - B. Product/service/design development
     1. Design quality progress reviews
     2. Design support activities
     3. Product design qualification test
     4. Service design qualification
     5. Field tests
   - C. Purchasing
     1. Supplier reviews
     2. Supplier rating
     3. Purchase order tech data reviews
     4. Supplier quality planning

*All detailed quality cost descriptions are from *Principles of Quality Costs*, John T. Hagan, editor, Milwaukee, WI: ASQ Quality Press, appendix B.

D. Operations (manufacturing or service)
   1. Operations process validation
   2. Operations quality planning
      a. Design and development of quality measurement and control equipment
   3. Operations support quality planning
   4. Operator quality education
   5. Operator SPC/process control
E. Quality administration
   1. Administrative salaries
   2. Administrative expenses
   3. Quality program planning
   4. Quality performance reporting
   5. Quality education
   6. Quality improvement
   7. Quality audits
   8. Other prevention costs

II. **Appraisal costs**—Costs incurred in measuring and controlling current production to assure conformance to requirements, such as
   A. Purchasing appraisal costs
      1. Receiving or incoming inspections and tests
      2. Measurement equipment
      3. Qualification of supplier product
      4. Source inspection and control programs
   B. Operations (manufacturing or service) appraisal costs
      1. Planned operations inspections, tests, audits
         a. Checking labor
         b. Product or service quality audits
         c. Inspection and test materials
      2. Set-up inspections and tests
      3. Special tests (manufacturing)
      4. Process control measurements
      5. Laboratory support
      6. Measurement equipment
         a. Depreciation allowances
         b. Measurement equipment expenses
         c. Maintenance and calibration labor
      7. Outside endorsements and certifications

- C. External appraisal costs
  1. Field performance evaluation
  2. Special product evaluations
  3. Evaluation of field stock and spare parts
- D. Review of tests and inspection data
- E. Miscellaneous quality evaluations

III. **Internal failure costs**—Costs generated before a product is shipped as a result of non-conformance to requirements, such as

- A. Product/service design failure costs (internal)
  1. Design corrective action
  2. Rework due to design changes
  3. Scrap due to design changes
- B. Purchasing failure costs
  1. Purchased material reject disposition costs
  2. Purchased material replacement costs
  3. Supplier corrective action
  4. Rework of supplier rejects
  5. Uncontrolled material losses
- C. Operations (product or service) failure costs
  1. Material review and corrective action costs
     - a. Disposition costs
     - b. Troubleshooting or failure analysis costs (operations)
     - c. Investigation support costs
     - d. Operations corrective action
  2. Operations rework and repair costs
     - a. Rework
     - b. Repair
  3. Reinspection/retest costs
  4. Extra operations
  5. Scrap costs (operations)
  6. Downgraded end product or service
  7. Internal failure labor losses
- D. Other internal failure costs

IV. **External failure costs**—Costs generated after a product is shipped as a result of non-conformance to requirements, such as

- A. Complaint investigation/customer or user service
- B. Returned goods
- C. Retrofit costs

D. Recall costs
E. Warranty claims
F. Liability costs
G. Penalties
H. Customer/user goodwill
I. Lost sales
J. Other external failure costs

QUALITY COST BASES

The guidelines for selecting a base for analyzing quality costs are:

- The base should be related to quality costs in a meaningful way
- The base should be well-known to the managers who will receive the quality cost reports
- The base should be a measure of business volume in the area where quality cost measurements are to be applied
- Several bases are often necessary to get a complete picture of the relative magnitude of quality costs

Some commonly used bases are (Campanella, 1999):

- A labor base (such as total labor, direct labor, or applied labor)
- A cost base (such as shop cost, operating cost, or total material and labor)
- A sales base (such as net sales billed, or sales value of finished goods)
- A unit base (such as the number of units produced, or the volume of output).

While actual dollars spent are usually the best indicator for determining where quality improvement projects will have the greatest impact on profits and where corrective action should be taken, unless the production rate is relatively constant, it will not provide a clear indication of quality cost improvement trends. Since the goal of the cost of quality program is improvement over time, it is necessary to adjust the data for other time-related changes such as production rate, inflation, etc. Total quality cost compared to an applicable base results in an index which may be plotted and analyzed using control charts, run charts, or one of the other tools described in chapter V.

For long-range analyses and planning, net sales is the base most often used for presentations to top management (Campanella, 1999). If sales are relatively constant over time, the quality cost analysis can be performed for relatively short spans of time. In other industries this figure must be computed over a longer time interval to smooth out large swings in the sales base. For example, in industries such as shipbuilding or satellite manufacturing, some periods may have no deliveries, while others have large dollar amounts. It is important that the quality costs incurred be related to the sales for the same period. Consider the sales as the "opportunity" for the quality costs to happen.

Some examples of cost of quality bases are (Campanella, 1999):

- Internal failure costs as a percent of total production costs
- External failure costs as an average percent of net sales
- Procurement appraisal costs as a percent of total purchased material cost
- Operations appraisal costs as a percent of total productions costs
- Total quality costs as a percent of production costs

An example of a cost of quality report that employs some of these bases is shown in Figure I.28.

QUALITY COST SUMMARY REPORT
FOR THE MONTH ENDING ______
(In Thousands of U.S. Dollars)

DESCRIPTION	CURRENT MONTH			YEAR TO DATE		
	QUALITY COSTS	AS A PERCENT OF		QUALITY COSTS	AS A PERCENT OF	
		SALES	OTHER		SALES	OTHER
1.0 PREVENTION COSTS						
1.1 Marketing/Customer/User						
1.2 Product/Service/Design Development						
1.3 Purchasing Prevention Costs						
1.4 Operations Prevention Costs						
1.5 Quality Administration						
1.6 Other Prevention Costs						
TOTAL PREVENTION COSTS						
PREVENTION TARGETS						
2.0 APPRAISAL COSTS						
2.1 Purchasing Appraisal Costs						
2.2 Operations Appraisal Costs						
2.3 External Appraisal Costs						
2.4 Review Of Test And Inspection Data						
2.5 Misc. Quality Evaluations						
TOTAL APPRAISAL COSTS						
APPRAISAL TARGETS						
3.0 INTERNAL FAILURE COSTS						
3.1 Product/Service Design Failure Costs						
3.2 Purchasing Failure Costs						
3.3 Operations Failure Costs						
3.4 Other Internal Failure Costs						
4.0 EXTERNAL FAILURE COSTS						
TOTAL FAILURE COSTS						
FAILURE TARGETS						
TOTAL QUALITY COSTS						
TOTAL QUALITY TARGETS						

BASE DATA	CURRENT MONTH		YEAR TO DATE		FULL YEAR	
	BUDGET	ACTUAL	BUDGET	ACTUAL	BUDGET	ACTUAL
Net Sales						
______ Other Base (Specify)						

Figure I.28. Quality costs summary report.
From Campanella (1999). Used by permission of the publisher.

QUALITY COST TREND ANALYSIS

As stated above, the purpose of collecting quality cost data is to provide a sound basis for taking the necessary action to eliminate the causes of these costs, and thereby eliminate the costs themselves. If the action taken is effective, the data will indicate a positive trend. Trend analysis is most often performed by presenting the data in run chart form and analyzing the runs (see section VI.E). It is common to combine all of the cost of quality data on a single graph, as shown in Figure I.29.

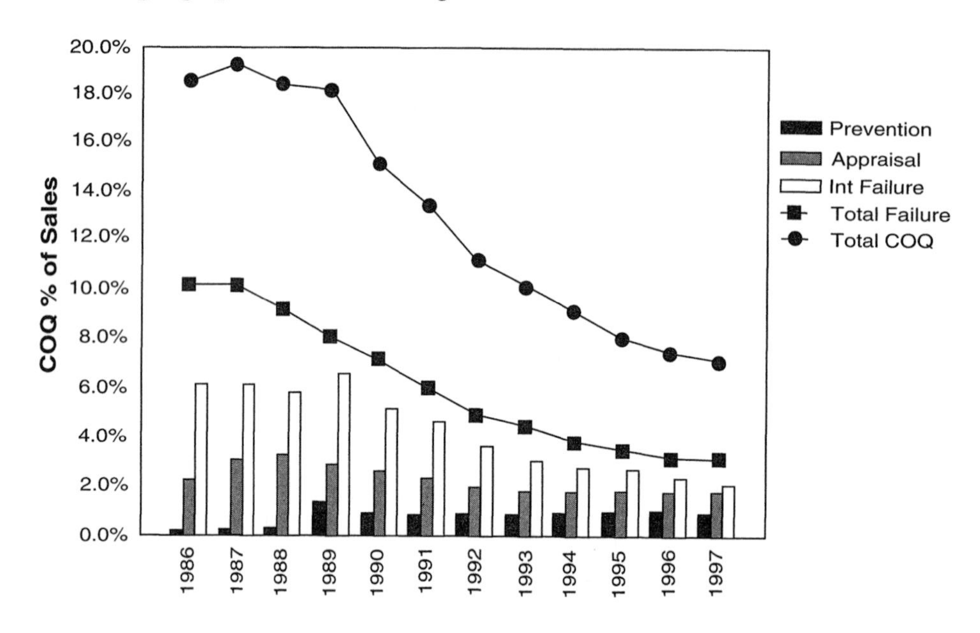

Figure I.29. Cost of quality history.

If the runs are subjected to the run tests described below, it can be shown that the total failure and total COQ (cost of quality) trends are statistically significant. However, for these data, the use of formal statistical rules is superfluous—the improvement is obvious.

While such aggregate analysis is useful for senior management, it is of little value to those engaged in more focused cost of quality projects. In these cases the trend data should be as specific as possible to the area being studied. Also, the measurement may be something more directly related to the work being done by the improvement team rather than dollars, and the time interval should be shorter. For example, if it has been determined that a major internal failure cost item is defective solder joints, then the team should plot a control chart of the solder defect rate and analyze the process in real-time. Obviously, reducing solder defects should reduce the cost associated with solder defects.

IMPLEMENTING THE QUALITY COST PROGRAM

Quality cost program introduction is a major project and should utilize the tools and techniques described in section I.B. Prior to implementation, a needs analysis should be performed to determine if, in fact, a cost of quality program can benefit the company. The needs assessment should also include a benefit/cost analysis and a plan for the implementation. The plan should include:

- the management presentation, designed to identify the overall opportunity and show an example of how the program will achieve its benefits
- a description of the pilot program
- material designed to educate and involve all functions in the program
- outline the internal cost of quality accounting procedures
- describe the data collection and analysis of cost of quality data at the highest level of aggregation
- describe the cost of quality reporting system and how the data will be used to improve quality

As with any major quality project, a sponsor should be found and management support secured. In the case of cost of quality, the sponsor should be the controller or one of her subordinates.

USE OF QUALITY COSTS

The principal use of quality cost data is to justify and support quality performance improvement. Quality cost data help identify problem areas and direct resources to these areas. To be effective, the cost of quality system has to be integrated with other quality information systems to assure that root causes will be addressed. Statistical analysis can be used to correlate quality cost trends with other quality data to help direct attention to problem causes.

One mission of the quality management function is to educate top management about the long-range effects of total quality performance on the profits and quality reputation of the company. Management must understand that strategic planning for quality is as important as strategic planning for any other functional area. When the strategic plan addresses cost issues, quality cost consideration should be prominent. Quality costs should be considered first because, since they are waste costs, their reduction is always taken from the "fat" of the organization. The role of the quality manager in this process should be to (Campanella, 1999)

- analyze major trends in customer satisfaction, defects or error rates, and quality costs, both generally and by specific program or project. These trends should also be used to provide inputs for setting objectives;
- assist the other functions to ensure that costs related to quality are included in their analyses for setting objectives;

- develop an overall quality strategic plan which incorporates all functional quality objectives and strategic action plans, including plans and budgets for the quality function.

BENEFITS OF QUALITY COST REDUCTION

Quality cost reductions can have a significant impact on a company's growth rate and bottom line. Research done by the Chicago Graduate School of Business showed that companies using TQM for an average of 6.5 years increased revenues at an annual rate of 8.3% annually, versus 4.2% annually for all U.S. manufacturers. Suminski (1994) reports that the average manufacturer's price of nonconformance is 25% of operating costs, for service businesses the figure is 35%. These costs represent a direct charge against a company's profitability. A New England heavy equipment manufacturer reports that their price of nonconformance was 31% of total sales when they undertook a quality cost reduction project. In just one year they were able to lower these costs to 9%. Among their accomplishments:

- Scrap and rework reduced 30%.
- Manufacturing cost variance reduced 20%.
- Late collections reduced 46%.
- Average turnaround on receivables reduced from 62 days to 35 days.

I.G QUALITY PHILOSOPHIES AND APPROACHES (e.g., JURAN, DEMING, TAGUCHI, ISHIKAWA)

The 1970s made one thing abundantly clear: a quality advantage could lead to improved market share and profitability. Whether the profit measure is return on sales or return on investment, businesses with a superior product or service offering clearly outperform those with inferior quality. Several key benefits accrue to businesses that offer superior perceived quality: stronger customer loyalty, more repeat purchases, less vulnerability to price wars, ability to command higher relative price without affecting market share, lower marketing costs, and market-share improvements. Firms with high quality receive a double benefit: they have lower costs than their low-quality competitors, and their superior offerings command higher prices in the marketplace. The net result is that high-quality firms enjoy greater market share and greater profitability, simultaneously.

Quality also has an impact on growth. Customers perceive that they receive greater value from their purchases of high quality goods or services. High quality tends to set off a chain reaction. Businesses that achieve superior quality tend to gain market share. The superior quality and market share lead to improved profitability. Capacity utilization increases, leading to better return on investment, higher employee productivity, and lower marketing expense per dollar of sales. Quality improvement is clearly not a zero-sum game!

Given these proven relationships, it is no surprise that business leaders seek to obtain a strategic quality advantage. Obtaining a quality advantage is a matter of leadership. The Profit Impact of Marketing Strategy (PIMS) studies indicate that those firms that enjoy a strategic advantage are usually pioneers rather than early followers (a finding that casts doubt on such practices as benchmarking).

HISTORICAL BACKGROUND

Garvin (1985) notes that, in the last century, quality has moved through four distinct "quality eras": inspection, statistical quality control, quality assurance, and strategic quality management. A fifth era is emerging: complete integration of quality into the overall business system. Managers in each era were responding to problems which faced them at the time. To fully comprehend the modern approach to quality, we must first understand how perceptions of quality have evolved over time.

THE INSPECTION ERA

Quality has been a matter of great concern to people throughout recorded history. Quality issues have been written into law since the Mesopotamian era. The Babylonian king Hammurabi (1792–1750B.C.) made Babylon the chief Mesopotamian kingdom and codified the laws of Mesopotamia and Sumeria. The Code of Hammurabi called for the death of builders whose buildings collapsed and killed its occupants. No doubt the builders of the time took great care to construct high quality structures!

Prior to the industrial revolution, items were produced by individual craftsmen for individual customers. Should quality problems arise, the customer took the issue directly to the producer. Products were truly "customized" and each item was unique. The entire process of material procurement, production, inspection, sale, and customer support was performed by the craftsman or his apprentices.

The American industrial revolution changed the situation dramatically. When Eli Whitney agreed to produce muskets for the new United States government, he proposed a radically new approach to manufacturing. Rather than handcrafting each musket, he would use special-purpose machinery to produce parts that were interchangeable. A preestablished sequence of operations would be performed by men working with the machines to produce parts that would be, to the extent possible, identical to one another. The advantages were obvious. If a soldier's weapon needed replacement parts they could be obtained from an inventory of spares, rather than needing to be remade to custom specifications. The new approach could be carried out by people with less skill than the craftsman, thus lowering labor costs. In addition, the new approach lead directly to a tremendous increase in productivity. The idea caught on quickly and spread beyond the military to consumer products such as Singer's sewing machines and McCormick's harvesting equipment. Mass production had been born.

The new approach rested on a key assumption: to be interchangeable the parts must be essentially identical. Much effort was expended towards this goal. The design process was directed towards producing a working prototype that performed its intended function well. Manufacturing processes were designed to reproduce the prototype with minimal variation. Tooling, jigs, fixtures, etc. were produced to exacting tolerances. Material quality was subjected to tight control. Rigorous inspection was performed at every important step of the process, using elaborate gaging systems. By the early 1900s, inspection had matured to such a state that Frederick W. Taylor listed it as one of the activities that should have its own functional boss (foreman).

The first book to discuss quality per se was G.S. Radford's *The Control of Quality in Manufacturing*, published in 1922. Radford presented quality as a distinct management function. He touched on such modern issues as the role of the designer, coordination among departments effecting quality, and quality's role in improving productivity and lowering costs. However, Radford's focus was clearly on inspection; 39% of the book's chapters were devoted exclusively to inspection. Quality was defined as "that evenness or uniformity which results when the manufacturer adheres to his established requirements." To Radford it was obvious that one measures quality by comparing inspection results with established requirements.

THE QUALITY CONTROL ERA

The inspection-based approach to quality was challenged by Walter A. Shewhart. Shewhart's landmark 1931 book *Economic Control of Quality of Manufacturing* introduced the modern era of quality management. In 1924, Shewhart was part of a group working at Western Electric's Inspection Engineering Department of Bell Laboratories. Other members of the group included Harold Dodge, Harry Romig, G.D. Edwards, and Joseph Juran, a veritable "who's who" of the modern quality movement.

The new concept of quality included ideas that were quite radical at the time. Shewhart recognized that variation could never be completely eliminated. Try as one might, no two things could ever be made exactly the same. Thus, he reasoned, attempts to eliminate variability were certain to fail. Then Shewhart took a huge conceptual leap: the central task of quality control was not to identify variation from requirements, it was to distinguish between variation that was a normal result of the process and variation that indicated trouble. This insight lead directly to Shewhart's now famous concept of statistical control. The concept is explained by Shewhart as follows:

> A phenomenon will be said to be controlled when, through the use of past experience, we can predict, at least within limits, how the phenomenon may be expected to vary in the future. Here it is understood that prediction means that we can state, at least approximately, the probability that the observed phenomenon will fall within the given limits. (Shewhart 1931, p. 6)

Shewhart's approach to quality was to identify the limits of variation that could be expected from a process operating in a "normal" state. To do this he developed simple statistical and graphical tools that could be used to study data obtained from a process. Unlike inspection, Shewhart's approach did not require 100% inspection or sorting; samples could be used. Furthermore, as long as the process variability was less than the design required, one could be assured that acceptable process quality was being maintained. Shewhart's approach is known today as statistical process control, or SPC. SPC remains one of the quality professional's most powerful tools, in a form largely unchanged from Shewhart's original presentation. One of the most common SPC tools is the control chart, such as the one shown in Figure I.30.

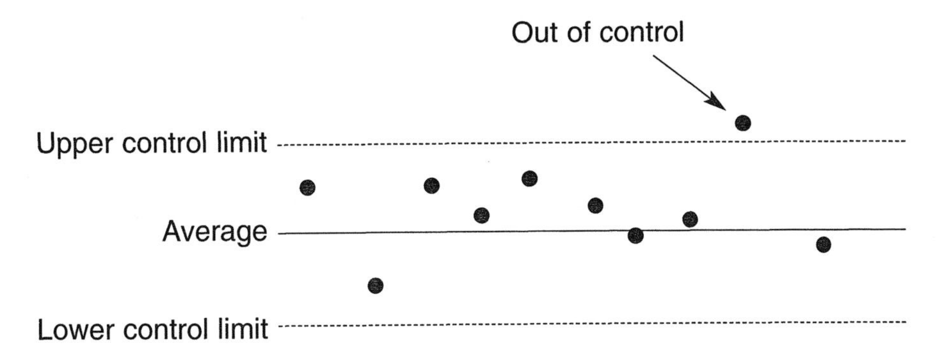

Figure I.30. Control chart.

While Shewhart pursued a process-focused approach, Dodge and Romig looked at the issue of product sampling, also known as acceptance sampling. The premise *was that* adequate information regarding delivered product quality could be obtained from a sample, rather than from inspecting all of the items. The approach developed by Dodge and Romig developed the science of acceptance sampling.

The new approach to quality was largely limited to Bell Labs until the outbreak of World War II. The War Department issued quality standards in 1941 and 1942 in an attempt to deal with the poor quality of ordinance. The War Department chose to pursue acceptance sampling rather than SPC. Thus, acceptance sampling was widely employed by the United States military in World War II. Acceptance sampling was successful in eliminating the inspection bottleneck in the production of war materials. Quality and productivity improvements were realized as well. Thousands of people were trained in the use of sampling tables and the concepts of the acceptance sampling approach to quality.

Quality continued to evolve after World War II. Initially, few commercial firms applied the new, statistical approach. However, those companies that did achieved spectacular results and the results were widely reported in the popular and business press. Interest groups, such as the Society of Quality Engineers (1945), began to form around the country. In 1946, the Society of Quality Engineers joined with other groups to form the American Society for Quality (ASQ). In July 1944, the Buffalo Society of Quality Engineers published *Industrial Quality Control*, the first journal devoted to the subject of quality. By 1950, quality was a widely recognized, if not widely practiced, management discipline.

THE QUALITY ASSURANCE ERA

One of the legacies of acceptance sampling is the idea of an "acceptable quality level" (AQL). Conceptually, the AQL is some less-than-perfect level of quality that management is willing to "live with" long term. Officially, the AQL is defined as

> The maximum percentage or proportion of variant units in a lot or batch that, for the purposes of acceptance sampling, can be considered satisfactory as a process average. (ASQ Statistics Division 1983, p 46)

The definition makes it clear that the authors intend the AQL idea to be applied only to statistical sampling problems. However, some believe that the message that came across was that there was a quality level that was "good enough" as long as both the supplier and producer agreed to that level. Quality goals became a matter of negotiation between buyer and seller and, once the goal was achieved, quality improvement often ceased.

Quality assurance is based on the premise that defects can be prevented, i.e., some of the losses from poor quality can be economically eliminated. Juran (1951) laid out his theory that quality costs could be divided into costs of prevention and costs of failure. By investing in quality improvement, management could reap substantial returns in reduced failures. Juran's model of quality costs is shown in Figure I.31.

In this model total quality costs are at an optimum at some quality level that is less than perfect. At quality levels worse than the optimum, management is spending too much on failure costs. Management is spending too much on *prevention* if quality levels exceed the optimum.

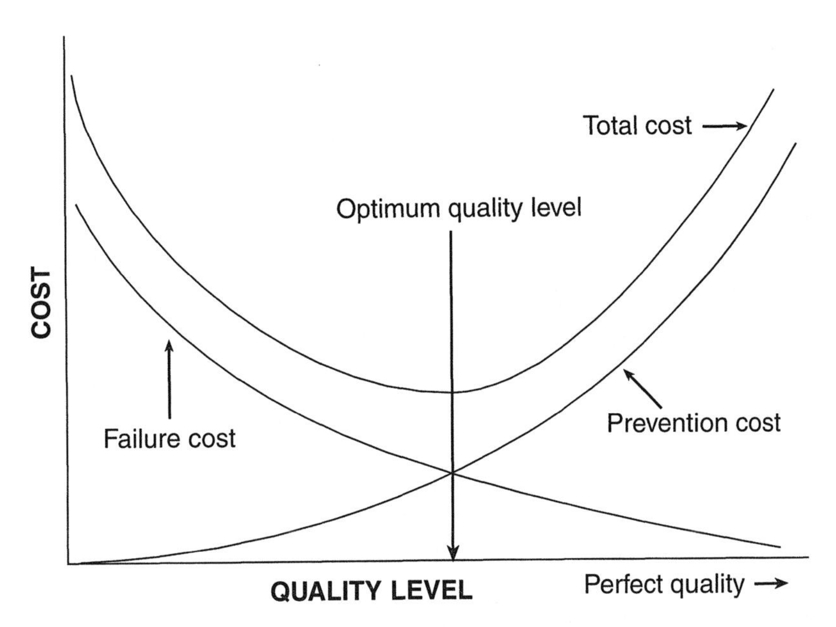

Figure I.31. Juran's original optimum quality cost model .* *(Figure I.26 repeated)*

Given this mindset, managers began the pursuit of identifying the correct amount to invest in quality. High failure rates were used to justify additional expenditures on quality improvement. Conversely, low failure rates were used to justify lower levels of spending on quality.

Juran's quality cost model had implications for management in general. The model showed that money spent in areas other than inspection had an impact on quality, e.g., design engineering decisions. In 1956 Armand Feigenbaum introduced the idea of *total quality control* (Feigenbaum 1956). Feigenbaum observed that manufacturing was only one element in the long chain of activities involved in producing a new product. Other activities included design, procurement of raw materials, storage, delivery, installation, and so on. In fact, said Feigenbaum, the first principle is that *quality is everyone's job*. Total quality control meant that every function played an important role in assuring product quality. Furthermore, this was not to be done in a haphazard manner. Feigenbaum presented elaborate matrices which linked the different functional areas with the appropriate quality control activity. Total quality control was to be implemented with rigor throughout the organization as well as in the organization's supplier base.

*In later years, Juran modified the model to show that the optimum cost is often found at 100% conformance.

Phillip Crosby popularized the zero defects approach to quality assurance. The zero defects philosophy started at the Martin company's Pershing missile manufacturing facility in Orlando, Florida. Prior to the zero defects program, no missiles had been delivered without extensive inspection, rework, and repair effort. This changed when Martin accepted a challenge from their customer, the U.S. Army, to deliver the first field Pershing a full month ahead of schedule. The result was to be accomplished by eliminating the normal pre-delivery inspection-repair-reinspection activities. Thus, employees had to assume responsibility for getting it right the first time. In February 1962, the employees of Martin delivered a defect-free missile to their customer. The missile was fully operational in less than 24 hours.

Martin's management was stunned by the result. After intense study of the incident, they concluded that defects were produced because management expected them to be produced. The one-time management demanded perfection; the employees delivered it. In other words, Martin management concluded that defects were a result of poor worker motivation.

The program developed by Martin to implement the zero defects philosophy was primarily an employee motivation program. As such, it was long on awareness and philosophy and short on methods and techniques. The fundamental message was very simple: the quality standard was to be zero defects, nothing less. This represented a dramatic philosophic departure from Juran's original optimum quality cost model. The debate over which approach is best continues to this day.

While broader in scope than its predecessors, quality assurance activities were still conducted primarily at the middle levels of management. Senior leadership was seldom actively involved with setting quality goals or monitoring progress towards the goals.

STRATEGIC QUALITY MANAGEMENT

The quality assurance perspective suffers from a number of serious shortcomings. Its focus is internal. Specifications are developed by the designers, often with only a vague idea of what customers really want. The scope of quality assurance is generally limited to those activities under the direct control of the organization; important activities such as transportation, storage, installation, and service are typically either ignored or given little attention. Quality assurance pays little or no attention to the competition's offerings. The result is that quality assurance may present a rosy picture, even while quality problems are putting the firm out of business. Such a situation existed in the United States in the latter 1970s.

The approaches taken to achieve the quality edge vary widely among different firms. Some quality leaders pioneer and protect their positions with patents or copyrights. Others focus on relative image or service. Some do a better job of identifying and meeting the needs of special customer segments. And others focus on value-added operations and technologies.

Once a firm obtains a quality advantage, it must continuously work to maintain it. As markets mature, competition erodes any advantage. Quality must be viewed from the customer's perspective, not as conformance to self-imposed requirements. Yet a quality advantage often cannot be obtained only by soliciting customer input, since customers usually are not aware of potential innovations.

W. EDWARDS DEMING

Dr. W. Edwards Deming is perhaps the best-known figure associated with quality. Dr. Deming is given much of the credit for the transformation of Japan from a war-shattered nation with a reputation for producing cheap junk to a global economic superpower.

Deming is probably best known for his theory of management as embodied in his 14 points. According to Deming, "The 14 points all have one aim: to make it possible for people to work with joy." Deming's 14 points follow.

Deming's 14 Points

1. Create constancy of purpose for the improvement of product and service, with the aim to become competitive, stay in business, and provide jobs.
2. Adopt the new philosophy of cooperation (win-win) in which everybody wins. Put it into practice and teach it to employees, customers, and suppliers.
3. Cease dependence on mass inspection to achieve quality. Improve the process and build quality into the product in the first place.
4. End the practice of awarding business on the basis of price tag alone. Instead, minimize total cost in the long run. Move toward a single supplier for any one item, on a long-term relationship of loyalty and trust.
5. Improve constantly and forever the system of production, service, planning, or any activity. This will improve quality and productivity and thus constantly decrease costs.
6. Institute training for skills.
7. Adopt and institute leadership for the management of people, recognizing their different abilities, capabilities, and aspirations. The aim of leadership should be to help people, machines, and gadgets do a better job. Leadership of management is in need of overhaul, as well as leadership of production workers.
8. Eliminate fear and build trust so that everyone can work effectively.
9. Break down barriers between departments. Abolish competition and build a win-win system of cooperation within the organization. People in research, design, sales, and production must work as a team to foresee problems of production and in use that might be encountered with the product or service.

10. Eliminate slogans, exhortations, and targets asking for zero defects or new levels of productivity. Such exhortations only create adversarial relationships, as the bulk of the causes of low quality and low productivity belong to the system and thus lie beyond the power of the work force.
11. Eliminate numerical goals, numerical quotas, and management by objectives. Substitute leadership.
12. Remove barriers that rob people of joy in their work. This will mean abolishing the annual rating or merit system that ranks people and creates competition and conflict.
13. Institute a vigorous program of education and self-improvement.
14. Put everybody in the company to work to accomplish the transformation. The transformation is everybody's job.

Deming also described a system of "profound knowledge." Deming's system of profound knowledge consists of four parts: appreciation for a system, knowledge about variation, theory of knowledge, and psychology.

A system is a network of interdependent components that work together to accomplish the aim of the system. The system of profound knowledge is, itself, a system. The parts are interrelated and cannot be completely understood when separated from one another. Systems must be managed: the greater the interdependence of the various system components, the greater the need for management. In addition, systems should be globally optimized; global optimization cannot be achieved by optimizing each component independent of the rest of the system.

Systems can be thought of as networks of intentional cause-and-effect relationships. However, most systems also produce unintended effects. Identifying the causes of the effects produced by systems requires understanding of variation—part two of Deming's system of profound knowledge. Without knowledge of variation, people are unable to learn from experience. There are two basic mistakes made when dealing with variation:

1. Reacting to an outcome as if it were produced by a special cause, when it actually came from common causes;
2. Reacting to an outcome as if it were produced by common causes, when it actually came from a special cause. The concept of special and common cause variation is discussed in detail in V.C.

Deming's theory of profound knowledge is based on the premise that management is prediction. Deming, following the teachings of the philosopher C.I. Lewis, believed that prediction is not possible without theory. Novelist/philosopher Ayn Rand defines *theory* as "A set of abstract principles purporting to be either a correct description of reality or a set of guidelines for man's actions. Correspondence to reality is the standard of value by which one estimates a theory." Deming points out that knowledge is acquired as one makes a rational prediction based on theory, then revises the theory based on comparison of prediction with observation. Knowledge is reflected in the new theory. Without theory, there is nothing to revise; i.e., there can be no new knowledge, no learning. The process of learning is operationalized by Deming's Plan-Do-Study-Act cycle (a modification of Shewhart's Plan-Do-Check-Act cycle). It is important to note that information is not knowledge. Mere "facts", in and of themselves, are not knowledge. Knowing what the Dow Jones Industrial Average is right now, or what it has been for the last 50 years, is not enough to tell us what it will be tomorrow.

Psychology is the science that deals with mental processes and behavior. In Deming's system of profound knowledge, psychology is important because it provides a theoretical framework for understanding the differences between people, and provides guidance in the proper ways to motivate them.

INTEGRATION OF QUALITY INTO SELF-DIRECTED SYSTEMS

The reader should note that the common thread in the evolution of quality management is that attention to quality has moved progressively further up in the organizational hierarchy. Quality was first considered a matter for the line worker, then the inspector, then the supervisor, the engineer, the middle manager and, today, for upper management. The question arises: what's next? The answer appears to be that quality will continue to increase in importance. And the fact is that the most important people in any organization are its customers. Ultimately, it is the customer's concern with quality that has been driving quality matters ever higher in the organization. This can be expected to continue. As Juran (1994) stated, the next century will be the century of quality.

The *American Heritage Dictionary* defines *management* as:

1. The act, manner, or practice of managing; handling, supervision, or control; management of a crisis; management of factory workers.
2. The person or persons who control or direct a business or other enterprise.
3. Skill in managing; executive ability.

As the definitions make clear, management is primarily a control-oriented activity. This was necessary in the past because the organization was designed in such a way as to create conflicting goals among its members. As a result managers had to oversee the operation to assure that the organization as a whole did not suffer. This created a "chain of command" system with layer upon layer of people watching one another. Cowen and Ellig (1993) liken the traditional approach to management with the central planning activities of the former Soviet Union. This author believes that, in the future, there will be a dramatic shift away from the traditional command-hierarchical approach to management and towards minimal-intervention systems of production.

The 1995 Baldrige criteria provide an indication of how quality will fit into the new business systems. The criteria were extensively revised to reflect a dramatic change of direction in the quality field. Simply put, quality is no longer a separate focus for the organization. Note that the word "quality" does not even appear in the Baldrige model!

However, the importance of quality is implicit in the model itself. The model presents quality as completely integrated into all of the important management systems. Customer satisfaction is a result of good quality, among other things. The focus of the organization is twofold: business results and customer satisfaction. Quality is important to the extent that it furthers these aims.

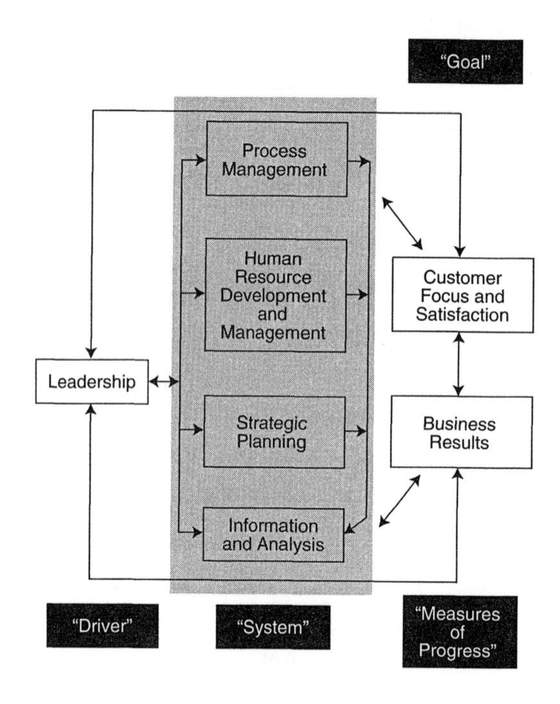

Figure I.32. 1995 Baldrige model of management.

Self-directed systems do not require active intervention to keep them moving towards their goals. In the new model of business, the interests of customers and other stakeholders (owners, employees, suppliers) are designed to be in harmony. By satisfying the demands of customers, stakeholders also meet their own objectives. This new model is analogous to natural ecosystems or, for that matter, Adam Smith's "invisible hand" free market system. Ideally, active management of such systems becomes unnecessary.

FUNDAMENTAL PRINCIPLES OF QUALITY

The first principle of modern quality improvement is that "The customer is king!" The 1980s have shown clearly that quality drives market share and when superior quality is combined with large market share, profitability is virtually guaranteed. But relative perceived quality is not the same thing as traditional "conformance" quality. Many a company has struggled to obtain a high conformance rate, only to be disappointed with the results in the marketplace. What matters is what your customers think about your quality. Of course, vague customer requirements must still be translated into meaningful specifications to guide internal activities.

DEFINITIONS OF QUALITY

While the importance of quality is now generally accepted, there is no single generally accepted definition of "quality." Garvin (1988) describes five principal approaches to defining quality:

1. **Transcendent**—"Quality cannot be defined, you know what it is." (Persig 1974, p. 213)
2. **Product-based**—"Differences in quality amount to differences in the quantity of some desired ingredient or attribute." (Abbott 1955, pp. 126–27)

   "Quality refers to the amounts of the unpriced attributes contained in each unit of the priced attribute." (Leffler 1982, p. 956)
3. **User-based**—"Quality consists of the ability to satisfy wants" (Edwards 1968, p. 37)

   In the final analysis of the marketplace, the quality of a product depends on how well it fits patterns of consumer preference." (Kuehn and Day 1962, p. 101)

   "Quality is fitness for use." (Juran 1974, p. 2-2)
4. **Manufacturing-based**—"Quality [means] conformance to requirements." (Crosby 1979, p. 15)

   "Quality is the degree to which a specific product conforms to a design or specification." (Gilmore 1974, p. 16)
5. **Value-based**—"Quality is the degree of excellence at an acceptable price and the control of variability at an acceptable cost." (Broh 1982, p. 3)

   "Quality means 'best for certain customer conditions.' These conditions are a) the actual use and b) the selling price of the product." (Feigenbaum 1961, p. 1)

Other quality experts have weighed in with their own definitions. Genichi Taguchi (1986) defines quality as "The loss a product causes to society after being shipped, other than any losses caused by its intrinsic functions." Taguchi points out that many of the customer-based definitions of quality are based on the concept of value to the customer. While this is an important point, it is a point to be addressed by marketing or product planning, while our interest is confined to engineering. However, Taguchi's objections notwithstanding, the distinction between quality engineering and other areas of the organization are rapidly blurring. Today the quality engineer must understand all aspects of quality.

I.H CUSTOMER RELATIONS, EXPECTATIONS, NEEDS, AND SATISFACTION

Although customers seldom spark true innovation (for example, they are usually unaware of state-of-the art developments), their input is extremely valuable. Obtaining valid customer input is a science itself. Market research firms use scientific methods such as critical incident analysis, focus groups, content analysis and surveys to identify the "voice of the customer." Noritaki Kano developed the following model of the relationship between customer satisfaction and quality.

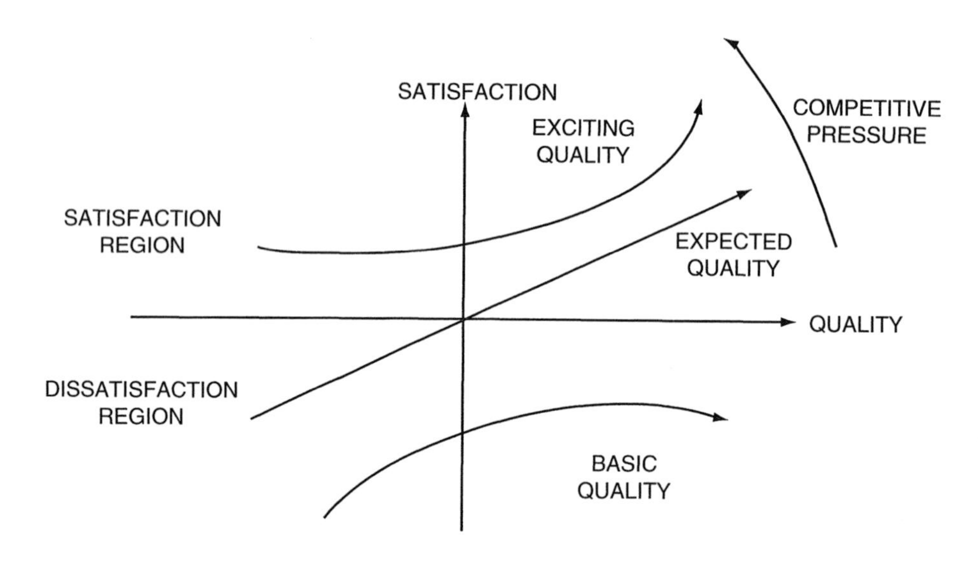

Figure I.33. Kano model.

The Kano model shows that there is a basic level of quality that customers assume the product will have. For example, all automobiles have windows and tires. If asked, customers don't even mention the basic quality items, they take them for granted. However, if this quality level *isn't* met the customer will be dissatisfied; note that the entire "Basic Quality" curve lies in the lower half of the chart, representing dissatisfaction. However, providing basic quality isn't enough to create a satisfied customer.

The expected quality line represents those expectations which customers explicitly consider. For example, the length of time spent waiting in line at a checkout counter. The model shows that customers will be dissatisfied if their quality expectations are not met; satisfaction increases as more expectations are met.

The exciting quality curve lies entirely in the satisfaction region. This is the effect of innovation. Exciting quality represents *unexpected* quality items. The customer receives more than they expected. For example, Cadillac pioneered a system where the headlights stay on long enough for the owner to walk safely to the door.

Competitive pressure will constantly raise customer expectations. Today's exciting quality is tomorrow's basic quality. Firms that seek to lead the market must innovate constantly. Conversely, firms that seek to offer standard quality must constantly research customer expectations to determine the currently accepted quality levels. It is not enough to track competitors since expectations are influenced by outside factors as well. For example, the quality revolution in manufacturing has raised expectations for service quality as well.

QUALITY FUNCTION DEPLOYMENT

Once information about customer expectations has been obtained, techniques such as quality function deployment (QFD) can be used to link the voice of the customer directly to internal processes.

Tactical quality planning involves developing an approach to implementing the strategic quality plan. One of the most promising developments in this area has been policy deployment. Sheridan (1993) describes policy deployment as the development of a measurement-based system as a means of planning for continuous quality improvement throughout all levels of an organization. Originally developed by the Japanese, American companies also use policy deployment because it clearly defines the long-range direction of company development, as opposed to short-term.

QFD is a customer-driven process for planning products and services. It starts with the voice of the customer, which becomes the basis for setting requirements. QFD matrices, sometimes called "the house of quality," are graphical displays of the result of the planning process. QFD matrices vary a great deal and may show such things as competitive targets and process priorities. The matrices are created by inter-departmental teams, thus overcoming some of the barriers which exist in functionally organized systems.

QFD is also a system for design of a product or service based on customer demands, a system that moves methodically from customer requirements to specifications for the product or service. QFD involves the entire company in the design and control activity. Finally, QFD provides documentation for the decision-making process. The QFD approach involves four distinct phases (King, 1987):

Organization phase—Management selects the product or service to be improved, the appropriate interdepartmental team, and defines the focus of the QFD study.

Descriptive phase—The team defines the product or service from several different directions such as customer demands, functions, parts, reliability, cost, and so on.

Breakthrough phase—The team selects areas for improvement and finds ways to make them better through new technology, new concepts, better reliability, cost reduction, etc., and monitors the bottleneck process.

Implementation phase—The team defines the new product and how it will be manufactured.

QFD is implemented through the development of a series of matrices. In its simplest form QFD involves a matrix that presents customer requirements as rows and product or service features as columns. The cell, where the row and column intersect, shows the correlation between the individual customer requirement and the product or service requirement. This matrix is sometimes called the "requirement matrix." When the requirement matrix is enhanced by showing the correlation of the columns with one another, the result is called the "house of quality." Figure I.34 shows one commonly used house of quality layout.

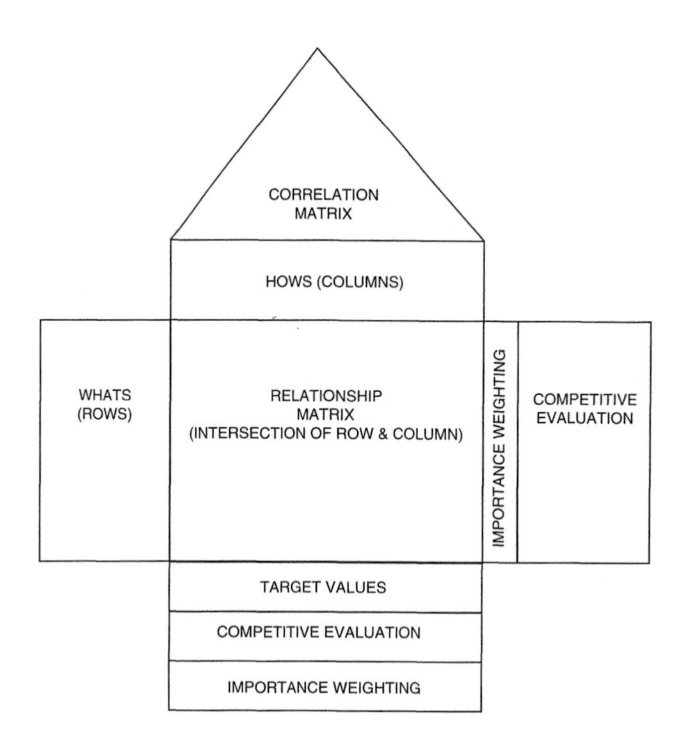

Figure I.34. The house of quality.

The house of quality relates, in a simple graphical format, customer requirements, product characteristics, and competitive analysis. It is crucial that this matrix be developed carefully since it becomes the basis of the entire QFD process. By using the QFD approach, the customer's demands are "deployed" to the final process and product requirements.

One rendition of QFD, called the Macabe approach, proceeds by developing a series of four related matrices (King, 1987): product planning matrix, part deployment matrix, process planning matrix, and production planning matrix. Each matrix is related to the previous matrix as shown in Figure I.35.

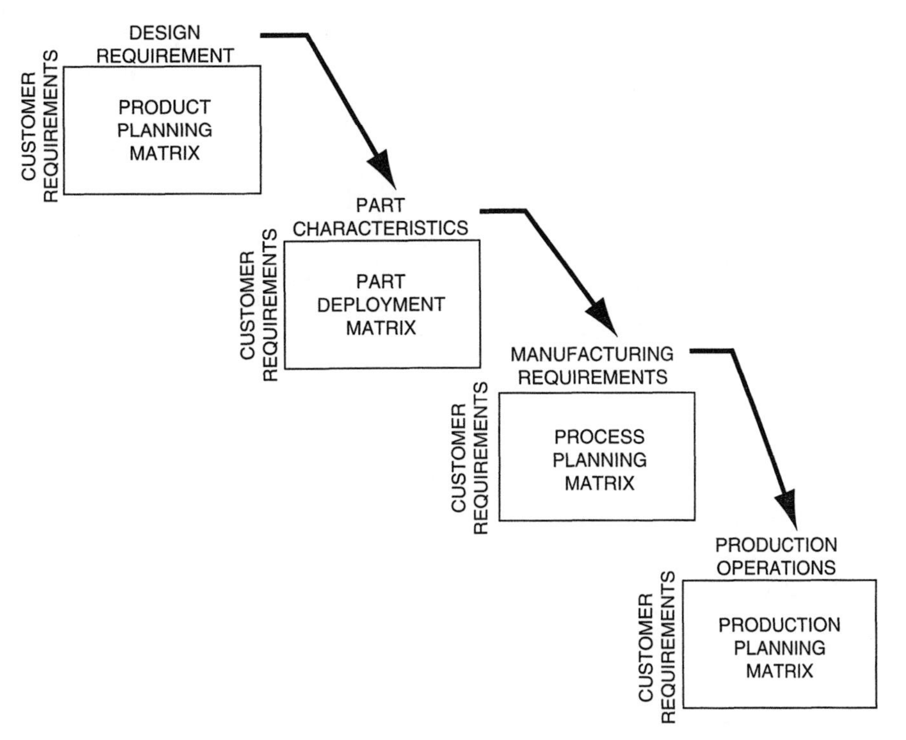

Figure I.35. QFD methodology: Macabe approach.

Figure I.36 shows an example of an actual QFD matrix.

T/R Module ManTech QFD

INTERACTIONS
⊕ STRONG POSITIVE
+ POSITIVE
– NEGATIVE
⊖ STRONG NEGATIVE

RELATIONSHIP
● STRONG = 9
O MEDIUM = 3
⊿ WEAK OR POSSIBLE = 1

DESIGN REQUIREMENTS / CUSTOMER DEMANDS	ITEM NUMBER / IMPORTANCE RATING	YIELD	PROCESS CAPABILITY	RATE	MODULE COST	MTBF	FAIL. FREE OP. PERIOD	SIZE	WEIGHT	PROCESS FLEXIBILITY	COST TO MAINTAIN	AVAILABILITY OF ARRAY
DIRECTION FROM TARGET VALUE		↑	↑	↑	↓	↑	↑	↓	↓	↑	↓	↑
		1	2	3	4	5	6	7	8	9	10	11
LOW COST	10	●	●	●	●	⊿	⊿	⊿	⊿	O	⊿	
REPRODUCIBLE	10	●	●	●	●	⊿	⊿	⊿				
FACTORY IMPLEMENTATION / DEMO.	10	●	●	●	●					O		
PROCESS VERIFICATION (PROOF)	10	●	O	●	●							
HIGH RELIABILITY (> 85,000 hrs.)	9	O	●	●	⊿	●	●	⊿			●	●
SMALL SIZE - LIGHT WEIGHT	8		O	⊿	⊿	⊿	⊿	●	●	⊿	⊿	
PERFORMANCE EXPANDABLE	6	⊿	●	⊿	⊿			⊿	⊿	●		
MEETS MODULE ELECT. PERFORMANCE	5	●	●	●	●	⊿	⊿	⊿	●	●	⊿	⊿
SYSTEM MAINTAINABILITY	1					●	●	⊿			●	●
ABSOLUTE WEIGHT		49	60	56	48	22	22	15	11	16	21	19
RELATIVE WEIGHT		438	504	500	428	123	123	113	88	122	113	95
KEY DESIGN REQUIREMENTS		X	X	X	X							
TARGET VALUES		80 %	1.33	1000/DAY	$200	85,000 HRS.	10,000 HRS.	4"X0.46"X0.2"	1 OZ.	TBD	TBD*	TBD*

Figure I.36. QFD matrix for an aerospace firm.
From Wahl, P.R. and Bersbach, P.L. (1991), "TQM Applied—Cradle to Grave," ASQ 45th Quality Congress Transactions. Reprinted with permission.

DATA COLLECTION AND REVIEW OF CUSTOMER EXPECTATIONS, NEEDS, REQUIREMENTS, AND SPECIFICATIONS

Another approach to QFD is based on work done by Yoji Akao. Akao (1990, pp. 7–8) presents the following 11-step plan for developing the quality plan and quality design, using QFD.

1. First, survey both the expressed and latent quality demands of consumers in your target marketplace. Then decide what kinds of "things" to make.
2. Study the other important characteristics of your target market and make a demanded quality function deployment chart that reflects both the demands and characteristics of that market.
3. Conduct an analysis of competing products on the market, which we call a competitive analysis. Develop a quality plan and determine the selling features (sales points).
4. Determine the degree of importance of each demanded quality.
5. List the quality elements and make a quality elements deployment chart.
6. Make a quality chart by combining the demanded quality deployment chart and the quality elements deployment chart.
7. Conduct an analysis of competing products to see how other companies perform in relation to each of these quality elements.
8. Analyze customer complaints.
9. Determine the most important quality elements as indicated by customer quality demands and complaints.
10. Determine the specific design quality by studying the quality characteristics and converting them into quality elements.
11. Determine the quality assurance method and the test methods.

CUSTOMER RELATIONSHIP MANAGEMENT AND COMMITMENT

While the principles of personal selling and negotiation are transaction-oriented, the sellers' dealings with customers should always be conducted with the larger concept of relationship management in mind. Relationship management is most appropriate for those customers who can most affect the company's future. For many companies, the top 20% of the customers account for a disproportionate share of the company's sales. Contact with such customers should not be limited to sales calls. These key accounts should be monitored to assure that their special needs are being addressed. The relationship should be viewed as a partnership, with the supplier proactively offering suggestions as to how the customer might improve their sales, quality, etc. For larger accounts, a formal relationship management program should be considered. Kotler (2002) presents the following steps for establishing a relationship management program within a company:

- Identify the key customers meriting relationship management—The company can choose the five or ten largest customers and designate them for relationship management. Additional customers can be added who show exceptional growth or who pioneer new developments in the industry, and so on.
- Assign a skilled relationship manager to each key customer or customer group—The salesperson who is currently servicing the customer should receive training in relationship management or be replaced by someone who is more skilled in relationship management. The relationship manager should have characteristics that match or appeal to the customer.
- Develop a clear job description for relationship managers—It should describe their reporting relationships, objectives, responsibilities, and evaluation criteria. The relationship manager is responsible for the client, is the focal point for all information about the client, and is the mobilizer of company services for the client. Each relationship manager will have only one or a few relationships to manage.
- Appoint an overall manager to supervise the relationship managers—This person will develop job descriptions, evaluation criteria, and resource support to increase relationship managers' effectiveness.
- Each relationship manager must develop long-range and annual customer-relationship plans—The annual relationship plan will state objectives, strategies, specific actions, and required resources.

Figure I.37 lists a number of relationship factors.

THINGS AFFECTING RELATIONSHIPS

Good Things	***Bad Things***
Initiate positive phone calls	Make only callbacks
Make recommendations	Make justifications
Can do in language	Accommodative language
Use phone	Use correspondence
Show appreciation	Wait for misunderstandings
Make service suggestions	Wait for service requests
Use "we" problem-solving language	Use "owe us" legal language
Get to problems	Only respond to problems
Use jargon/shorthand	Use long-winded communication
Personality problems aired	Personality problems hidden
Talk of "our future together"	Talk about making good on the past
Routinize responses	Fire drill/emergency responsiveness
Accept responsibility	Shift blame
Plan the future	Rehash the past

Figure I.37. Actions affecting buyer-seller relationships.
From "After the sale is over," *Harvard Business Review* (September–October 1983), p. 119.

Complaints

This section will describe frameworks for resolving customer complaints in the context of preserving customer relationships and honoring the expressed and implied commitments to the customer. Techniques for complaint analysis and prevention are discussed a little later in this chapter.

The term which describes the organization's response to a customer complaint is recovery. Effective recovery is an important element in customer retention; the subject of customer retention and loyalty is discussed a little later in this chapter.

Edosomwan proposes the complaint resolution model shown in Figure I.38. The model shows that complaint resolution is one of the elements in the customer satisfaction model. Complaints arise from a variety of sources. The source most commonly encountered by the quality manager is product defects or service errors. Another source of complaints is poor communication between the customer and the company. A common cause of communication breakdown are complex or rigid procedures for resolving customer problems. Company employees who are rude, poorly trained, or unable to answer the customer's questions will generate complaints. In addition, company systems and procedures can lead to complaints.

Not all problems result in complaints. It has been shown that the percentage of defects or service failures that actually result in a complaint is related to the seriousness of the problem and the price paid for the product or service. Minor problems are reported less than major problems; problems are reported more often when higher prices are paid. Also, complex systems for processing complaints will reduce the number of complaints logged, but it won't reduce the number of dissatisfied customers.

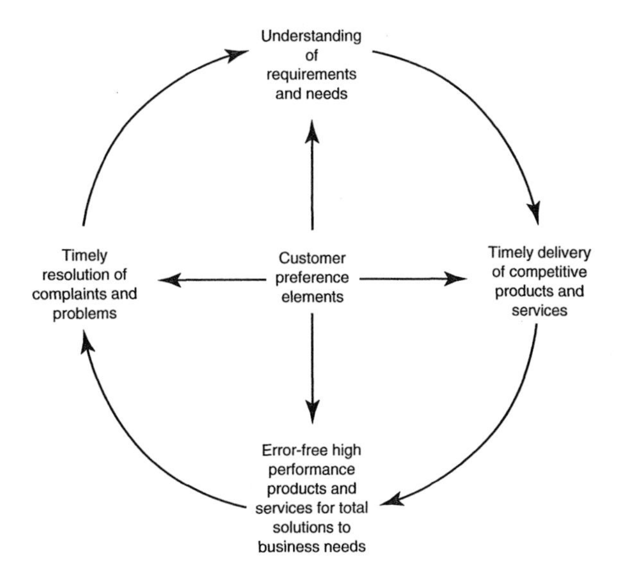

Figure I.38. Complaint resolution model.

From *Customer and Market-Driven Quality Management,* Figure 3.12, p. 70, by Johnson A. Edosomwan, copyright © 1993 by ASQ. Reprinted by permission.

Given a suitable company policy and procedure for complaints, complaint processing becomes largely a communications matter. The most important activity in the process is to listen to the customer. Listening skills should be routinely taught to all customer service employees (CSE); these skills are absolutely essential to complaint processors. The CSE should attempt to understand the nature, magnitude, and potential impact of the complaint on the customer. They should agree with the customer that the complaint is valid, and offer the company's apology for the inconvenience. CSEs should scrupulously avoid arguing or becoming defensive. Every effort should be made to get the facts. If possible, an immediate solution should be offered. If resolving the problem cannot be accomplished on the spot, the customer should be told precisely what steps will be taken to resolve their problem. During the resolution process the customer should be kept informed of the status of their complaint.

Many complaints are not covered under legally binding agreements, such as warranties. Instead, the emphasis is on maintaining good relations with the customer by honoring implicit "guarantees." In these situations, the spirit of the understanding between the company and its customer is more important than the letter of a legal agreement. The message conveyed to the customer should be "You are a very important person to us." After all, it's the truth.

PARTNERSHIP AND ALLIANCES BETWEEN CUSTOMERS AND SUPPLIERS*

Research shows that purchased items account for:

- 60% of sales
- 50% of all quality problems
- 75% of all warranty claims

These facts alone should be sufficient to make the importance of a good customer-supplier relationship obvious. But even these impressive figures understate the importance of suppliers to a firm's success. The emphasis on just-in-time inventory management systems (JIT) has created a situation where any slippage in quality or delivery commitments causes an immediate detrimental impact on production schedules and the firm's ability to ship finished product. The interdependence of the supplier and the purchaser is now painfully obvious. This has lead many firms to reduce the number of suppliers in an effort to better manage supplier relationships. The new approach is to treat suppliers as partners rather than as adversaries. Suppliers are given larger and longer contracts and, in turn, they are expected to work closely with the purchaser to assure that the purchaser's customers are satisfied.

*The subject of supplier management is also discussed in I.I.

Supplier interaction types

Interactions between supplier and purchaser fall into a few common types.

- **Inspection**—Inspection focuses on various forms of product inspection. The relationship is adversarial.
- **Prevention**—Prevention is based on the premise that quality must be built by the supplier with the purchaser's help. The relationship is not adversarial, but it is still arm's-length.
- **Partnership**—This involves not just quality, but other business issues. Purchasers and suppliers work closely on such issues as joint training, planning, sharing confidential information (e.g., sales forecasts). Communications channels are wide-open and include such new forms as designer-to-designer or quality engineer-to-quality engineer. This approach results in parallel communication, a dramatic change from the purchasing agent-to-sales representative approach used in the past (serial communication). A research study showed that, for serial communications channels, quality levels stayed flat or declined, while, for parallel communication, quality levels improved dramatically.
- **Teamwork**—Very early in the relationship, supplier/purchaser teams are formed. The teams are process improvement teams, formed and chartered as described in I.C. The process to be improved is the supplier/purchaser communication process. These teams meet at supplier's and purchaser's facilities to set goals for the relationship and to develop plans for achieving their goals. Contact personnel and methods of contact are established, including a schedule of meetings. These meetings are used to update each other on progress and new plans, not merely for discussion of problems. Purchasers show suppliers how the product is designed, where the supplier's materials are used, and the manufacturing processes involved.

Joint technological planning

With joint technological planning (JTP), detailed plans are developed cooperatively by suppliers and purchasers. The plans include specific performance requirements, the meaning of which are discussed and agreed to by both parties. Quality, reliability, and maintainability requirements are specified, and the supplier's role in meeting those requirements is defined.

JTP also encompasses the processes to be used by suppliers and purchasers. Process control plans are prepared, including the identification of key process variables and how they are to be controlled and monitored. Special tasks to be done by the supplier (e.g., in-process inspections, SPC) are described. Classification of characteristics and defects is performed. Purchasers work with suppliers to prepare aesthetic and sensory standards. Test and measurement methods are standardized.

Finally, finished product inspection and audit requirements are established. Sampling requirements are specified as well as acceptable quality levels. Lot identification and traceability systems are developed. Corrective action and follow-up systems acceptable to supplier and purchaser are instituted.

Contract administration is a cooperative effort. Both supplier and purchaser participate in evaluation of initial samples of product. Systems are implemented to assure that design information and key information about changes in requirements is rapidly and accurately communicated. Provisions are made for routine surveillance of supplier processes. All parties are involved in evaluating delivered product, and the data collected are used for improvement and learning, not for judgment. Action on nonconforming product is rational and based on what is best for the end-user. Supplier quality personnel are part of the team.

Keiretsu

In Japan, "Keiretsu" is the name for the close coordination of companies, suppliers, banks, and many other companies that work together for the good of the whole. Control is based on equity holdings of their suppliers. The Japanese system, commonly called a cartel, is precluded by the current U.S. antitrust laws. However, as described by Burt and Doyle (1993), American firms have created Keiretsu-type relationships with suppliers through strategic supply management. The focus is on value-added outsourcing relationships. Significant efficiencies can be gained through these approaches to strategic supply management. The basis for the American Keiretsu-type relationships includes shared long-term objectives and commitments. The key ingredients that must be present are strategic fit and mutual trust.

American Keiretsu-type relationships are compatible and interdependent with other corporate strategies, such as total quality management, strategic cost management, just-in-time manufacturing, simultaneous engineering, flexible manufacturing, core competencies, and value-chain management. The approach can be applied to procurement of hardware or nonproduction and service requirements, such as legal services, consulting services, and maintenance, repair, and operating supplies.

In these relationships, it is important that companies assure protection of a company's core competencies through early agreements and during subsequent discussions and actions. Disciplines to control interim contacts at all levels are needed to ensure that the relationships stay focused within pre-established boundaries and do not flow over into product or technical activities that are beyond the scope of the agreed cooperation. The focus is on cooperation, trust, and synergism; however, this does not mean that executives can be careless in their business practices. They should not contribute to the development of a new, stronger competitor.

COMMUNICATION TECHNIQUES

There are any number of reasons why a firm may wish to communicate with its customers. A primary reason is the evaluation of the customer's perception of the firm's product and service quality and its impact on customer satisfaction. The purpose may be to get an idea of the general condition of quality and satisfaction, or a comparison of the current levels with the firm's goals. To be effective, the communication will provide sufficient detail to help the firm determine how they could improve quality and satisfaction.

Strategies for communicating with customers

There are four primary strategies commonly used to obtain information from or about customers:

- sample surveys
- case studies
- field experiments
- available data

With sample surveys, data are collected from a sample of a universe to estimate the characteristics of the universe, such as their range or dispersion, the frequency of occurrence of events, or the expected values of important universe parameters. The reader should note that these terms are consistent with the definition of enumerative statistical studies described in VI.A. This is the traditional approach to such surveys. However, if survey results are collected at regular intervals, the results can be analyzed using the quality control tools described in sections V.C and VI.G to obtain information on the underlying process. The quality manager should not be reticent in recommending that survey budgets be allocated to conducting small, routine, periodic surveys rather than infrequent "big studies." Without the information available from time-ordered series of data, it will not be possible to learn about processes which produce changes in customer satisfaction or perceptions of quality.

A case study is an analytic description of the properties, processes, conditions, or variable relationships of either single or multiple units under study. Sample surveys and case studies are usually used to answer descriptive questions ("How do things look?") and normative questions ("How well do things compare with our requirements?"). A field experiment seeks the answer to a cause-and-effect question ("Did the change result in the desired outcome?"). Use of available data as a strategy refers to the analysis of data previously collected or available from other sources. Depending on the situation, available data strategies can be used to answer all three types of questions: descriptive, normative, and cause-and-effect. Original data collection strategies such as mail questionnaires are often used in sample surveys, but they may also be used in case studies and field experiments.

Surveys

Survey development consists of the following major tasks (GAO, 1986, 15):

1. initial planning of the questionnaire
2. developing the measures
3. designing the sample
4. developing and testing the questionnaire
5. producing the questionnaire
6. preparing and distributing mailing materials
7. collecting data
8. reducing the data to forms that can be analyzed
9. analyzing the data

Figure I.39 shows a typical timetable for the completion of these tasks.

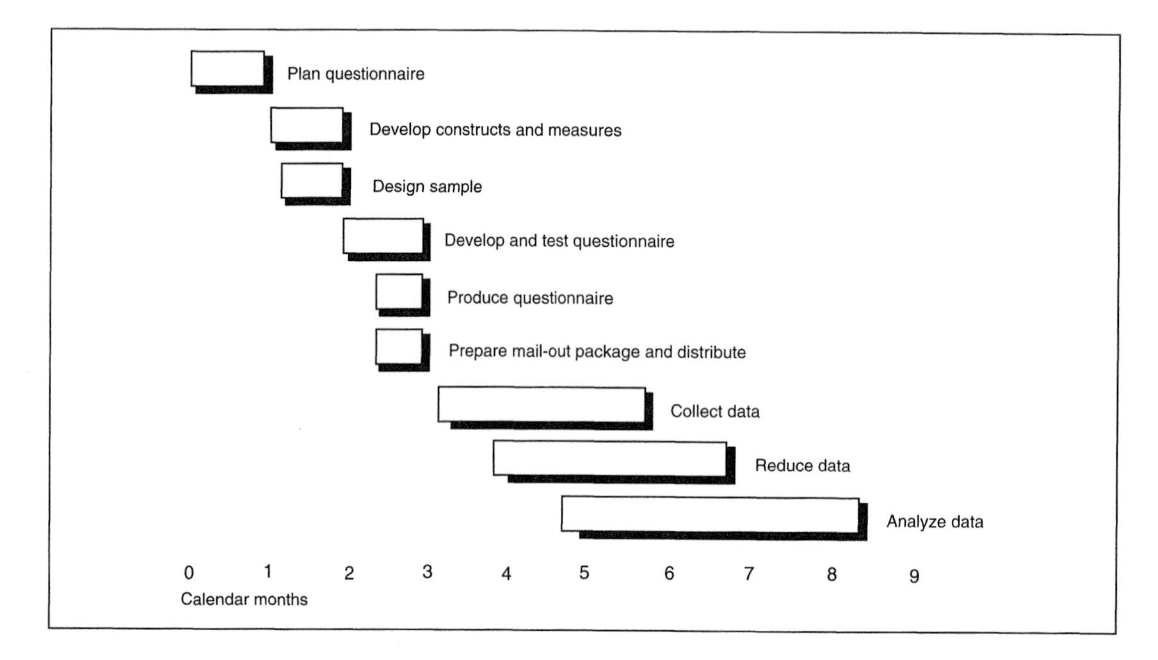

Figure I.39. Typical completion times for major questionnaire tasks.
From *Developing and Using Questionnaires*, Transfer Paper 7, p. 15, Washington, DC: GAO Program Evaluation and Methodology Division.

Guidelines for developing questions

The axiom that underlies the guidelines shown below is that the question-writer(s) must be thoroughly familiar with the respondent group and must understand the subject matter from the perspective of the respondent group. This is often problematic for the quality professional when the respondent group is the customer; methods for dealing with this situation are discussed below. There are eight basic guidelines for writing good questions:

1. Ask questions in a format that is appropriate to the questions' purpose and the information required.
2. Make sure the questions are relevant, proper, and qualified as needed.
3. Write clear, concise questions at the respondent's language level.
4. Give the respondent a chance to answer by providing a comprehensive list of relevant, mutually exclusive responses from which to choose.
5. Ask unbiased questions by using appropriate formats and item constructions and by presenting all important factors in the proper sequence.
6. Get unbiased answers by anticipating and accounting for various respondent tendencies.
7. Quantify the response measures where possible.
8. Provide a logical and unbiased line of inquiry to keep the reader's attention and make the response task easier.

The above guidelines apply to the *form* of the question. Using the critical incident technique to develop good question *content* is described below.

Response types

There are several commonly used types of survey responses.

- **Open-ended questions**—These are questions that allow the respondents to frame their own response without any restrictions placed on the response. The primary advantage is that such questions are easy to form and ask using natural language, even if the question writer has little knowledge of the subject matter. Unfortunately, there are many problems with analyzing the answers received to this type of question. This type of question is most useful in determining the scope and content of the survey, not in producing results for analysis or process improvement.
- **Fill-in-the-blank questions**—Here the respondent is provided with directions that specify the units in which the respondent is to answer. The instructions should be explicit and should specify the answer units. This type of question should be reserved for very specific requests, e.g., "What is your age on your last birthday? (age in years)."
- **Yes/No questions**—Unfortunately, yes/no questions are very popular. Although they have some advantages, they have many problems and few uses. Yes/no questions are

ideal for dichotomous variables, such as defective or not defective. However, too often this format is used when the measure spans a range of values and conditions, e.g., "Were you satisfied with the quality of your new car (yes/no)?" A yes/no response to such questions contains little useful information.

- **Ranking questions**—The ranking format is used to rank options according to some criterion, e.g., importance. Ranking formats are difficult to write and difficult to answer. They give very little real information and are very prone to errors that can invalidate all the responses. They should be avoided whenever possible in favor of more powerful formats and formats less prone to error, such as rating. When used, the number of ranking categories should not exceed five.
- **Rating questions**—With this type of response, a rating is assigned on the basis of the score's absolute position within a range of possible values. Rating scales are easy to write, easy to answer, and provide a level of quantification that is adequate for most purposes. They tend to produce reasonably valid measures. Here is an example of a rating format:

  For the following statement, check the appropriate box:
  The workmanship standards provided by the purchaser are
  ☐ *Clear*
  ☐ *Marginally adequate*
  ☐ *Unclear*

- **Guttman format**—In the Guttman format, the alternatives increase in comprehensiveness; that is, the higher-valued alternatives include the lower-valued alternatives. For example,

  Regarding the benefit received from training in quality improvement:
  ☐ *No benefit identified*
  ☐ *Identified benefit*
  ☐ *Measured benefit*
  ☐ *Assessed benefit value in dollar terms*
  ☐ *Performed cost/benefit analysis*

- **Likert and other intensity scale formats**—These formats are usually used to measure the strength of an attitude or an opinion. For example,

  Please check the appropriate box in response to the following statement:
  "The quality auditor was knowledgeable."

  ☐ *Strongly disagree*
  ☐ *Disagree*
  ☐ *Neutral*
  ☐ *Agree*
  ☐ *Strongly agree*

  Intensity scales are very easy to construct. They are best used when respondents can agree or disagree with a statement. A problem is that statements must be worded to present a single side of an argument. We know that the respondent agrees, but we must infer what he believes. To compensate for the natural tendency of people to agree, statements are usually presented using the converse as well, e.g., "The quality auditor was not knowledgeable."

  When using intensity scales, use an odd-numbered scale, preferably with five categories. If there is a possibility of bias, order the scale in a way that favors the hypothesis you want to disprove and handicaps the hypothesis you want to confirm. This way you will confirm the hypothesis with the bias against you—a stronger result. If there is no bias, put the most undesirable choices first.
- **Semantic differential format**—In this format, the values that span the range of possible choices are not completely identified; only the end points are labeled. For example,

  Indicate the number of times you initiated communication with your customer in the past month.

  few   *many*
  (2 or less) ☐ ☐ ☐ ☐ ☐ ☐ ☐ *(20 or more)*

  The respondent must infer that the range is divided into equal intervals. The range seems to work better with seven categories rather than the usual five.

  Semantic differentials are very useful when we do not have enough information to anchor the intervals between the poles. However, they are very difficult to write well and if not written well the results are ambiguous.

Survey development case study

This actual case study involves the development of a mail survey at a community hospital. The same process has been used by the author to develop customer surveys for clientele in a variety of industries.

The study of service quality and patient satisfaction was performed at a 213 bed community hospital in the southwestern United States. The hospital is a non-profit, publicly funded institution providing services to the adult community; pediatric services are not provided. The purpose of the study was to:

1. Identify the determinants of patient quality judgments.
2. Identify internal service delivery processes that impacted patient quality judgments.
3. Determine the linkage between patient quality judgments and intent-to-patronize the hospital in the future or to recommend the hospital to others.

To conduct the study, the author worked closely with a core team of hospital employees, and with several ad hoc teams of hospital employees. The core team included the Nursing Administrator, the head of the Quality Management Department, and the head of Nutrition Services.*

The team decided to develop their criteria independently. It was agreed that the best method of getting information was directly from the target group, in-patients. Due to the nature of hospital care services, focus groups were not deemed feasible for this study. Frequently, patients must spend a considerable period of time convalescing after being released from a hospital, making it impossible for them to participate in a focus group soon after discharge. While the patients are in the hospital, they are usually too sick to participate. Some patients have communicable diseases, which makes their participation in focus groups inadvisable.

Since memories of events tend to fade quickly (Flanagan, 1954, p. 331), the team decided that patients should be interviewed within 72 hours of discharge. The target patient population was, therefore, all adults treated as in-patients and discharged to their homes. The following groups were not part of the study: families of patients who died while in the hospital, patients discharged to nursing homes, patients admitted for psychiatric care.**

The team used the Critical Incident Technique (CIT) to obtain patient comments. The CIT was first used to study procedures for selection and classification of pilot candidates in world war II (Flanagan, 1954). A bibliography assembled in 1980 listed over seven hundred studies about or using the CIT (Fivars, 1980). Given its popularity, it is not surprising that the CIT has also been used to evaluate service quality .

---

*The nutrition services manager was very concerned that she get sufficient detail on her particular service. Thus, the critical incident interview instrument she used included special questions relating to food service.

**The team was unable to obtain a Spanish-speaking interviewer, which meant that some patients that were candidates were not able to participate in the survey.

CIT consists of a set of specifically defined procedures for collecting observations of human behavior in such a way as to make them useful in addressing practical problems. Its strength lies in carefully structured data collection and data classification procedures that produce detailed information not available through other research methods. The technique, using either direct observation or recalled information collected via interviews, enables researchers to gather firsthand patient-perspective information. This kind of self-report preserves the richness of detail and the authenticity of personal experience of those closest to the activity being studied. Researchers have concluded that the CIT produces information that is both reliable and valid.

This study attempted to follow closely the five steps described by Flanagan as crucial to the CIT: 1) establishment of the general aim of the activity studied; 2) development of a plan for observers or interviewers; 3) collection of data; 4) analysis (classification) of data; and 5) interpretation of data.

Establishment of the general aim of the activity studied

The general aim is the purpose of the activity. In this case the activity involves the whole range of services provided to in-patients in the hospital. This includes every service activity between admission and discharge.* From the service provider's perspective the general aim is to create and manage service delivery processes in such a way as to produce a willingness by the patient to utilize the provider's services in the future. To do this the service provider must know which particular aspects of the service are remembered by the patient.

Our general aim was to provide the service provider with information on what patients remembered about their hospital stay, both pleasant and unpleasant. This information was to be used to construct a new patient survey instrument that would be sent to recently discharged patients on a periodic basis. The information obtained would be used by the managers of the various service processes as feedback on their performance, from the patient's perspective.

Interview plan

Interviewers were provided with a list of patients discharged within the past 3 days. The discharge list included all patients. Non-psychiatric patients who were discharged to "home" were candidates for the interview. Home was defined as any location other than the morgue or a nursing home. Interviewers were instructed to read a set of predetermined statements. Patients to be called were selected at random from the discharge list. If a patient could not be reached, the interviewer would try again later in the day. One interview form was prepared per patient. To avoid bias, 50% of the interview forms asked the patient to recall unpleasant incidents first and 50% asked for pleasant incidents first. Interviewers were instructed to record the patient responses using the patient's own words.

*Billing was not covered in the CIT phase of the study because patients had not received their bills within 72 hours.

Collection of data

Four interviewers participated in the data collection activity, all were management level employees of the hospital. Three of the interviewers were female, one was male. The interviews were conducted when time permitted during the interviewer's normal busy work day. The interviews took place during the September 1993 time period. Interviewers were given the instructions recommended by Hayes (1997) for generating critical incidents.

A total of 36 telephone attempts were made and 23 patients were reached. Of those reached, three spoke only Spanish. In the case of one of the Spanish-speaking patients a family member was interviewed. Thus, 21 interviews were conducted, which is slightly greater than the 10 to 20 interviews recommended by Hayes (1997). The 21 interviews produced 93 critical incidents.

Classification of data

The Incident Classification System required by CIT is a rigorous, carefully designed procedure with the end goal being to make the data useful to the problem at hand while sacrificing as little detail as possible (Flanagan, 1954, p. 344). There are three issues in doing so: 1) identification of a general framework of reference that will account for all incidents; 2) inductive development of major area and subarea categories that will be useful in sorting the incidents; and 3) selection of the most appropriate level of specificity for reporting the data.

The critical incidents were classified as follows:

1. Each critical incident was written on a 3x5 card, using the patient's own words.
2. The cards were thoroughly shuffled.
3. Ten percent of the cards (10 cards) were selected at random, removed from the deck and set aside.
4. Two of the four team members left the room while the other two grouped the remaining 83 cards and named the categories.
5. The ten cards originally set aside were placed into the categories found in step 3.
6. Finally, the two members not involved in the initial classification were told the names of the categories. They then took the reshuffled 93 cards and placed them into the previously determined categories.

The above process produced the following dimensions of critical incidents:

- Accommodations (5 critical incidents)
- Quality of physician (14 critical incidents)
- Care provided by staff (20 critical incidents)
- Food (26 critical incidents)
- Discharge process (1 critical incident)
- Attitude of staff (16 critical incidents)
- General (11 critical incidents)

Interpretation of data

Interjudge agreement, the percentage of critical incidents placed in the same category by both groups of judges, was 93.5%. This is well above the 80% cutoff value recommended by experts. The setting aside of a random sample and trying to place them in established categories is designed to test the comprehensiveness of the categories. If any of the withheld items were not classifiable it would be an indication that the categories do no adequately span the patient satisfaction space. However, the team experienced no problem in placing the withheld critical incidents into the categories.

Ideally, a critical incident has two characteristics: 1) it is *specific* and 2) it describes the service provider in *behavioral terms* or the service product with *specific adjectives* (Hayes, 1997). Upon reviewing the critical incidents in the General category, the team determined that these items failed to have one or both of these characteristics. Thus, the 11 critical incidents in the General category were dropped. The team also decided to merge the two categories "Care provided by staff" and "Attitude of staff" into the single category "Quality of staff care." Thus, the final result was a five dimension model of patient satisfaction judgments: Food, Quality of physician, Quality of staff care, Accommodations, and Discharge process.

A rather obvious omission in the above list is billing. This occurred because the patients had not yet received their bill within the 72 hour time frame. However, the patient's bill was explained to the patient prior to discharge. This item is included in the Discharge process dimension. The team discussed the billing issue and it was determined that billing complaints do arise after the bills are sent, suggesting that billing probably is a satisfaction dimension. However, the team decided not to include billing as a survey dimension because 1) the time lag was so long that waiting until bills had been received would significantly reduce the ability of the patient to recall the details of their stay; 2) fear that the patient's judgments would be overwhelmed by the recent receipt of the bill and 3) a system already existed for identifying patient billing issues and adjusting the billing process accordingly.

Survey item development

As stated earlier, the general aim was to provide the service provider with information on what patients remembered about their hospital stay, both pleasant and unpleasant. This information was then to be used to construct a new patient survey instrument that would be sent to recently discharged patients on a periodic basis. The information obtained would be used by the managers of the various service processes as feedback on their performance, from the patient's perspective.

The core team believed that accomplishing these goals required that the managers of key service processes be actively involved in the creation of the survey instrument. Thus, ad hoc teams were formed to develop survey items for each of the dimensions determined by the critical incident study. The teams were given brief instruction by the author in the characteristics of good survey items. Teams were required to develop items that, in the opinion of the core team, met five criteria: 1) relevance to the dimension being measured; 2) concise; 3) unambiguous; 4) one thought per item; and

5) no double negatives. Teams were also shown the specific patient comments that were used as the basis for the categories and informed that these comments could be used as the basis for developing survey items.

Writing items for the questionnaire can be difficult. The process of developing the survey items involved an average of three meetings per dimension, with each meeting lasting approximately two hours. Ad hoc teams ranged in size from four to eleven members. The process was often quite tedious, with considerable debate over the precise wording of each item.

The core team discussed the scale to be used with each ad hoc team. The core team's recommended response format was a five point Likert-type scale. The consensus was to use a five point *agree-disagree* continuum as the response format. Item wording was done in such a way that agreement represented better performance from the hospital's perspective.

In addition to the response items, it was felt that patients should have an opportunity to respond to open-ended questions. Thus, the survey also included general questions that invited patients to comment in their own words. The benefits of having such questions is well known. In addition, it was felt that these questions might generate additional critical incidents that would be useful in expanding the scope of the survey.

The resulting survey instrument contained 50 items and three open-ended questions and is included in the Appendix.

Survey administration and pilot study

The survey was to be tested on a small sample. It was decided to use the total design method (TDM) to administer the survey (Dillman, 1983). Although the total design method is exacting and tedious, Dillman indicated that its use would assure a high rate of response. Survey administration would be handled by the Nursing Department.

TDM involves rather onerous administrative processing. Each survey form is accompanied by a cover letter, which must be hand-signed in blue ink. Follow up mailings are done 1, 3 and 7 weeks after the initial mailing. The 3 and 7 week follow ups are accompanied by another survey and another cover letter. No "bulk processing" is allowed, such as the use of computer-generated letters or mailing labels. Dillman's research emphasizes the importance of viewing the TDM as a completely integrated approach (Dillman, 1983, p. 361).

Because the hospital in the study is small, the author was interested in obtaining maximum response rates. In addition to following the TDM guidelines, he recommended that a $1 incentive be included with each survey. However, the hospital administrator was not convinced that the additional $1 per survey was worthwhile. It was finally agreed that to test the effect of the incentive on the return rate $1 would be included in 50% of the mailings, randomly selected.

The hospital decided to perform a pilot study of 100 patients. The patients selected were the first 100 patients discharged to home starting April 1, 1994. The return information is shown in the table below.

Table I.14. Pilot patient survey return information.

A. NUMBERS

Surveys mailed: 100
Surveys delivered: 92
Surveys returned as undeliverable: 8
Survey returned, needed Spanish version: 1

Total surveys returned: 45
Percentage of surveys delivered returned: 49%

Number delivered that had $1 incentive: 47
Number returned that had $1 incentive: 26
Percentage returned that had $1 incentive: 55%
Number delivered that had no $1 incentive: 45
Number returned that had no $1 incentive: 19
Percentage returned that had no $1 incentive: 42%

B. SURVEY RESPONSES BY MAILING

Number of surveys returned after:

Initial mailing: 12
One week follow up: 16
Three week follow up: 8
Seven week follow up: 9

Although the overall return rate of 49% is excellent for normal mail-survey procedures, it is substantially below the 77% average and the 60% "minimum" reported by Dillman. As possible explanations, the author conjectures that there may be a large Spanish-speaking constituency for this hospital. As mentioned above, the hospital is planning a Spanish version of the survey for the future.

The survey respondent demographics were analyzed and compared to the demographics of the non-respondents to assure that the sample group was representative. A sophisticated statistical analysis was performed on the responses to evaluate the reliability and validity of each item. Items with low reliability coefficients or questionable validity were reworded or dropped.

Focus groups

The focus group is a special type of group in terms of purpose, size, composition, and procedures. A focus group is typically composed of seven to ten participants who are unfamiliar with each other. These participants are selected because they have certain characteristic(s) in common that relate to the topic of the focus group.

The researcher creates a permissive environment in the focus group that nurtures different perceptions and points of view, without pressuring participants to vote, plan, or reach consensus. The group discussion is conducted several times with similar types of participants to identify trends and patterns in perceptions. Careful and systematic analysis of the discussions provide clues and insights as to how a product, service, or opportunity is perceived.

A focus group can thus be defined as a carefully planned discussion designed to obtain perceptions on a defined area of interest in a permissive, non-threatening environment. The discussion is relaxed, comfortable, and often enjoyable for participants as they share their ideas and perceptions. Group members influence each other by responding to ideas and comments in the discussion.

In quality management, focus groups are useful in a variety of situations:

- prior to starting the strategic planning process
- generate information for survey questionnaires
- needs assessment, e.g., training needs
- test new program ideas
- determine customer decision criteria
- recruit new customers

Focus group advantages

The focus group is a socially oriented research procedure. The advantage of this approach is that members stimulate one another, which may produce a greater number of comments than would individual interviews. If necessary, the researcher can probe for additional information or clarification. Focus groups produce results that have high face validity, i.e., the results are in the participant's own words rather than in statistical jargon. The information is obtained at a relatively low cost, and they can be obtained very quickly.

Focus group disadvantages

There is less control in a group setting than with individual interviews. When group members interact, it is often difficult to analyze the resulting dialogue. The quality of focus group research is highly dependent on the qualifications of the interviewer. Trained and skilled interviewers are hard to find. Group-to-group variation can be considerable, further complicating the analysis. Finally, focus groups are often difficult to schedule.

Other customer information systems

- Complaint and suggestion systems typically provide all customers with an easy-to-use method of providing favorable or unfavorable feedback to management. Due to selection bias, these methods do not provide statistically valid information. However, because they are a census rather than a sample, they provide opportunities for individual customers to have their say. These are moments of truth that can be used to increase customer loyalty. They also provide anecdotes that have high face validity and are often a source of ideas for improvement.
- Customer panels are composed of a representative group of customers who agree to communicate their attitudes periodically via phone calls or mail questionnaires. These panels are more representative of the range of customer attitudes than customer complaint and suggestion systems. To be effective, the identity of customers on the panel must be withheld from the employees serving them.
- Mystery shoppers are employees who interact with the system as do real customers. The identity of the mystery shopper is withheld from employees.

Once customer feedback has been obtained, it must be used to improve process and product quality. A system for utilizing customer feedback is shown in Figure I.40.

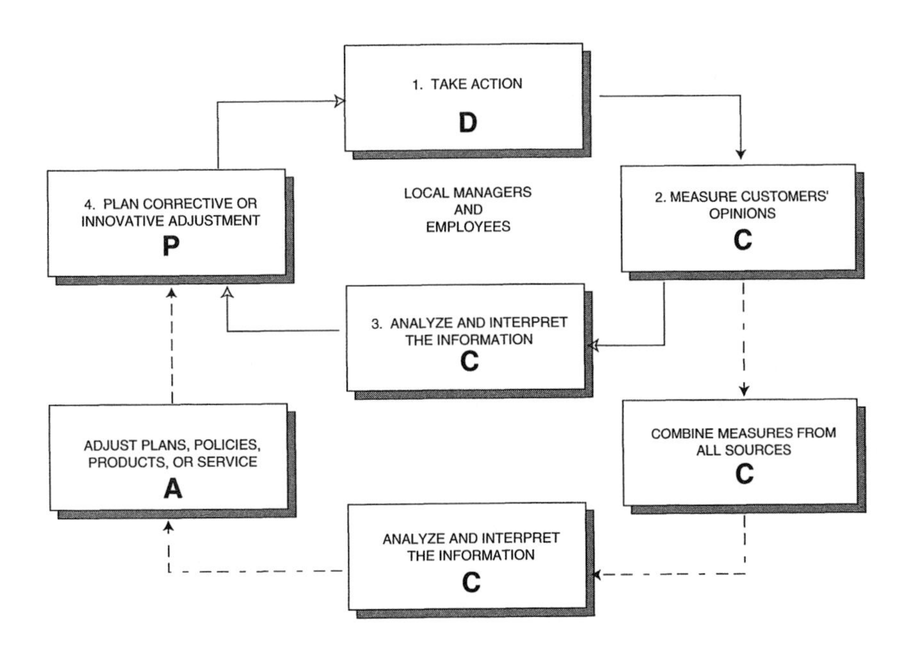

Figure I.40. System for utilizing customer feedback.

From Daltas, A.J. (1977). "Protecting service markets with consumer feedback," *Cornell Hotel and Restaurant Administration Quarterly*, May, pp. 73–77.

1. Local managers and employees serve customers' needs on a daily basis, using locally modified procedures along with general corporate policies and procedures.
2. By means of a standardized and locally sensitive questionnaire, determine the needs and attitudes of customers on a regular basis.
3. Comparing financial data, expectations, and past attitude information, determine strengths and weaknesses and their probable causes.
4. Determine where and how effort should be applied to correct weaknesses and preserve strengths. Repeat the process by taking action—step 1—and maintain it to attain a steady state or to evolve in terms of customer changes.
5. A similar process can take place at higher levels, using aggregated data from the field and existing policy flows of the organization.

Although this system was developed by marketing specialists, note that it incorporates a variation of the classical Shewhart quality improvement PDCA cycle.

MULTIPLE-CUSTOMER MANAGEMENT AND CONFLICT RESOLUTION

For additional information on customer relationship management see earlier this chapter. For additional information on conflict resolution principles and techniques see section I.C.

The principal issues involved when dealing with multiple customers is privacy of the customer's proprietary information. Firms wish to protect those trade secrets that give them a competitive edge in the marketplace. This imposes a responsibility on you, in your role as supplier, to develop protocols and procedures that assure each customer of the desired level of confidentiality. Of course, this adds cost to your operation and complexity to your processes. Such costs are one reason why Deming and others argue for single-source suppliers. The same costs arise here, and for the same reason: competition.

In previous sections of this chapter a great deal of time was spent discussing the importance of supplier-purchaser partnerships and close cooperation. Such cooperation becomes more difficult when multiple customers are involved. This is especially the case when your customers are competitors. Such proven successful methods as cooperative contract administration, joint technological planning, customer-supplier teams, and Keiretsu-type arrangements are often difficult when multiple competing customers are involved.

Multiple-customer relations

The principles of customer relationship management discussed earlier in this chapter also apply here. Paying attention to the "Do's and Don'ts" listed in Figure I.37 (Actions Affecting Buyer-Seller Relationships) is vital. Inevitably, despite your best efforts, conflicts will arise when multiple customers are involved. The quality system requirements of the different customers will vary, rumors of "leaks" of customer information will start mysteriously, one of your key people will be hired by one of the customers, etc. In such situations the quality manager will need to have a firm grasp of the basics of customer relationship management and conflict resolution skills.

The best approach is to begin building trust at the outset. The most important work is performed *before* a problem arises. As with the rest of quality management, the prevention approach is better than inspection. If all customers are convinced that you are making an honest effort, conflict resolution will be far easier. All customers should be involved in planning of the security arrangements. Consensus agreement that the arrangements are acceptable is absolutely vital. Written records of all communication with customers should be carefully maintained, including logbooks of phone conversation, copies of letters (including email), meeting minutes, etc. The security of these communications records should be covered by the security plan. Customer satisfaction with the multiple-customer management program should be a topic at relationship management meetings.

While security arrangements must be adequate, information security by its nature conflicts with open communications. Care must be taken to avoid problems due to excessive security. The author recommends that the security system provide different levels of security access (e.g., unrestricted, confidential, secret). Quality managers should consult security experts for advice in this area.

Legal issues

Customer privacy concerns go beyond causing ill will. If customer patents are infringed upon, or if the loss of intellectual property leads to significant loss, legal action may result. Reasonable measures must be taken to provide a defense against these lawsuits. This is an area of immense importance and one which involves myriad abstruse legal concepts. The quality manager is again urged to seek the counsel of experts in this field. Legal protection in contracts is desirable.

CUSTOMER RETENTION/LOYALTY

Customers have value. This simple fact is obvious when one looks at a customer making a single purchase. The transaction provides revenue and profit to the firm. However, when the customer places a demand on the firm, such as a return of a previous purchase or a call for technical support, there is a natural tendency to see this as a loss. At these times it is important to understand that customer value must not be viewed on a short-term transaction-by-transaction basis. Customer value must be measured *over the lifetime of the relationship*. One method of calculating the lifetime value of a loyal customer, based on work by Frederick Reichheld of Bain and Co. and the University of Michigan's Claes Fornell, is as follows (Stewart, 1995):

1. Decide on a meaningful period of time over which to do the calculations. This will vary depending on your planning cycles and your business: A life insurer should track customers for decades, a disposable diaper maker for just a few years, for example.
2. Calculate the profit (net cash flow) customers generate each year. Track several samples—some newcomers, some old-timers—to find out how much business they gave you each year, and how much it cost to serve them. If possible, segment them by age, income, sales channel, and so on. For the first year, be sure to subtract the cost of acquiring the pool of customers, such as advertising, commissions, back-office costs of setting up a new account. Get specific numbers—profit per customer in year one, year two, etc.—not averages for all customers or all years. Long-term customers tend to buy more, pay more (newcomers are often lured by discounts), and create less bad debt.
3. Chart the customer "life expectancy," using the samples to find out how much your customer base erodes each year. Again, specific figures are better than an average like "10% a year"; old customers are much less likely to leave than freshmen. In retail banking, 26% of account holders defect in the first year; in the ninth year, the rate drops to 9%.
4. Once you know the profit per customer per year and the customer-retention figures, it's simple to calculate net present value. Pick a discount rate—if you want a 15% annual return on assets, use that. In year one, the NPV will be profit ÷ 1.15. Next year, NPV = (year-two profit × retention rate) ÷ 1.152. In year n, the last year in your figures, the NPV is the n year's adjusted profit ÷ 1.15n. The sum of the years one through n is how much your customer is worth—the net present value of all the profits you can expect from his tenure.

This is very valuable information. It can be used to find out how much to spend to attract new customers, and which ones. Better still, you can exploit the leverage customer satisfaction offers. Take your figures and calculate how much more customers would be worth if you increased retention by 5%. Figure I.41 shows the increase in customer NPV for a 5% increase in retention for three industries.

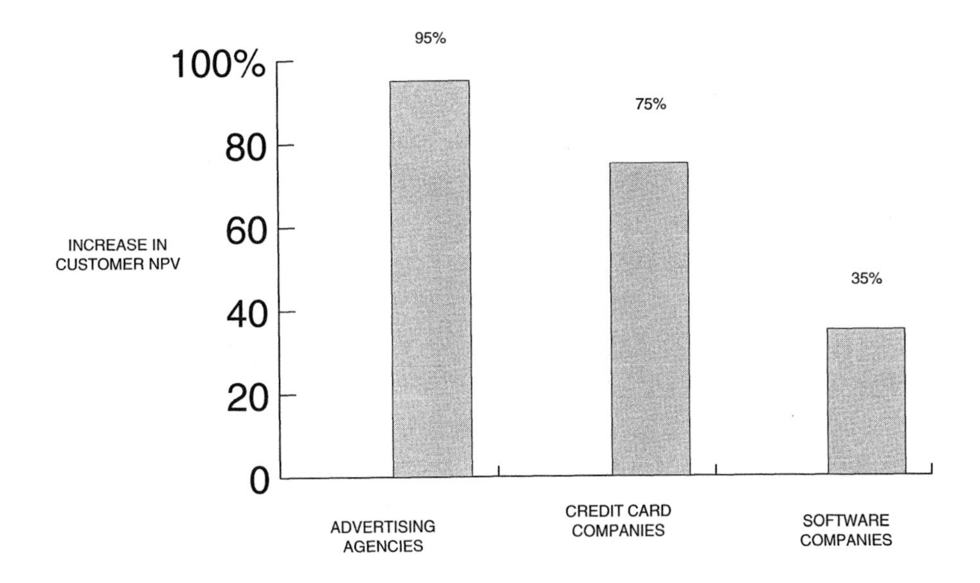

Figure I.41. Increase in customer NPV for a 5% increase in customer retention.

Once the lifetime value of the customer is known, it forms the basis of loyalty-based management[SM] of the customer relationship. According to Reichheld (1996), loyalty-based management is the practice of carefully selecting customers, employees, and investors, and then working hard to retain them. There is a tight, cause-and-effect connection between investor, employee and customer loyalty. These are the human assets of the firm.

Complaint handling

When a customer complaint has been received it represents an opportunity to increase customer loyalty, and a risk of losing the customer. The way the complaint is handled is crucial. The importance of complaint handling is illustrated in Figure I.42. These data illustrate that the decision as to whether a customer who complains plans to repurchase is highly dependent on how well they felt their complaint was handled. Add to this the fact that customers who complain are likely to tell as many as 14 others of their experience, and the importance of complaint handling in customer relations becomes obvious.

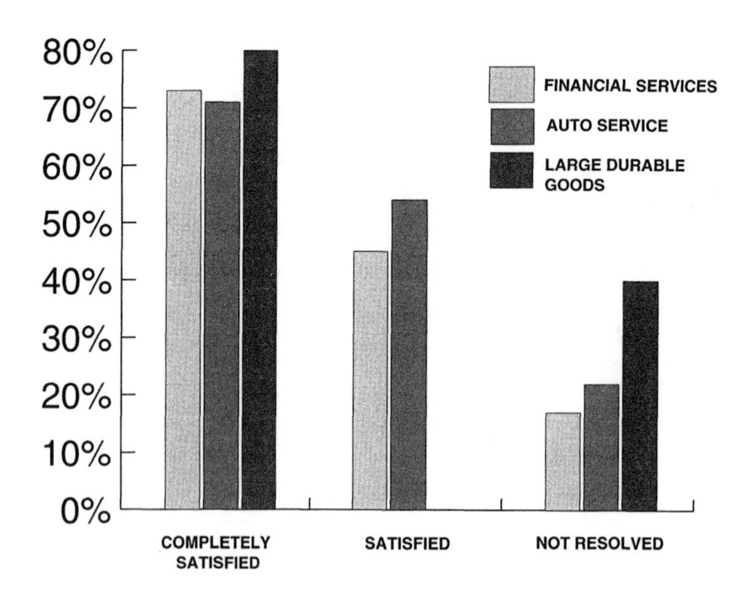

Figure I.42. Percent planning to repurchase vs. how complaint was handled. (*Note: The large durable goods survey did not include a response category of "satisfied".)

Despite the impressive nature of Figure I.42, even these figures dramatically understate the true extent of the problem. Complaints represent people who were not only unhappy, they notified the company. Research indicates that up to 96% of unhappy customers never tell the company. This is especially unfortunate since it has been shown that customer loyalty is *increased* by proper resolution of complaints. Given the dramatic impact of a lost customer, it makes sense to maximize the opportunity of the customer to complain. Complaints should be actively sought, an activity decidedly against human nature. This suggests that a system must be developed and implemented to force employees to seek out customer complaints. In addition to actively soliciting customer complaints, the system should also provide every conceivable way for an unhappy customer to contact the company on their own, including toll-free hotlines, email, comment cards, etc.

I.I SUPPLIER RELATIONS AND MANAGEMENT METHODOLOGIES

The major cost of most manufactured product is in purchased materials. In some cases, the percentage is 80% or more, it is seldom less than 50%. The importance of consistently high levels of quality in purchased materials is clear. This section will examine important aspects of vendor quality control systems.

It is important to remember that dealings between companies are really dealings between people. People work better together if certain ground rules are understood and followed. Above all, the behavior of both buyer and seller should reflect honesty and integrity. This is especially important in quality control, where many decisions are "judgment calls." There are certain guidelines that foster a relationship based on mutual trust:

- Don't be too legalistic. While it is true that nearly all buyer-seller arrangements involve a contract, it is also true that unforeseen conditions sometimes require special actions be taken. If buyer and seller treat each other with respect, these situations will present no real problem.
- Maintain open channels of communication. This involves both formal and informal channels. Formal communication includes such things as joint review of contracts and purchase order requirements by both seller and buyer teams, on-site seller and buyer visits and surveys, corrective action request and follow-up procedures, record-keeping requirements, and so on. Informal communications involve direct contact between individuals in each company on an ongoing and routine basis. Informal communications to clarify important details, ask questions, gather background to aid in decision-making, etc., will prevent many problems.
- The buyer should furnish the seller with detailed product descriptions. This includes drawings, workmanship standards, special processing instructions, or any other information the seller needs to provide product of acceptable quality. The buyer should ascertain that the seller understands the requirements.
- The buyer should objectively evaluate the seller's quality performance. This evaluation should be done in an open manner, with the full knowledge and consent of the seller. The buyer should also keep the seller informed of his relative standing with respect to other suppliers of the same product. However, this should be done in a manner that does not compromise the position of any other seller or reveal proprietary information.
- The buyer should be prepared to offer technical assistance to the seller, or vice-versa. Such assistance may consist of on-site visits by buyer or seller teams, telephone assistance, or transfer of documents. Of course, both parties are obligated to protect the trade secrets and proprietary information they obtain from one another.
- Seller should inform buyer of any known departure from historic or required levels of quality.
- Buyer should inform seller of any change in requirements in a timely fashion.
- Seller should be rewarded for exceptional performance. Such rewards can range from plaques to increased levels of business.

The basic principles of ethical behavior have been very nicely summarized in the code of ethics for members of the American Society for Quality, which is discussed in section I.A of this chapter.

SCOPE OF VENDOR QUALITY CONTROL

Most companies purchase several different types of materials. Some of the materials are just supplies, not destined for use in the product to be delivered to the customer. Traditionally, vendor quality control does not apply to these supplies. Of those items destined for the product, some are simple items that have loose tolerances and an abundant history of acceptable quality. The quality of these items will usually be controlled informally, if at all. The third category of purchased goods involves items that are vital to the quality of the end-product, complex, and with limited or no history. Purchase of these items may even involve purchase of the vendor's "expertise"; e.g., designs, application advice, etc. It is the quality of this category of items that will be the subject of subsequent discussions.

VENDOR QUALITY SYSTEMS

One early question is always "who is responsible for vendor quality?" When asked in this way, the result is usually much pointless discussion. Instead of beginning with such a broad question, it is usually better to break down the tasks involved in assuring vendor quality, and then assign responsibility for the individual tasks. Table I.15 shows one such breakdown.

Table I.15. Responsibility matrix vendor relations.

From Juran, J.M. and Gryna, F.M., Jr., *Quality Planning and Analysis*, p. 229, Copyright© 1980 by McGraw-Hill, Inc., reprinted by permission.

	PARTICIPATING DEPARTMENTS		
ACTIVITY	Product Design	Purchasing	Quality Control
Establish a vendor quality policy	X	X	XX
Use multiple vendors for major procurements		XX	
Evaluate quality capability of potential vendors	X	X	XX
Specify requirements for vendors	XX		X
Conduct joint quality planning	X		X
Conduct vendor surveillance		X	XX
Evaluate delivered product	X		XX
Conduct improvement programs	X	X	XX
Use vendor quality ratings in selecting vendors		XX	X

X = shared responsibility XX = primary responsibility

It is important to recognize that, in the end, the responsibility for quality always remains with the supplier. The existence of vendor quality systems for "assuring vendor quality" in no way absolves the vendor of his responsibility. Rather, these systems should be viewed as an aid to the vendor.

MULTIPLE VENDORS

The subject of whether or not to use multiple vendors is one that arouses strong feelings, both pro and con. The conventional wisdom in American quality control was, for decades, that multiple vendors would keep all suppliers "on their toes" through competition. Also, having multiple vendors was looked at as a hedge against unforeseen problems like fire, flood, or labor disputes. These beliefs were so strong that multiple vendors became the de facto standard for most firms and required by major government agencies, including the Department of Defense.

In the 1980s, the consensus on multiple sources of supply began to erode. Japan's enormous success with manufacturing in general and quality in particular inspired American businessmen to study the Japanese way of doing things. Among the things Japanese businesses do differently is that they *discourage* multiple-source purchases whenever possible. This is in keeping with the philosophy of W. Edwards Deming, the noted American quality expert who provided consulting and training to many Japanese firms. The advocates of single-source procurement argue that it encourages the supplier to take long-term actions on your behalf and makes suppliers more loyal and committed to your success. A statistical argument can also be made; minimum variability in product can be obtained if the sources of variation are minimized and different suppliers are an obvious source of variation.

Since both sides have obvious strong points, the decision regarding single-source versus multiple-source is one that must be made on a case-by-case basis. Your company will have to examine its unique set of circumstances in light of the arguments on both sides. In most cases, the author has found that a policy of using single-sources, except under unusual conditions, works well.

EVALUATING VENDOR QUALITY CAPABILITY

When making an important purchase, most companies want some sort of an advance assurance that things will work out well. When it comes to quality, the vendor quality survey is the "crystal ball" used to provide this assurance. To some degree this approach has been replaced by third-party audits such as ISO 9000. The vendor quality survey usually involves a visit to the vendor by a team from the buyer prior to the award of a contract; for this reason, it is sometimes called a "pre-award survey." The team is usually composed of representatives from the buyer's design engineering, quality control, production, and purchasing departments. The quality control elements of the survey usually include, at a minimum, the following:

- Quality management
- Design and configuration control
- Incoming material control
- Manufacturing and process control
- Inspection and test procedures

- Control of non-conforming material
- Gage calibration and control
- Quality information systems and records
- Corrective action systems

The evaluation typically involves use of a checklist and some numerical rating scheme. A simplified example of a supplier evaluation checklist is shown in Figure I.43. Since this type of evaluation is conducted at the supplier's facility, it is known as a physical survey.

CRITERIA	PASS	FAIL	SEE NOTE
The quality system has been developed	❑	❑	❑
The quality system has been implemented	❑	❑	❑
Personnel are able to identify problems	❑	❑	❑
Personnel recommend and initiate solutions	❑	❑	❑
Effective quality plans exist	❑	❑	❑
Inspection stations have been identified	❑	❑	❑
Management regularly reviews quality program status	❑	❑	❑
Contracts are reviewed for special quality requirements	❑	❑	❑
Processes are adequately documented	❑	❑	❑
Documentation is reviewed by quality	❑	❑	❑
Quality records are complete, accurate, and up-to-date	❑	❑	❑
Effective corrective action systems exist	❑	❑	❑
Non-conforming material is properly controlled	❑	❑	❑
Quality costs are properly reported	❑	❑	❑
Changes to requirements are properly controlled	❑	❑	❑
Adequate gage calibration control exists	❑	❑	❑

Figure I.43. Vendor evaluation checklist.

The above checklist is very simple compared to those used in practice. Many checklists run on to 15 pages or more. The author's experience is that these checklists are very cumbersome and difficult to use. If you are not bound by some government or contract requirement, it is recommended that you prepare a brief checklist similar to the one above and supplement the checklist with a report which documents your personal observations. Properly used, the checklist can help guide you without tying your hands.

Bear in mind that a checklist can never substitute for the knowledge of a skilled and experienced evaluator. Numerical scores should be supplemented by the observations and interpretations of the evaluator. The input of vendor personnel should also be included. If there is disagreement between the evaluator and the vendor, the position of both sides should be clearly described.

In spite of their tremendous popularity, physical vendor surveys are only one means of evaluating the potential performance of a supplier. Studies of their accuracy suggest that they should be taken with a large grain of salt. One such study by Brainard (1966) showed that 74 of 151 vendor surveys resulted in incorrect predictions, i.e., either a good supplier was predicted to be bad or vice versa; a coin flip would've been as good a predictor, and a lot cheaper! Since no contrary data has been published, it is recommended that physical vendor surveys be avoided whenever possible. One option is to use so-called "desk surveys." With desk surveys, you have the prospective suppliers themselves submit lists of test equipment, certifications, organization charts, manufacturing procedures, the quality manual, etc. Check these documents over for obvious problems.

Regardless of the survey method used or the outcome of the survey, it is important to keep close tabs on the actual quality history. Have the vendor submit "correlation samples" with the first shipments. The samples should be numbered and each should be accompanied by a document showing the results of the vendor's quality inspection and test results. Verify that the vendor has correctly checked each important characteristic and that your results agree, or "correlate" with his. Finally, keep a running history of quality performance (see "Quality records" later in this chapter). The best predictor of future good performance seems to be a record of good performance in the past. If you are a subscriber to GIDEP, the Government, Industry Data Exchange Program, you have access to a wealth of data on many suppliers. (GIDEP subscribers must also contribute to the data bank.) Another data bank is the Coordinated Aerospace Supplier Evaluation (CASE). If relevant to your application, these compilations of the experience and audits of a large number of companies can be a real money and timesaver.

VENDOR QUALITY PLANNING

Vendor quality planning involves efforts directed toward preventing quality problems, appraisal of product at the vendor's plant as well as at the buyer's place of business, corrective action, disposition of non-conforming merchandise, and quality improvement. The process usually begins in earnest after a particular source has been selected. Most pre-award evaluation is general in nature; after the vendor is selected, it is time to get down to the detail level.

A first step in the process is the transmission of the buyer's requirements to the vendor. Even if the preliminary appraisal of the vendor's capability indicated that the vendor could meet your requirements, it is important that the requirements be studied in detail again before actual work begins. Close contact is required between the buyer and the vendor to assure that the requirements are clearly understood. The vendor's input should be solicited; could a change in requirements help them produce better quality parts?

Next it is necessary to work with the vendor to establish procedures for inspection, test, and acceptance of the product. How will the product be inspected? What workmanship standards are to be applied? What in-process testing and inspection is required? What level of sampling will be employed? These and similar questions must be answered at this stage. It is good practice to have the first few parts completely inspected by both the vendor and the buyer to assure that the vendor knows which features must be checked as well as how to check them. The buyer may want to be at the vendor's facility when production first begins.

Corrective action systems must also be developed. Many companies have their own forms, procedures, etc., for corrective action. If you want your system to be used in lieu of the vendor's corrective action system, the vendor must be notified. Bear in mind that the vendor may need additional training to use your system. Also, the vendor may want some type of compensation for changing his established way of doing things. If at all possible, let the vendor use his own systems.

At an early stage, the vendor should be made aware of any program you have that would result in his being able to reduce inspection. At times it is possible to certify the vendor so that *no* inspection is required by the buyer and shipments can go directly to stores, bypassing receiving inspection completely. Vendor statistical process control can be used to certify vendor processes.

If special record-keeping will be required, this needs to be spelled out. Many companies have special requirements imposed upon them by the nature of their business. For example, most major defense items have traceability and configuration control requirements. Government agencies, such as the FDA, often have special requirements. Automotive companies have record-keeping requirements designed to facilitate possible future recalls. The vendor is often in the dark regarding your special record-keeping requirements; it is your job to keep him informed.

POST-AWARD SURVEILLANCE

Our focus up to this point has been to develop a process that will minimize the probability of the vendor producing items that don't meet your requirements. This effort must continue after the vendor begins production. However, after production has begun, the emphasis can shift from an evaluation of systems to an evaluation of *actual* program, process, and product performance.

Program evaluation is the study of a supplier's facilities, personnel, and quality systems. While this is the major thrust during the pre-award phase of an evaluation, program evaluation doesn't end when the contract is awarded. Change is inevitable, and the buyer should be kept informed of changes to the vendor's program. Typically this is accomplished by providing the buyer with a registered copy of the vendor's quality manual, which is updated routinely. Periodic follow-up audits may also be required, especially if product quality indicates a failure of the quality program.

A second type of surveillance involves surveillance of the vendor's process. Process evaluations involve a study of methods used to produce an end-result. Process performance can usually be best evaluated by statistical methods, and it is becoming common to require that statistical process control (SPC) be applied to critical process characteristics. Many large companies require that their suppliers perform statistical process control studies, called process capability studies, as part of the pre-award evaluation (see section VI.H).

The final evaluation, product evaluation, is also the most important. Product evaluation consists of evaluating conformance to requirements. This may involve inspection at the vendor's site, submission of objective evidence of conformance by the vendor, inspection at the buyer's receiving dock, or actual use of the product by the buyer or the end-user. This *must* be the final proof of performance. If the end-product falls short of requirements, it matters little that the vendor's program looks good, or that all of the in-process testing meets established requirements.

Bear in mind that the surveillance activity is a communications tool. To be effective, it must be conducted in an ethical manner, with the full knowledge and cooperation of the vendor. The usual business communications techniques, such as advanced notification of visits, management presentations, exit briefings, and follow up reports, should be utilized to assure complete understanding.

VENDOR RATING SCHEMES

As the discussion so far makes clear, evaluating vendors involves comparing a large number of factors, some quantitative and some qualitative. Vendor rating schemes attempt to simplify this task by condensing the most important factors into a single number, the vendor rating, that can be used to evaluate the performance of a single vendor over time, or to compare multiple sources of the same item.

Most vendor rating systems involve assigning weights to different important measures of performance, such as quality, cost, and delivery. The weights are selected to reflect the relative importance of each measure. Once the weights are determined, the performance measure is multiplied by the weight and the results totalled to get the rating. For example, we might decide to set up the following rating scheme:

PERFORMANCE	MEASURE	WEIGHT
Quality	% lots accepted	5
Cost	lowest cost/cost	300
Delivery	% on-time shipments	2

The ratings calculations for 3 hypothetical suppliers are shown below.

VENDOR	% LOTS ACCEPTED	PRICE	% ON-TIME DELIVERIES
A	90	$60	80
B	100	70	100
C	85	50	95

VENDOR	QUALITY RATING	PRICE RATING	DELIVERY RATING	OVERALL RATING
A	450	250	160	860
B	500	214	200	914
C	425	300	190	915

As you can see in the example shown above, even simple rating schemes combine reject rates, delivery performance, and dollars into a single composite number. Using this value, we would conclude that vendors B and C are approximately the same. What vendor B lacks in the pricing category, he makes up for in quality and delivery. Vendor A has a rating much lower than B or C.

PROBLEMS AND PITFALLS OF VENDOR RATING SCHEMES

The engineer should note that all rating schemes have "holes" in them that will require subjective interpretation; many quality experts believe that when subjective judgments are allowed, it defeats the purpose of the rating. However, disallowing the subjective evaluations often leads to serious problems. Also, there are very few cases where the ratings can be usefully applied, e.g. multiple sources of the same important item over a time period sufficient to provide a valid basis for comparison. Many ratings have values that are highly questionable; for example, a best-selling book on quality control shows a model rating scheme that is based in part on "percent of promises kept." Ratings are usually easy for vendors to manipulate by varying lot sizes, delivery dates, or selectively sorting lots. Finally, rating schemes tend to draw attention from the true objective, never-ending improvement in quality through cooperative effort between buyer and supplier.

A valid question at this point would be, "is all of this meaningful?" Obviously, one of the problems is the mixture of units. As one quality control text put it, "APPLES + ORANGES = FRUIT SALAD." Other questions involve the selection of weights, disposition of rejected lots (scrap, rework, use-as-is, return to vendor, etc.), the number of lots involved, cost of defects, types of defects, how late were the deliveries, should early deliveries and late deliveries both count against the delivery score, etc. It is easier to answer such questions if some guidelines are provided. Here are some characteristics of good rating schemes:

- The scheme is clearly defined and understood by both the buyer and the seller.
- Only relevant information is included.
- The plan should be easy to use and update.
- Rating schemes should be applied only where they are needed.
- The ratings should "make sense" when viewed in light of other known facts.

SPECIAL PROCESSES

We will define a special process as one that has an effect that can't be readily determined by inspection or testing subsequent to processing. The difficulty may be due to some physical constraint, such as the difficulty in verifying grain size in a heat-treated metal, or the problem may simply be economics, such as the cost of 100% X-ray of every weld in an assembly. In these cases, special precautions are required to assure that processing is carried out in accordance with requirements.

The two most common approaches to control of special processes are certification and process audit. Certification can be applied to the skills of key personnel, such as welder certification, or to the processes themselves. With processes the certification is usually based on some demonstrated capability of the process to perform a specified task. For example, a lathe may machine a special test part designed to simulate product characteristics otherwise difficult or impossible to measure. The vendor is usually responsible for certification. Process audit involves establishing a procedure for the special process, then reviewing actual process performance for compliance to the procedure. A number of books exist to help with the evaluation of special processes. In addition, there are inspection service companies that allow you to hire experts to verify that special processes meet established guidelines. These companies employ retired quality control professionals, as well as full-time personnel. In addition to reducing your costs, these companies can provide a level of expertise you may not otherwise have.

QUALITY RECORDS

Reference was made earlier in this chapter to the importance of tracking actual quality performance. In fact, this "track record" is the most important basis for evaluating a vendor, and the best predictor of future quality performance. Such an important task should be given careful consideration. The purpose of vendor quality records is to provide a basis for action. The data may indicate a need to reverse an undesirable trend, or it may suggest a cost savings from reduced (or no) inspection. Also, the records can provide an objective means of evaluating the effectiveness of corrective action.

An early step is to determine what data will be required. Data considerations include analyzing the level of detail required to accomplish the objectives of the data collection described above, i.e., how much detail will you need to determine an appropriate course of action? Some details normally included are part number, vendor identification, lot identification number, name of characteristic(s) being monitored, quantity shipped, quantity inspected, number defective in the sample, lot disposition (accept or reject), and action taken on rejected lots (sort, scrap, rework, return to vendor, use as is). These data are usually first recorded on a paper form, and later entered into a computerized data base. The original paper documents are kept on file for some period of time, the length of time being dictated by contract, warranty, tax, and product liability considerations. Electronic archiving is becoming increasingly common as technology makes it possible to store massive amounts of data at low cost. Electronic storage offers numerous other advantages, too, such as accessibility.

The data collection and data entry burden can be significant, especially in larger companies. Some companies have resorted to modern technology to decrease the workload. For example, common defects may have bar codes entered on menus for the inspector. The vendor will apply a bar code sticker with important information like the purchase order number, lot number, etc., which the computer can use to access a database for additional details on the order. Thoughtful use of automated data capture systems can substantially reduce, or even eliminate the data input chore. Many systems have been designed for use on desktop computers, bringing the technology within the reach of nearly everyone.

Although the raw data contain all of the information necessary to provide a basis for action, the information will typically be in a form that lacks meaning to decision-makers. To be meaningful, the raw data must be analyzed and reported in a way that makes the information content obvious. Reports should be designed with the end-reader in mind. The data reported should be carefully considered and only the necessary information should be provided. In most cases it will take several reports to adequately distribute the information to the right people. Typically, the level of detail provided will decrease as the level of management increases. Thus, senior management will usually be most interested in summary trends, perhaps presented as graphs and charts rather than numbers. Operating personnel will receive detailed reports that will provide them with information on specific parts and characteristics. At the lowest level, audit personnel will receive reports that allow them to verify the accuracy of the data entered into the system. Report recipients are customers to those generating the reports. As with all customers, their input should be actively solicited using a wide variety of forums, e.g., surveys, focus groups, face-to-face interviews. Their input should be used to drive the report generation and distribution processes as described in section I.B above.

PROCUREMENT STANDARDS AND SPECIFICATIONS*

Quality control of purchased materials is an area that has been well explored. As a result, many of the basics of supplier quality control have been drafted into standards and specifications that describe many of the details common to most buyer and seller transactions. We will discuss some examples of commercial specifications, as well as government specifications.

Most large companies have formalized at least some of their purchased materials quality control requirements. Traditionally, their specifications have covered the generic requirements of their supplier's quality system, covering the details of any given purchase in the purchase order or contract. These supplier quality specifications typically cover such things as the seller's quality policy, quality planning, inspection and test requirements, quality manual, audits, the *supplier's* material control system, special process control, defective material control, design change control, and corrective action systems. The scope of the different specifications varies considerably from company to company, as does the level of detail.

A good example of a third-party standard is ANSI/ASQC Z1.15, "Generic guidelines for quality systems" available from the American Society for Quality. The ISO 9000 series is also widely used. While the Department of Defense Military series (including Mil-Q-9858, Mil-I-45208 amongst many others) had previously served as industry-wide standards for quality, they are now nearly completely superceded by corresponding ISO or NIST standards. The auto industry has adopted QS 9000, which is built upon an ISO 9000 base with added auto-industry-specific and company-specific requirements.

*See II.C for additional information.

CHAPTER

II

Quality Systems Development, Implementation, and Verification

II.A ELEMENTS OF A QUALITY SYSTEM

QUALITY FUNCTION MISSION

Traditional organizations are functionally specialized, e.g., marketing, engineering, purchasing, manufacturing. Conceptually, each function performs an activity essential in delivering value to the customer. In the past, these activities have performed sequentially. As shown in Figure II.1, Shewhart, Deming, and Juran all depict these activities as forming a circle or a spiral, where each cycle incorporates information and knowledge acquired during the previous cycle.

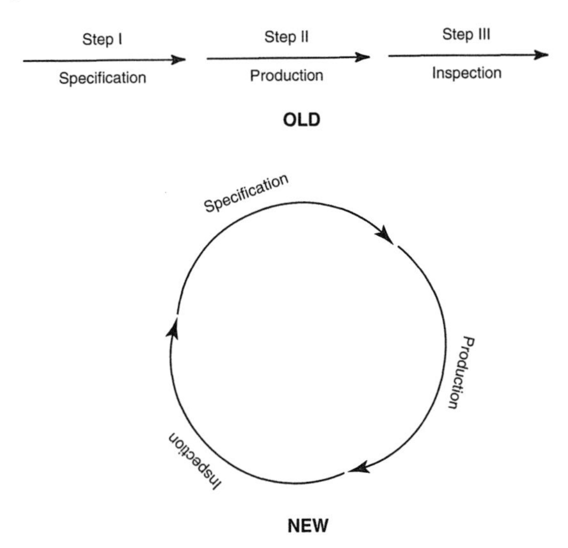

Figure II.1a. Representation of quality activities in the organization.
From Shewhart, W.A. (1939). *Statistical Method from the Viewpoint of Quality Control,* p. 45. New York: Dover Publications, Inc.

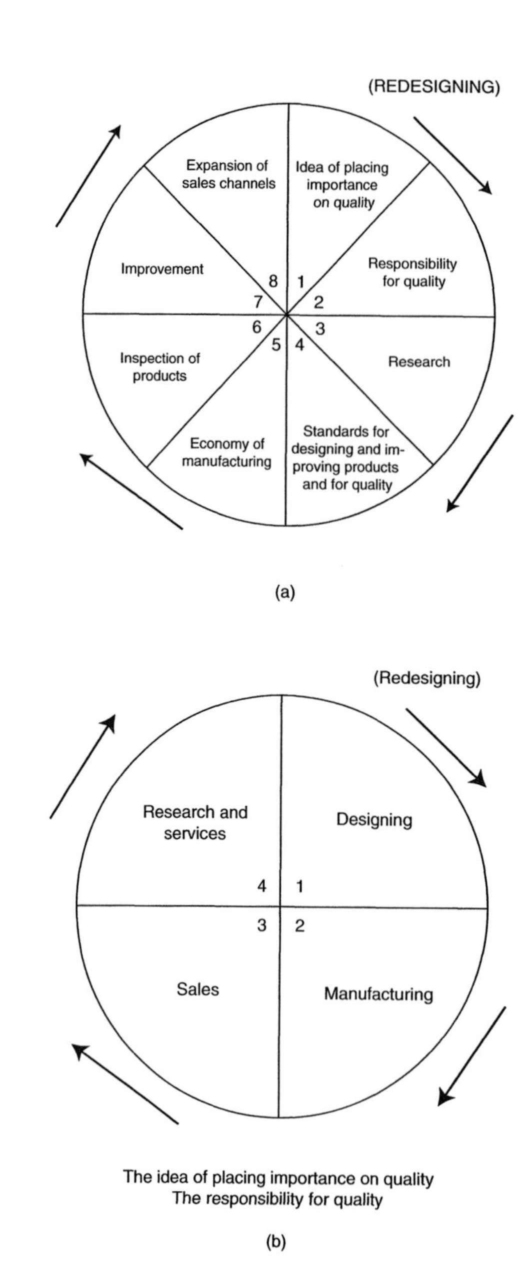

Figure II.1b. Deming's wheel of quality control (1951).

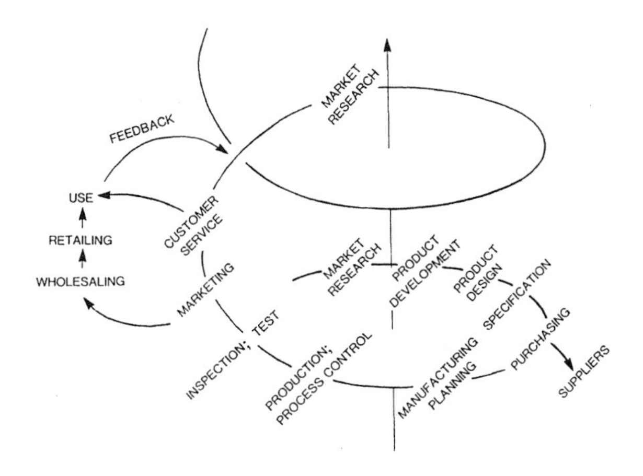

Figure II.1c. Juran's spiral of progress in quality.
From Juran, J.M. and Gryna, F.M. (1988). *Juran's Quality Control Handbook, Fourth Edition*. p. 2.5. New York: McGraw-Hill.

Juran and Gryna (1988, p. 2.6) defines the quality function as "the entire collection of activities through which we achieve fitness for use, no matter where these activities are performed." As noted throughout this book, quality is the responsibility of many different departments. In most cases, the quality department's role is a secondary, supporting one. The quality department is a specialized function. But, quality activities are dispersed throughout the organization. The term "quality function" applies to those activities, departmental and companywide, that collectively result in product or service quality. An analogy can be made with the finance department. Even though many specialized finance and accounting functions are managed by the finance department, every employee in the organization is expected to practice responsible management of their budgets and expenditures.

JURAN'S QUALITY TRILOGY

Quality activities can be conveniently divided into three categories: planning, control, and improvement. Juran (1988) defines these categories as follows:

Quality planning is the activity of developing the products and processes required to meet customers' needs. It involves a number of universal steps:

1. Determine who are the customers.
2. Determine the needs of the customers.
3. Develop product features which respond to customers' needs.
4. Develop processes which are able to produce those product features.
5. Transfer the resulting plans to the operating forces.

Quality control is the process used by the operating forces as an aid to meeting the product and process goals. It is based on the feedback loop and consists of the following steps:

1. Evaluating actual operating performance
2. Comparing actual performance to goals
3. Act on the difference

Quality improvement aims to attain levels of performance which are unprecedented—levels which are significantly better than any past level. The methodologies used in quality improvement are described in sections I.G and V.A.

PRIORITY OF THE QUALITY FUNCTION WITHIN THE ORGANIZATION

The mission of the quality function is companywide quality management. Quality management is the process of identifying and administering the activities necessary to achieve the organization's quality objectives. These activities will fall into one of the three categories of Juran's quality trilogy. The quality function, therefore, transcends the quality department and extends to all of the activities throughout the company that affect quality. The quality department plays a major role in the quality efforts of the organization, but it is not the only player.

The primary role in managing the quality function is played by the senior leadership. Since quality by its nature crosses functional boundaries, it cannot be effectively managed by a single functional department. Because the traditional, functionally specialized hierarchy creates a "silo mentality," each function tends to focus on its own "silo," to the detriment of cross-functional concerns like quality. This is not a failing of the people in the company, it is a result of the system in which these people work. The situation will not be corrected by exhortations to think differently or act differently, it can only be changed by modifying the system itself. The cross-functional organization is, as of this writing, the most widespread alternative structure. Briefly, cross-functional organizations are teams of senior executives brought together to deal with significant issues that span several functional areas. Quality "councils" are cross-functional teams that look specifically at quality issues that affect the entire company. The quality council sets quality policy and, to a great extent, determines the role to be played by the quality department in achieving the policy goals. The quality council also makes decisions regarding company resources to be devoted to quality planning, improvement, and control. Quality councils direct the quality activities of other functions as well, with the objective being the accomplishment of the organization's quality goals, rather than the goals of a particular department.

Of course, quality is not the only concern of the company. Quality concerns must be balanced with cost, cycle time, and other customer issues. The customer's concerns must be balanced with the concerns of investors and employees. The senior leadership, consisting of top management and the board of directors, must weigh all of these concerns and arrive at a resource allocation plan that meets the needs of all stakeholders in the organization. The unifying principle for all stakeholders is the organization's purpose, which remains constant over time.

QUALITY AND BUSINESS PERFORMANCE

Deming's first point tells organizations to "become competitive." There are two basic ways to do this. One is to achieve superior perceived quality by developing a set of product specifications and service standards that more closely meet customer needs than competitors. The other is to achieve superior conformance quality by being more effective than your competitors in conforming to the appropriate product specifications and service standards. These are not mutually exclusive; excellent companies do both simultaneously.

Research findings indicate that achieving superior *perceived quality*, as perceived by customers, gives the business three options—all of them good (Buzzell and Gale, 1987):

1. You can charge a higher price for your superior quality and increase profitability.
2. You can charge a higher price and invest the premium in R&D, thus ensuring higher perceived quality and greater market share in the future.
3. You can charge the same price as your competitor for your superior product, building market share. Increased market share, in turn, means volume growth, rising capacity utilization (or capacity expansion) allowing you to lower costs.

Achieving superior *conformance quality* provides two key benefits:

1. Lower cost of quality than competitors, which translates to lower over-all cost.
2. Conformance quality is a factor in achieving perceived quality, thus it is a link to the benefits mentioned for perceived quality above.

Other benefits also accrue to companies that provide superior perceived quality:

- higher customer loyalty
- more repeat purchases
- less vulnerability to price wars
- ability to command higher relative price without affecting market share
- lower marketing costs
- improved market share

METRICS AND GOALS THAT DRIVE ORGANIZATIONAL PERFORMANCE

Assessment metrics

The choice of what to measure is crucial to the success of the organization. Improperly chosen metrics leads to suboptimal behavior and can lead people away from the organization's goals instead of towards them. Joiner (1994) suggests three systemwide measures of performance: overall customer satisfaction (see section I.H), total cycle time (see V.A), and first-pass quality. An effective metric for quantifying first pass quality is total cost of poor quality (see section I.F). Once chosen, the metrics must be communicated to the members of the organization. To be useful, the employee must be able to influence the metric through his performance, and it must be clear precisely how the employee's performance influences the metric.

Rose (1995) lists the following attributes of good metrics:

- They are customer centered and focused on indicators that provide value to customers, such as product quality, service dependability, and timeliness of delivery, or are associated with internal work processes that address system cost reduction, waste reduction, coordination and teamwork, innovation, and customer satisfaction.
- They measure performance across time, which shows trends rather than snapshots.
- They provide direct information at the level at which they are applied. No further processing or analysis is required to determine meaning.
- They are linked with the organization's mission, strategies, and actions. They contribute to organizational direction and control.
- They are collaboratively developed by teams of people who provide, collect, process, and use the data.

Rose also presents a performance measurement model consisting of eight steps:

Step 1: performance category—This category is the fundamental division of organizational performance that answers the question: What do we do? Sources for determining performance categories include an organization's strategic vision, core competencies, or mission statement. An organization will probably identify several performance categories. These categories define the organization at the level at which it is being measured.

Step 2: performance goal—The goal statement is an operational definition of the desired state of the performance category. It provides the target for the performance category and, therefore, should be expressed in explicit, action-oriented terms. An initial goal statement might be right on the mark, so complex that it needs further division of the performance category, or so narrowly drawn that it needs some combination of performance categories. It might be necessary to go back and forth between the performance goals in this step and the performance categories in step 1 before a satisfactory result is found for both.

Step 3: performance indicator—This is the most important step in the model because this is where progress toward the performance goal is disclosed. Here irrelevant measures are swept aside if they do not respond to an organizational goal. This is where the critical measures—those that communicate what is important and set the course toward organizational success—are established. Each goal will have one or more indicators, and each indicator must include an operational definition that prescribes the indicator's intent and makes its role in achieving the performance goal clear. The scope of the indicator might be viewed differently at various levels in the organization.

Step 4: elements of measure—These elements are the basic components that determine how well the organization meets the performance indicator. They are the measurement data sources—what is actually measured—and are controlled by the organization. Attempting to measure things that are beyond organizational control is a futile diversion of resources and energy because the organization is not in a position to respond to the information collected. This would be best handled in the next step.

Step 5: parameters—These are the external considerations that influence the elements of measure in some way, such as context, constraint, and boundary. They are not controlled by the organization but are powerful factors in determining how the elements of measure will be used. If measurement data analysis indicates that these external considerations present serious roadblocks for organizational progress, a policy change action could be generated.

Step 6: means of measurement—This step makes sense out of the preceding pieces. A general, how-to action statement is written that describes how the elements of measure and their associated parameters will be applied to determine the achievement level in the performance indicator. This statement can be brief, but clarifying intent is more important than the length.

Step 7: notional metrics—In this step, conceptual descriptions of possible metrics resulting from the previous steps are put in writing. This step allows everyone to agree on the concept of how the information compiled in the previous steps will be applied to measuring organizational performance. It provides a basis for validating the process and for subsequently developing specific metrics.

Step 8: specific metrics—In this final step, an operational definition and functional description of the metrics to be applied are written. The definition and description describe the data, how they are collected, how they are used, and, most important, what the data mean or how they affect organizational performance. A prototype display of real or imaginary data and a descriptive scenario that shows what actions might be taken as a result of the measurement are also made. This last step is the real test of any metric. It must identify what things need to be done and disclose conditions in sufficient detail to enable subsequent improvement actions.

Rose presents an application of his model used by the U.S. Army Material Command, which is shown in Figure II.2.

Application of the Performance Measurement Model

Organizational performance metrics

Performance category: Material acquisition

Performance goal: Provide technologically superior equipment in the shortest possible time, while ensuring best value in all procurements.

Performance indicator	Elements of measure	Parameters	Means of measurement	Notional metrics	Specific metrics
Industrial base management (assessing and assuring adequate industrial facilities to meet military needs)	Source of supply for material and services	Material and services requirements	Determine the level of industrial base support for required material and services	Number and capacity of existing and potential sector suppliers vs. material requirements	(Separate package containing operational definitions; functional descriptions of data collection, use, and meaning; and prototype display)
	Quality of suppliers	Procurement policy	Certify the quality of existing and potential suppliers	Percentage of quality-certified sector suppliers	
(Other performance indicators)					

(Other performance categories and goals)

(Optional: subordinate organizational performance metrics)

Figure II.2. Organizational performance metrics.

From "A performance measurement model," by Kenneth H. Rose. *Quality Progress*, February 1995, p. 65. Reprinted by permission.

FORMULATION OF QUALITY PRINCIPLES AND POLICIES

Ishikawa described what he called total quality control (TQC) as a "thought revolution in management." TQC is a movement towards greater rationality in management. The thought revolution involves a transformation in six categories (Ishikawa, 1988):

1. Quality first—not short-term profit first.

   Management that stresses "quality first" can gain customer confidence step-by-step and the company's sales will increase gradually. In the long run, profits will be substantial, and will permit the company to have stable management. If a company follows the principle of "profit first," it may obtain a quick profit, but it cannot sustain competitiveness for a long period of time.

2. Consumer orientation—not producer orientation.

   A logical reaction to the consumer orientation approach is always to think in terms of another party's position. This means to listen to their opinions and to act in a way that will take their views into account.

3. The next process is your customer—breaking down the barrier of sectionalism.

   The phrase "the next process is your customer" could be incorporated into the previous passage where we discussed consumer orientation. However, in a company where sectionalism is strong, this approach can be so important that it deserves to be treated separately. Companywide quality cannot be accomplished unless sectionalism, i.e., each department looking after its own best interests, is broken down. The company must look at its processes for delivering customer value, rather than a department or section separately.
4. Using facts and data to make presentations—utilization of statistical methods.

   The importance of facts must be clearly recognized. Facts, in contrast to opinions, can be translated into data. If data are accurate, they can be analyzed using statistical methods and engineering or management models. This, in turn, forms the basis of decisions. Such decisions will, in the long run, prove to be better than decisions made without this process, i.e., the decisions will produce results that more closely match management's goals.
5. Respect for humanity as a management philosophy—full participatory management.

   When management decides to make companywide quality its goal, it must standardize all processes and procedures and then boldly delegate authority to subordinates. The fundamental principle of successful management is to allow subordinates to make full use of their ability. The term humanity implies autonomy and spontaneity. People are different from animals or machines. They have their own minds, and they are always thinking. Management based on humanity is a system of management that lets the unlimited potential of human beings blossom.
6. Cross-functional management.

   From the perspective of companywide goals, the main functions are quality assurance, cost control, quantity control, and personnel control. The company must establish cross-functional committees to address these section-spanning issues. The committee chair should be a senior managing director. Committee members are selected from among those who hold the rank of director or above.

These fundamental principles are the drivers behind all of the quality techniques and practices described elsewhere in this book.

THE MANAGEMENT CYCLE

The management cycle consists of five primary functions: planning, organizing, commanding, coordinating, and controlling. As applied to quality, elements of the management cycle include (Juran and Gryna, 1993):

Quality policies—A policy is a broad guide to action, a statement of principles. Policies are statements of "whats"; they specify what will be done. Procedures accompany the policies and specify the "how." Procedures detail how a given activity is to be accomplished. For example, a quality policy might specify that customer-perceived quality is to be measured; the corresponding procedure would specify the measurement process in detail. Here are some examples of quality policy items:

- It is the policy of XXX corporation to manufacture, efficiently and economically, products which consistently meet or exceed customer expectations.
- Quality standards will be developed through the coordinated efforts of engineering, manufacturing, and quality control.

Quality goals—A goal is a statement of the desired result to be achieved within a specified time. Goals can be created to achieve stability or unprecedented new levels of performance. Goals can be either short-term or long-term. The use of benchmarking in goal-setting is discussed later in this chapter. Other methods of goal-setting include inspiration, competitive evaluations, reengineering, customer requirements, new regulations, etc. Pareto analysis is another technique that is often used to set quality improvement goals (see section V.C).

RESOURCE REQUIREMENTS TO MANAGE THE QUALITY FUNCTION

The modern approach to quality requires an investment of time and resources throughout the entire organization—for many people, the price is about 10% of their time (Juran and Gryna, 1993, p. 129). Eventually, this investment yields time-savings that become available for quality activities and other activities. In the short run, however, these resources are diverted from other, urgent organizational priorities.

Upper management has the key role in providing resources for quality activities. This is accomplished through the quality councils mentioned earlier, as well as through routinely funding the activities of the quality department. One alternative is to add resources, but this is seldom feasible in a highly competitive environment. More commonly, priorities are adjusted to divert existing resources to quality planning, control, and improvement. This means that other work must be eliminated or postponed until the payback from quality efforts permits its completion.

Before resources can be requested, their usage must be determined. This should be done using a rational process which can be explained to others. The exercise will also provide a basis for budgeting. Figure II.3 illustrates the approach used by a large integrated healthcare organization to determine their quality resource needs. The approach is used by the entire organization, not just the quality department. In fact, the organization doesn't have a "quality department" in the traditional sense of a centralized quality control activity.

The following questions have been outlined in an effort to provide a recipe to enable each area of HealthPartners of Southern Arizona to identify, illustrate and detail the service requirements and costs of delivering services and/or products to our customers. Please answer each of the following questions in as much detail as necessary.

I. Customer/Customer requirements
   A. What are the key products or services you provide to your customers?
   B. Who are your customers? (Who receives the output of your products or services?)
   C. What are your customers' expectations of the products or services you provide to them?
   D. What attributes of these services or products are most important to your customers? (E.g., timely, comfortable, cost effective)
   E. What measurement techniques do you use to confirm your customers' requirements?
   F. What are the major factors that make your products or services necessary?

II. Structure and processes
   A. Create a top-down or detailed flowchart of your key processes. (Key processes are defined as those processes that deliver services or products directly to your customers.)
   B. Identify skill levels required for each step of your process. (E.g., an RN starts an IV or a Health Records Analyst completes the coding process for the chart.)
      1. Identify the number of skill levels required at each step. (Steps may involve more than one skill level and/or multiples of one skill level.)
      2. How much time is required by each skill level to complete each step of the process?
      3. What is the average time to complete one cycle of the process? Does an industry standard time to complete this process exist? How does your process compare to any industry standards? If there is no comparable standard, how long should this process take as compared to the actual time required to complete the process? (Hint: if you were to measure this process assuming there were no delays, bottlenecks, shortage of resources, etc., how long would it take?)
      4. Identify any other resources required to support the entire process. (E.g., reusable equipment, disposable supplies, purchased services.)

Figure II.3—Process for identifying quality resource needs. HealthPartners of Southern Arizona; Decision Package. *Continued on next page . . .*

Figure II.3—*Continued . . .*

III. Cost assignment
   A. What are the total costs each time the process is completed and the product or service is delivered? (I.e., total costs per unit of service.)
   B. What are the cycle time costs for each step of the process?

IV. Process improvements
   A. What are the causes, delays or roadblocks in your process that create non-value added cost?
   B. What level of variation and complexity exist in your process? How do you measure the variation and complexity of your process? (Hint: What does a control chart tell you about the process variability?)
   C. If you could do one or two things to reduce time delays, waste, or complexity, what would you do?
   D. If you were successful in reducing time delays, waste, errors and complexity how would this improve the product or service in the eyes of your customer? (What improved level of satisfaction would result?)

Courtesy of HealthPartners of Southern Arizona. Used with permission.

Figure II.3. Process for identifying quality resource needs.
HealthPartners of Southern Arizona; Decision Package

People assigned to quality teams should be aware of the amount of time that will be required while they are on a team. If time is a problem, they should be encouraged to propose changes in their other priorities before the team activity starts. Resources for project teams will be made available only if pilot teams demonstrate benefits by achieving tangible results. As teams compile records of success, it will become easier to secure resources for the quality function. Ideally, quality improvement funding will become a routine part of the budgeting process, much like R&D.

Quality cost reports provide a valuable source of information for securing resources for quality improvement. Quality costs indicate the total impact of nonconformance on the company's bottom line. If the allocation for preventing these costs is small relative to total quality costs, management can see for themselves the potential impact of spending in the area of quality cost reduction. Likewise, if external failure costs are high relative to appraisal costs additional appraisal expenditures may be justified.

A shortcoming of using quality costs to determine resource requirements for the quality function is that the highest costs are difficult to determine with a high degree of accuracy. What is the cost of someone *not* buying your product? As difficult as it is to measure the value of keeping an existing customer (see I.H), it is more difficult to know when a prospective customer didn't consider your product because of a poor reputation for quality. It is also hard to estimate the cost of future failures with new products or processes. For example, a company decided to begin manufacturing their own gear boxes. The quality department's request for an expensive gear-checker was turned down because they couldn't precisely quantify the benefit of the equipment (the cost was easy to determine, $50,000). Two years later a field problem with the gears lead to a $1.5 million field repair program. Afterwards, the purchase of the gear-checker was quickly approved.

Another problem with quality costs is that they measure negatives, rather than the lack of positives. Quality is not only the absence of negatives (e.g., defects), it is the presence of desirable features. Quality costs only measure the cost of the former. While techniques exist for estimating the cost of lost business due to a lack of desirable features (see I.H), they are not as well-defined or standardized as quality costs. Thus, it is difficult to make a general statement regarding their use in securing resources for the quality function.

QUALITY ORGANIZATION

As the importance of quality has increased, the position of the quality function within the organization has moved up in the organizational hierarchy. Figure II.4 presents a prototypical modern organization chart for a hypothetical large manufacturing organization.

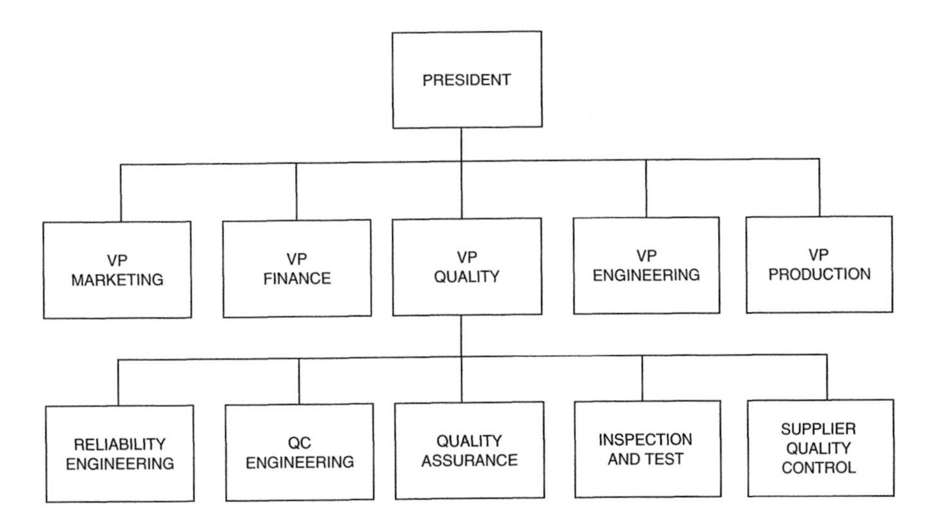

Figure II.4. Quality organization chart.

WORK ELEMENTS IN THE QUALITY SYSTEM

The organization chart and discussion above is misleading in that it suggests that quality is the responsibility of specialists in the quality department. In fact, this is decidedly not the case. The specialists in the quality department have no more than a secondary responsibility for most of the important tasks that impact quality. Table II.1 lists the major work elements normally performed by quality specialists.

Table II.1. Work elements.

RELIABILITY ENGINEERING	Establishing reliability goals Reliability apportionment Stress analysis Identification of critical parts Failure mode, effects, and criticality analysis (FMECA) Reliability prediction Design review Supplier selection Control of reliability during manufacturing Reliability testing Failure reporting and corrective action system
QUALITY ENGINEERING	Process capability analysis Quality planning Establishing quality standards Test equipment and gage design Quality troubleshooting Analysis of rejected or returned material Special studies (measurement error, etc.)
QUALITY ASSURANCE	Write quality procedures Maintain quality manual Perform quality audits Quality information systems Quality certification Training Quality cost systems

Continued on next page . . .

Table II.1—*Continued . . .*

INSPECTION AND TEST	In-process inspection and test Final product inspection and test Receiving inspection Maintenance of inspection records Gage calibration
VENDOR QUALITY	Pre-award vendor surveys Vendor quality information systems Vendor surveillance Source inspection

The role of others in the company can be better understood if we look at a listing of the basic tasks as part of a system, which we will refer to as the Total Quality System (TQS). TQS can be viewed as a process that assures continual improvement while implementing the policy established by top management. Figure II.5 depicts the TQS process. The TQS makes it clear that the quality specialist shares responsibility for quality with many others in the company.

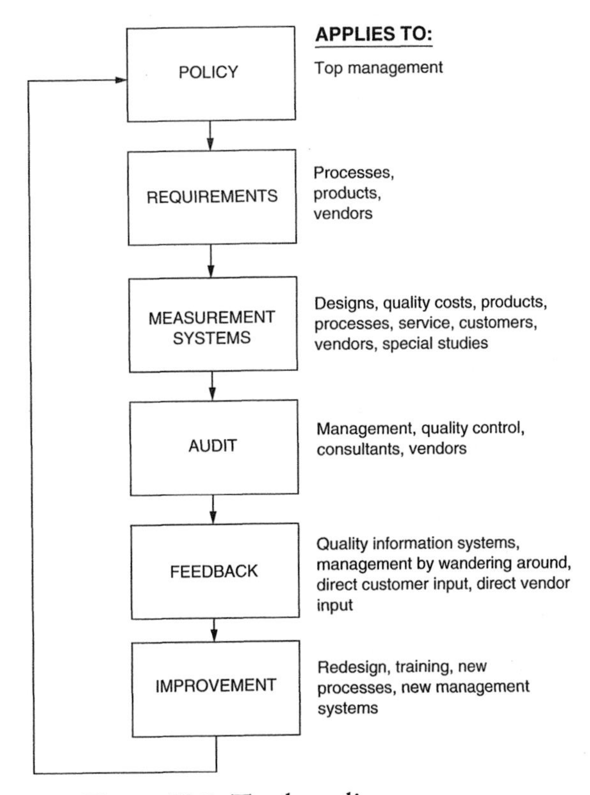

Figure II.5. Total quality system.

TQS POLICY

As discussed earlier, top management is responsible for establishing the company policy on quality. This policy, derived in turn from the organization's mission and customers, forms the basis of the requirements.

TQS REQUIREMENTS

Requirements are established for key products, processes, and vendors. Product requirements are based on market research regarding the customer's needs and the ways which competitor's products meet or fail to meet these needs. Depending on the policy of the company, a product design must either compete with existing products or improve upon them. Process requirements assure that designs are reproduced with a minimum of variability, where "minimum" is interpreted consistent with established policy. Process design is usually the responsibility of process engineering specialists. Vendor selection is typically based on reliability, price, and quality. Purchasing bears the primary responsibility, with quality and engineering providing assistance.

TQS MEASUREMENT SYSTEMS

The term "measurement system" is used here in the broadest sense. The purpose of a measurement system is to determine the degree to which a product, service, vendor, process, or management system conforms to the requirements established. At times the measurement is not straightforward. For example, design conformance is accomplished by a design review. Quality, marketing, and manufacturing evaluate the design to determine if it meets the customer's expectations and can be economically and consistently reproduced by manufacturing. This is a subjective evaluation but the parties involved should reach a consensus.

Quality cost measurement is discussed in I.F. The design of quality cost systems is usually the primary responsibility of the accounting department, with assistance from quality.

Product and process measurements are designed to assure that they conform to requirements. This usually involves obtaining one or more observations and comparing the observation to operational standards. While the quality department may be responsible for obtaining measurements on certain products and processes, the responsibility for acting on the results remains with the operating department.

Measurement of service quality is woefully inadequate. In fact, many companies fail to recognize their responsibility to the customer in this area. In most cases of neglect it takes an external force to create a change, such as a government agency or a competitor. Unlike most of the other areas, responsibility for measuring service quality hasn't been established by precedent. However, it seems logical that quality and marketing share a joint responsibility.

Measurement of vendor quality is discussed later in this chapter. Quality and purchasing are the primary players in developing the measurement systems. Of course, the final responsibility for quality remains with the vendors themselves.

Special studies are different from the other measurement systems described here in that they are not an ongoing activity. Studies may be commissioned to answer an important question or to obtain vital facts for a particular purpose. For example, a competitor may have just released a revolutionary product that requires close examination. The responsibility for special studies is usually stipulated in the charter which commissions the study.

Finally, remember that early in this chapter we said that the proper emphasis was to be results, not task completion. The most important measurement is, therefore, customer satisfaction. Marketing is usually given responsibility for developing systems that measure customer satisfaction.

TQS AUDIT

The purpose of a TQS audit is to seek verification that an adequate system exists and is being followed. The primary responsibility for TQS auditing is top management's. In addition, the manager of each activity must perform audits within his area to assure that the requirements are being met. Finally, "third-party" audits by the quality department or auditors from outside the company are also common. All third-party audits should be commissioned by management. However, the third-party audit should never be the primary source of detailed information for management. There is no substitute for first-hand observation. Audits are discussed in greater detail in II.D.

TQS FEEDBACK

All of the information collection in the world is absolutely useless without an effective feedback mechanism. The feedback loop of the total quality system is closed by a variety of means. The most visible is the formal Quality Information System (QIS) of the company. QIS encompasses all of the formal, routine activities associated with quality data collection, analysis, and dissemination. QIS is discussed in greater detail in I.B.

Formal systems, while necessary, provide only part of the feedback. Human beings may get the facts from these systems, but they can't get a feeling for the situation from them. This level of feedback comes only from firsthand observation. One company encourages their senior managers, in fact all of their managers, to "wander around" their work areas to observe and to interact with the people doing the work. The term they give this activity is Management By Wandering Around, or MBWA. The author can personally attest to the effectiveness of this approach, having done it and seen it work. There are many "facts" that never appear in any formal report. Knowing that quality costs are 10% of sales is not the same as looking at row upon row of inspectors or a room full of parts on hold for rework or a railroad car of scrap pulling away from the dock.

Just as MBWA provides direct feedback from the workplace, a company needs direct personal input from its customers and suppliers. Company personnel should be encouraged to visit with customers. All managers should be required to periodically answer customer complaints (after being properly trained, of course). A stint on the customer complaint "hot line" will often provide an insight that can't be obtained in any other way. Vendors should be asked to visit your facility. More formal methods of getting customer and vendor input are also important. Vendors should be made part of the process from design review to post-sale follow-up. Often a slight design modification can make a part much easier to produce, and the vendor, who knows his processes better than you do, will often suggest this modification if he's invited to do so. Customer input should be solicited through opinion surveys, customer review panels, etc. Of course, the QIS will provide some of this input.

IMPROVEMENT

One reason feedback is obtained is to facilitate improvement. Why bother collecting, analyzing, and distributing information if it will not be used? This simple fact is often overlooked and many companies devote vast resources to developing evermore sophisticated management information systems that produce little or no improvement. Discovering this situation is one of the primary objectives of quality audits.

Continuous improvement should be a part of everyone's routine, but making this actually happen is extremely difficult. Most operating precedents and formal procedures are designed to maintain the status quo. Systems are established to detect negative departures from the status quo and react to them. Continuous improvement implies that we constantly attempt to change the status quo for the better. Doing this wisely requires an understanding of the nature of cause systems. Systems will always exhibit variable levels of performance, but the nature of the variation provides the key to what type of action is appropriate. If a system is "in control" in a statistical sense, then all of the observed variability is from common causes of variation that are inherent in the system itself. Improving performance when this situation exists calls for fundamental changes to the system. Other systems will exhibit variability that is clearly non-random in nature. Variation of this sort is said to be due to "special causes" of variation. When this is true it is usually best to identify the special cause rather than to take any action to change the system itself. Changing the system when variability is from special causes is usually counterproductive, as is looking for "the problem" when the variability is from common causes. Determining whether variability is from special causes or common causes requires an understanding of statistical methods.

Quality improvement must be companywide in scope. The entire cycle from marketing and design through installation and service should be geared toward improvement.

This holds regardless of the policy of the company. Companies that aspire to be cost leaders have as much need for improvement within their market segment as those who aim for quality leadership. A key part of this effort must be continuous improvement of the management systems themselves. See also V.A.

II.B DOCUMENTATION SYSTEMS

Many firms create quality manuals to document their formal quality systems. Figure II.6 shows portions of the table of contents of a quality manual for a large defense contractor. Smaller firms usually have less elaborate systems and less extensive documentation.

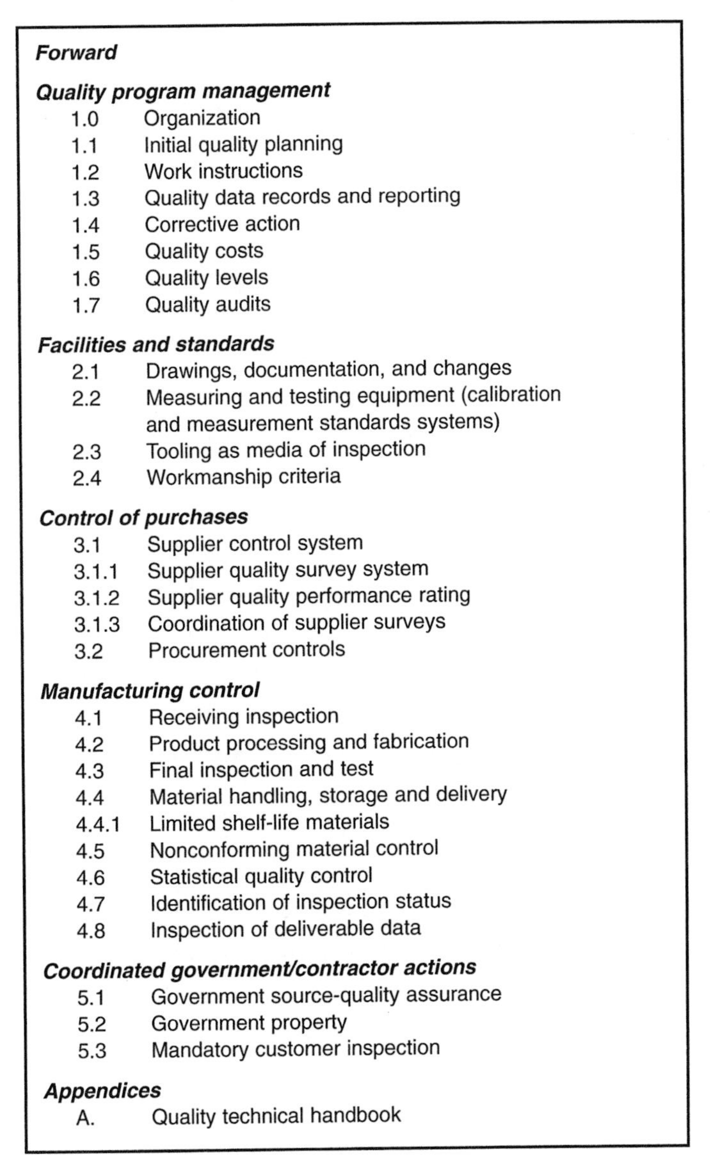

Forward

Quality program management

1.0	Organization
1.1	Initial quality planning
1.2	Work instructions
1.3	Quality data records and reporting
1.4	Corrective action
1.5	Quality costs
1.6	Quality levels
1.7	Quality audits

Facilities and standards

2.1	Drawings, documentation, and changes
2.2	Measuring and testing equipment (calibration and measurement standards systems)
2.3	Tooling as media of inspection
2.4	Workmanship criteria

Control of purchases

3.1	Supplier control system
3.1.1	Supplier quality survey system
3.1.2	Supplier quality performance rating
3.1.3	Coordination of supplier surveys
3.2	Procurement controls

Manufacturing control

4.1	Receiving inspection
4.2	Product processing and fabrication
4.3	Final inspection and test
4.4	Material handling, storage and delivery
4.4.1	Limited shelf-life materials
4.5	Nonconforming material control
4.6	Statistical quality control
4.7	Identification of inspection status
4.8	Inspection of deliverable data

Coordinated government/contractor actions

5.1	Government source-quality assurance
5.2	Government property
5.3	Mandatory customer inspection

Appendices

A.	Quality technical handbook

Figure II.6. Sample quality manual table of contents.

The size and complexity of the quality manual has little to do with the quality of the firm's products and services. Good quality manuals, like the effective systems they document, are well thought out and customized to the particular needs of the firm. The emphasis is on clarity and common sense, not on an effort to document every conceivable eventuality. For example, the quality manual of a very large defense contractor listed *68 ways to identify scrap items* (red paint, stamped with an "R," a scrap tag, etc., etc.). This approach led to constant difficulty because there were always new cases not on the list. In one instance, an auditor complained because some defective microchips were not properly identified. The chips were in a sealed plastic bag with a completed scrap tag. However, the auditor wanted the parts stamped with an R, never mind that the parts were physically too small for such identification. This company's quality manual was hundreds of pages thick, few people had ever read it and no one understood it completely.

In contrast, when a medium-sized electronics company that supplied this defense contractor was studied because of their excellent quality record it was discovered that their entire manual contained only 45 pages. The manual was to the point and clearly written. Scrap identification was described by a single sentence: "Scrap material will be clearly and permanently identified." The details were deliberately left out. Employees and auditors had no difficulty confirming that the requirement was met, and there were no known instances of scrap material being inadvertently shipped. Employees who were registered holders of the manual had read it, in fact most of them helped to write it. All of the manuals were up to date and employees were able to answer auditor questions about the contents of the manual. In other words, the manual served its intended purpose.

WORK INSTRUCTIONS

Eventually the engineering drawing has to be made into physical product and the management plan has to be implemented to provide service to the customer. How this is done is just as important as the drawing or the plan, perhaps more so. In other words, *how* a thing is to be done is sometimes just as important as *what* is to be done. Work instructions provide the answer to specifically how the work is to be done. Mil-Hbk-H50 describes work instructions as follows:

> The quality program shall assure that all work affecting quality . . . shall be prescribed in clear and complete documented instructions of a type appropriate to the circumstances. Such instructions shall provide the criteria for performing the work functions and they shall be compatible with acceptance criteria for workmanship. The instructions are intended also to serve for supervising, inspecting and managing work. The preparation and maintenance of and compliance with work instructions shall be monitored as a function of the quality program.

The instructions must establish quantitative or qualitative means for determining that each operation has been done satisfactorily These criteria must also be suitable for use with related inspections or tests, because work instructions serve operating personnel, supervisors, inspectors, managers, and even customers. Of course, compliance with instructions is subject to review and audit.

Work instructions consist of documented procedures that define how production, installation, or servicing will take place. The instructions describe the operating environmental conditions as well as the activities necessary to assure that the finished product meets all of the customer's requirements. This includes such things as "cheat sheets" and "crib notes" and other tidbits that people frequently keep to remind them of the way "it's really done." In the past, this information was kept on scraps of paper or in personal notebooks, but ISO 9000 makes these informal notes part of the official documentation of the process.

Just how far one should go in documenting a process is a highly debatable issue. In this author's opinion, a reasonable limit is soon reached beyond which the documentation becomes so massive that no one has time to read it all. Some work instructions grow to huge size (e.g., one work instruction for a welding operation of a small part of a missile wing runs well over 100 pages) and are filled with masses of "tips" which result from problems encountered in the operation of the process over a period of years. Many of the situations described in the documentation are responses to rare events unlikely to be encountered in the future. By documenting them, one makes it that much more difficult to locate the truly useful information. Consider, for example, your daily trip to work. Simple documentation might list the streets that you take under normal conditions. However, one day you find a traffic jam and take an alternate route. Should you write this down? Well, if the traffic jam is caused by a long-term construction project, perhaps. But if it's due to a rare water main rupture, probably not. As technology improves, it may be possible to develop databases that can be quickly and effectively searched for information that is relevant to the task at hand. This will effectively increase the amount of data that can be useful to the process operator. Until then, the documentation must be contained within human cognitive limits. The guiding principle should be minimum size, subject to being reasonably complete and accessible to those who will use it.

Work instructions should cover the following items:

- The manner in which the work will be done.
- The equipment needed to do the work.
- The working environment.
- Compliance with other procedures and documents.
- Process parameters to be monitored and how they will be monitored (e.g., checklists, control charts).

- Product characteristics to be monitored and how they will be monitored.
- Workmanship criteria.
- Maintenance procedures.
- Verification methods (process qualification).

Work instructions should be written in clear, simple terms, using the language that is easiest for the person doing the work to understand. The people doing the work should be intimately involved in preparing the work instructions. Pictures, diagrams, graphics, and illustrations should be used to make the documentation easier to understand and apply. If the instructions are voluminous, they should include such aids as indexes, tables of contents, tables of figures, tabs, etc., to assist in locating relevant information. Of course, to assure that they are up-to-date, the documentation should be cross-indexed to the engineering drawings, purchase orders, or other documents which they implement. Work instructions should be part of the overall document control system of a firm.

CONFIGURATION CONTROL

It has been said that the only constant is change. As time passes, new and better ways of doing things are discovered, mistakes are found, new technology is introduced. Quality improvement demands that these changes for the better be incorporated into new designs and new processes. At other times, it is in the best interest of all parties to waive certain requirements temporarily. However, doing so is anything but simple. When a change is made in a part, for example, the new part may no longer work properly with existing equipment. Part of the solution is to maintain lot traceability (see section III.B), but traceability addresses only some of the parts. A broader solution is needed to deal with routine changes. This solution is configuration control.

Configuration control is the systematic evaluation, coordination, approval or disapproval, and implementation of all approved changes in the configuration of an item after formal establishment of its configuration identification.

The intent of configuration management is to control the form, fit, and function of configuration items. Configuration control is primarily concerned with managing engineering changes. For configuration control purposes, engineering changes are divided into two classes, Class I and Class II. A Class I change is any change that has an impact on the form, fit, or function while Class II changes involve non-substantive changes such as correcting documentation errors or cosmetic items.

The easiest change to manage is no change at all. Thus, the first line of defense in configuration control is to restrict changes to those things which offer a significant benefit to the customer. This includes improving safety, improved compatibility, correcting serious deficiencies, improving performance, reducing the cost of purchase and ownership or helping maintain schedule. The configuration management procedure should include change-justification review as part of the change-approval process.

PRIORITY OF CHANGES

Priorities should be assigned to engineering changes to reflect the relative speed at which the ECP is to be reviewed and evaluated, and at which the engineering change is ordered and implemented. The proposed priority is assigned by the originator. The following priorities are recommended:

Emergency—An emergency priority is assigned to an engineering change which corrects a hazardous condition which may result in death or injury to personnel or in extensive damage to the equipment. Emergency priorities generally require that the product be removed from service pending the engineering change.

Urgent—This priority is applied to engineering changes which correct potentially hazardous conditions. Under this designation, limited use of the product is sometimes allowed under carefully defined conditions and with adequate warning to the user. Urgent priority is also assigned to those changes that will result in large savings, major performance improvements, or the prevention of significant schedule slippage.

Routine—Any engineering change other than emergency or urgent.

Of course, the configuration control procedure should specify how to process engineering changes with different priorities.

II.C DOMESTIC AND INTERNATIONAL STANDARDS AND SPECIFICATIONS

It can be said that quality is all about eliminating unwanted variation. This applies to human behavior as well as to management systems, manufacturing processes, products, and so on. Standards are documents used to define acceptable conditions or behaviors and to provide a baseline for assuring that conditions or behaviors meet the acceptance criteria. In most cases standards define *minimum* criteria; world-class quality is, by definition, beyond the standard level of performance. Standards can be written or unwritten, voluntary or mandatory. Unwritten quality standards are generally not acceptable. Quality standards serve the following purposes (Sullivan 1983):

Standards educate—They set forth ideals or goals for the guidance of manufacturers and users alike. They are invaluable to the manufacturer who wishes to enter a new field and to the naive purchaser who wants to buy a new product.

Standards simplify—They reduce the number of sizes, the variety of processes, the amount of stock, and the paperwork that largely accounts for the overhead costs of making and selling.

Standards conserve—By making possible large-scale production of standard designs, they encourage better tooling, more careful design, and more precise controls, and thereby reduce the production of defective and surplus pieces. Standards also benefit the user through lower costs.

Standards provide a base upon which to certify—They serve as hallmarks of quality which are of inestimable value to the advertiser who points to proven values, and to the buyer who sees the accredited trademark, nameplate, or label.

In general, standards define requirements for systems, products, or processes. The assumption is that when a producer meets the requirements outlined in the standard, the customer can be assured of at least minimally acceptable performance. This assumption, like all assumptions, should not be taken for granted. Complying with standards involves considerable cost and the organization undertaking the process of certification should determine for themselves that there is sound business logic in their pursuit. In addition to the initial cost of designing, documenting, and implementing the changes required by the standard, the organization should consider the costs of obtaining certification and of maintaining the certification as the standard and the organization change.

There are many organizations which produce standards (Sullivan lists 118 in his book). In the quality field, the best known organizations are those shown in Table II.2.

Table II.2. Standards organizations.

ORGANIZATION	ADDRESS
International Organization for Standardization (ISO)	1, rue de Varembe, 1211 Geneva 20, Switzerland
American Society for Quality (ASQ)	600 North Plankinton Ave., Milwaukee, WI 53203
American National Standards Institute (ANSI)	1819 L Street, NW, Suite 600, Washington, D.C. 20036
Department of Defense (DOD)	Naval Publications and Forms Center, 5801 Tabor Ave., Philadelphia, PA 19120
American Society for Testing and Materials (ASTM)	100 Barr Harbor Drive, West Conshohocken, PA 19428–2959

The best known series of quality standards is, by far, the ISO 9000 series. ISO 9000 is a set of five standards for quality systems. The standards were originally published in 1987, and updated in 1994 and 2000. The 1994 ISO series numbering scheme is shown in Table II.3.

Table II.3. 1994 ISO 9000 system standards.

STANDARD	TITLE
ISO 9000-1	Part 1: Guidelines for selection and use
-2	Part 2: Generic guidelines for application of ISO 9001, 9002 and 9003
-3	Part 3: Guidelines for application of ISO 9001 to development, supply, and maintenance of software
-4	Part 4: Application for dependability management
ISO 9001	Quality Systems—Model for Quality Assurance in Design, Development, Production, Installation, and Servicing
ISO 9002	Quality Systems—Model for Quality Assurance in Production, Installation, and Servicing
ISO 9003	Quality Systems—Model for Quality Assurance in Final Inspection and Test

Continued on next page . . .

Table II.3—*Continued . . .*

STANDARD	TITLE
ISO 9004-1	Part 1: Guidelines
-2	Part 2: Guidelines for services
-3	Part 3: Guidelines for processed materials
-4	Part 4: Guidelines for quality improvement
-5	Part 5: Guidelines for quality plans
-6	Part 6: Guidelines for project management
-7	Part 7: Guidelines for configuration management
-8	Part 8: Guidelines on quality principles

The 2000 revisions integrated the ISO 9001, 9002, and 9003 standards into a single new ISO 9001:2000 standard. ISO 9000 was repaced with *ISO 9000:2000 Fundamentals and vocabulary*. ISO 9001:2000 is based on eight quality management principles (defined in ISO 9000:2000):

- Principle 1: Customer focus
- Principle 2: Leadership
- Principle 3: Involvement of people
- Principle 4: Process approach
- Principle 5: Systems approach to management
- Principle 6: Continual improvement
- Principle 7: Factual approach to decision making
- Principle 8: Mutually beneficial supplier relationships

Figure II.7 shows the correspondence between the twenty main elements of the 1994 standard and the five major clauses of the 2000 standard.

The use of ISO 9000 is extremely widespread all over the world. In addition to carefully reading what is written here, the reader is encouraged to obtain a set of the standards and to read them carefully.

Development of the ISO 9000 series standards started with the formation of ISO Technical Committee 176 (TC176) in 1979. Prior to that time various national and international quality standards existed, but they were too inconsistent with one another to be used for international trade. TC176 was charged with developing a set of quality system standards which utilized consistent terminology and described an approach that could be applied to different industries in different countries. The ISO 9000 series is the result of the efforts of TC176. The series embodies comprehensive quality management concepts and provides guidance for implementing the principles.

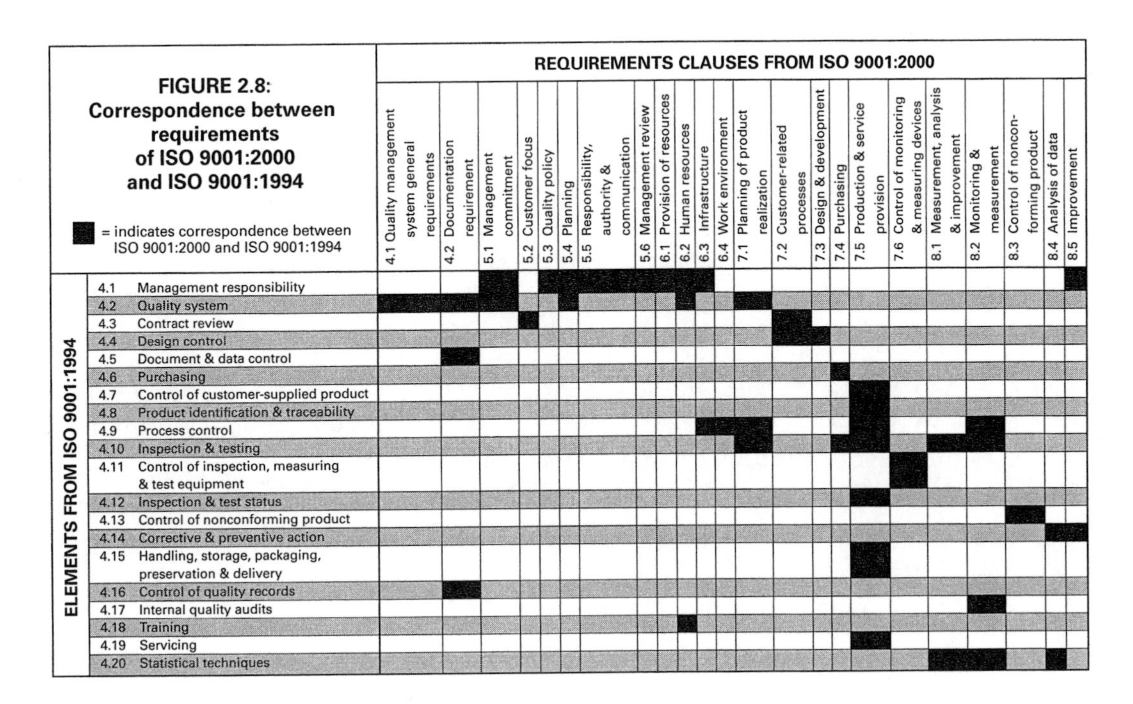

Figure II.7. Correspondence between the twenty main elements of the 1994 standard and the five major clauses of the 2000 standard.

Stapp, Eric H., (2001). ISO 9000:2000, An essential guide to the new standard. Tucson, AZ: QA Publishing, pg. 29.

A major advantage of the ISO 9000 system is the elimination of the need for multiple quality systems audits by the various customers of a firm. ISO 9000 registration is achieved by third-party certification, i.e., audits are not performed by customers but by specially trained, independent, third-party auditors. In the past many firms had to deal with auditors from many different customers; furthermore, the customers usually had their own specific requirements. The result was that parallel systems had to be developed to meet the various requirements. This was costly, confusing, and ineffective. While some industry groups, notably aerospace and automotive, made progress in coordinating their requirements, ISO 9000 has greatly improved upon and extended the scope of acceptable quality systems standards. ISO 9000 has been approved by the "Big Three" auto manufacturers (see "Other quality standards", below), the US Department of Defense, NASA, and most other large government and industry groups. However, since the various groups often impose additional requirements, ISO 9000 registration isn't always sufficient.

While ISO 9000 applies to any organization and to all product categories, it does not specify *how* the requirements are to be implemented. Also, the series specifies requirements at a very high level; ISO 9000 does not replace product, safety, or regulatory requirements or standards. The concept which underlies the ISO 9000 standards is that consistently high quality is best achieved by a combination of technical product specifications and management system standards. ISO 9000 standards provide only the management systems standards.

ISO 9000 REGISTRATION

Many firms, especially those doing business internationally, pursue registration. A supplier seeking to be registered starts the process by completing an application form and sending it, with a fee, to the appropriate registrar. There are several organizations that provide quality system registration services in the United States. When the registrar accepts the application, it sometimes performs a preassessment visit. During the visit the registrar determines the supplier's readiness for assessment and the type of expertise needed on the assessment team.

Using the 1994 standards, firms were registered to one of the ISO 9000 series standards as appropriate, either ISO 9001, 9002, or 9003. With ISO 9000:2000, all firms are registered to the single 9001:2000 standard. Since some organizations do not perform all the functions outlined in Clause 7 of the ISO 9001:2000 standard (see Figure II.7), specific aspects of this clause may be excluded from the organization's requirements for registration (Stapp, 2001). The registrar begins by appraising the supplier's quality documentation for conformance to the standard, e.g., their quality manual, procedures, etc. Deviations are reported and the supplier is informed regarding any deviations to be corrected. This phase of the registration process may or may not be conducted at the supplier's site.

Once the supplier reports that the corrections have been made, an on-site assessment is scheduled. The registrar selects an assessment team and informs the supplier of the team membership. Suppliers have the right to object to team members on the grounds of conflict of interest. At least one of the team members must be familiar with the supplier's technology. Each team has a designated team leader, who has overall responsibility for the assessment.

On arrival, the assessment team, which usually consists of two to four members, meets with the supplier's management. In this meeting the team leader explains the assessment process. The supplier selects someone to be the liaison between the team and management. All information obtained during the assessment is confidential.

Assessment involves evaluating all areas of the operation that are relevant to the standard. Since the standard is written at a high level, the assessment includes support operations such as purchasing, engineering, shipping, and transportation in addition to manufacturing. Assessors look for objective evidence that the supplier is conforming with its own documentation. They also assure that the documentation complies with the standard.

After the on-site assessment, the team again meets with the supplier's management. The leader summarizes the assessment team's findings and answers any questions that the supplier might have. Findings are presented verbally and confirmed in writing. The team also identifies deficiencies and notifies the supplier of any time limits on correcting the deficiencies.

The supplier sends the registrar written descriptions of how it plans to correct the deficiencies. The registrar evaluates the supplier's response and decides how it will assure that effective action will be taken, including, possibly, reassessment in whole or in part. Once the registrar is satisfied that the supplier has complied with the standard, a registration document is issued to the supplier. If the registrar is accredited, they also notify the RAB, which keeps a master directory of all suppliers registered by accredited registrars. Suppliers are reassessed by surveys conducted quarterly; full reassessments are conducted every three or four years. If the supplier changes the scope of its operations beyond what it has been assessed and registered for, it must inform the registrar. The registrar then decides whether or not reassessment (complete or partial) is required.

Suppliers should approach registration in a planned, orderly fashion. The supplier should understand the standards in general, and how the standards apply to the operations in particular. Personnel should be trained in their own systems and procedures. At least one individual should be familiar with the standards.

A full self-assessment should be conducted and the findings reported to key personnel. Many firms have arrangements with other firms or consultants to conduct unofficial assessments of one another. Based on the results of the self-assessment, the supplier should prepare a plan to correct the deficiencies discovered. The documentation should be brought into conformance with actual (accepted) practices.

It is important to note that the term "documentation" applies not only to the quality manual, but to management policies and work instructions. Documentation should be prepared by those actually doing the work, not by the quality department. The systems being documented should conform to the appropriate ISO standard. Once the documentation is ready, the supplier should assure that everyone understands what the documentation says, and that the policies and procedures are actually being followed. When this has been accomplished, the supplier is ready to begin the registration process.

Registration is not for everyone. The process can be expensive; Lofgren (1991) conducted an informal survey of several registrars and asked them to price an initial registration of a supplier with 200–300 employees producing a single product line at a single site to ISO 9002. The cost estimates for a single-pass registration ranged from $10,000 to $15,000 plus travel expenses for the team. Cost estimates did not include ongoing surveillance, preparation by the supplier, corrective action, or additional assessments. Larger firms report proportionately higher registration costs.

Costly or not, many firms feel that registration is worth it. ISO 9000 registration reduces the number of second-party audits and it is viewed as providing a marketing advantage over unregistered competitors. Also, registration provides access to markets not open to unregistered companies. More important than any of these reasons is that the registration process helped suppliers improve their operations and quality.

OTHER QUALITY STANDARDS

Because of its general nature, ISO 9000 doesn't fit the specific needs of any one industry, or any one supplier. Consequently, the automotive industry, the largest industry, has developed its own industry-specific standard which is designated QS 9000. QS 9000 starts with ISO 9000 as a base and adds additional customer-specific requirements. The QS 9000 standard essentially integrates ISO 9000 requirements with the requirements found in the quality systems manuals of Ford, General Motors, and Chrysler. Thus, a firm that wishes to supply all of the "Big Three" automakers will need to conform to the generic requirements of the appropriate ISO 9000 standard, the industry-specific requirements of QS 9000, and the customer-specific requirements of Ford, GM and/or Chrysler.

There are some who feel that efforts to make ISO 9000 more specific undermines the primary benefit of the program, namely, a centralized approach to quality systems certification (Zuckerman 1995). In addition to the automotive industry, there is pressure from the medical-device industry, the electronics industry and others to develop more specific criteria for quality systems. It seems likely that this pressure will continue into the foreseeable future.

Prior to ISO 9000 there was a proliferation of quality system standards. Chief among these was the United States Department of Defense (DOD) Mil-Q-9858 (Quality Program Requirements) and Mil-I-45208 (Inspection System Requirements). Mil-Q-9858 was a standard used by the DOD to evaluate the quality systems of prime defense contractors. While there is a great deal of overlap between Mil-Q-9858 and the ISO 9000 systems, the military standard has a more restricted scope. The DOD formally adopted ISO 9000 as their quality systems standard, abandoning Mil-Q-9858A as a requirement effective October 1995 (Morrow 1995).

Standards are such an important part of the quality profession that the journals *Quality Engineering* and *The Journal of Quality Technology* both have regular standards departments.

In addition to quality systems standards, there are also standards for just about every other aspect of quality engineering. Of course, standards exist for many different types of products, processes, and raw materials. But standards also exist for many commonly used statistical procedures. For example, ASTM has a standard on the proper use of control charts [Manual 7A, Manual on the Presentation of Data and Control Chart Analysis, Seventh edition, (2002)], as does ANSI/ASQ (B1–B3 1996, Guide for Quality Control Charts, Control Chart Method of Analyzing Data, Control Chart Method for Controlling Quality During Production). The quality engineer can easily be overwhelmed by the sheer number and scope of standards. However, there are a few standards that are well known and widely used and knowledge of these will cover 99% of the situations. Two widely-used standards (from a historical perspective) are Mil-Std-105E (Sampling Procedures and Tables for Inspection by Attributes) and Mil-Std-414 (Sampling Procedures and Charts for Inspection by Variables), which have been replaced by ANSI/ASQ Z1.4-1993 and ANSI/ASQ Z1.9-1993, respectively. These standards are discussed further in III.C.

Many calibration systems are based on Mil-Std-45662, Calibration Systems Requirements, or its replacement ANSI/ASQ M1-1996. These standards are discussed in greater detail in the measurement systems section of this book (III.D).

II.D QUALITY AUDITS

Basically, an audit is simply a comparison of observed activities and/or results with documented requirements. More formally, ISO 10011 *Guidelines for Auditing Quality Systems* (p.1), defines a quality audit as

> A systematic and independent examination to determine whether quality activities and related results comply with planned arrangements and whether these arrangements are implemented effectively and are suitable to achieve objectives.

By implication, this definition makes it clear that quality audits must involve several characteristics. First, the audit is to be a systematic evaluation. Informal walk-throughs do not qualify. This is not to say such walk-throughs are not important, just that they are not audits per se. Examinations by employees who report to the head of the function being examined are also important, but not audits. According to the definition, the audit is also concerned with "planned arrangements." This implies the existence of written procedures and other types of documentation such as quality manuals. Undocumented quality systems are not proper subject matter for quality audits. Objectives are a third aspect of quality audits. Implementation is audited by comparing the planned arrangement to observed practices, with an eye towards whether or not 1) the plan is properly implemented and 2) if so, will it accomplish the stated objectives.

BENEFITS OF CONDUCTING QUALITY AUDITS

Audits are conducted to determine if systems, processes, and products comply with documented requirements. The evidence provided from audits forms the basis of improvement in either the element audited, or in the requirements. The benefit derived depends on which type of audit is being conducted: system, process, or product (see II.D).

AUDIT STANDARDS

It is no surprise to find that an activity as important and as common as quality audits are covered by a large number of different standards. It is in the best interest of all parties that the audit activity be standardized to the extent possible. One of the fundamental principles of effective auditing is "no surprises," something easier to accomplish if all parties involved use the same rule book. Audit standards are, in general, guidelines that are voluntarily adopted by auditor and auditee. Often the parties make compliance mandatory by specifying the audit standard as part of a contract. When this occurs, it is common practice to specify the revision of the applicable standard to prevent future changes from automatically becoming part of the contract.

Willborn (1993) reviews eight of the most popular quality audit standards and provides a comparative analysis of these standards in the following areas:

- General features of audit standards
- Auditor (responsibilities, qualifications, independence, performance)
- Auditing organizations (auditing teams, auditing departments/groups)
- Client and auditee
- Auditing (initiation, planning, implementation)
- Audit reports (drafting, form, content, review, distribution)
- Audit completion (follow-up, record retention)
- Quality assurance

It can be seen from the above list that auditing standards exist to cover virtually every aspect of the audit. The reader is encouraged to consult these standards, or Willborn's summaries, to avoid reinventing the wheel.

TYPES OF QUALITY AUDITS

There are three basic types of quality audits: systems, products, and processes. Systems audits are the broadest in terms of scope. The system being audited varies, but the most commonly audited system is the quality system. The quality system is that set of activities designed to assure that the product or service delivered to the end-user complies with all quality requirements. Product audits are performed to confirm that the system produced the desired result. Process audits are conducted to verify that the inputs, actions, and outputs of a given process match the requirements. All of these terms are formally defined in several audit standards; several definitions for each type of audit are provided in the Appendix 2.

Product audits

Product audits are generally conducted from the customer's perspective. ISO 9000:2000 divides products into four generic categories:

1. Hardware
2. Software
3. Processed materials
4. Services

The quality system requirements are essentially the same for all product categories. There are four facets of quality: quality due to defining the product to meet marketplace requirements; quality due to design; quality due to conformance with the design; and quality due to product support. Traditionally, product quality audits were conducted primarily to determine conformance with design. However, modern quality audit standards (e.g., the ISO 9000 family) are designed to determine all four facets of quality.

One purpose of product audit is to estimate the quality being delivered to customers, thus product audits usually take place after inspections have been completed. The audit also provides information about how product quality can be improved. In addition, the audit provides another level of assurance beyond routine inspection.

Product audits differ from inspection in a number of ways. Two fundamental differences exist: 1) audits are broader in scope than inspections and 2) audits go "deeper" than inspections. Inspection normally focuses on a small number of important product characteristics. Inspection samples are selected at random in sizes large enough to produce statistically valid inferences regarding lot quality. Audits, on the other hand, are concerned with the quality being produced by the *system*. Thus, the unit of product is viewed as representing common result of the system which produced it. Audit samples are sometimes seemingly quite small, but they serve the purpose of showing a system snapshot.

Audit samples are examined in greater depth than are product samples, i.e., more information is gathered per unit of product. The sample results are examined from a systems perspective. The examination goes beyond mere conformance to requirements. Minor discrepancies are noted, even if they are not serious enough to warrant rejection. A common practice is to use a weighting scheme to assign "demerits" to each unit of product. Esthetics can also be evaluated by the auditor (e.g., paint flaws, scratches, etc.). Table II.4 presents an example of a publisher's audit of their books.

Table II.4. Book audit results.

SAMPLE SIZE = 1,000 BOOKS				
PROBLEM	SERIOUSNESS	WEIGHT	FREQUENCY	DEMERITS
Cover bent	Major	5	2	10
Page wrinkled	Minor	3	5	15
Light print	Incidental	1	15	15
Binding failure	Major	5	1	5
		TOTAL	23	45

These audit scores are presented on histograms and control charts to determine their distribution and to identify trends. Product audit results are compared to the marketing requirements (i.e., customer requirements) as well as the engineering requirements.

Product audits are often conducted in the marketplace itself. By obtaining the product as a customer would, the auditor can examine the impact of transportation, packaging, handling, storage, and so on. These audits also provide an opportunity to compare the condition of the product to that being offered by competitors.

Process audit

Process audits focus on specific activities or organizational units. Examples include engineering, marketing, calibration, inspection, discrepant materials control, corrective action, etc. Processes are organized, value-added manipulations of inputs which result in the creation of a product or service. Process audits compare the actual operations to the documented requirements of the operations. Process audits should begin with an understanding of how the process is supposed to operate. A process flowchart is a useful tool in helping to reach this understanding (see V.C).

It has been said that a good reporter determines the answer to six questions: who? what? when? where? why? and how? This approach also works for the process auditor. For each important process task ask the following questions:

- Who is supposed to do the job? (Are any credentials required?)
- What is supposed to be done?
- When is it supposed to be done?
- Where is it supposed to be done?
- Why is this task done?
- How is this task supposed to be done?

The documentation should contain the answers to every one of these questions. If it doesn't, the auditor should suspect that the process isn't properly documented. Of course, the actual process should be operated in conformance to documented requirements.

Systems audit

Systems are actually just arrangements of processes. The dictionary definition of a system is a group of interacting, interrelated, or interdependent elements forming a complex whole. Systems audits differ from process audits primarily in their scope. Where process audits focus on an isolated aspect of the system, systems audits concentrate on the relationships *between* the various parts of the system. In the case of quality audits, the concern is with the quality system. The quality system is the set of all activities designed to assure that all important quality requirements are determined, documented, and followed.

The level of requirements for quality systems varies with the type of organization being audited and, perhaps, with the size of the organization. Not every organization is required to have every quality system element. Table II.5 lists the twenty elements of ANSI/ASQ 9001-1994, along with their corresponding five clauses of ANSI/ASQ 9000:2000.

Table II.5. Summary of elements in Q9001-9004. *(Figure II.7 repeated.)*

Stapp, Eric H., (2001). ISO 9000:2000, An essential guide to the new standard. Tucson, AZ: QA Publishing, pg. 29.

FIGURE 2.8: Correspondence between requirements of ISO 9001:2000 and ISO 9001:1994

■ = indicates correspondence between ISO 9001:2000 and ISO 9001:1994

ELEMENTS FROM ISO 9001:1994 \ REQUIREMENTS CLAUSES FROM ISO 9001:2000	4.1 Quality management system general requirements	4.2 Documentation requirement	5.1 Management commitment	5.2 Customer focus	5.3 Quality policy	5.4 Planning	5.5 Responsibility, authority & communication	5.6 Management review	6.1 Provision of resources	6.2 Human resources	6.3 Infrastructure	6.4 Work environment	7.1 Planning of product realization	7.2 Customer-related processes	7.3 Design & development	7.4 Purchasing	7.5 Production & service provision	7.6 Control of monitoring & measuring devices	8.1 Measurement, analysis & improvement	8.2 Monitoring & measurement	8.3 Control of nonconforming product	8.4 Analysis of data	8.5 Improvement
4.1 Management responsibility			■		■	■	■	■	■	■	■												■
4.2 Quality system	■	■	■			■				■			■										
4.3 Contract review				■										■									
4.4 Design control														■	■								
4.5 Document & data control		■																					
4.6 Purchasing																■							
4.7 Control of customer-supplied product																	■						
4.8 Product identification & traceability																	■						
4.9 Process control											■	■	■				■			■			
4.10 Inspection & testing													■			■	■		■	■			
4.11 Control of inspection, measuring & test equipment																		■					
4.12 Inspection & test status																	■						
4.13 Control of nonconforming product																					■		
4.14 Corrective & preventive action																						■	■
4.15 Handling, storage, packaging, preservation & delivery																	■						
4.16 Control of quality records		■																					
4.17 Internal quality audits																				■			
4.18 Training										■													
4.19 Servicing																	■						
4.20 Statistical techniques																			■	■		■	

The ISO 9000 approach determines the applicable requirements by using either a "stakeholder motivated" approach or a "management motivated" approach. In either case, the supplier may use the ANSI/ASQ 9001-1994 and ANSI/ASQ 9000:2000 to understand the basic concepts and elements of a complete quality system. Note that the twenty elements of the 1994 (and preceding) standards are dispersed throughout the 2000 standard, as indicated by Table II.5.

In the stakeholder-motivated approach, the supplier implements a quality system in response to demands by customers, owners, subsuppliers or society at large. Supplier management plays a leadership role, but the effort is driven by external stakeholders. In the management-motivated approach, the supplier's management initiates the effort in anticipation of emerging marketplace needs and trends. ANSI/ASQ 9000 standards are used to guide a quality management approach to installing a quality system that will enhance quality. The management-driven approach normally produces more comprehensive and beneficial quality systems. This realization is reflected in the ANSI/ASQ 9000:2000 standard's requirement for management commitment and leadership of the Quality System (Stapp).

Internal audits

Considering the benefits which derive from quality auditing, it is not surprising that most quality audits are internal activities conducted by organizations interested in self-improvement. Of course, the same principles apply to internal audits as to external audits (e.g., auditor independence). Ishikawa describes four types of internal audits:

1. Audit by the president
2. Audit by the head of the unit (by division head, factory manager, branch office manager, etc.)
3. QC audit by QC staff
4. Mutual QC audit

President's audits are similar to what Tom Peters has called "management by walking around" (MBWA). The president personally goes into different areas of the organization to make firsthand observations of the effectiveness of the quality system. Audit by the head of the unit is the equivalent to the president's audit, except the audit is limited to functional areas under the jurisdiction of the head person. Quality control (QC) audits are conducted by the quality department in various parts of the organization. Unlike presidents and unit heads, who are auditing their own areas, quality department auditors must obtain authorization before conducting audits. In mutual QC audits, separate divisions of the company exchange their audit teams. This provides another perspective from a team with greater independence.

Two-party audits

Most audits are conducted between customers and suppliers. In this case, suppliers usually provide a contact person to work with the customer auditor. In addition, suppliers usually authorize all personnel to provide whatever information the auditor needs, within reason, of course. Two-party audits are generally restricted to those parts of the quality system of direct concern to the parties involved. The customer will evaluate only those processes, products, and system elements that directly or indirectly impact upon their purchases.

Third-party audits

One problem with two-party audits is that a supplier will be subject to audits by many different customers, each with their own (sometimes conflicting) standards. Likewise, customers must audit many different suppliers, each with their own unique approach to quality systems design. Third-party audits are one way of overcoming these difficulties.

In a third-party audit, the auditing organization is not affiliated with either the buyer or the seller. The audit is conducted to a standard that both the buyer and seller accept. By far, the most common standard used in third-party audits is the ISO 9000 series. The ISO 9000 approach to certification of quality systems is discussed in this chapter. As the use of ISO 9000 becomes more widespread, the incidence of third-party audits will continue to increase. However, ISO 9000 audits are conducted at a high system level. Product and process audits will continue to be needed to address specific issues between customers and suppliers.

Quality systems for small businesses

Historically, for small business the basic understanding of process control, sampling, calibration techniques, material traceability and corrective-action has been less critical than in major industry. However, this situation is changing rapidly. Even very small businesses are increasingly expected to demonstrate the existence of sound quality systems. There are special considerations involved when implementing quality systems in small businesses. The small business manager must develop a quality program that meets customer requirements while controlling and maintaining product conformance.

Evaluation of the small suppliers' quality system begins with a review of the quality manual. In the small company, it may be only a few typed pages, but it must accurately represent the quality system being used. Basic items for a small supplier audit include (ASQC 1981) the following:

- Drawing and specification control
- Purchased material control/receiving inspection
- Manufacturing control/in-process inspection/final inspection
- Gage calibration and test equipment control
- Storage, packaging and shipment control
- Nonconforming material control
- Quality system management

During the audit the supplier should demonstrate his procedures in each of these areas.

AUDITOR QUALIFICATIONS

In this section we discuss some of the qualifications of quality auditors. Willborn (1993, 11–23) provides an extensive discussion of auditor qualifications.

The first requirement for any auditor is absolute honesty and integrity. Auditors are often privy to information of a proprietary or sensitive nature. They sometimes audit several competing organizations. The information an auditor obtains must be used only for the purpose for which it was intended. It should be held in strict confidence. No amount of education, training, or skill can compensate for lack of ethics.

The auditor must be independent of the auditee. In addition, auditors must comply with professional standards, possess essential knowledge and skills, and must maintain technical competence. Auditors must be fair in expressing opinions and should inspire the confidence of both the auditee and the auditor's parent organization.

The auditor acts as only an auditor and in no other capacity, such as management consultant or manager. Managers of audit organizations should have a working knowledge of the work they are supervising.

An auditor's qualifications must conform to any applicable standards and they must be acceptable to all parties. The auditing organization should establish qualifications for auditors and provide training for technical specialists. Some auditing activities, such as for Nuclear Power Plants, require special certification. Lead auditors require additional training in leadership skills and management. Third parties provide certification of auditors.

Auditors should be able to express themselves clearly and fluently, both orally and in writing. They should be well-versed in the standards in which they are auditing. Where available, this knowledge should be verified by written examination and/or certification.

Auditors should master the auditing techniques of examining, questioning, evaluating and reporting, identifying methods, following up on corrective action items, and closing out audit findings. Industrial quality auditors should have knowledge of design, procurement, fabrication, handling, shipping, storage, cleaning, erection, installation, inspection, testing, statistics, nondestructive examinations, maintenance, repair, operation, modification of facilities or associated components, and safety aspects of the facility/process. In a specific audit assignment, the knowledge of individual auditors might be complemented by other audit team members.

SCHEDULING OF AUDITS

Most quality audits are pre-announced. There are many advantages to pre-announcing the audit. A pre-announced audit is much less disruptive of operations. The auditee can arrange to have the right people available to the auditor. The auditor can provide the auditee with a list of the documentation he or she will want to review so the auditee can make it available. Much of the documentation can be reviewed prior to the on-site visit. The on-site visit is much easier to coordinate when the auditee is informed of the audit. Finally, pre-announced audits make it clear that the audit is a cooperative undertaking, not a punitive one.

Of course, when deliberate deception is suspected, surprise audits may be necessary. Surprise audits are usually very tightly focused and designed to document a specific problem. In most cases, quality auditors are not trained to conduct adversarial audits. Such audits are properly left to accounting and legal professionals trained in the handling of such matters.

Audits can be scheduled at various points in the buying cycle. The following timing of audits are all quite common:

Pre-award audit—Conducted to determine if the prospective supplier's quality system meets the customer's requirements.

Surveillance audit—Conducted to assure that an approved supplier's quality system continues to comply with established requirements.

Contract renewal—Conducted to determine if a previously approved supplier continues to meet the quality system requirements.

Problem resolution—A tightly focused audit conducted to identify the root cause of a problem and to assure that effective corrective action is taken to prevent future occurrences of the problem.

In-process observation—On-site audits performed to assure that processes are performed according to established requirements. These audits are often performed when it is difficult or impossible to determine whether or not requirements have been met by inspecting or testing the finished product.

At times, periodic audits are automatically scheduled. For example, to maintain certification to the ISO 9000 series standards, the organization is periodically reassessed (see II.C for additional information on the registration process).

DESK AUDITS

The emphasis of the discussion above has been on the on-site visit. However, a significant portion of the auditing activity takes place between the auditor and auditee each working at their respective organizations. A great deal of the audit activity involves examination of documentation. The documentation reveals whether or not a quality system has been developed. It describes the system as the supplier wants it to be. From a documentation review the auditor can determine if the quality system, as designed, meets the auditor's requirements. If not, a preliminary report can inform the auditee of any shortcomings. Corrective action can be taken to either modify the documentation or to develop new system elements. Once the documentation is in a form acceptable to the auditor, an on-site visit can be scheduled to determine if the system has been properly implemented. Properly done, desk audits can save both auditor and auditee significant expense and bother.

STEPS IN CONDUCTING AN AUDIT

Most quality systems audits involve similar activities. The list below is an adaptation of the basic audit plan described by Keeney (1995).

- Choose the audit team. Verify that no team member has a conflict of interest.
- Meet with the audit team and review internal audit procedures.
- Discuss forms to be used and procedures to be followed during the audit.
- Perform a desk audit of the quality manual and other documentation to verify the scope of the audit and provide an estimate of the duration of the audit.
- Assign audit subteams to their respective audit paths.
- Contact the auditee and schedule the audit.
- Perform the audit.
- Write corrective action requests (CARs) and the audit summary report, listing the CARs in the audit summary.

- Conduct a closing meeting (exit briefing).
- Issue the audit summary report.
- Present the complete audit findings, including all notes, reports, checklists, CARs, etc., to the quality manager.
- Prepare a final audit report.
- Follow up on CARs.

Keeney also provides a "kit" to aid the auditor in planning and conducting the audit, including forms, worksheets, procedures, and so on.

AUDIT REPORTING PROCESS

Audit results are reported while the audit is in progress and upon completion of the audit. The principle is simple: the auditor should keep the auditee up-to-date at all times. This is a corollary of the "no surprises" principle*. In addition, it helps the auditor** avoid making mistakes by misinterpreting observations; in general, the auditee is better informed about internal operations than the auditor.

Auditees should be informed before, during, and after the audit. Prior to the audit, the auditee is told the scope, purpose, and timing of the audit and allowed to play an active role in planning the audit. Upon arrival, the auditor and auditee should meet to review plans and timetables for the audit. Oral daily briefings should be made, presenting the interim results and tentative conclusions of the auditor. At these meetings, the auditee is encouraged to present additional information and clarification to the auditor. Written minutes of these meetings should be maintained and published. Upon completion of the audit an exit briefing is recommended.

As a matter of courtesy, an interim report should be issued as soon as possible after completion of the audit. The interim report should state the main findings of the auditor and the auditor's preliminary recommendations.

Formal audit reports are the ultimate product of the audit effort. Formal audit reports usually include the following items:

- Audit purpose and scope
- Audit observations
- Conclusions and recommendations
- Objectives of the audit
- Auditor, auditee, and third-party identification
- Audit dates
- Audit standards used
- Audit team members
- Auditee personnel involved

*This discussion presumes that we are discussing a typical audit, not an audit where deliberate wrongdoing is suspected.

**The term "auditor" is taken to include individual auditors and members of audit teams as well.

- Statements of omission
- Qualified opinions
- Issues for future audits
- Auditee comments on the report (if applicable)
- Supplementary appendices

The audit report is the "official document" of the audit. Audit reports should be prepared in a timely fashion. Ideally, the deadline for issuing the report should be determined in the audit plan prepared beforehand. Audit reports should describe the purpose and scope of the audit, the entity audited, the membership of the audit team including affiliation and potential conflicts of interest, the observations of the audit, and recommendations. Detailed evidence supporting the recommendations should be included in the report. Audit reports may include recommendations for improvement. They may also include acknowledgment of corrective action already accomplished. The formal report may draw upon the minutes of the meetings held with the auditee. It should also note the auditee's views of previously reported findings and conclusions.

Auditor opinions are allowed, but they should be clearly identified as opinions and supporting evidence should be provided.

Anyone who has been an auditor for any time knows that it is sometimes necessary to report unpleasant findings. These results are often the most beneficial to the auditee, providing information that in-house personnel may be unwilling to present. When presented properly by an outside auditor, the "bad news" may act as the catalyst to long-needed improvement. However, the auditor is advised to expect such reactions as denial, anger, and frustration when the findings are initially received. In general, unpleasant findings should be supported more extensively. Take care in wording the findings so that the fewest possible emotional triggers are involved. The sooner the auditee and auditor can begin work on correcting the problems, the better for both parties.

If a report is prepared by more than one person, one person will be designated as senior auditor. The senior auditor will be responsible for assembling and reviewing the completed report. If the audit was conducted by a second party, distribution of the report will be limited to designated personnel in the auditor and auditee's organizations, usually senior management. If a third-party audit was conducted, the audit report will also be sent to the client, who in turn is responsible for informing the auditee's senior management. In certain cases (e.g., government audits, financial audits), the results are also made available to the public. However, unless otherwise specified, audit reports are usually considered to be private, proprietary information that cannot be released without the expressed written permission of the auditee.

Bases of measurement

Simple counts of discrepancies and problems are generally meaningless unless placed in a context. The measurement base provides this context by providing an indication of the number of opportunities for problems to appear, the counts are then compared to the base. For example, if records are being audited, the base might be the percentage of records with problems. Here are some other problems/bases:

- Data entry errors per 1,000 records entered
- Planning errors per 100 pages of planning
- Clerical errors per 100 discrepancy reports
- Percentage of material batches incorrectly identified
- Percentage of out-of-date quality manuals

Audit reports often contain descriptions of problems and discrepancies encountered during the audit. However, not all problems are equal, some are more serious than others. A well-written audit report will classify the problems according to how serious they are. Product defect seriousness classification schemes are discussed in section III.A. Some organizations also apply seriousness classification to discrepancies found in planning, procedures and other areas. These seriousness classifications (e.g., "critical," "major," "minor") should be explicitly defined and understood by all parties. Generally, some sort of weighting scheme is used in conjunction with the classification scheme and an audit score is computed (e.g., Table II.4). Although the audit score contains information, it should not be the sole criterion for deciding whether or not the audit was passed or failed. Such "mindlessness" is generally counterproductive. Instead, consider the numerical data as additional input to assist in the decision-making process.

POST-AUDIT ACTIVITIES (E.G., CORRECTIVE ACTION, VERIFICATION)

Most audits are not pass/fail propositions. Rather, they represent an effort on the part of buyers and sellers to work together for the long term. When viewed in this light, it is easy to see that identifying a problem is just the first step. Solving the problem requires locating the root cause of the problem. This is not a trivial matter. Identifying the root cause(s) of problems is challenging work. Many times the problem is treated as if *it* were a cause, i.e., action is taken to "manage the problem" rather than addressing its cause. Examples of this are inspection to remove defects or testing software to catch bugs. Wilson, Dell and Anderson (1993, 9) define *root cause* as that most basic reason for an undesirable condition or problem which, if eliminated or corrected, would have prevented it from existing or occurring. Root causes are usually expressed in terms of specific or systematic factors. A root cause usually is expressed in terms of the least common organizational, personal, or activity denominator.

In most cases the auditor is not capable of identifying the root cause of a problem. The auditee is expected to perform the necessary analysis and to specify the action taken to address the cause(s) of the problem. At this point the auditor can sometimes determine that the root cause has not been identified and can assist the auditee in pursuing the problem at a deeper level. At other times there is no choice but to validate the corrective action by additional audits, tests or inspections. The final proof of the effectiveness of any corrective action must be in achieving the desired result.

Preventing problems

A major benefit of quality auditing is the prevention of problems. Audits often uncover situations that are still acceptable but trending toward an eventual problem. Product audits, for example, monitor minor and even incidental problems as well as more serious problems. Without the attention of management brought on by an unfavorable audit report, noncompliance tends to increase. In quality cost accounting, quality audits are properly classified as an appraisal cost, but preventing problems is the objective.

Quality audit versus quality survey

Evaluation of processes, products, and quality systems should not be limited to the performance of quality audits. Quality audits are formal, structured evaluations involving independent auditors. Quality surveys are less formal reviews of quality systems, products, or processes. Unlike audits, quality surveys are often conducted by non-independent survey personnel who communicate their findings to their own management. The purpose of a quality survey is informational, rather than being focused on supplier selection or certification. Formal reports are generally not prepared. Rather, the survey results are presented in information-sharing sessions with concerned personnel. Quality surveys are often conducted prior to quality audits to determine if the organization is ready for the audit.

Planning, Controlling, and Assuring Product and Process Quality

III.A PROCESSES FOR PLANNING PRODUCT AND SERVICE DEVELOPMENT

All but the most simple products contain very large numbers of features. In theory, every feature of every unit produced could be inspected and judged against the requirements. However, this would add considerable cost to the product while, for most features, adding little or no value to the customer. The producer is faced with the need for establishing a hierarchy of importance for the various characteristics of the product. Which features are so important that they deserve a great deal of attention? Which need only a moderate amount of attention? Which need only a cursory inspection? The activity of arriving at this determination is known as classification of characteristics.

In practice, characteristics are usually classified into the categories *critical, major*, and *minor*. The terms can be defined in simple terms as follows:

Critical characteristic—Any feature whose failure can reasonably be expected to present a safety hazard either to the user of the productor to anyone depending on the product functioning properly.

Major characteristic—Any feature, other than critical, whose failure would likely result in a reduction of the usability of the product.

Minor characteristic—Any feature, other than major or critical, whose failure would likely be noticeable to the user.

Incidental characteristic—Any feature other than critical, major, or minor.

Of course, it is possible to develop classification schemes that are more detailed. However, the above definitions suffice for the vast majority of applications. Most often classifications of critical characteristics are noted on the drawing as well as in the manufacturing plan, as well as in such other ways as to give the user ample warning of potential hazards.

One excellent method of identifying characteristic classifications is failure mode, effects, and criticality analysis (FMECA), see IV.F.

CLASSIFICATION OF CHARACTERISTICS VERSUS CLASSIFICATION OF DEFECTS

A classification of defects is the enumeration of possible defects of the unit of product classified according to their seriousness.

The following definitions are from Mil-Std-105E:

Defect—Any nonconformance of the unit of product with specified requirements.

Defective—A product with one or more defects.

Critical defect—A critical defect is a defect that judgment and experience indicate would result in hazardous or unsafe conditions for individuals using, maintaining, or depending upon the product or a defect that judgment and experience indicate is likely to prevent performance of the tactical function of a major end item such as a ship, aircraft, tank, missile, or space vehicle.

Critical defective—A critical defective is a unit of product which contains one or more critical defects and may also contain major and/or minor defects.

Major defect—A major defect is a defect, other than critical, that is likely to result in failure or to reduce materially the usability of the unit of product for its intended purpose.

Major defective—A major defective is a unit of product which contains one or more major defects and may also contain minor defects but contains no critical defects.

Minor defect—A minor defect is a defect that is not likely to reduce materially the usability of the unit of product for its intended purpose or is a departure from established standards having little bearing on the effective use or operation of the unit.

Minor defective—A minor defective is a unit of product which contains one or more minor defects but contains no critical or major defect.

DESIGN REVIEW AND QUALIFICATION

A great deal of what we learn comes from experience. The more we do a thing, the more we learn about doing it better. As a corollary, when something is new or untried we tend to make more mistakes. Design review and qualification is performed to apply the lessons learned from experience with other products and projects to the new situation. The objective is to introduce the new item with a minimum of startup problems, errors, and engineering changes. This involves such activities as:

- Locating qualified suppliers
- Identifying special personnel, equipment, handling, storage, quality and regulatory requirements

- Providing information to marketing for forecasting, promotional, and public-relations purposes.

The design review and qualification activity is usually performed after the development of an acceptable prototype and before full-scale production.

Design review often takes place in formal and informal meetings involving manufacturing, quality, and engineering personnel. In some cases, customer personnel are also present. The meetings involve the discussion of preliminary engineering drawings and design concepts. The purpose is to determine if the designs can be produced (or procured) and inspected within the cost and schedule constraints set by management. If not, one of two courses of action must be taken: 1) change the design or 2) acquire the needed production or inspection capabilities. The design review is commonly where critical and major characteristics are identified. This information is used to design functional test and inspection equipment, as well as to focus manufacturing and quality efforts on high-priority items. Formal Failure Mode, Effects and Criticality Analysis (FMECA) and Fault Tree Analysis (FTA) is also performed to assist in identification of important features.

When feasible, a pilot run will be scheduled to confirm readiness for full-scale production. Pilot runs present an excellent opportunity for process capability analysis (PCA) to verify that the personnel, machines, tooling, materials and procedures can meet the engineering requirements. The pilot run usually involves a small number of parts produced under conditions that simulate the full-scale production environment. Parts produced in the pilot run are subject to intense scrutiny to determine any shortcomings in the design, manufacturing, or quality plans. Ideally, the pilot run will encompass the entire spectrum of production, from raw materials to storage to transportation, installation and operation in the field.

Properly done, design review and qualification will result in a full-scale production plan that will minimize startup problems, errors, and engineering changes after startup. The production plan will include error-free engineering drawings, a manufacturing plan and a quality plan. See also chapter II.

PROCESS QUALIFICATION AND VALIDATION METHODS

Process qualification and validation are primarily control issues. One objective is to identify those processes that are capable of meeting management and engineering requirements, if properly controlled. Another objective is to assure that processes are actually performing at the level which they are capable of performing. This requires that process capability be analyzed using statistical methods and that products are produced only on those processes capable of holding the required tolerances. These issues are discussed in detail in VI.G, and VI.H. Process validation is accomplished by using SPC (see VI.G), quality audit (see II.D), acceptance sampling (see III.C) and inspection.

INTERPRETATIONS OF DIAGRAMS, SCHEMATICS, DRAWINGS, OR BLUEPRINTS

The quality engineer will be called upon to deal with technical issues relating to product quality. In most cases, the scope of the quality engineer's job is broad, extending to many different areas of manufacturing. Many quality engineers move from industry to industry as their careers progress. Therefore quality engineers should be able to demonstrate some familiarity with the technical "language" of the various trades. What is presented here is a brief overview of the different types of engineering drawings.

Reading standard engineering drawings and blueprints

Engineering drawings are a means used by engineers and designers to communicate their requirements to others. Engineering drawings are akin to a legal document and they are often used as such. A blueprint is a copy of an engineering drawing, so named because the process used in the past to produce them resulted in blue lines or a blue background. Blueprint "reading" involves visualizing and interpreting the information contained on the blueprint. Visualization is the ability to mentally "see" the three-dimensional object represented by the drawing. Interpretation involves the ability to understand the various lines, symbols, dimensions, notes and other information on the drawing.

Blueprints are idealized, two-dimensional pictures of something to be created as a physical, three-dimensional reality. The fundamental problem of mechanical drafting is to completely describe the three-dimensional object using only two dimensions. (In fact, it is not even possible to see a real three-dimensional object all at once without the use of mirrors or some other aid.) Engineering drawings deal with this by presenting multiple "views" of the object, and by establishing conventions for showing parts of the object that are "hidden" from view.

To understand the different views on a blueprint, visualize a box placed before you on a table. Each side of the box has a word printed on it which describes the view from your perspective, i.e., the front of the box is labeled "Front" and so on. This approach is shown in Figure III.1.

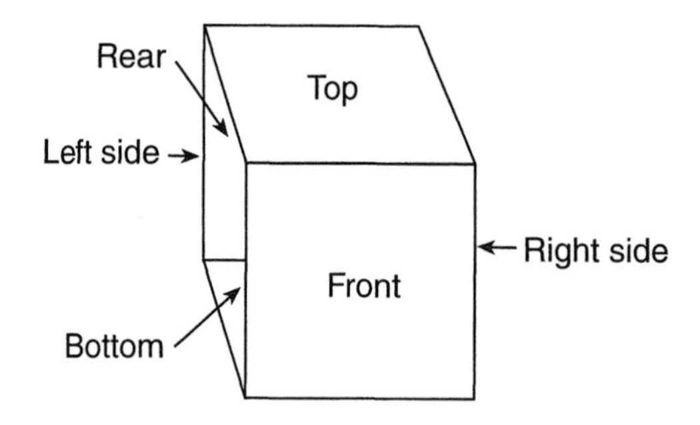

Figure III.1. Blueprint views #1.

Now, imagine cutting this box and folding the various sides out flat. It will then be possible to see every surface in two dimensions. Essentially, this is how, on a blueprint a 3-D object is converted to a series of two-dimensional views, which are known as plan views. This is illustrated in Figure III.2.

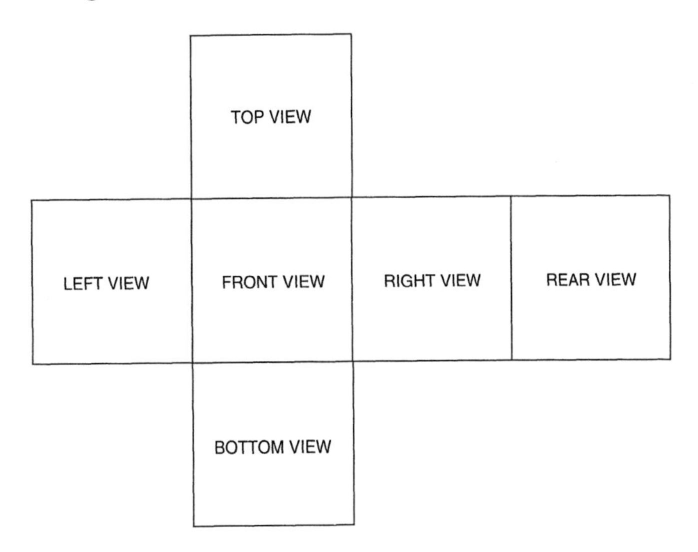

Figure III.2. Blueprint views #2.

Figure III.3 shows how this approach works with a part.

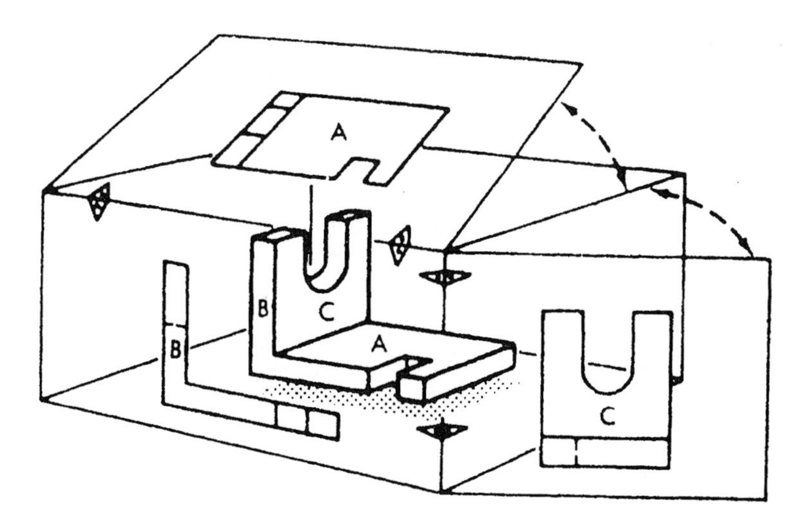

Figure III.3. Plan views of a part.

The language of lines

A great deal of blueprint information is conveyed by using a variety of lines. There are 12 types of lines commonly used on blueprints.

1. **Visible lines**—A visible line represents all edges or surfaces that are visible in a given view. Visible lines are shown as thick, continuous lines.

THICK

Figure III.4. Visible line.

2. **Hidden lines**—When, in the opinion of the engineer or draftsman, there is a need to show a surface or edge that cannot be seen in a given view, hidden lines are used. Hidden lines are shown as a dashed line of medium weight.

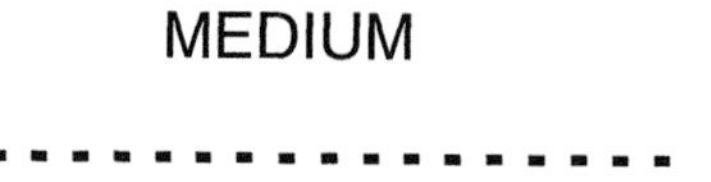

Figure III.5. Hidden line.

3. **Section lines**—Section lines are drawn to "fill" a surface that represents a sectional view. A sectional view is obtained when you are looking at a part as if it were cut open in a particular location.

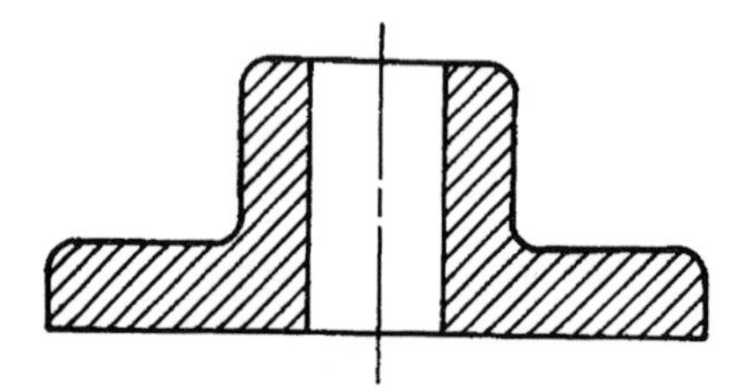

Figure III.6. Section line and sectional view.

From *Blueprint Reading for Industry*, Figure 3-2. Copyright © 1979 by Goodheart-Willcox Co., Inc.

4. **Center lines**—Center lines are thin lines used to designate centers of holes, arcs, and symmetrical objects.

THIN

Figure III.7. Center line.

5. **Dimensions and notes**—Dimensions and notes are shown by using a combination of three different types of lines: dimension lines, extension lines and leaders.

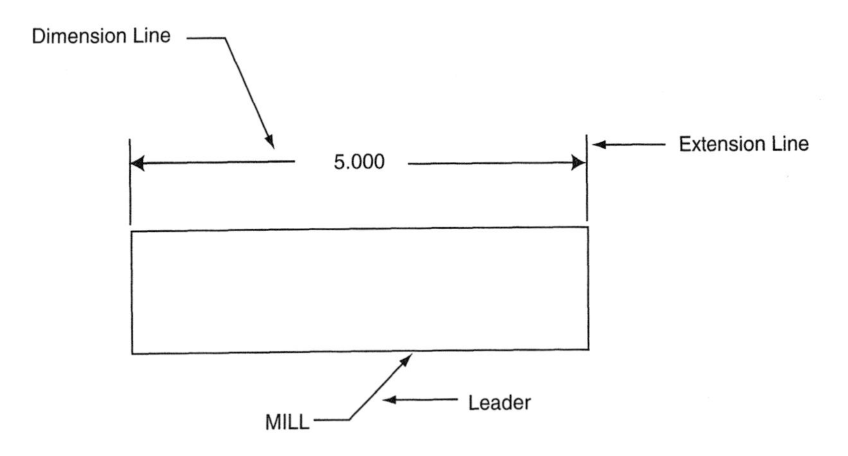

Figure III.8. Dimension, extension, and leader lines.

6. **Dimension and extension lines**—Dimension and extension lines are thin lines used to indicate dimensions of parts. Extension lines show the surfaces which are being measured. Dimension lines are normally terminated with arrowheads which point to the extension lines. Leaders are thin lines used to indicate drawing areas to which dimensions or notes apply.

7. **Cutting plane lines**—Also called viewing plane lines, cutting plane lines are used to indicate views obtained by mentally "cutting" a part with a plane. Cutting planes are used because it is sometimes impossible otherwise to properly show all of the various perspectives needed to understand the designer's intent.

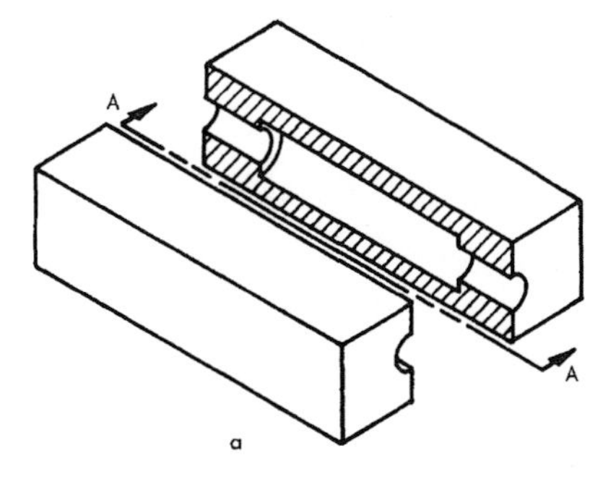

Figure III.9. Cutting plane lines.

Continued on next page . . .

Figure III.9—*Continued . . .*

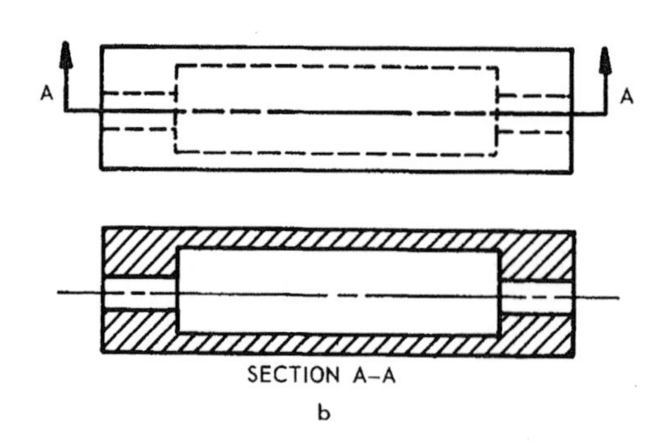

Figure III.9. Cutting plane lines.

8. **Break lines**—Break lines are used for shortening the representation of parts with large identical segments. By using break lines, the draftsman can preserve the scale of the remainder of the part and clearly show details drawn to scale.
9. **Long breaks**—Long breaks are shown as thin lines with sharp angular portions.

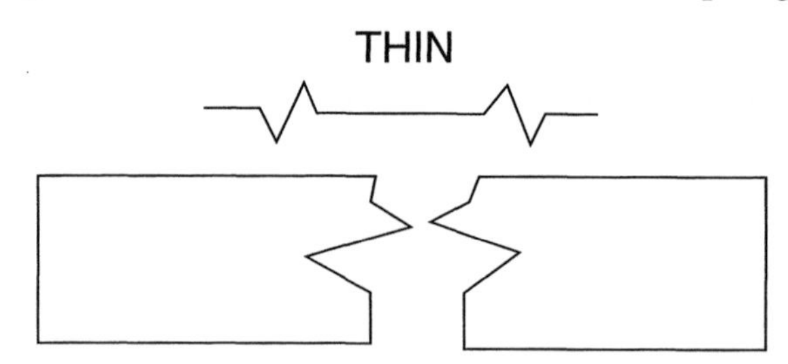

Figure III.10. Long break lines.

10. **Short breaks**—short breaks are shown as thick, wavy lines. Note that some draftsmen use this convention for both short and long breaks.

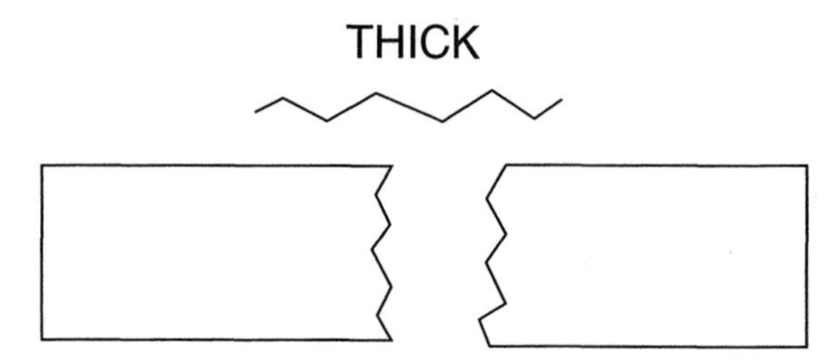

Figure III.11. Short break lines.

11. **Cylindrical break lines**—Break lines in round stock are shown using curved lines.

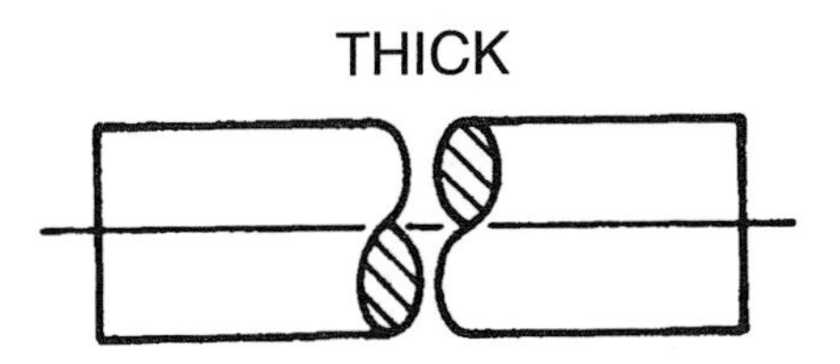

Figure III.12. Cylindrical break lines.

From *Blueprint Reading for Industry*, Figure 3-6c. Copyright © 1979 by Goodheart-Willcox Co., Inc.

12. **Phantom lines**—Phantom lines are thin lines composed of long dashes alternating with pairs of short dashes. Phantom lines show alternate positions of moving parts, adjacent positions of related parts, or repeated detail. Phantom lines are also used to show reference lines, points, or surfaces from which dimensions are measured. The references shown by phantom lines may not actually exist on the physical part.

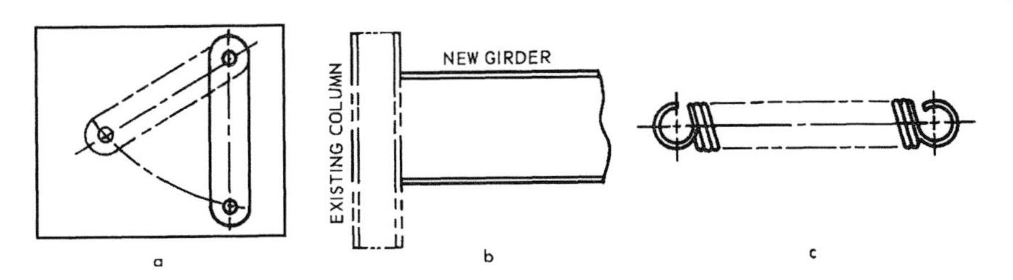

Figure III.13. Phantom lines.

From *Blueprint Reading for Industry*, Figure 3-7. Copyright © 1979 by Goodheart-Willcox Co., Inc.

The drawing below illustrates the conventions described above.

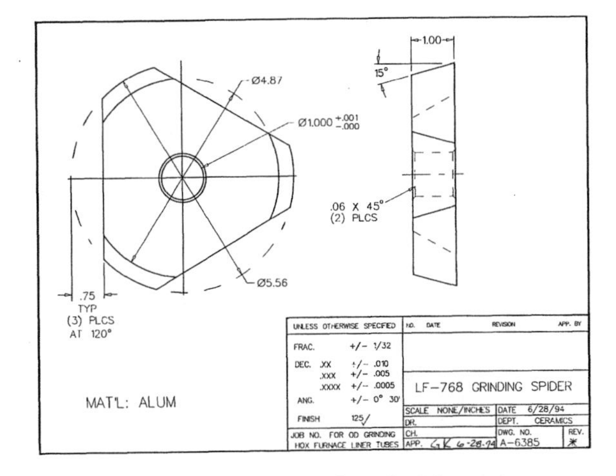

Figure III.14. Example of a blueprint.

Geometric dimensioning and tolerancing

The traditional method of indicating tolerances on engineering drawings suffers from a number of problems. For example, if a round pin is designed to fit through a round hole, the traditional drawing method cannot provide all of the available tolerance because it produces tolerance zones which are either square or rectangular rather than circular. An alternative method of dimensioning and tolerancing was developed to address these shortcomings. Known as *geometric dimensioning and tolerancing*, the new method is a radical departure from the traditional method. Geometric dimensioning and tolerancing incorporates new concepts which are operationalized with a special set of drawing conventions and symbols. The approach is documented in the standard *ANSI Y14.5M: Dimensioning and Tolerancing*. A complete discussion of this standard is beyond the scope of this book; however, a brief overview of the most relevant subjects will be covered.

ANSI Y14.5M describes standardized methods of showing size, form and location of features on mechanical drawings. It does not provide information on how to inspect these features. An understanding of the standard begins with knowing the following terms: (ANSI Y14.5M–1994).

Dimension—A numerical value expressed in appropriate units of measure and indicated on a drawing and in other documents along with lines, symbols, and notes to define the size or geometric characteristic, or both, of a part or part feature.

Basic dimension—A numerical value used to describe the theoretically exact size, profile, orientation, or location of a feature or datum target. It is the basis from which permissible variations are established by tolerances on other dimensions, in notes, or in feature-control frames. Basic dimensions are shown enclosed in boxes, as shown in Figure III.15.

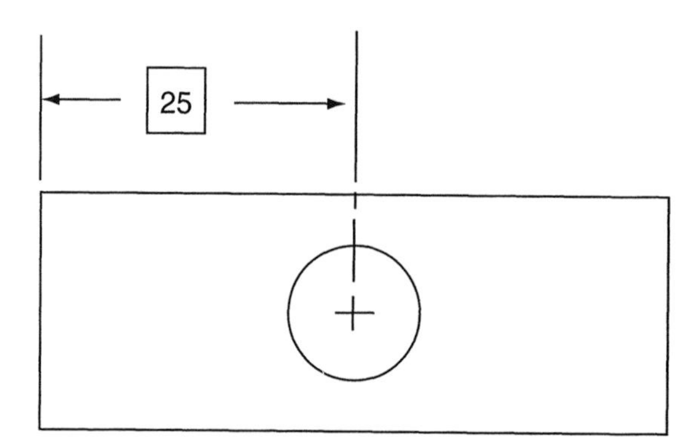

Figure III.15. Basic dimension.

True position—The theoretically exact location of a feature established by basic dimensions.

Reference dimension—A dimension, usually without tolerance, used for information purposes only. It is considered auxiliary information and does not govern production or inspection operations. A reference dimension is a repeat of a dimension or is derived from other values shown on the drawing or on related drawings.

Datum—A theoretically exact point, axis, or plane derived from the true geometric counterpart of a specified datum feature. A datum is the origin from which the location or geometric characteristics of features of a part are established.

Datum target—A specified point, line, or area on a part used to establish a datum.

Feature—The general term applied to a physical portion of a part, such as a surface, hole, or slot.

Feature of size—One cylindrical or spherical surface, or a set of two plane parallel surfaces, each of which is associated with a size dimension.

Datum feature—An actual feature of a part that is used to establish a datum.

Actual size—The measured size.

Limits of size—The specified maximum and minimum sizes.

Maximum Material Condition (MMC)—The condition in which a feature of size contains the maximum amount of material within the stated limits of size; for example, minimum hole diameter, maximum shaft diameter.

Least Material Condition (LMC)— The condition in which a feature of size contains the minimum amount of material within the stated limits of size; for example, maximum hole diameter, minimum shaft diameter.

Regardless of Feature Size (RFS)—The term used to indicate that a geometric tolerance or datum reference applies at any increment of size of the feature within its size tolerance.

Virtual condition—The boundary generated by the collective effects of the specified MMC limit of size of a feature and any applicable geometric tolerances.

Tolerance—The total amount by which a specific dimension is permitted to vary. The tolerance is the difference between the maximum and minimum limits.

Unilateral tolerance—A tolerance in which variation is permitted in one direction from the specified dimension.

Bilateral tolerance—A tolerance in which variation is permitted in both directions from the specified dimension.

Geometric tolerance—The general term applied to the category of tolerances used to control form, profile, orientation, location, and runout.

Full Indicator Movement (FIM)—The total movement of an indicator when appropriately applied to a surface to measure its variations.

ANSI Y14.5M uses the symbols shown in Figure III.16.

	TYPE OF TOLERANCE	CHARACTERISTIC	SYMBOL	SEE:
FOR INDIVIDUAL FEATURES	FORM	STRAIGHTNESS	—	6.4.1
		FLATNESS	▱	6.4.2
		CIRCULARITY (ROUNDNESS)	○	6.4.3
		CYLINDRICITY	⌭	6.4.4
FOR INDIVIDUAL OR RELATED FEATURES	PROFILE	PROFILE OF A LINE	⌒	6.5.2 (b)
		PROFILE OF A SURFACE	⌓	6.5.2 (a)
FOR RELATED FEATURES	ORIENTATION	ANGULARITY	∠	6.6.2
		PERPENDICULARITY	⊥	6.6.4
		PARALLELISM	//	6.6.3
	LOCATION	POSITION	⌖	5.2
		CONCENTRICITY	◎	5.11.3
	RUNOUT	CIRCULAR RUNOUT	↗*	6.7.2.1
		TOTAL RUNOUT	⌰*	6.7.2.2

*Arrowhead(s) may be filled in.

TERM	SYMBOL
AT MAXIMUM MATERIAL CONDITION	Ⓜ
REGARDLESS OF FEATURE SIZE	Ⓢ
AT LEAST MATERIAL CONDITION	Ⓛ
PROJECTED TOLERANCE ZONE	Ⓟ
DIAMETER	Ø
SPHERICAL DIAMETER	SØ
RADIUS	R
SPHERICAL RADIUS	SR
REFERENCE	( )
ARC LENGTH	⌒

Figure III.16. Geometric characteristics symbols.

From *ANSI Y14.5M: Dimensioning and Tolerancing.* 1994.

The symbols are generally shown on the engineering drawing in a feature control frame, as shown in Figure III.17.

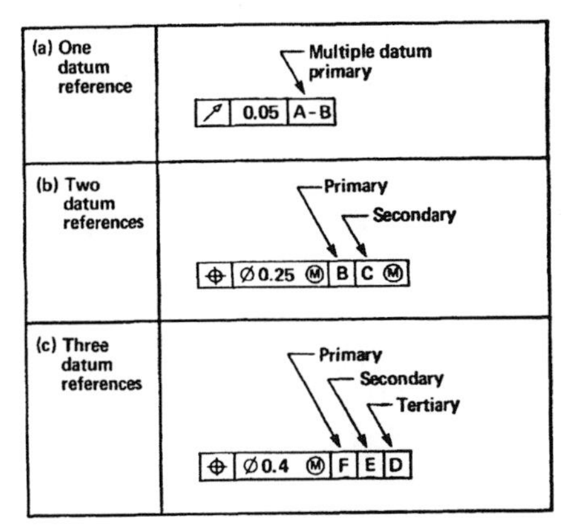

Figure III.17. Feature control frame.
From *ANSI Y14.5M: Dimensioning and Tolerancing.* 1994.

Feature control frames are read as shown in Figure III.18.

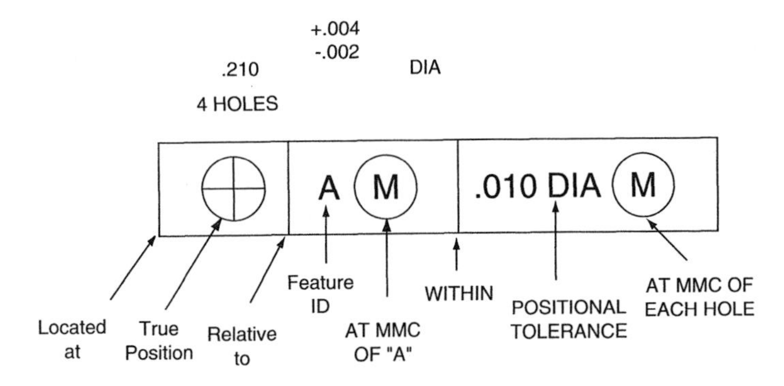

Figure III.18. Feature control frame example.

The above is interpreted as follows: a pattern of 4 holes (whose size can be between .208 and .214 inches in diameter) are to be positioned, as a group, relative to a feature identified as datum A. The tolerance applies at the MMC of datum A. The positional tolerance of .010 is applicable to each of the 4 holes, relative to each other, at the MMC of each hole.

Schematics

Schematics are engineering drawings of electrical circuits. Normally, schematics show the layout of printed circuit boards (PC boards). PC board schematics are essentially two-dimensional drawings; multi-layer boards exist, but schematics of these boards usually show only the discrete components on the surface(s) of the boards. While the quality engineer working in an electronic industry will need to know a great deal about schematics, the typical quality engineer should, at a minimum, know the symbols shown in Figure III.19.

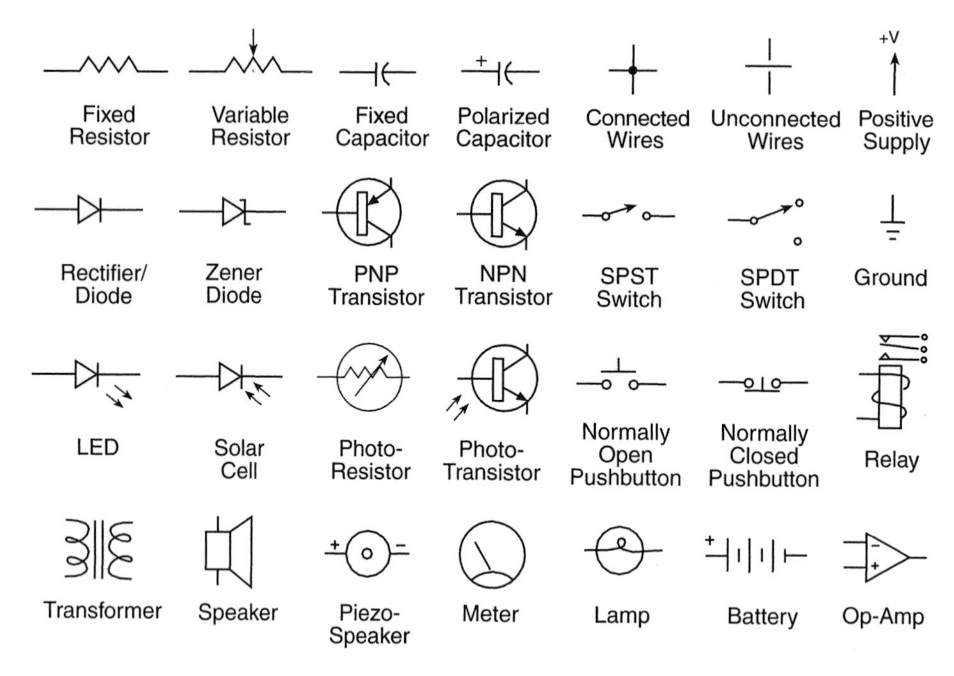

Figure III.19. Circuit symbols.

III.B MATERIAL CONTROL

IDENTIFICATION OF MATERIALS AND STATUS

Has this been inspected? If so, was it accepted? Rejected? Does it require rework? Reinspection? Retest? Obtaining clear answers to these questions is a primary task in quality control. Virtually all quality systems standards and specifications require that systems which identify the status of purchased materials, customer-supplied materials, production materials, work-in-progress, and finished goods be developed, well-documented and fully implemented.

Sample quality manual entry

4.0 Manufacturing control
  [Sections 4.01–4.07 omitted]

4.08 Identification of inspection status
  [subsections 1–2 omitted]

3.0 Procedure

- 3.1 Inspection and test status will be indicated by means of the inspection acceptance stamp.
- 3.2 Inspection stamps are designed to identify both Acme Corporation and individual inspectors and engineers to whom the stamp is assigned.
- 3.3 The stamp impressions shown are maintained by the Quality Control Department.
  - a. Inspection acceptance stamp is used to indicate inspection acceptance of articles that conform to applicable requirements. [stamp shown]
  - b. Inspection rejection stamp is used to indicate rejection of non-conforming articles. [stamp shown]
  - c. Materials review stamp—Withheld—is used to indicate discrepant material to be referred to the material review board. [stamp shown]
  - d. Material review stamp—Accepted—is used to indicate acceptance of previously withheld material. [stamp shown]
  - e. Material review stamp—Rejected—is used to indicate unusable discrepant material to be scrapped or salvaged. [stamp shown]
  - f. Functional test stamp—when required by contract or purchase order, the Functional Test Stamp shall be applied to each article or its identification tag. This indicates that Inspection accepts the article as having successfully met the applicable functional test requirements.

PURCHASED MATERIALS

Proper identification of purchased materials begins, of course, with the supplier. Supplier quality management was discussed in I.I. A key part of the supplier's quality system must include the identification of materials and status discussed below. Once received, the quality status of purchased materials should be identified in accordance with documented procedures. The procedures should cover how purchased material will be identified (e.g., status tags), where the materials are to be stored until conformance to requirements has been established, how non-conforming material will be identified and stored, and how to process non-conforming purchased materials.

Sample quality manual entry

4.0 Manufacturing control

4.01 Receiving inspection

[subsections 1–4 omitted]

5.0 Receiving inspection control center

5.1 Receiving inspector will determine that all information required accompanies material or lot.

5.2 Material or lots found defective will be logged in, on Vendor Conformance Record, Form xxx, and determination of type of inspection will be made.

5.3 Parts and materials having characteristics that are unable to be examined during receiving inspection, but which otherwise conform and which can be determined to be acceptable during subsequent stages of assembly, can be forwarded to stock after stamping all documents including the Identification Tag with the following stamp: [stamp is shown]

CUSTOMER-SUPPLIED MATERIALS

Procedures must be developed and documented for the control of verification, storage, and maintenance of customer-supplied product provided for incorporation into the supplies or for related activities. The procedures must assure that product that is lost, damaged, or otherwise unsuitable for use is recorded and reported to the customer.

Sample quality manual entry

5.0 Coordinated customer/contractor actions

5.01 Control of customer furnished property

[subsections1–3 omitted]

4.0 Customer furnished property

4.1 Receiving Inspection will assure material is examined for completeness and identification. Visually examine for damage caused during transit.

4.2 Rejected material shall be identified as Customer-furnished material.

4.3 Customer-furnished material shall only be used on the contract for which it is furnished.

WORK-IN-PROGRESS (WIP)

Procedures for the identification of the inspection and test status of all WIP should be developed and documented. The identification of inspection and test status should be part of the quality plan covering the entire cycle of production. The purpose of the procedures is to assure that only product that has passed the necessary inspection and test operations is delivered. WIP procedures should also include any in-process observations, verifications, and tests which are required. For example, some products must undergo certain interim processing which cannot be verified except by direct observation as the processing is taking place.

Sample quality manual entry

4.0 Manufacturing control

[Sections 4.01–4.02 omitted]

4.03 In-process inspection

[subsection 1 omitted]

2.0 General procedure

In-process inspection is the process of measuring, examining, testing, gaging, or otherwise comparing a product during and at the conclusion of a prescribed stage of its manufacture with the requirements established for that product at that stage.

2.1 In-process inspection is performed by Quality Control inspectors located strategically throughout the plant.

2.2 Surveillance over the manufacturing process is the primary duty of the inspector assigned to an area.

2.3 In-process inspection is accomplished on an individual piece basis during a process as well as on a batch basis at the end of the process.

2.4 In-process steps and sequences are established by Engineering and Quality Control and are shown on each product drawing or on Form xxx, Station Inspection Requirement.
2.5 Statistical Process Control shall be in accordance with NAVORD OD 42565.
2.6 Evidence of inspection is found on the Work Order Router front and back.
2.7 Discrepant parts are removed from the production run or batch, grouped according to defect, tagged by groups with a Discrepant Material Tag xxx, stamped or otherwise individually identified as a reject and held for disposition.

FINISHED GOODS

The quality plan should include procedures which document the tests or inspections required prior to the release of product for delivery to the customer. The procedures should specify how the inspection and test status of finished goods will be shown, where the goods will be stored while awaiting shipment, and proper methods of packaging, handling, and loading the goods for final delivery.

Sample quality manual entry

4.0 Manufacturing control
[subsections 4.01–4.03 omitted]
4.04 Final inspection and test
[subsections 1–3 omitted]
4.0 General procedures
4.1 Statistical sampling will be in accordance with Mil-Std-105E.
a. Sampling inspection will not be required if evidence of statistical control is documented for the process. When statistical control exists, the disposition of the lot will be determined in accordance with the procedure described in the Quality Control Instruction Manual (QCI).
4.2 Evidence of inspection is found on the Work Order Router front and back.
4.3 If required by contract, the Final Inspector's stamp will be applied to individual items, tags, or containers.
4.4 Final Inspection will pay particular attention to physical appearance, dimensions, weight, marking, and ascertain that all in-process inspections have been performed and documented according to procedures.
4.5 When required, units and documentation will be submitted to customer inspection personnel for final approval.

LOT TRACEABILITY

Documented procedures should be prepared to assure that, when required, lot traceability is maintained. Traceability is largely a matter of recordkeeping. The system should assure the following:

1. The units in the lot and the lot itself are identified.
2. The integrity of the lot is maintained (i.e., every unit that is part of the lot remains in the lot).

Lot traceability is generally required when there is reason to believe that the unit in question may need to be located at some time in the future. There are many reasons why this might be necessary, but the primary reason is that a safety defect might be discovered. The manufacturer should be able to quickly communicate with all those who are at risk from the defect. Items whose failure would cause an unsafe condition to exist are known as *critical components* or *critical items*. The FDA defines a critical component as follows:*

> Any component of a medical device whose failure to perform can be reasonably expected to cause the safety and effectiveness of a device to be altered in such a manner as to result in significant user injury, death, or a clearly diminished capacity to function as an otherwise normal human being. Those failures which are clearly a result of improper use due to failure to follow published operating instructions are not to be considered in the classification process. Those failures which result in interruption of the desired function, but do not in any way cause harm as described above, are not to be considered critical.

Case study of lot traceability implementation

A widespread practice among manufacturers of medical equipment is to label the products as sterile, then send them to outside contractors for sterilization; i.e., for a time during the process, there are items which are labeled "Sterile," but they are actually not sterile. Without traceability, if such items found their way into the stream of commerce, the result could be catastrophic. Not only would there be a danger from usage of unsterilized items, but there would be no way of knowing which items were sterilized and which were not. This would lead to the inability to use *any* of the affected items. On the other hand, it is often an economic necessity that the items be made in one location and sterilized in another. If tight control of the process can be maintained, it is possible to deliver safe product to the consumer at a reasonable cost. This example describes a procedure for assuring traceability of a critical medical item; it is excerpted from Subpart E of 21 Code of Federal Regulations (CFR) 801, along with an FDA memo which explains the FDA position on the application of this regulation.

---

* *GMP: A Workshop Manual*, October 1982, FDA.

As it is common practice to manufacture and/or assemble, package, and fully label a device as sterile at one establishment and then ship such device in interstate commerce to another establishment or to a contract sterilizer for sterilization, the Food and Drug Administration will initiate no regulatory action against the device as misbranded or adulterated when the nonsterile device is labeled sterile, provided all the following conditions are met:

1. There is in effect a written agreement which:
   a. Contains the names and post office addresses of the firms involved and is signed by the person authorizing such shipment and the operator or person in charge of the establishment receiving the devices for sterilization.
   b. Provides instructions for maintaining proper records or otherwise accounting for the number of units in each shipment to insure that the number of units shipped is the same as the number of units received and sterilized.
   c. Acknowledges that the device is nonsterile and is being shipped for further processing, and
   d. States in detail the sterilization process, the gaseous mixture or other media, the equipment, and the testing method or quality controls to be used by the contract sterilizer to assure that the device will be brought into full compliance with the Federal Food, Drug and Cosmetic Act.
2. Each pallet, carton, or other designated unit is conspicuously marked to show its nonsterile nature when it is introduced into and is moving in interstate commerce, and while it is being held prior to sterilization. Following sterilization, and until such time as it is established that the device is sterile and can be released from quarantine, each pallet, carton, or other designated unit is conspicuously marked to show that it has not been released from quarantine, e.g., "sterilized—awaiting test results" or an equivalent designation.

The following is Good Manufacturing Practice (GMP)-related information that should be included in the written agreement for contract sterilization.

A. **Information transfer**—The manufacturer and contract sterilizer must designate the individual(s) at each facility responsible for coordinating the flow of information between establishments and for approving changes in procedures. All technical, procedural, and associated activities must pass through these designees.
B. **Recordkeeping**—Documentation such as procedures, records, etc., to be used and maintained should be specified. If changes are made to the documentation, both parties should agree on the manner in which the changes are to be made. These documentation changes must comply with GMP requirements 820.100a(2) and b(3) for manufacturing and process specification changes, and with 820.181 for changes in device master records.
C. **Process qualification (validation)**—The manufacturer has primary responsibility for assuring the sterility of the device and, therefore, must work with the contract sterilizer to assure the facilities, equipment, instrumentation, sterilization process, and other related activities. The agreement should specify all parameters to be qualified by the contractor and the criteria for

requalification. Cycle parameters should be qualified and documented for each specific device or family of devices. The agreement should list the measurements to be taken and the records to be kept for each lot of devices sterilized during the qualification process.

D. **Bioindicators and dosimeters**—The agreement should state who is responsible for, and must define, placement recovery, handling, and processing of product samples and any biological, chemical, or physical indicators. Instructions should include instructions for packaging and shipment of indicators and samples to test laboratories for analysis.

.
.
.

The requirement goes on to cover various other items involved in traceability of the items throughout processing and delivery.

Discussion of lot traceability case study

While the details will vary from item-to-item, the case study provides a detailed implementation of certain basic principles of traceability:

1. The traceability procedure should be in writing.
2. Control points should be identified. The control points specify who is responsible for maintaining records showing the location of the controlled items.
3. Checkpoints should be established to confirm that the lot integrity is being maintained and that necessary processing has been completed.
4. When items are transferred between physical locations, the entire transfer process (loading, transportation, delivery, storage, etc.) should be described.
5. All concerned parties should be trained in the procedure.
6. Responsibility for obtaining and maintaining records should be clearly specified.
7. Audits should be conducted to verify that the procedures are implemented.

The factors to be considered in setting up a traceability program as part of a larger product recall program are the following:

1. **Product category**—A unique designation such as a part name or part number is required.
2. **Product life**—The longer the product life, the greater the number of similar items that are present in the field. Also, as modifications are made in design or manufacturing, the traceability plan must be able to identify which units of product embody which changes.
3. **Unit cost**—The greater the unit cost, the greater the incentive to recall only the most important problem items.
4. **Recall or modify in the field**—If the product is repairable, then a decision must be made whether to a) bring the product in from the field and have it corrected by the

manufacturer; b) send a manufacturer's representative to the field to correct the problem; or c) provide the user with parts and instructions to make the necessary modification.

5. **Product complexity**—If a design consists of many components, then a decision must be made to determine how far down the product hierarchy to extend the traceability.
6. **Level of downstream traceability**—This refers to the reason for recall. For each failure mode it must be decided what level of traceability is needed to segregate the suspect items from the total product population. For example, if a recall results from a manufacturing problem, then the manufacturing date, facility, lot identification, etc., may be required.
7. **Documents providing traceability**—An analysis of the failure modes will indicate the types of information necessary.
8. **Type of identification**—The cost of identification should be consistent with the product price. For expensive products, serial numbers are often used; for lower-priced products, dates are used.
9. **Should identification be coded or uncoded?**—An uncoded identification may have an adverse effect on sales, but it makes it easier to identify the defective product and thereby finish the recall sooner.
10. **Method of identification**—This includes tags, name plates, ink stamps, and other means.

SAMPLE INTEGRITY AND CONTROL

In quality engineering sampling is the science that guides the quantitative study of quality. Conceptually, a sample is merely a surrogate for a larger population of interest. The fundamental idea is that we can use a subset to obtain information on the whole. The incentive for doing this is usually economics, but it is sometimes a practical necessity (e.g., destructive testing). When samples are used as the basis of decision-making, a fundamental question is "Does the sample represent the population?" This issue is independent of statistical questions of validity, which are primarily dependent on sample size and measurement error. The issue here is more fundamental: if the sample is not representative of the population to which the decision applies, no statistical technique can improve the quality of the decision.

Types of samples

In quality engineering samples are drawn for a number of purposes. The systems needed to control sample integrity depend on the purpose for which the sample was drawn.

Acceptance samples—are drawn to determine whether or not an acceptable proportion of a defined inspection lot conforms to requirements. Sample integrity requires that the sample be drawn at random from the inspection lot. With discrete parts this can be accomplished by assigning each unit in the lot a unique serial number and selecting the sample using a table of random numbers. In practice, this is seldom done because of the difficulty of serialization. However, the quality engineer should attempt to make the actual sampling procedure as close as possible to this ideal. The unfortunate tendency to sample the easiest-to-reach units can result in a very biased sample.

Statistical process control samples—are drawn to determine if the process is stable, i.e., if there are special causes of variation acting on the process. Sample integrity requires that a "rational subgroup" be formed. A rational subgroup is one where every item in the subgroup was produced under essentially the same process conditions. When possible, this is usually accomplished by creating a subgroup from consecutive units.

Process validation samples—are taken to show that the proper processing conditions were met during a production run. For example, test coupons might be heat-treated along with piece-parts. The test coupons can then be evaluated using test methods that would destroy or damage actual product. Or, a punch-press operation might be validated for size by saving the first and last piece of the production run. If both items meet the requirement, then the operation is verified. The primary concern is the identification of the first piece.

Measurement system correlation samples—are taken to allow a comparison between two different methods of measurement. Here every item in the sample must be identified and the appropriate measurements recorded for each item. The sample must be reinspected under essentially the same conditions and the results compared statistically.

Procedures

Systems for sample integrity and control are usually quite specific to the situation. Of course, the systems should be documented and periodically audited for compliance to the documentation.

Deviations and waivers

While an Engineering Change Proposal (ECP) involves a permanent change to the engineering design, a deviation is a temporary departure from an established requirement. Deviation requests should be formally evaluated and approved only when they result in significant benefit. Repeated deviations should be investigated; if the underlying requirements are too stringent, they should be modified.

MATERIALS SEGREGATION PRACTICES

The previous sections describe various activities relating to the identification of various types of materials, e.g., by type of defective, by processing status. Once a "special" classification has been made (e.g., material to be scrapped or reworked), the procedure specifies how the affected material will be identified. Next, provision must often be made to physically remove the material from the normal processing stream. Formal, written procedures should be developed to describe the control, handling, and disposition of nonconforming materials to assure that such materials are adequately identified and prevented from becoming mixed with acceptable materials.

The physical control of nonconforming materials varies widely from firm-to-firm and by type of problem. Some organizations require that discrepant critical components be immediately removed to a locked storage area and require authorization from designated individuals for release. For typical quality problems the procedure described below is representative.

Sample quality manual entry

4.0 Manufacturing control
[subsections 4.01–4.05 omitted]

4.06 Control of non-conforming material
[subsections 1–2 omitted]

3.0 Requirements

3.1 It is mandatory that immediately upon detection, discrepant material be positively identified as provided in this manual and segregated from normal manufacturing.

3.2 The removal of a discrepant material tag or other identification by other than authorized personnel is not permitted and may be cause for dismissal.

4.0 Procedure for use of discrepant material tag, xxx

4.1 Upon finding a non-conforming item, the inspector shall complete a discrepant material tag, xxx, and attach it to the rejected item.

4.2 All rejected material will be moved to a clearly identified location for storage.

4.3 Rejected material will be reviewed on a daily basis by representatives from departments who possess the expertise necessary to determine how to dispose of the material.

4.4 Material to be repaired or reworked will be identified with the reject tag until it is confirmed that the reworked item conforms to requirements.

4.5 Material to be scrapped will be clearly and conspicuously identified and it will be immediately removed to an area separate from normal material and disposed of quickly.

MATERIALS REVIEW BOARD CRITERIA AND PROCEDURES

The material review board (MRB) activity has two basic responsibilities:

1. disposing of nonconforming materials.
2. assuring that effective action is taken to prevent future occurrences of the problem.

MRB membership consists of technical and management personnel who possess the necessary expertise and authority to accomplish these basic objectives. MRB membership generally includes representatives from production, engineering and quality. Large customers are sometimes represented on MRBs. For small businesses, there may be a single MRB for the entire organization. Large organizations may have several MRBs representing different product lines, customers, etc.

In the past, MRBs focused almost entirely on disposing of non-conforming materials. This generally meant reviewing rejected materials and deciding if they could be reworked or if they should be scrapped. For rework dispositions, the MRB prepared a rework and reinspection plan. Some products which could not be reworked were downgraded, or the requirement waived. Corrective action often involved little more than a cursory statement, e.g., "Cause: operator, corrective action: operator will be instructed to take greater care in processing." Although such activities still occur, the modern approach to corrective action is much more rigorous. Control charts are used to determine if the problem is due to special or common causes; statistical evidence is required to show a link between the problem and its cause; then corrective action plans are reviewed in light of the statistical evidence.

Not all nonconformances are taken to the MRB. Most of the time the disposition is handled at the Preliminary Review level. Preliminary Review involves dispositions of nonconforming materials when the amount involved is not large and there is no question as to the status of the material or the method of rework or repair.

When the amount of potential loss is very large, an upper-level MRB is sometimes convened. This group acts in the same capacity as the usual MRB, but it is composed of individuals with the authority to take actions involving large dollar consequences.

Example of an MRB procedure

4.0 Manufacturing control

[subsections 4.01–4.08 omitted]

4.09 Materials review

[subsections 1–4 omitted]

5.0 Materials review board

Material referred to the MRB will be examined to determine if it will be:

a. Scrapped

b. Used "as is" with the approval of the customer's representative.
c. Reworked. Rework instructions will be indicated on the back of the MRB Tag Form xxx.
d. Repaired. Repair instructions will be indicated on the back of the MRB Tag Form xxx.
e. Returned to the vendor.

6.0 Materials review board membership

MRB personnel will include qualified representatives from Engineering and Quality. Manufacturing Engineering concurrence is required for any rework or repair dispositions. Customer concurrence is required for any "use as-is" dispositions.

Note that this example illustrates the rather dated perspective of MRB as solely devoted to judging the fitness of nonconforming materials. As was mentioned above, MRB responsibility now generally extends to identifying and correcting the root causes of repeated problems. Also note that while most quality systems standards require a system be in place for dealing with nonconforming materials, MRBs are not always required. In those cases the producer is allowed to develop alternative systems so long as they assure that the *intent* of disposing of the materials and preventing recurrence is met.

III.C ACCEPTANCE SAMPLING

Acceptance sampling was once the mainstay of statistical quality control. In recent years its use has become somewhat controversial. While not completely ignoring the controversy, we will focus on the fundamental ideas behind acceptance sampling.

GENERAL COMMENTS ON ACCEPTANCE SAMPLING

- It can be mathematically shown that acceptance sampling should never be used for processes in statistical control. Acceptance sampling is a *detection* approach to quality control, while SPC is an approach that *prevents problems*. Thus, concentrate on getting statistical control first—then you will need acceptance sampling only for isolated lots or other special situations, such as purchasing from a central distribution warehouse.
- Some information can be obtained by plotting acceptance sampling results on control charts. However, the information is not as valuable as that obtained from true statistical process control because 1) the time-lag in obtaining the results makes it more difficult to identify any special causes of variation, 2) the control limits may

be too close together or too far apart since there is no relationship between the sample size required for SPC and that required for acceptance sampling, 3) process variable information isn't available to correlate with sample results, 4) characteristics checked by acceptance sampling are usually not the best choices for process control, 5) the sampling interval is often too long, and 6) acceptance sampling requires random samples, but SPC requires non-random rational subgroups.*

- There are many occasions when rejected lots are re-inspected. This is poor practice and it creates a sampling plan that has very high consumer risks.
- It is often difficult to draw random samples, but it is also very important. If you don't plan to draw your samples at random, don't dignify the inspection process by pretending that a sampling plan will somehow make it "scientific."
- If you are sampling for critical defects, the accept number should always be zero. Also, except for destructive tests, sampling should be used for critical defects only to verify that a prior 100% inspection or test was effective. The reasons for this are both ethical and legal.
- If, while sampling, you observe defects that are not part of your sample, you must identify the defective units and remove them from the inspection lot. You can't knowingly ship defective product, regardless of whether or not it is part of your sample. Of course, the acceptability of the lot is determined by those units in the sample only. Keep in mind that you can't knowingly exclude a unit from your sample because it is defective; i.e., don't remove obvious defectives before drawing the sample.

DEFINITIONS

Acceptance sampling—is defined as making a decision regarding the acceptability of a lot or batch based on the results from one or more samples rather than from 100% inspection.

Acceptable Quality Level (AQL)—The worst quality level that is considered acceptable as a long-term process average for acceptance sampling purposes.

Rejectable Quality Level (RQL)—A quality level with a very high probability of being rejected by the sampling plan. Also called the lot tolerance percentage defective (LTPD), lot tolerance fraction defective (LTFD), or limiting quality (LQ).

Average outgoing quality—The quality level that results when a particular sampling *scheme* (a sampling plan combined with repair or replacement of defectives in the sample and/or lot) is applied to a series of lots from a process.

Average Outgoing Quality Limit (AOQL)—For a given sampling plan and repair or replacement scheme, the maximum outgoing defective rate for *any* incoming defective rate. Figure III.20 illustrates the AOQ and AOQL concepts.

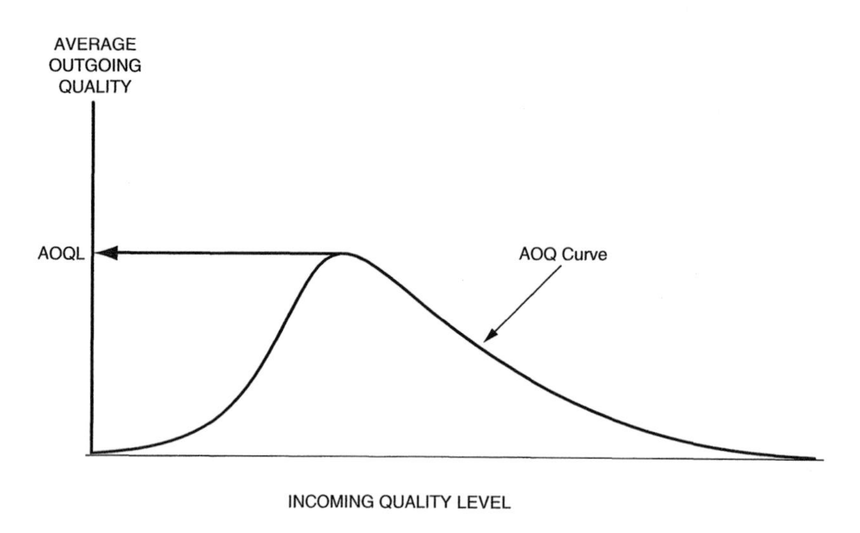

Figure III.20. AOQ curve and AOQL.

Average Total Inspected (ATI)—The average number in a lot inspected as part of the sample plus the screening of rejected lots.

Acceptance number (c)—A defect or defective count that, if not exceeded, will cause acceptance of the lot.

Reject number (r)—A defect or defective count that will cause rejection of the lot.

Single sampling plan—A sampling plan where the acceptance decision is reached after the inspection of a single sample of n items. Single sampling plans work as follows:

$$\text{Sample size} = n$$
$$\text{Accept number} = c$$
$$\text{Reject number} = c + 1$$

For example, if n=50 and c=0, we will inspect a sample of 50 units and accept the lot if there are 0 defective units in the sample. If there are one or more defectives in the sample, the lot will be rejected.

GENERAL THEORY

Acceptance sampling is a popular quality control technique that is applied to discrete lots or batches of product. A *lot* is a collection of physical units; the term *batch* is usually applied to chemical materials. The lot or batch is typically presented to the inspection department by either a supplier or a production department. The inspection department then inspects a sample from the lot or batch and based on the results of the inspection they determine the acceptability of the lot or batch.

Acceptance sampling schemes generally consist of three elements:

1. The sampling plan. How many units should be inspected? What is the acceptance criteria?
2. The action to be taken on the current lot or batch. Actions include accept, sort, scrap, rework, downgrade, return to vendor, etc.
3. Action to be taken in the future. Future action includes such options as switching to reduced or tightened sampling, switching to 100% inspection, shutting down the process, etc.

OPERATING CHARACTERISTICS CURVE (OC CURVE)

An OC curve is a graph that, for a given sampling plan, gives the probability of accepting the lot as a function of the quality of the lot or process. Figure III.21 illustrates a perfect OC curve and an actual OC curve. A perfect OC curve will accept all lots of acceptable quality, and reject all other lots.

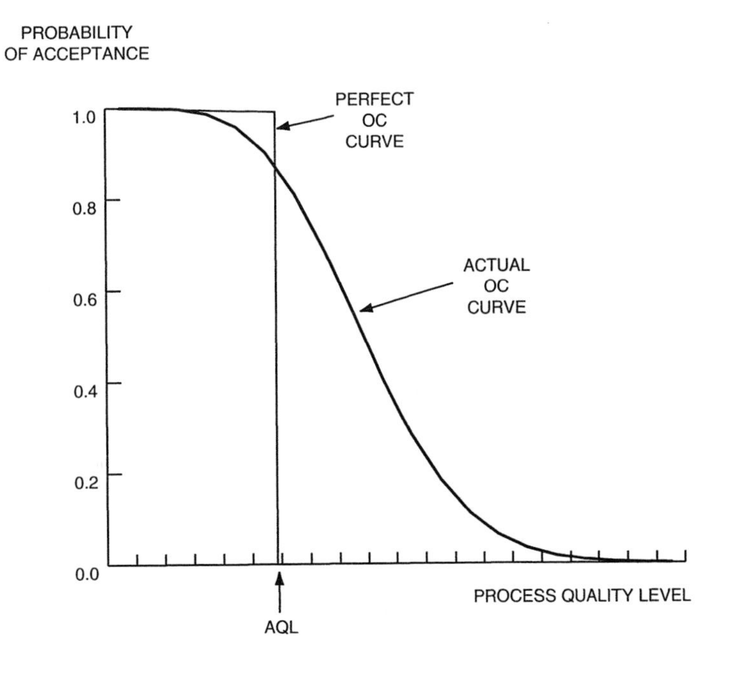

Figure III.21. The OC curve for a "perfect" acceptance sampling plan and an actual sampling plan.

Indifference Quality Level (IQL)—A quality level that a particular sampling plan will accept 50% of the time. The IQL can be found on the OC curve.

Producer's risk—The probability that an acceptable lot, that is, a lot of AQL or better quality, will be rejected; also known as the α risk. The α risk can be found using the OC curve.

Consumer's risk—The probability that a lot of rejectable quality will be accepted by a particular sampling plan. The consumer's risk can be found using the OC curve.

SAMPLING TYPES

Single sampling plans

As mentioned earlier, acceptance sampling involves making a decision regarding the acceptability of a lot or batch based on the results from one or more samples rather than from 100% inspection.

Calculating operating characteristics

Operating characteristics, and the OC curve, are the most important tools for evaluating a particular sampling plan. Many of the items of concern to both producer and consumer can be obtained directly from the OC curve, such as AQL, LTPD, producer's risk, and consumer's risk. The remaining items can be calculated from the OC curve. We will learn a method for calculating operating characteristics that uses tables of the Poisson (pronounced "pwä-son") distribution. The worksheet and instructions in Figure III.22 can be used for this purpose.

Instructions for using worksheet

1. Record the sample size (n), lot size (N), and accept number (c) in the appropriate box. Since this is a single sampling plan the reject number is $r = c + 1$.
2. Enter the lot quality levels of interest in column 1. When evaluating defectives, use fractions, not percentages.
3. Multiply the values in column 1 by the sample size, n, and record the results in column 2.
4. Refer to Appendix 9, where the accept number c determines the row, and the value in column 2 determines the column μ. The numbers in the table are the probability of acceptance, P_A. Find the values for column 3 from the table.
5. Complete column 4, AOQ, by using Equation III.1.

$$AOQ = p \times P_A \times \left(1 - \frac{\text{sample size}}{\text{lot size}}\right) \qquad \text{(III.1)}$$

   where p is the value in column 1, P_A is the value in column 3, and the sample size and lot size are given.
6. Plot the OC curve with column 1 along the bottom axis and column 3 on the vertical axis.

SAMPLE SIZE	
ACCEPT NUMBER	
LOT SIZE	

(1) Fraction defective or defects-per-unit	(2) *np* or *nu*	(3) Probability of acceptance	(4) Average outgoing quality

Figure III.22. Worksheet for computing operating characteristics for single sampling plans.

7. Plot the AOQ curve with column 1 along the bottom axis and column 4 on the vertical axis. Note that the AOQL can be approximately found as the largest value in column 4. The AOQL can also be calculated as

$$AOQL = y\left(\frac{1}{\text{sample size}} - \frac{1}{\text{lot size}}\right) \qquad \text{(III.2)}$$

Where y is a factor from Table 12 in the Appendix.

Example of evaluating an acceptance sampling plan

You've been talking to a supplier who tells you that they plan to use the following sampling plan for your product:

1. Sample 100 units from every lot.
2. Accept the lot if there are 1 or fewer defectives found in the sample.

Your lots are typically 1000 units each. You are concerned about the performance of this sampling plan for quality levels ranging from 0.5% (p=.005) to 5% defective (p=.05). The completed worksheet is shown in Figure III.23.

SAMPLE SIZE	100
ACCEPT NUMBER	1
LOT SIZE	1,000

(1) Fraction defective or defects-per-unit	(2) *np* or *nu*	(3) Probability of acceptance	(4) Average outgoing quality
0.005	0.5	0.910	0.0041*
0.010	1.0	0.736	0.0066
0.015	1.5	0.558	0.0075
0.020	2.0	0.406	0.0073
0.024	2.4	0.308	0.0067
0.030	3.0	0.199	0.0054
0.036	3.6	0.126	0.0041
0.040	4.0	0.092	0.0033
0.046	4.6	0.056	0.0023
0.050	5.0	0.040	0.0018

$$* \; AOQ = 0.005 \times 0.910\left(1 - \frac{100}{1,000}\right) = 0.0041$$

$$AOQL = 0.841\left(\frac{1}{100} - \frac{1}{1,000}\right) = 0.0076$$

Figure III.23. Example of evaluating a single sampling plan.

Note that this AOQL is very close to the largest value in column 4, 0.0075 found in the table at p = 0.015. Assuming perfect inspection, the worst average outgoing quality of 0.76% will occur when the process average is near 1.5% defective.

The information in Figure III.23 is shown graphically in Figure III.24.

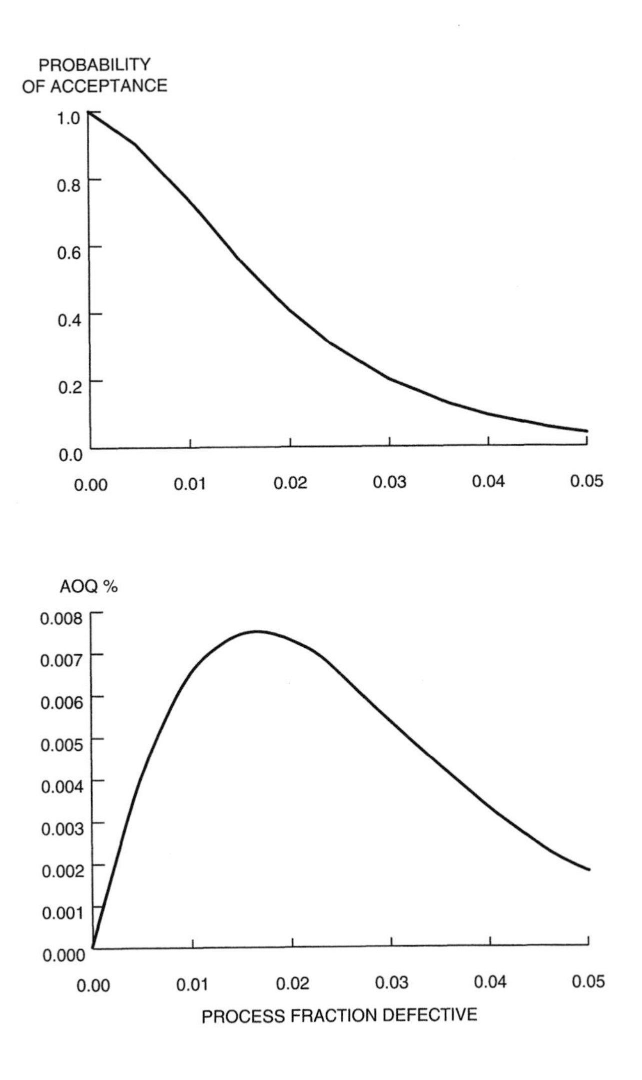

Figure III.24. The OC and AOQ curves for the sampling plan in Figure III.23.

Discussion of sampling plan evaluation

Producers are usually interested in the quality level they must achieve to have the vast majority of their lots accepted, i.e., the AQL. Typically, we look for quality levels accepted 95% of the time or more. We can see that for this plan the acceptable quality level is something better than 0.5% defective. A more precise acceptable quality level can be found by entering the Appendix 9 Poisson Tables in the row for c=1 and finding the column corresponding to a probability of acceptance=95%. Doing this we find P_A=0.951 in the column for *np*=0.35. The AQL is found by dividing the np number by the sample size, e.g., AQL=0.35/100=0.0035=0.35%. The consumer will be more interested in how the plan protects against poor quality. The operating characteristics curve shows that this sampling plan provides good protection from individual lots that are 4% defective. Also, if this supplier provides a steady flow of lots, the average quality after sorting of rejected lots will be no worse than 0.76% defective.

Double sampling plan

Using the double sampling plan the acceptance decision may not be reached until two samples have been inspected. The samples may or may not be the same size. Double sampling plans work as follows:

Table III.1. Double sampling plan layout.

SAMPLE NUMBER	SAMPLE SIZE	ACCEPT NUMBER	REJECT NUMBER
1	n_1	c_1	$r_1 > c_1+1$
2	n_2	c_2	c_2+1

Note that for the first sample, r_1 is at least 2 greater than c_1. However, the accept decision must be made after the second sample since the reject number is just 1 greater than the accept number.

An example of a double sampling plan is shown in Table III.2:

Table III.2. Double sampling plan example.

SAMPLE NUMBER	SAMPLE SIZE	ACCEPT NUMBER	REJECT NUMBER
1	20	1	3
2	25	2	3

The above double sampling plan works as follows: select 20 units from the lot at random and inspect them. If there are 0 or 1 defectives in the sample, accept the lot. If there are 3 or more defectives in the sample, reject the lot. If there are exactly 2 defectives in the sample, draw another random sample of 25 units. Add the number defective in the first sample to the number defective found in the second sample. If the *total* number defective is 2 or less, accept the lot. If there are 3 or more defectives, reject the lot.

The motivation to use double sampling is economics; on average double sampling plans require less total inspection than their single sampling counterparts. The reason to not use them is administrative: double sampling plans are a bit more confusing than single sampling plans. Also, the fact that the second sample is sometimes selected and sometimes not inspected makes the inspection workload difficult to predict, which leads to scheduling problems.

Multiple sampling plan

Using the multiple sampling plan the acceptance decision may not be reached until 3 or more samples have been inspected. The samples are usually the same size. A multiple sampling plan with *k* stages works as shown in Table III.3.

Table III.3. Multiple sampling plan layout.

SAMPLE NUMBER	SAMPLE SIZE	ACCEPT NUMBER	REJECT NUMBER
1	n_1	c_1	$r_1>c_1+1$
2 ⋮	n_2 ⋮	c_2 ⋮	$r_2>c_2+1$ ⋮
k	n_k	c_k	$r_k=c_k+1$

The above sampling plan works in the same way as the double sampling plan described above except that there are more than 2 stages; ANSI/ASQ Z1.4-1993 multiple sampling plans have 7 stages. In general, multiple sampling requires less total inspection than either single or double sampling, at the cost of additional administrative confusion.

Sequential sampling plan

Using the sequential sampling plan an acceptance decision is made upon the inspection of each unit in the sample. Figure III.25 illustrates a sequential plan.

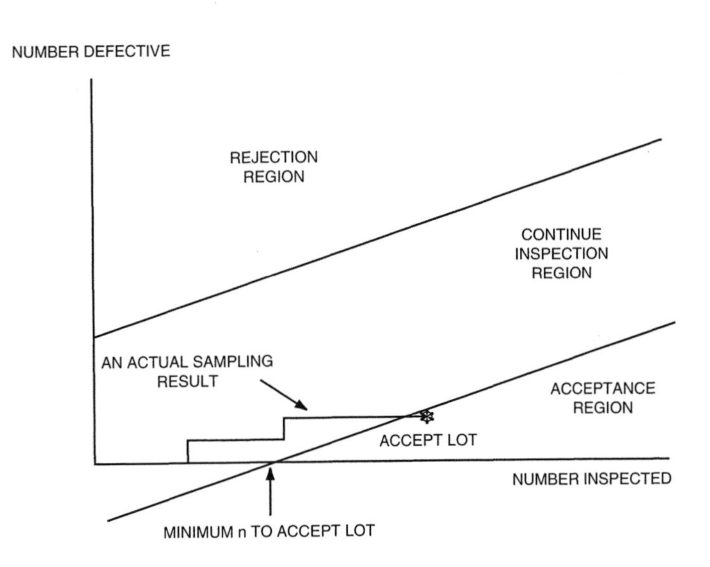

Figure III.25. Sequential sampling plan.

Continuous sampling plan

Acceptance sampling can be applied in a variety of different situations. When acceptance sampling is applied to a process in continuous production it is possible to design an acceptance sampling scheme that does not require assembling a discrete lot or batch. Instead, these plans call for the inspection of some fraction of units selected at random as the process is running. These plans mix 100% inspection with sampling in such a way that the consumer is guaranteed some maximum fraction non-conforming in the outgoing product. Such plans are called continuous sampling plans.

Deming's all–or–none rule

If a process is in continuous production it is generally better to use a p chart for process control than to apply a continuous sampling plan. Based on the stable p chart you can determine the process average fraction defective, from which you can determine whether to sort the output or ship it by applying Deming's all-or-none rule.

$$\text{If } \overline{\text{p}} < \frac{K_1}{K_2} \text{ then ship, otherwise sort} \qquad \text{(III.3)}$$

where K_1 is the cost of inspecting one piece and K_2 is the cost of shipping a defective, including lost customer goodwill. For example, if K_1=\$1 and K_2=\$100, then output from a process with an average fraction defective of less than 1% would be shipped without additional inspection; if the process average were 1% or greater, the output would be sorted.

Note that this discussion does *not* apply to critical defects or defectives, since sampling is only done in these cases to confirm a previous 100% inspection or test procedure.

Orsini (See Deming 1986, pp. 415–416) has developed an approach to be used when a process is in a state of chaos, i.e., when no stable process exists. Orsini's rules are essentially a modification of Deming's all-or-none rule which produce a substantial savings over 100% inspection when the process average fraction defective is not known. Her rules are as follows:

For $k_2 \geq 1{,}000k_1$	Inspect 100% of incoming lots
For $1{,}000k_1 > k_2 \geq 10k_1$	Test a sample of n = 200. Accept the remainder if you find no defective in the sample. Screen the remainder if you encounter a defective in the sample.
For $k_2 < 10k_1$	No inspection.

SAMPLING PLANS

The work of creating sampling plans can be tedious and error-prone, especially for multi-level sampling plans. Thus, several standards have been developed which simplify the selection and use of acceptance sampling. The three methods described below, ANSI/ASQ Z1.4-1993 (formerly MIL-STD-105), ANSI/ASQ Z1.9-1993 (formerly MIL-STD-414), and Dodge-Romig sampling tables, integrate sampling into a scheme with provisions for problem follow-up and corrective action.

SAMPLING PLANS: ANSI/ASQ Z1.4

ANSI/ASQ Z1.4 is a standard for sampling inspection by attributes. Sampling plans in ANSI/ASQ Z1.4-1993 are intended for use with an ongoing series of lots or batches produced in a stable environment.* ANSI/ASQ Z1.4-1993 can be applied to either defectives or defects.

How to use ANSI/ASQ Z1.4

1. Determine the AQL. The AQL must be one of the "preferred AQL's" in the master tables.
2. Determine the inspection level. Use general inspection level II unless there is a good reason to select some other level.
3. Determine the type of inspection to be used: single, double, or multiple.
4. Determine the switching level: reduced, normal, or tightened. You always begin at normal and follow the switching procedures described in ANSI/ASQ Z1.4-1993.
5. Determine the size of the inspection lot.
6. Using Table 1 in Z1.4, find the sample size code letter for your lot size and level of inspection.
7. Depending on the severity and type of inspection, enter one of the master tables and find the sampling plan. A guide to the master tables is shown in Table III.4. Note that there are cases where you will encounter an arrow in a master table. When this occurs, follow the arrow to the first row in the table that contains an accept (Ac) or a reject number (Re) and use the sampling plan in that row. Be sure to read the notes at the bottom of the master table for special instructions.
8. Draw the samples at random from the lot and inspect each item in the sample. Determine the acceptability of the lot by comparing the results to the acceptance/rejection criteria given in the master table.

---

*Refer to the discussion of Deming's all-or-none rule described earlier.

9. Dispose of the lot. This may mean acceptance, rework, scrap, etc., depending on your sample results and acceptance criteria. Be sure that you have a record of your disposition. The record will be used for switching, as well as for quality control purposes.
10. Where appropriate, take action on the process to effect a quality improvement.

Table III.4. Master table selection guide for ANSI/ASQ Z1.4-1993.

TYPE OF	SEVERITY OF INSPECTION		
INSPECTION	Reduced	Normal	Tightened
Single	II-C	II-A	II-B
Double	III-C	III-A	III-B
Multiple	IV-C	IV-A	IV-B

SAMPLING PLANS: ANSI/ASQ Z1.9

The previous discussion covered acceptance sampling plans for attributes; i.e., the accept/reject decision was based on a count of defects or defectives. It is also possible to base the decision on statistics computed from samples. This is known as acceptance sampling by variables. The best-known approach to acceptance sampling by variables is ANSI/ASQ Z1.9-1993 (formerly Mil-Std-414 and interchangeable with ISO/DIS 3951).

ANSI/ASQ Z1.9-1993 is not nearly as widely used as ANSI/ASQ Z1.4-1993, its attribute sampling counterpart. Compared to ANSI/ASQ Z1.4-1993, ANSI/ASQ Z1.9-1993 sampling plans provide equivalent protection with far fewer samples. However, the application of ANSI/ASQ Z1.9-1993 involves mathematical calculations that are complex, and the acceptance criterion is not as easy to understand. It is this author's experience that when using ANSI/ASQ Z1.9-1993, explaining reject decisions to production managers and suppliers is no easy task. By comparison, ANSI/ASQ Z1.4-1993 reject decisions (or, for that matter, acceptance decisions) are understood without the need for explanation.

ANSI/ASQ Z1.9-1993 is based on a strategy of estimating the lot percentage defective. This is approached in two different ways. If Form 1 is used, statistics are computed from samples and then compared to constants found in tables. If Form 2 is used, the computed statistics are used to obtain estimated percentages of non-conformances (both high and low); these percentages are then compared to critical values found in tables. The procedures are summarized in Table III.5.

Table III.5. Summary of ANSI/ASQ Z1.9-1993 equations and acceptance criteria.

PURPOSE	MIL-STD-414 SEC.	FORM 1 ACTIVITY	FORM 2 ACTIVITY
Preparation	—	Find n, k from tables	Find n, M from tables
	B (s method, σ unknown)	$Z_U = \frac{U - \overline{X}}{s}$ $Z_L = \frac{\overline{X} - L}{s}$	$Q_U = \frac{U - \overline{X}}{s}$ $Q_L = \frac{\overline{X} - L}{s}$
Calculation	C (R method, σ unknown)	$Z_U = \frac{U - \overline{X}}{\overline{R}}$ $Z_L = \frac{\overline{X} - L}{\overline{R}}$	$Q_U = \frac{(U - \overline{X})c}{\overline{R}}$ $Q_L = \frac{(\overline{X} - L)c}{\overline{R}}$
	D (s method, σ known)	$Z_U = \frac{U - \overline{X}}{\sigma}$ $Z_L = \frac{\overline{X} - L}{\sigma}$	$Q_U = \frac{(U - \overline{X})v}{\sigma}$ $Q_L = \frac{(\overline{X} - L)v}{\sigma}$
Find % low, % high and compute % out	—	Not applicable	Get P_U and P_L from table
	One-sided specification	Accept if $Z_U \geq k$ or $Z_L \geq k$	Accept if $P_U \leq M$ or $P_L \leq M$
	Two-sided Specification	Not applicable	Accept if $P_U + P_L \leq M$

Form 2 is more complicated to apply than Form 1. However, if two-sided specifications are involved, only Form 2 can be used. When Form 2 is used, the procedure produces an estimate of the lot non-conformance, which is valuable as a communication tool. Form 1 produces a statistic that, while useful in deciding acceptance, provides no information on the non-conformance rate.

SAMPLING PLANS: DODGE-ROMIG SAMPLING TABLES

Dodge-Romig tables are attribute acceptance sampling tables designed for use with ongoing processes that have known process averages. There are tables for both single sampling or double sampling. Sampling plans in the Dodge-Romig tables are designed to minimize the average total inspected. There are two classes of sampling plans: AOQL constrained and LTPD constrained. To use the Dodge-Romig tables, the following information must be provided:

- The AOQL for AOQL constrained plans, or the LTPD for LTPD constrained plans. For the LTPD constrained plans, you have a choice of probabilities of acceptance at the LTPD quality level: 5% or 10%.
- The process average fraction defective.
- The lot size.
- The type of sampling to be used (single or double sampling).

III.D MEASUREMENT SYSTEMS

A good measurement system possesses certain properties. First, it should produce a number that is "close" to the actual property being measured; i.e., it should be *accurate*. Second, if the measurement system is applied repeatedly to the same object, the measurements produced should be close to one another; i.e., it should be *repeatable*. Third, the measurement system should be able to produce accurate and consistent results over the entire range of concern; i.e., it should be *linear*. Fourth, the measurement system should produce the same results when used by any properly trained individual; i.e., the results should be *reproducible*. Finally, when applied to the same items, the measurement system should produce the same results in the future as it did in the past; i.e., it should be *stable*. The remainder of this section is devoted to discussing ways to ascertain these properties for particular measurement systems. In general, the methods and definitions are consistent with those described by the Automotive Industry Action Group (AIAG) [Automotive Industry Action Group (2002)].

DEFINITIONS

Bias—The difference between the average measured value and a reference value is referred to as *bias*. The reference value is an agreed-upon standard, such as a standard traceable to a national standards body (see below). When applied to attribute inspection, bias refers to the ability of the attribute inspection system to produce agreement on inspection standards. Bias is controlled by *calibration*, which is the process of comparing measurements to standards. The concept of bias is illustrated in Figure III.26.

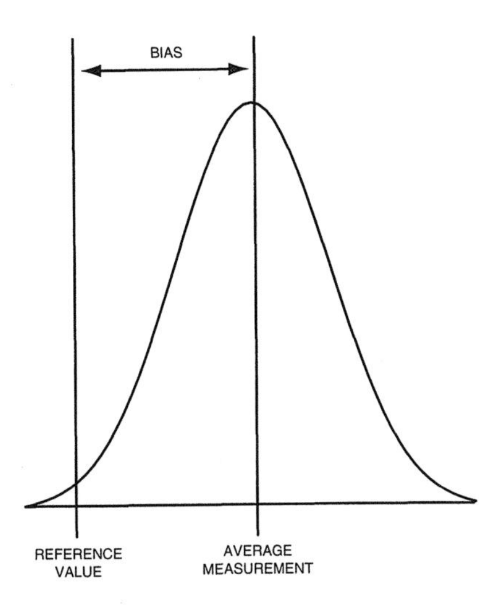

Figure III.26. Bias illustrated.

Repeatability—AIAG defines repeatability as the variation in measurements obtained with one measurement instrument when used several times by one appraiser while measuring the identical characteristic on the same part. Variation obtained when the measurement system is applied repeatedly under the same conditions is usually caused by conditions inherent in the measurement system.

ASQ defines *precision* as "The closeness of agreement between randomly selected individual measurements or test results. Note: The standard deviation of the error of measurement is sometimes called 'imprecision.' " This is similar to what we are calling repeatability. Repeatability is illustrated in Figure III.27.

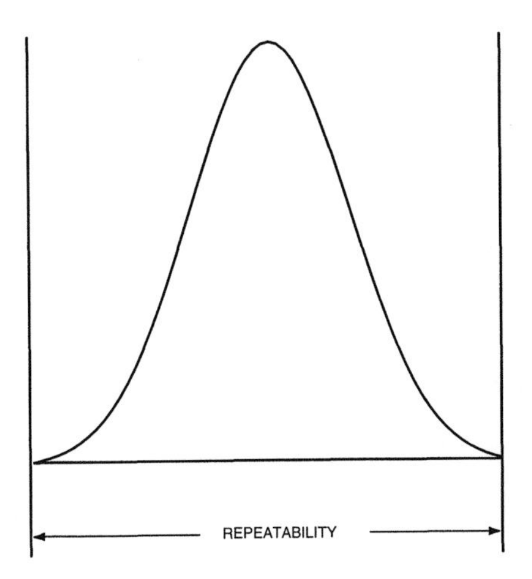

Figure III.27. Repeatability illustrated.

Reproducibility—Reproducibility is the variation in the average of the measurements made by different appraisers using the same measuring instrument when measuring the identical characteristic on the same part. Reproducibility is illustrated in Figure III.28.

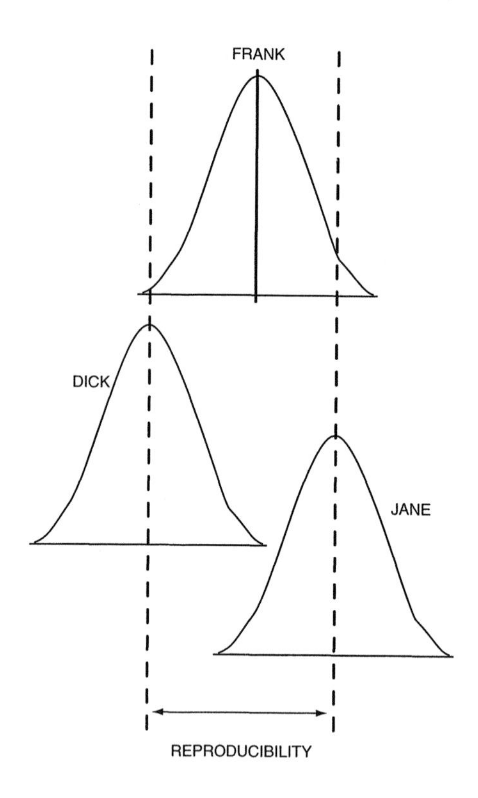

Figure III.28. Reproducibility illustrated.

Stability—Stability is the total variation in the measurements obtained with a measurement system on the same master or parts when measuring a single characteristic over an extended time period. A system is said to be stable if the results are the same at different points in time. Stability is illustrated in Figure III.29.

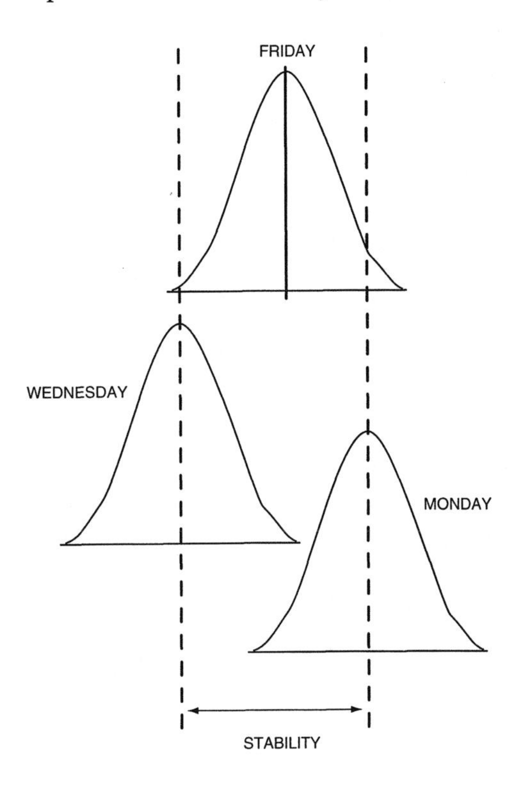

Figure III.29. Stability illustrated.

Linearity—the difference in the bias values through the expected operating range of the gage. Linearity is illustrated in Figure III.30.

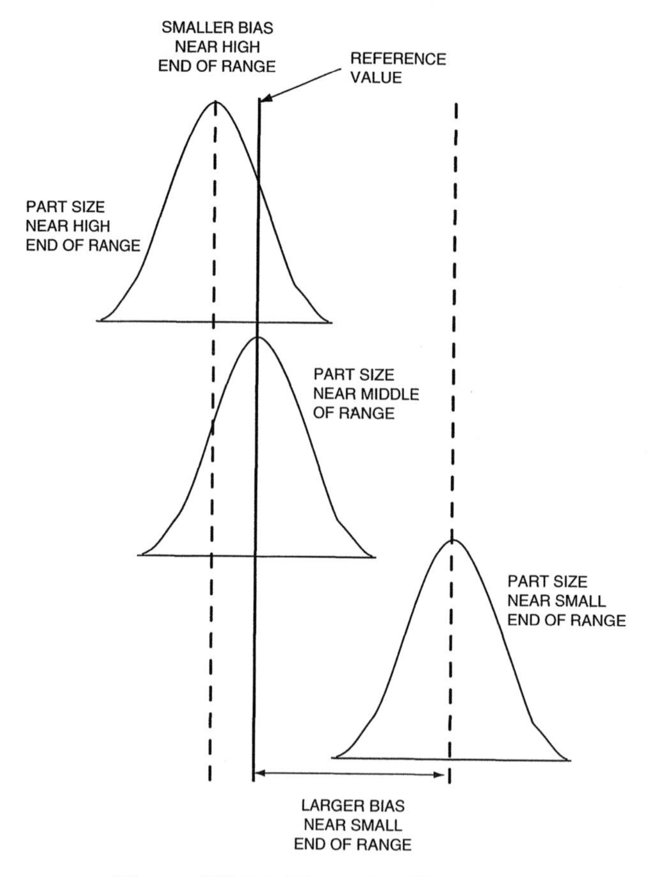

Figure III.30. Linearity illustrated.

DESTRUCTIVE TESTING

Destructive testing refers to analyses that result in the destruction of the item being analyzed. This includes corrosion tests, tensile-strength tests, impact testing, shear-strength tests, bonding strength tests, and many chemical tests. An important consideration in destructive testing is test cost. In general, the cost to obtain a measurement is higher when the test is destructive. Obviously, sampling is a must; 100% destructive tests would not leave any items to ship! However, destructive sampling favors the statistical methods that provide the greatest information at the lowest cost. This usually means that variables sampling or SPC methods are preferred over attribute sampling or SPC methods. The added complexity of the variables analyses is usually justified by the reduced sampling costs. Destructive testing planning is also emphasized. By making maximum use of engineering and process knowledge, testing can be designed to obtain maximum information from each test item. It is often possible to obtain test specimens that represent the production lot at a lower cost, e.g., a hardness test specimen might be processed instead of a part when the surface of a part might be damaged. It is also sometimes possible to eliminate destructive testing by certifying that the process was performed according to an agreed-upon procedure.

NON-DESTRUCTIVE TESTING

Non-Destructive Testing (NDT) is the name applied to a family of testing methods that determine the physical properties of materials or products. Typical NDT methods are described below (courtesy of the American Society for Non-Destructive Testing, 1–800–222–2768):

Radiography—Radiography involves the use of penetrating X or gamma radiation to examine parts and products for imperfections. An X-ray machine or radioactive isotope is used as a source of radiation. Radiation is directed through a part and onto film. When the film is developed, a shadowgraph is obtained that shows the internal soundness of a part. Possible imperfections show up as density changes in the film, in much the same manner as an X-ray can show broken bones.

Magnetic particle—Magnetic particle testing is done by inducing a magnetic field in a ferro-magnetic material and dusting the surface with iron particles (either dry or suspended in a liquid). Surface imperfections will distort the magnetic field and concentrate the iron particles near imperfections, thus indicating their presence.

Ultrasonics—Ultrasonic testing uses the transmission of high frequency sound waves into a material to detect imperfections within the material or changes in material properties. The most commonly used ultrasonic testing technique is pulse echo, wherein sound is introduced into the test object and reflections (echoes) are returned to a receiver from internal imperfections or from geometrical surfaces of the part.

Liquid penetrant—Liquid penetrant testing is probably the most widely used NDT method. The test object or material is coated with a visible or fluorescent dye solution. The excess dye is removed from the surface, and then a developer is applied. The developer acts like a blotter and draws penetrant out of imperfections which open to the surface. With visible dyes, the vivid color contrast between the penetrant and the developer makes the "bleedout" easy to see. With fluorescent dyes, an ultraviolet lamp is used to make the "bleedout" fluoresce brightly, thus allowing the imperfection to be seen readily.

Eddy current—In eddy current testing, electrical currents are generated in a conductive material by an induced magnetic field. Interruptions in the flow of the electric currents (eddy currents), which are caused by imperfections or changes in a material's conductive properties, will cause changes in the induced magnetic field. These changes, when detected, indicate the presence of a change in the test object.

Leak testing—A variety of techniques are used to detect and locate leaks in pressure containment parts and structures. Leaks can be detected by using electronic listening devices, pressure gauge measurements, liquid and gas penetrant techniques, and/or a simple soap bubble test.

Acoustic emission—When a solid material is stressed, imperfections within the material emit short bursts of energy called "emissions." In much the same manner as ultrasonic testing, these acoustic emissions can be detected by special receivers. The source of emissions can be evaluated through study of their strength, rate, and location.

Visual examination—Probably the oldest and most common method of NDT is visual examination, which has numerous industrial and commercial applications. Examiners follow procedures ranging from simple to very complex, some of which involve comparison of workmanship samples with production parts. Visual techniques are used with all other NDT methods.

Special NDT methods—NDT engineers and technicians also use microwaves, ultrasonic imaging, lasers, holography, liquid crystals, and infrared-thermal testing techniques in addition to many other specialized methods.

Table III.6. Objectives of nondestructive testing methods.

OBJECTIVES	ATTRIBUTES MEASURED OR DETECTED
Discontinuities	
Surface anomalies	roughness, scratches, gouges, crazing, pitting, inclusions and imbedded foreign material
Surface connected anomalies	cracks, porosity, pinholes, laps, seams, folds, inclusions
Internal anomalies	cracks, separations, hot tears, cold shuts, shrinkage, voids, lack of fusion, pores, cavities, delaminations, disbonds, poor bonds, inclusions, segregations
Structure	
Microstructure	molecular structure, crystalline structure and/or strain, lattice structure, strain, dislocation, vacancy, deformation
Matrix structure	grain structure, size, orientation and phase, sinter and porosity, impregnation, filler and/or reinforcement distribution, anisotropy, heterogeneity, segregation
Small structural anomalies	leaks (lack of seal or through-holes), poor fit, poor contact, loose parts, loose particles, foreign objects
Gross structural anomalies	assembly errors, misalignment, poor spacing or ordering, deformation, malformation, missing parts

Continued on next page . . .

Table III.6—*Continued . . .*

OBJECTIVES	ATTRIBUTES MEASURED OR DETECTED
Dimensions and metrology	
Displacement, position	linear measurement, separation, gap size, discontinuity size, depth, location and orientation
Dimensional variations	unevenness, nonuniformity, eccentricity, shape and contour, size and mass variations
Thickness, density	film, coating, layer, plating, wall and sheet thickness, density or thickness variations
Physical and mechanical properties	
Electrical properties	resistivity, conductivity, dielectric constant and dissipation factor
Magnetic properties	polarization, permeability, ferromagnetism, cohesive force
Thermal properties	conductivity, thermal time constant and thermoelectric potential
Mechanical properties	compressive, shear and tensile strength (and moduli), Poisson's ratio, sonic velocity, hardness, temper and embrittlement
Surface properties	color, reflectivity, refraction index, emissivity
Chemical composition and analysis	
Elemental analysis	detection, identification, distribution and/or profile
Impurity concentrations	contamination, depletion, doping and diffusants
Metallurgical content	variation, alloy identification, verification and sorting
Physiochemical state	moisture content, degree of cure, ion concentrations and corrosion, reaction products
Stress and dynamic response	
Stress, strain, fatigue	heat-treatment, annealing and cold-work effects, residual stress and strain, fatigue damage and life (residual)
Mechanical damage	wear, spalling, erosion, friction effects
Chemical damage	corrosion, stress corrosion, phase transformation
Other damage	radiation damage and high frequency voltage breakdown
Dynamic performance	crack initiation and propagation, plastic deformation, creep, excessive motion, vibration, damping, timing of events, any anomalous behavior
Signature analysis	
Electromagnetic field	potential, strength, field distribution and pattern
Thermal field	isotherms, heat contours, temperatures, heat flow, temperature distribution, heat leaks, hot spots
Acoustic signature	noise, vibration characteristics, frequency amplitude, harmonic spectrum and/or analysis, sonic and/or ultrasonic emissions
Radioactive signature	distribution and diffusion of isotopes and tracers
Signal or image analysis	image enhancement and quantization, pattern recognition, densitometry, signal classification, separation and correlation, discontinuity identification, definition (size and shape) and distribution analysis, discontinuity mapping and display

MECHANICAL INSPECTION*

The subject of mechanical inspection is too big to cover here completely. In fact, it is possible to become certified in the area of mechanical inspection. Although the quality technician need not be a full-fledged mechanical inspector, some familiarity with the subject is required. This section covers the basic ideas most commonly encountered by quality technicians in mechanical manufacturing.

All methods of measuring length consist essentially of comparing the length of the object to be measured with a calibrated standard of length, which is used as the norm.

First measurements will be considered where the precision required is not high (where units smaller than one one-hundredth of an inch are not considered) that is, so-called non-precision measurements. The calibrated standard employed in such work consists generally of a strip of metal with lines ruled along one edge at a right angle to the edge, the lines being a known distance apart. This known distance is called the interval of graduation, and is rarely less than 0.01". If the standard is designed to measure distances less than four or five feet, it is usually made of a fairly rigid strip, and it is known as a *rule* or *scale*. If it is designed to measure larger distances, it is made of a flexible strip which may be rolled into a compact form, and it is known as a *tape*. The interval of graduation of tapes varies with their use and may be one-sixteenth of an inch, one-tenth of an inch, or one-hundredth of a foot.

Sometimes it is impossible to bring the scale into contact with the object to be measured. It is then necessary that some intermediate device be used to transfer the unknown distance to the scale. *Calipers* are used for this purpose. Two sorts of calipers are shown in Figures III.31 and III.32, one for measuring internal dimensions, the other for external dimensions. When they are in use, the distance between the tips of the calipers is adjusted until it is equal to the dimension to be measured. The distance is then measured by comparing the calipers with a scale (see Figure III.33).

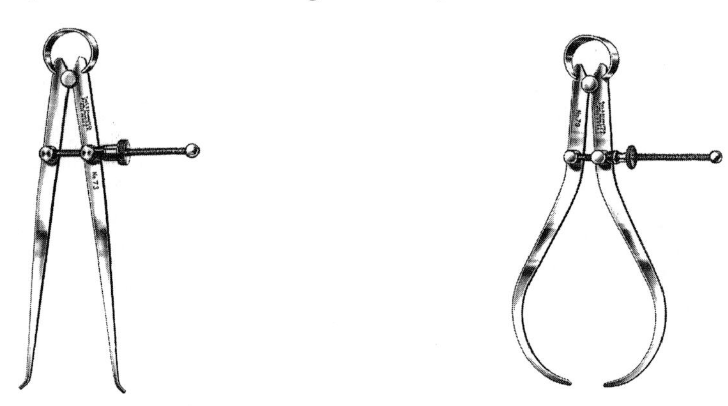

Figure III.31. Internal dimension caliper. **Figure III.32.** External dimension caliper.

*From Baker, R. (1945). "Mensuration," in *Practical Mathematics: Theory and Practice with Applications to Industrial, Business and Military Problems*, R. S. Kimball, editor. New York: National Educational Alliance.

Figure III.33. Pre-setting outside calipers to a rule for a transfer measurement.
(Figures III.31–III.33: Photos used with permission from L.S. Starrett Company.)

Precision instruments

The heart of many precision measuring instruments is the *vernier*. It is a small supplementary scale which enables the user to read with certainty a value which he would otherwise have to estimate on the main scale. One instrument which employs a vernier scale is the vernier caliper, shown in Figure III.34. The small scale engraved on the movable jaw of the instrument is the vernier scale, and it is so placed that, when the jaws are closed, the zero line of the vernier scale coincides with the zero line of the main scale. In use, the jaws are opened to the desired width and clamped. The reading of both scales indicates the spacing between the jaws and hence the dimension. The first reading is taken on the main scale and is that division which falls directly opposite, or most closely precedes, the vernier zero line.

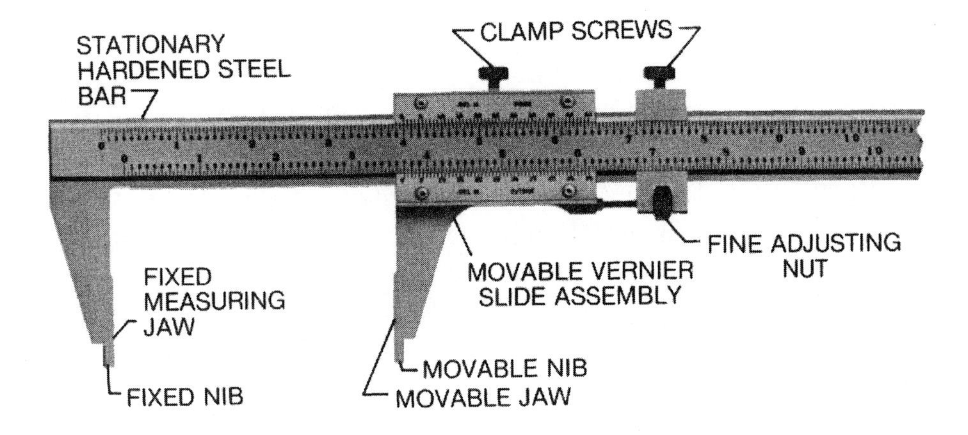

Figure III.34. Vernier caliper.
Photo used with permission from L.S. Starrett Company.

A decimal main scale with a vernier demonstrates the layout of the vernier scale. In Figure III.35, the zero line of the vernier falls at 1.3 on the main scale. Since, in this case, the vernier zero-mark lines up exactly with a graduation on the main scale, the second decimal place is 0 and the reading is 1.30. In Figure III.36, the zero line of the vernier scale falls just beyond 1.3 on the main scale. Here the vernier division, 1, lines up with a graduation on the main scale, and the reading is 1.31. In Figure III.37, the setting is similarly read as 1.36.

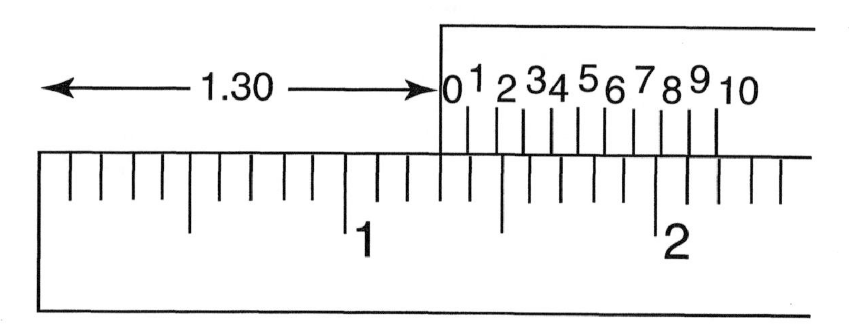

Figure III.35. Vernier reading 1.30.

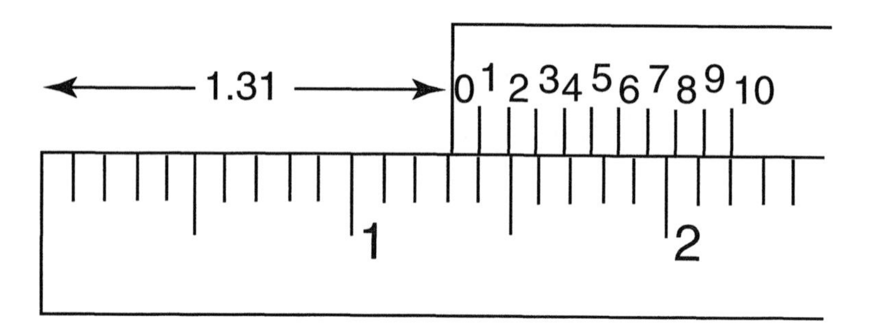

Figure III.36. Vernier reading 1.31.

Figure III.37 is a special case where none of the lines on the vernier corresponds with the graduations on the main scale. In this case, two adjacent vernier divisions very nearly correspond with two main-scale divisions, and the correct reading falls between these two vernier divisions. Here the reading is 1.365. (Note: the vernier scales shown in Figures III.35–III.37 are not designed to represent an instrument as they are sketches simplified in detail to illustrate the reading of the vernier scale.)

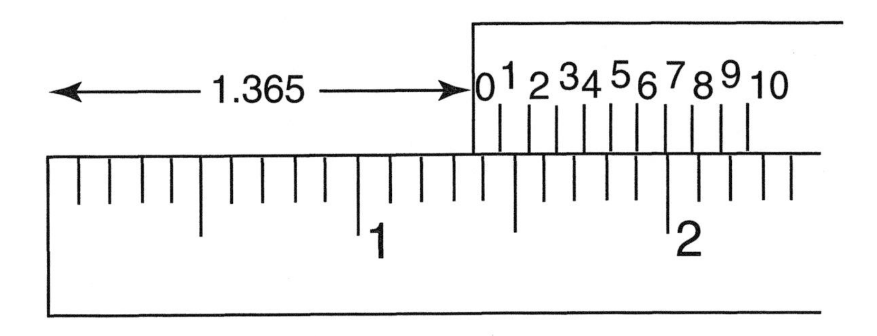

Figure III.37. Vernier reading 1.365.

The explanation of the vernier is not difficult. In this case, one vernier interval equals nine one-hundredths (0.09) of an inch, whereas one main-scale division equals ten one-hundredths (0.10) of an inch. If the vernier zero (0) is opposite a main-scale division, therefore, the vernier 1 is 0.01" from the next main-scale division (Figure III.36). If the vernier scale is moved 0.01" (Figure III.36) the vernier division, 1, lines up with the next main-scale division. Similarly, if the vernier scale is moved another 0.01" (or a total of 0.02"), the vernier division, 2, lines up with the following main-scale division. If the vernier division lines up with a division on the main scale, therefore, it indicates that the vernier 0 has moved that many one-hundredths from the original main-scale division. Thus, the actual dimension is shown on the main scale by the position of the vernier 0, but the precise reading to hundredths of an inch must be taken from the vernier scale.

The main scale of the vernier caliper just described was graduated in one-hundredths of an inch. The main scales of most vernier calipers are graduated to 0.025" and the vernier scales have 25 divisions, which equals 24 divisions of the main scale. The principle of reading is exactly the same as described in the above examples.

The *least count* of a vernier is the change in measurement which will cause a shift of one division on the vernier scale. In the cases illustrated, it is 0.01". In the common vernier caliper described above, where each main-scale division is 0.025", the least count is 0.001".

Micrometers

Another precision measuring instrument which is in universal use is the *micrometer* (Figure III.38). The spacing between the jaws may seldom be varied more than an inch on any single micrometer because the micrometer head would be long and unwieldy. An assortment of frames is often supplied, with one interchangeable head; the smallest frame is used to measure dimensions from 0" to 1", the next from 1" to 2", the next from 2" to 3", and so on. Hence, six micrometer frames of different sizes would be required to cover a range of measurement for which one vernier caliper would usually be sufficient.

The micrometer is built about a screw thread (generally of forty threads per inch) very accurately machined on a steel spindle. The spindle is moved in and out by the revolution of the screw inside the nut on the micrometer frame. A hollow barrel is part of the micrometer frame and holds the nut firmly in place. As the spindle moves inside the nut, the thimble moves outside the barrel. The barrel carries a scale divided into tenths, with four subdivisions for each tenth. One turn of the thimble advances the spindle one-fortieth of an inch (0.025") along the barrel. It is evident that, if the thimble turns through a fraction of a revolution, the spindle will advance the same fraction of one-fortieth of an inch. A graduated scale is etched on the edge of the thimble, dividing the circumference into 25 parts.

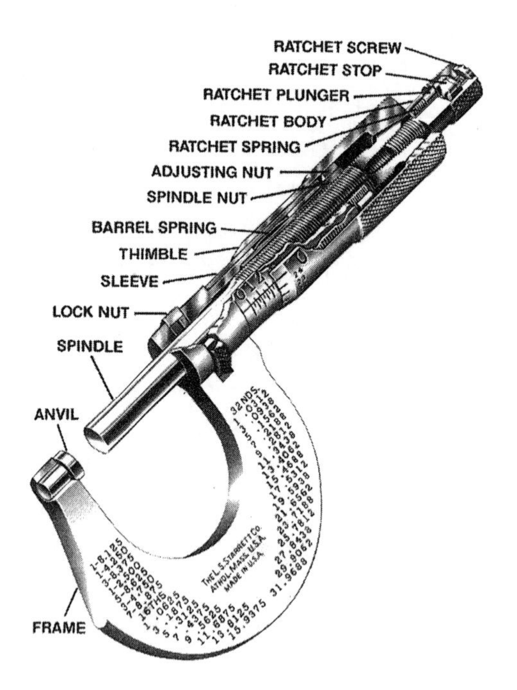

Figure III.38. Micrometer.

Photo used with permission from L.S. Starrett Company

Hence, turning the thimble one division (1/25 revolution) advances the spindle $\left(\frac{1}{25}\times 0.025 = 0.001''\right)$ one one-thousandth of an inch. To enable accurate reading to the fourth decimal place (ten-thousandths), a vernier scale is often ruled on the barrel to be used in conjunction with the thimble scale. The instrument is then known as the *vernier micrometer*. Figure III.39 shows the scale of such an instrument. Modern micrometers usually have digital readouts that replace the graduated thimbles and vernier scales.

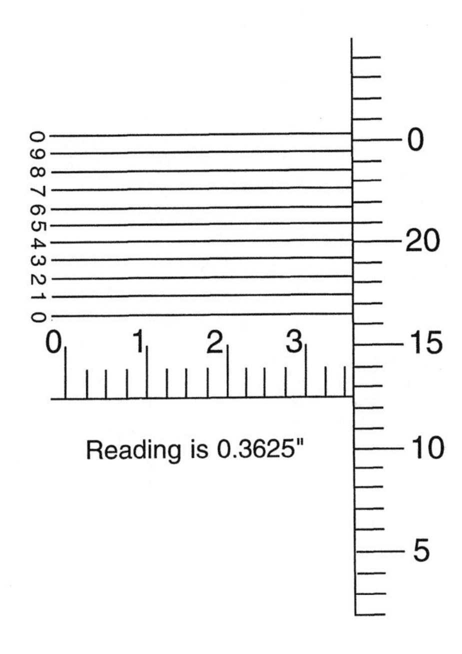

Figure III.39. Micrometer vernier scale.

Micrometers can also be obtained for measuring in the metric system. These have a screw or spindle with 20 threads per cm., so that one complete revolution of the thimble moves the spindle 0.5 mm. Apart from this, the form of construction and method of use are the same as for an instrument calibrated in fractions of an inch. Modern metric micrometers also have digital readouts.

Indicator gages

Another precision device for the measurement of length, and one which has a very broad application, is the *dial gage*, or *dial indicator* or simply *indicator gage* (Figure III.40). It consists of a dial and a pointer (or digital readout) which is connected by a magnifying linkage with a sliding rod projecting from the casing of the instrument. The dial gage is arranged in such a way that a definite displacement of the rod causes the pointer to make one revolution. Commercial dial gages read to one-thousandth or one ten-thousandth of an inch. Dial gages are seldom used to find over-all dimensions, but have a wide use in inspection and control work, such as gaging the thickness of metal sheeting.

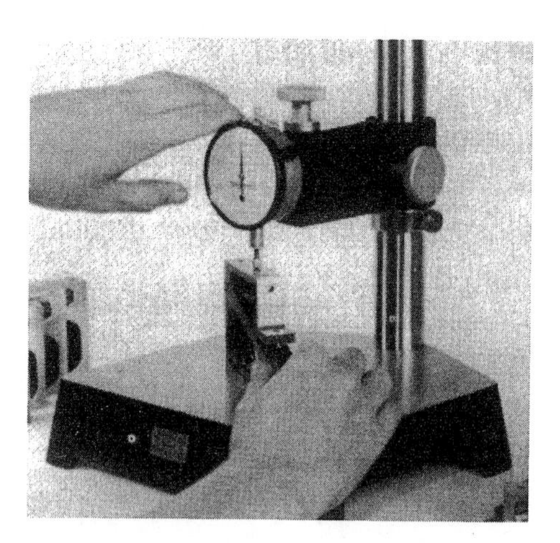

Figure III.40. Dial indicator gage on height stand.
Photo used with permission from L.S. Starrett Company.

Gage blocks

Gage blocks are one of the most accurate measuring standards in common use. Gage blocks are metal blocks, rectangular in all sections, which have been machined and polished until their dimensions between two opposite faces at standard temperature conditions are accurate to a few millionths of an inch. The blocks are made in various sizes and arranged in sets so that common fractional dimensions may be equaled to the nearest ten-thousandth by some combination of blocks.

To get a particular linear dimension with gage blocks, they are built up, one on the other, using a suitable selection of various size blocks, until a "stack" is obtained. For example, assume that a set of gage blocks contains blocks in the following sizes:

0.1001" to 0.1009" in steps of 0.0001"
0.1010" to 0.1090" in steps of 0.0010"
0.1100" to 0.1900" in steps of 0.0100"
0.1000" to 0.5000" in steps of 0.1000"
1.0000" to 4.0000" in steps of 1.0000"

To create a gage block stack of a certain height one begins by evaluating the last digit of the desired size and working right-to-left. For example, if one wishes to build a gage block stack of 5.4321" from the set shown, the following procedure would be employed:

5.4321	
<u>–0.1001</u>	(block 1)
5.332	
<u>–0.102</u>	(block 2)
5.23	
<u>–0.13</u>	(block 3)
5.1	
<u>–0.1</u>	(block 4)
5	
<u>–4</u>	(block 5)
1	
<u>–1</u>	(block 6)

Thus, the stack would consist of six gage blocks measuring 1", 4", 0.1", 0.13", 0.102", and 0.1001".

Gage blocks are stacked using a technique called wringing. Wringing consists of four careful steps:

1. Two properly cleaned blocks are brought together flat and oscillated slightly under very light pressure.
2. While in contact with the lower block, the top block is slid about half way across the lower block.
3. The top block is then rotated and simultaneously brought into approximate alignment with the bottom block.
4. Finally, the edges of the two blocks are precisely aligned.

The procedure requires only very light pressure. The final stack will be held together by vacuum force due to the highly polished surfaces of the two blocks. The stack will be as a single unit rather than as two separate blocks; it will take considerable force to pull the blocks apart. Additional blocks may be added to the stack by repeating the procedure.

Measuring angles

The precision measurement of angles is achieved by the use of a very accurately ruled protractor with a vernier scale attached at the periphery in such a way that it may be moved to any position (see Figure III.41). It is known as a *vernier protractor*. The center of the protractor circle is placed over the vertex of the angle to be measured, the zero line of the protractor is brought in line with a side of the angle, and the vernier is moved around to the position where the zero line of the vernier coincides with the other side of the angle.

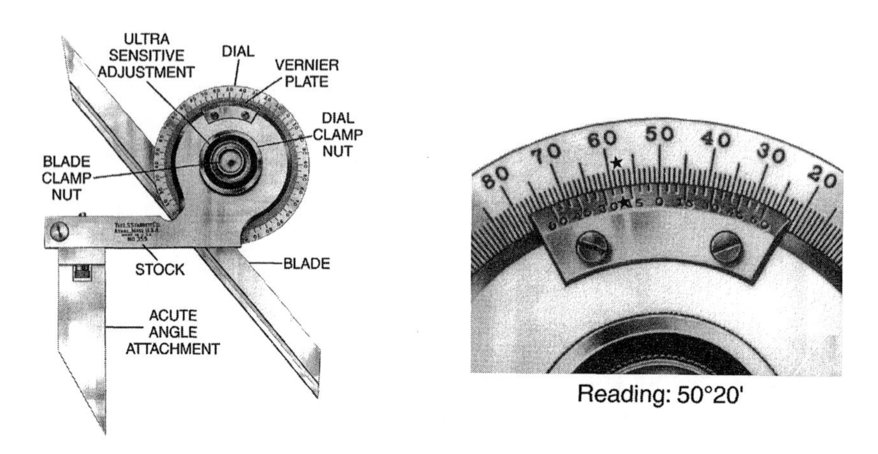

Figure III.41. Vernier protractor scale.
Photos used with permission from L.S. Starrett Company.

Other gaging

There are many other mechanical inspection gages in addition to those discussed here. Table III.7 lists a number of these instruments.

Table III.7. Accuracy and application of various measuring instruments.

TYPE OF GAGE	ACCURACY	APPLICATION
Steel ruler or scale	No better than 0.15 inch	Used to measure heights, depths, diameters, etc., where the tolerance specified is not exact.
Combination square	Within one degree	Used to make miscellaneous angular checks.
Vernier calipers	Usually no better than .001 inch or the increment of graduation	Used to check inside and outside diameters and also thickness.
Vernier depth gage	Usually no better than .001 inch or the increment of graduation	Normally used to check depths but can be used in conjunction with a surface plate to check thickness.

Continued on next page . . .

Table III.7—*Continued . . .*

TYPE OF GAGE	ACCURACY	APPLICATION
Micrometer (outside)	Usually no better than .001 inch or the increment of graduation	Normally used to check diameter or thickness. Special application permits checking thread pitch diameters.
Micrometer (inside)	Usually no better than .001 inch or the increment of graduation	Normally used for checking large bore or hole diameters.
Split sphere and telescope	Within .001 inch but no better than .0005 inch using an accurate micrometer graduated in .0001".	Normally used for measuring small hole diameters.
Fixed snap gages	No set accuracy determined by checking with this gage other than the part is in or out of the specification limit.	Normally used for checking diameters on a production basis where the only criteria is to be within the specification.
Adjustable snap gages	This type of gage is usually accurate within 10% of the tolerance for any given reading.	Measuring diameter on a production basis where an exact measurement is needed, as adjusting for tool wear.
Ring gages	Will only discriminate against diameters larger or smaller than the print specification. Will not check for an out-of-roundness condition.	Best application is to resemble a mating part in assembly.
Plug gages	Accuracy very good for checking the largest or smallest hole or bore diameter. Will not check for an out-of-roundness condition.	Checking the diameter of drilled or reamed holes.
Radius and template gages	Accuracy will be no better than .015 inch.	Used to check small radii, contours, and corner breaks.
Precision straight edge	Visual—no closer than 0.10 inch. Feeler gage—no closer than .003 inch.	Used to check flatness of a face, waviness of a face, or squareness of a face to a reference plane.
Flush pin gages	Accuracy no better than .002 inch.	Are designed to check virtually any characteristic, but are used only on single purpose applications for a continuing operation on a mass production line.
Tapered parallels	Used with an accurate micrometer the accuracy will be no better than .0005 inch.	Used to measure bore sizes. Cannot be used to measure blind hole sizes. Best adapted to a situation where production is low.
Gage blocks	Accuracy of the gage block depends upon the grade used; working sets which are the least precise are manufactured within 10 millionth of the specification.	Gage blocks are best adapted for limited production runs, tool room machining, and as a master for comparison purposes.

Continued on next page . . .

Table III.7—*Continued . . .*

TYPE OF GAGE	ACCURACY	APPLICATION
Surface plates	Flatness expected to be no better than .0005 inch between any two points on the surface.	Best application is to measure overall flatness of a piece or to establish a reference plane from which measurement can be obtained.
Height verniers	Accuracy in reading can be expected no better than .0001 inch.	Best application is in conjunction with a surface plate to check dimensional tolerances.
Tool makers flat	No better than .0005 inch. Accuracy depends upon the instrument used to measure the height obtained with the tool makers flat.	Best application is with a surface plate. Use gage blocks and dial indicator or micrometer to measure height dimensions.
Dial indicator	Accuracy depends upon the type of indicator used. Some indicators will produce measurements within .0001 inch of the specification.	Variety of applications when used correctly, such as: 1. To measure flatness 2. To measure taper 3. To measure concentricity 4. To measure diameters 5. To make machine and height adjustments.
Dial bore gages	Dial bore gages are commonly manufactured accurately within .0001 inch; however, great care must be used to obtain this accuracy.	Used to measure bore diameters, will also give a reading of an out-of-roundness or tapered condition within a bore.
Electronic	Accuracies to .00001 inch, or to micro inches; .000005 inch and even .000001 inch can be obtained.	Best use is for application where the allowable tolerance is less than .0001 inch. Depending upon the design of the gage, this type can be adapted to measure an unlimited number of characteristics.
Automatic sorting gages	Possible to design in accuracy within .0001 inch	Best application is for sorting parts by a dimensional characteristic into different classes.
Air gages	Accuracy depends upon the design of the gage. Air gages have accurately measured tolerances less than 50 millionths of an inch.	Primary application is measuring the diameter of a bore or hole. Other applications include measuring taper, ovality, triangular out of round, flatness, parallelism, squareness, hole location, outside diameters, width, and thickness.
Optical flat	May be accurate to within a few millionths of an inch; however, the accuracy obtained depends upon the skill of the user.	Used only for very precise work such as gage manufacture and tool room work. Best used for checking flatness.

Continued on next page . . .

Table III.7—*Continued . . .*

TYPE OF GAGE	ACCURACY	APPLICATION
Optical comparator	Properly designed for a particular detail, the accuracy can be within .0002 inch.	Best applications are for measuring contours and part configurations which are not easily measured by other methods.
Internal thread gages; External thread gages	No accuracy relating to an exact reading but will discriminate at a given specification limit.	Used for measuring the inside and outside pitch diameters on threaded items.
Coordinate measuring	Accuracy of this type of machine depends upon the part it is designed to check. Coordinate measuring machines have accuracies within 35 millionths of an inch on an axis and within 5 millionths of an inch for T.I.R.	Can be used to measure a variety of characteristics, such as coordinate dimensions on small and large parts, contour, taper, radii, roundness, squareness, etc.

Direct reading vs. differential measurements

There are two different methods for obtaining a measurement: direct measurement and differential measurement. Both methods provide essentially the same information. A direct measurement involves a technique which provides the desired measurement without any additional need for calculation. For example, a part is measured with a caliper and the resulting reading read directly from the caliper's main and vernier scale. A differential reading is obtained by first setting the gage to a specified dimension using gage blocks, a master part, or some other standard. Then the deviations of actual piece parts are measured relative to the standard. For example, a dial indicator might be "zeroed" at a nominal dimension of 0.4250" using a gage block stack on a height stand. Parts would be placed on the height stand and the deviations from nominal, plus or minus, would register on the dial indicator. Dial indicating micrometers also employ this principle. When using differential measurements it is common to record only those digits that actually change as integers, and to record them as deviations from the nominal dimension. For example a reading of 0.4231" would be recorded as –19 (0.4250" – 0.4231" = 0.0019" = 19 ten-thousandths below the nominal = –19).

Differential measurement instruments are often used when extremely high precision is required. Since they reduce the amount of travel in the instrument's sensing mechanism, the effects of drift are minimized. For example, an air gage with 1" of travel and a linearity error of 0.00005" (50 millionths) per inch would exhibit an error of 50 millionths if it were used to check dimensions over its entire range. If it were set with a master and used to check parts that varied over a range of 0.005", then the error over the range of measurements would only be 0.005 x 50 millionths = 0.25 millionths.

Analog versus digital measurement

Dials and scales are examples of *analog* readouts. Analog readouts are the traditional forms of providing information obtained by measuring instruments. There are certain problems encountered in the use of analog readouts. One problem is *parallax error.* Parallax is an observational error caused by the change of position of the observer. For example, one common analog readout is the automobile speedometer. It is not uncommon for the driver and the front-seat passenger to see two different speeds on the speedometer. Another problem is *interpolation error.* When the indicator needle does not point precisely to a mark on the dial, the inspector is required to either *round* the reading to the nearest mark, or to estimate the precise reading using *interpolation.* Either way, the opportunity for error exists.

Digital readouts overcome the problems with analog readouts. As the name implies, a digital readout provides the actual *digits* of measurement. This completely eliminates parallax and interpolation errors; however, the problem of rounding still occurs with some instruments. The ability of the instrument to resolve its rightmost digit, its *resolution,* determines the amount of rounding that occurs. For example, some digital micrometers round to the nearest 0.0005".

One should be careful to note that *readouts* are being discussed, not the accuracy or precision of the measuring instrument. The mere fact that an instrument's readout provides a measurement to a specific number of decimal places does not in any way imply that the instrument is *correct.*

REPEATABILITY AND REPRODUCIBILITY STUDIES

Modern measurement system analysis goes well beyond calibration. A gage can be perfectly accurate when checking a standard and still be entirely unacceptable for measuring a product or controlling a process. This section illustrates techniques for quantifying discrimination, stability, bias, repeatability, reproducibility and linearity for a measurement system. We also show how to express measurement error relative to the product tolerance or the process variation. For the most part, the methods shown here use control charts. Control charts provide graphical portrayals of the measurement process that enable the engineer to detect special causes that numerical methods alone would not detect.

Measurement system discrimination

Discrimination, sometimes called *resolution,* refers to the ability of the measurement system to divide measurements into "data categories." All parts within a particular data category will measure the same. For example, if a measurement system has a resolution of 0.001-inches, then items measuring 1.0002, 1.0003, 0.9997 would all be placed in the data category 1.000; i.e., they would all measure 1.000-inches with this particular measurement system. A measurement system's discrimination should enable it to divide the region of interest into many data categories. In quality engineering, the region of interest is the smaller of the engineering tolerance (the high specification minus the low specification) or six standard deviations. A measurement system should be able to divide the region of interest into at least five data categories. For example, if a process was capable (i.e., six sigma is less than the engineering tolerance) and $\sigma = 0.0005$, then a gage with a discrimination of 0.0005 would be acceptable (six data categories), but one with a discrimination of 0.001 would not (three data categories). When unacceptable discrimination exists, the range chart shows discrete "jumps" or "steps." This situation is illustrated in Figure III.42.

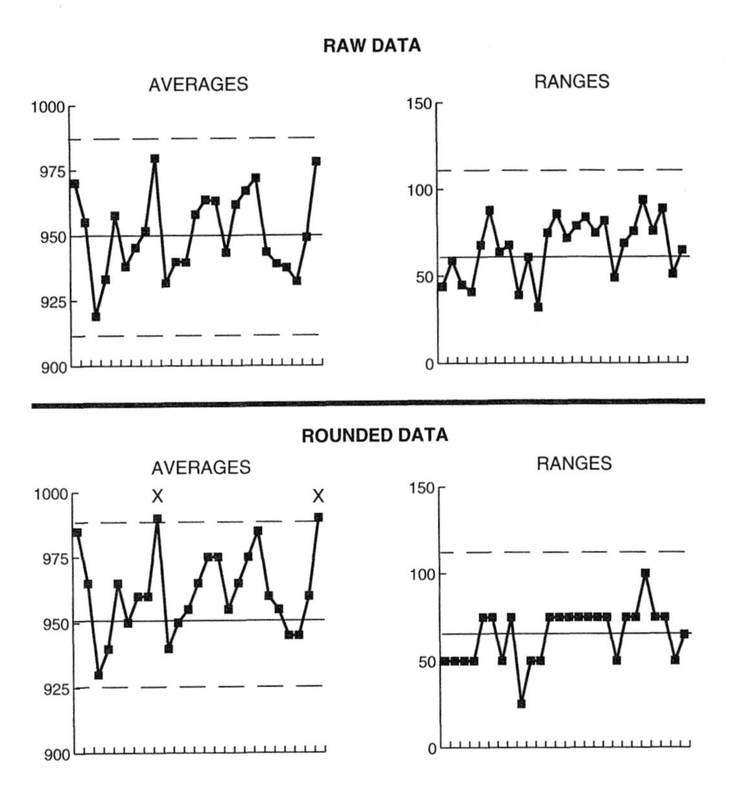

Figure III.42. Inadequate gage discrimination on a control chart.

Note that on the control charts shown in Figure III.42, the data plotted are the same, except that the data on the bottom two charts were rounded to the nearest 25. The effect is most easily seen on the R chart, which appears highly stratified. As sometimes happens (but not always), the result is to make the X-bar chart go out of control, even though the process is in control, as shown by the control charts with unrounded data. The remedy is to use a measurement system capable of additional discrimination, i.e., add more significant digits. If this cannot be done, it is possible to adjust the control limits for the round-off error by using a more involved method of computing the control limits. See Pyzdek (1992, 37–42) for details.

Stability

Measurement system stability is the change in bias over time when using a measurement system to measure a given master part or standard. *Statistical stability* is a broader term that refers to the overall consistency of measurements over time, including variation from *all causes*, including bias, repeatability, reproducibility, etc. A system's statistical stability is determined through the use of control charts. Averages and range charts are typically plotted on measurements of a standard or a master part. The standard is measured repeatedly over a short time, say an hour; then the measurements are repeated at predetermined intervals, say weekly. Subject matter expertise is needed to determine the subgroup size, sampling intervals and measurement procedures to be followed. Control charts are then constructed and evaluated. A (statistically) stable system will show no out-of-control signals on an X-control chart of the averages readings. No "stability number" is calculated for statistical stability; the system either is or is not statistically stable.

Once statistical stability has been achieved, but not before, measurement system stability can be determined. One measure is the process standard deviation based on the R or *s* chart. The equations are given in VI.H and are repeated here for reference.

R chart method:

$$\hat{\sigma} = \overline{R}/d_2$$

s chart method:

$$\hat{\sigma} = \overline{s}/c_4$$

The values d_2 and c_4 are constants from Table 13 in the Appendix.

Bias

Bias is the difference between an observed average measurement result and a reference value. Estimating bias involves identifying a standard to represent the reference value, then obtaining multiple measurements on the standard. The standard might be a master part whose value has been determined by a measurement system with much less error than the system under study, or by a standard traceable to NIST. Since parts and processes vary over a range, bias is measured at a point within the range. If the gage is non-linear, bias will not be the same at each point in the range (see the definition of linearity above).

Bias can be determined by selecting a single appraiser and a single reference part or standard. The appraiser then obtains a number of repeated measurements on the reference part. Bias is then estimated as the difference between the average of the repeated measurement and the known value of the reference part or standard.

Example of computing bias

A standard with a known value of 25.4mm is checked 10 times by one mechanical inspector using a dial caliper with a resolution of 0.025mm. The readings obtained are:

25.425	25.425	25.400	25.400	25.375
25.400	25.425	25.400	25.425	25.375

The average is found by adding the 10 measurements together and dividing by 10,

$$\overline{X} = \frac{254.05}{10} = 25.405$$

The bias is the average minus the reference value, i.e.,

$$\begin{aligned} bias &= average - reference\ value \\ &= 25.405mm - 25.400mm = 0.005mm \end{aligned}$$

The bias of the measurement system can be stated as a percentage of the tolerance or as a percentage of the process variation. For example, if this measurement system were to be used on a process with a tolerance of ±0.25mm then

$$\begin{aligned} \%\ bias &= 100 * |bias| / tolerance \\ &= 100 * 0.005 / 0.5 = 1\% \end{aligned}$$

This is interpreted as follows: this measurement system will, on average, produce results that are 0.005mm larger than the actual value. This difference represents 1% of the allowable product variation. The situation is illustrated in Figure III.43.

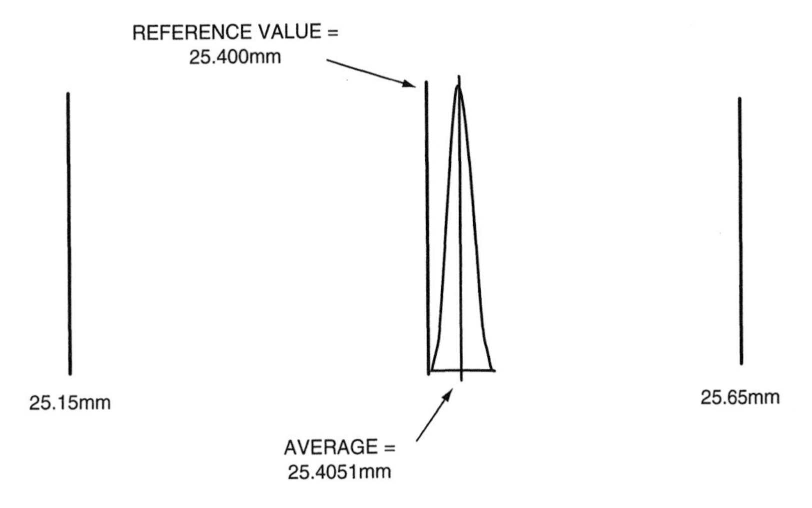

Figure III.43. Bias example illustrated.

Repeatability

A measurement system is repeatable if its variability is consistent. Consistent variability is operationalized by constructing a range or sigma chart based on repeated measurements of parts that cover a significant portion of the process variation or the tolerance, whichever is greater. If the range or sigma chart is out of control, then special causes are making the measurement system inconsistent. If the range or sigma chart is in control then repeatability can be estimated by finding the standard deviation based on either the average range or the average standard deviation. The equations used to estimate sigma are shown above.

Example of estimating repeatability

The data in Table III.8 are from a measurement study involving two inspectors. Each inspector checked the surface finish of five parts; each part was checked twice by each inspector. The gage records the surface roughness in μ-inches (micro-inches). The gage has a resolution of 0.1 μ-inches.

Table III.8. Measurement system repeatability study data.

PART	READING #1	READING #2	AVERAGE	RANGE
	INSPECTOR #1			
1	111.9	112.3	112.10	0.4
2	108.1	108.1	108.10	0.0
3	124.9	124.6	124.75	0.3
4	118.6	118.7	118.65	0.1
5	130.0	130.7	130.35	0.7
	INSPECTOR #2			
1	111.4	112.9	112.15	1.5
2	107.7	108.4	108.05	0.7
3	124.6	124.2	124.40	0.4
4	120.0	119.3	119.65	0.7
5	130.4	130.1	130.25	0.3

We compute:

Ranges chart

$$\overline{R} = 0.51$$

$$UCL = D_4\overline{R} = 3.267 \times 0.51 = 1.67$$

Averages chart

$$\overline{\overline{X}} = 118.85$$

$$LCL = \overline{\overline{X}} - A_2\overline{R} = 118.85 - 1.88 \times 0.51 = 117.89$$

$$UCL = \overline{\overline{X}} + A_2\overline{R} = 118.85 + 1.88 \times 0.51 = 119.81$$

The data and control limits are displayed in Figure III.44. The R chart analysis shows that all of the R values are less than the upper control limit. This indicates that the measurement system's variability is consistent; i.e., there are no special causes of variation.

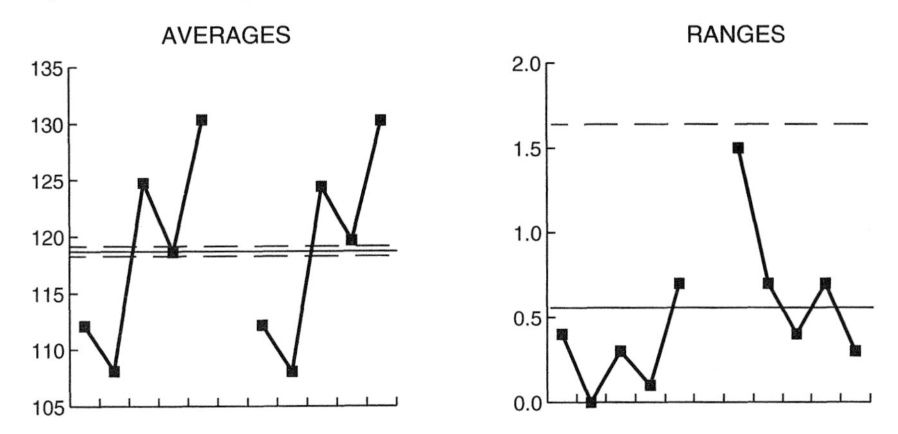

Figure III.44. Repeatability control charts.

Note that many of the averages are outside of the control limits. This is the way it should be! Consider that the spread of the X-bar chart's control limits is based on the average range, which is based on the repeatability error. If the averages *were* within the control limits it would mean that the part-to-part variation was less than the variation due to gage repeatability error, an undesirable situation. Because the R chart is in control we can now estimate the standard deviation for repeatability or gage variation:

$$\sigma_e = \frac{\overline{R}}{d_2^*} \tag{III.4}$$

where d_2^* is obtained from Table 16 in the Appendix. Note that we are using d_2^* and not d_2. The d_2^* values are adjusted for the small number of subgroups typically involved in gage R&R studies. Table 16 is indexed by two values: m is the number of repeat readings taken (m=2 for the example), and g is the number of parts times the number of inspectors ($g=5x2=10$ for the example). This gives, for our example

$$\sigma_e = \frac{\overline{R}}{d_2^*} = \frac{0.51}{1.16} = 0.44$$

The repeatability from this study is calculated by $5.15\sigma_e$=5.15x0.44=2.26. The value 5.15 is the Z ordinate which includes 99% of a standard normal distribution (see VI.C).

Reproducibility

A measurement system is reproducible when different appraisers produce consistent results. Appraiser-to-appraiser variation represents a bias due to appraisers. The appraiser bias, or reproducibility, can be estimated by comparing each appraiser's average with that of the other appraisers. The standard deviation of reproducibility (σ_o) is estimated by finding the range between appraisers (R_o) and dividing by d_2^*. Reproducibility is then computed as $5.15\sigma_o$.

Reproducibility example (AIAG method)

Using the data shown in the previous example, each inspector's average is computed and we find the following:

Inspector #1 average = *118.79 μ-inches*
Inspector #2 average = *118.90 μ-inches*
Range = R_o = *0.11 μ-inches*

Looking in Table 16 in the Appendix for one subgroup of two appraisers we find d_2^*=1.41 (m=2, g=1), since there is only one range calculation g=1. Using these results we find $R_o/\ d_2^*$=0.11/1.41=0.078.

This estimate involves averaging the results for each inspector over all of the readings for that inspector. However, since each inspector checked each part repeatedly, this reproducibility estimate includes variation due to repeatability error. The reproducibility estimate can be adjusted using the following equation, where n is the number of parts, and r is the number of operators:

$$\sqrt{\left(5.15\frac{R_0}{d_2^*}\right)^2-\frac{(5.15\sigma_e)^2}{nr}}=\sqrt{\left(5.15\times\frac{0.11}{1.41}\right)^2-\frac{(5.15\times0.44)^2}{5\times2}}=\sqrt{0.16-0.51}=0$$

As sometimes happens, the estimated variance from reproducibility exceeds the estimated variance of repeatability + reproducibility. When this occurs the estimated reproducibility is set equal to zero, since negative variances are theoretically impossible. Thus, we estimate that the reproducibility is zero.

The measurement system standard deviation is

$$\sigma_m=\sqrt{\sigma_e^2+\sigma_o^2}=\sqrt{(0.44)^2+0}=0.44 \qquad \text{(III.5)}$$

and the measurement system variation, or gage R&R, is $5.15\sigma_m$. For our data gage R&R = 5.15x0.44 = 2.27.

Reproducibility example (alternative method)

One problem with the above method of evaluating reproducibility error is that it does not produce a control chart to assist the engineer with the evaluation. The method presented here does this. This method begins by rearranging the data in Table III.8 so that all readings for any given part become a single row. This is shown in Table III.9.

Table III.9. Measurement error data for reproducibility evaluation.

	INSPECTOR #1		INSPECTOR #2			
Part	Reading 1	Reading 2	Reading 1	Reading 2	X bar	R
1	111.9	112.3	111.4	112.9	112.125	1.5
2	108.1	108.1	107.7	108.4	108.075	0.7
3	124.9	124.6	124.6	124.2	124.575	0.7
4	118.6	118.7	120	119.3	119.15	1.4
5	130	130.7	130.4	130.1	130.3	0.7
				Averages →	118.845	1

Observe that when the data are arranged in this way, the R value measures the combined range of repeat readings plus appraisers. For example, the smallest reading for part #3 was from inspector #2 (124.2) and the largest was from inspector #1 (124.9). Thus, R represents two sources of measurement error: repeatability and reproducibility.

The control limits are calculated as follows:

Ranges chart

$$\overline{R} = 1.00$$

$$UCL = D_4\overline{R} = 2.282 \times 1.00 = 2.282$$

Note that the subgroup size is 4.

Averages chart

$$\overline{\overline{X}} = 118.85$$

$$LCL = \overline{\overline{X}} - A_2\overline{R} = 118.85 - 0.729 \times 1 = 118.12$$

$$UCL = \overline{\overline{X}} + A_2\overline{R} = 118.85 + 0.729 \times 1 = 119.58$$

The data and control limits are displayed in Figure III.45. The R chart analysis shows that all of the R values are less than the upper control limit. This indicates that the measurement system's variability due to the combination of repeatability and reproducibility is consistent; i.e., there are no special causes of variation.

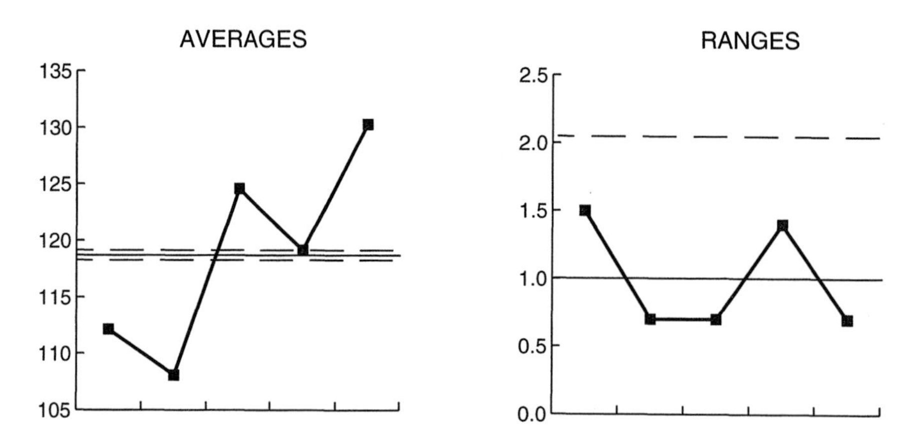

Figure III.45. Reproducibility control charts.

Using this method, we can also estimate the standard deviation of reproducibility plus repeatability as $\sigma_m = R_o/d_2^* = 1/2.08 = 0.48$. Now we know that variances are additive, so

$$\sigma^2_{repeatability+reproducibility} = \sigma^2_{repeatability} + \sigma^2_{reproducibility} \tag{III.6}$$

which implies that

$$\sigma_{reproducibility} = \sqrt{\sigma^2_{repeatability+reproducibility} - \sigma^2_{repeatability}}$$

In a previous example, we computed $\sigma_{repeatability} = 0.44$. Substituting these values gives Using this we estimate reproducibility as $5.15 \times 0.19 = 1.00$.

$$\sigma_{reproducibility} = \sqrt{\sigma^2_{repeatability+reproducibility} - \sigma^2_{repeatability}} = \sqrt{(0.48)^2 - (0.44)^2} = 0.19$$

Part-to-part variation

The X-bar charts show the part-to-part variation. To repeat, if the measurement system is adequate, *most of the parts will fall outside of the X-bar chart control limits.* If fewer than half of the parts are beyond the control limits, then the measurement system is not capable of detecting normal part-to-part variation for this process.

Part-to-part variation can be estimated once the measurement process is shown to have adequate discrimination and to be stable, accurate, linear (see below), and consistent with respect to repeatability and reproducibility. If the part-to-part standard deviation is to be estimated from the measurement system study data, the following procedures are followed:

1. Plot the average for each part (across all appraisers) on an averages control chart, as shown in the reproducibility error alternate method.
2. Confirm that at least 50% of the averages fall outside the control limits. If not, find a better measurement system for this process.
3. Find the range of the part averages, R_p.
4. Compute $\sigma_p = \frac{R_p}{d_2^*}$, the part-to-part standard deviation. The value of d_2^* is found in Table 16 in the Appendix using *m*= the number of parts and *g*=1, since there is only one R calculation.
5. The 99% spread due to part-to-part variation (PV) is found as $5.15\sigma_p$.

Once the above calculations have been made, the overall measurement system can be evaluated.

1. The total process standard deviation is found as $\sigma_t = \sqrt{\sigma_m^2 + \sigma_p^2}$. Where σ_m=the standard deviation due to measurement error.
2. Total variability (TV) is $5.15\sigma_t$.
3. The percent repeatability and reproducibility (R&R) is $100\frac{\sigma_m}{\sigma_t}\%$.
4. The number of distinct data categories that can be created with this measurement system is 1.41x(PV/R&R).

Example of measurement system analysis summary

1. Plot the average for each part (across all appraisers) on an averages control chart, as shown in the reproducibility error alternate method.
   Done above, see Figure III.45.
2. Confirm that at least 50% of the averages fall outside the control limits. If not, find a better measurement system for this process.
   4 of the 5 part averages, or 80%, are outside of the control limits. Thus, the measurement system error is acceptable.
3. Find the range of the part averages, R_p.
   $$R_p = 130.3 - 108.075 = 22.23.$$
4. Compute $\sigma_p = \frac{R_p}{d_2^*}$, the part-to-part standard deviation. The value of d_2^* is found in Table 16 in the Appendix using *m*= the number of parts and *g*=1, since there is only one R calculation.
   $$m = 5,\ g = 1,\ d_2^* = 2.48,\ \sigma_p = 22.23/2.48 = 8.96.$$

5. The 99% spread due to part-to-part variation (PV) is found as $5.15\sigma_p$.

$$5.15x8.96 = PV = 46.15$$

Once the above calculations have been made, the overall measurement system can be evaluated.

1. The total process standard deviation is found as $\sigma_t = \sqrt{\sigma_m^2 + \sigma_p^2}$

$$\sigma_t = \sqrt{\sigma_m^2 + \sigma_p^2} = \sqrt{(0.44)^2 + (8.96)^2} = \sqrt{80.5} = 8.97$$

2. Total variability (TV) is $5.15\sigma_t$.

$$5.15x8.97 = 28.72$$

3. The percent R&R is $100\frac{\sigma_m}{\sigma_t}\%$

$$100\frac{\sigma_m}{\sigma_t}\% = 100\frac{0.44}{8.97} = 4.91\%$$

4. The number of distinct data categories that can be created with this measurement system is 1.41x(PV/R&R).

$$1.41 \times \frac{46.15}{2.27} = 28.72 = 28$$

Since the minimum number of categories is five, the analysis indicates that this measurement system is more than adequate for process analysis or process control.

Linearity

Linearity can be determined by choosing parts or standards that cover all or most of the operating range of the measurement instrument. Bias is determined at each point in the range and a linear regression analysis is performed. Linearity is defined as the slope times the process variance (PV) or the slope times the tolerance, whichever is greater. A scatter diagram should also be plotted from the data.

Linearity example

The following example is taken from *Measurement Systems Analysis*, published by the Automotive Industry Action Group.

A plant foreman was interested in determining the linearity of a measurement system. Five parts were chosen throughout the operating range of the measurement system based upon the process variation. Each part was measured by a layout inspection to determine its reference value. Each part was then measured twelve times by a single appraiser. The parts were selected at random. The part average and bias average were calculated for each part as shown in Figure III.46. The part bias was calculated by subtracting the part reference value from the part average.

PART →	1	2	3	4	5
Average	2.49	4.13	6.03	7.71	9.38
Ref. Value	2.00	4.00	6.00	8.00	10.00
Bias	+0.49	+0.13	+0.03	–0.29	–0.62

Figure III.46. Gage data summary.

A linear regression analysis was performed using the methods described in VI.E. In the regression, x is the reference values and y is the bias. The results are shown below:

SUMMARY OUTPUT					
Regression statistics					
Multiple R	0.98877098				
R Square	0.97766805				
Adjusted R Square	0.97022407				
Standard Error	0.07284687				
Observations	5				
ANOVA					
	df	*SS*	*MS*	*F*	*Significance F*
Regression	1	0.69696	0.69696	131.336683	0.00142598
Residual	3	0.01592	0.00530667		
Total	4	0.71288			
	Coefficients	*Standard error*	*t Stat*	*P-value*	
Intercept	0.74	0.07640244	9.68555413	0.0023371	
Ref. Value	–0.132	0.0115181	-11.460222	0.00142598	

Figure III.47. Regression analysis of linearity summary data.

The p-values indicate that the result is statistically significant; i.e., there is actually a bias in the gage. The slope of the line is –0.132, and the intercept is 0.74. R^2=0.98, indicating that the straight line explains about 98% of the variation in the bias readings. The results can be summarized as follows:

Bias	b + ax = 0.74 – 0.132 (Reference Value)
Linearity	|slope| x Process Variation = 0.132 x 6 = 0.79, where 6 is the tolerance
% linearity	100% x |slope| = 13.2%

Note that the zero bias point is found at

$$x = -\left(\frac{\text{intercept}}{\text{slope}}\right) = -\left(\frac{0.74}{-0.132}\right) = 5.69$$

In this case, this is the point of least bias. Greater bias exists as you move further from this value.

This information is summarized graphically in Figure III.48.

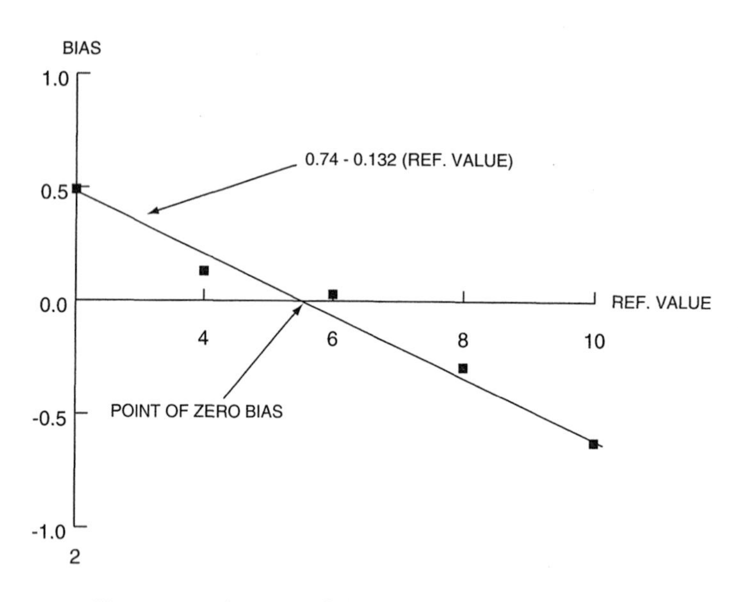

Figure III.48. Graphical analysis of linearity.

METROLOGY

The term *metrology* derives from the Greek word "metron" (measurement) and "logos" (science). Thus, quite literally, metrology is the science of measurement. In practice, the term is usually taken to mean the science of (1) providing, maintaining, and disseminating a consistent set of measurement units, (2) providing support for the fair and equitable transaction of trade, and (3) providing data for quality control of manufacturing.

Traceability to standards

Standards form the basis of metrology. They provide a well-defined trail from measurement of objects or chemicals with physical properties (size, weight, purity, etc.) that are universally accepted as accurate or "true," to measurements of objects or chemicals with unknown physical properties. In the United States, the principal standards organization is the National Institute of Standards and Technology (NIST), formerly the National Bureau of Standards. NIST serves as the repository for most of the nation's physical and chemical measurement standards. NIST coordinates its measurement standards with the standards of other countries and develops and distributes measurement and calibration procedures. One of NIST's functions is to transfer measurements from its standards to other measurement systems with a hierarchical system of transfers using an approved calibration procedure. The transfer system is illustrated in Figure III.49. A measurement is said to be "traceable to NIST" if it can be connected to NIST through an unbroken chain of measurements on standards in the hierarchy, obtained using proper calibration procedures.

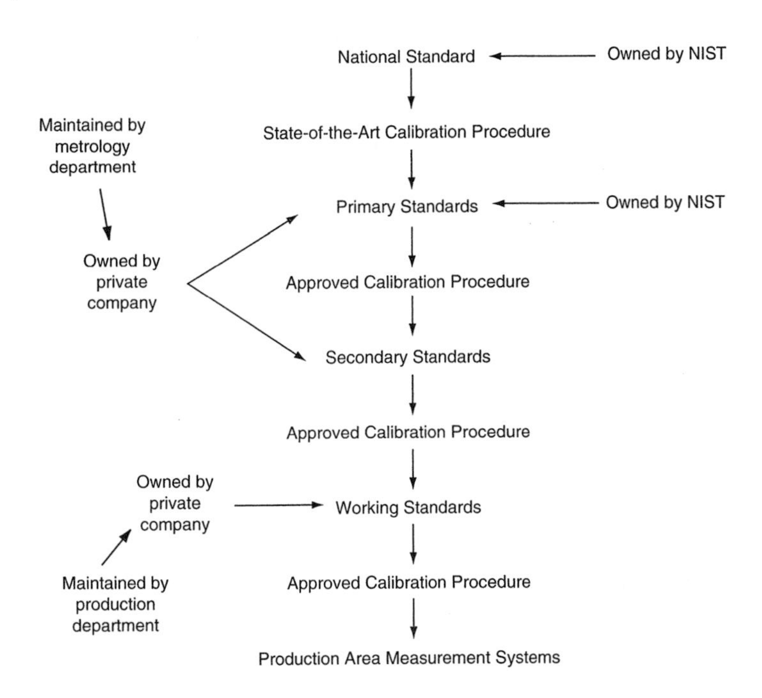

Figure III.49. Hierarchical system of transfers.

Measurement error

The difference between an actual property and a measurement of that property is termed *measurement error.* In quality engineering, the actual property is usually taken to be the standard, which is exact by definition. Measurement error is usually determined for a particular measurement system. A measurement system is more than simply the measuring instrument; it includes the entire process of measurement. A measurement process consists of inputs, actions, and outputs. The inputs include the item to be measured, the measurement equipment, the environmental controls, the written procedures to be followed, the people involved, and so on. The actions include the way in which the procedures are followed, the storage of the items to be measured, calibration procedures, analysis of the data, and so on. The outputs are the measurements obtained. As mentioned above, measurement error is typically described in terms of precision, resolution, accuracy, bias, repeatability, and linearity. Specific methods for quantifying these terms are described earlier in this chapter.

Calibration systems

The historical standard for the design of calibration systems is Mil-Std-45662, although more modern systems generally reference the updated ANSI/ASQ M1-1996 or ISO/DIS 10012: *Quality assurance requirements for measuring equipment.* Excerpts from Mil-Std-45662 are included in the Appendix. Electronic copies of the ANSI/ASQ M1-1996 may be obtained for a fee from <http://www.ansi.org>. Most calibration systems, including those of commercial as well as military contractors, are designed to meet the criteria described in these standards.

Traditionally the term *calibration* was taken to mean that the gage was accurate; i.e., it was able to measure the appropriate standard to within a prescribed maximum error. As can be seen, calibration defined in this way equates to accuracy. Furthermore, the gage is usually taken out of the work area and into a laboratory for calibration. This, of course, removes variation due to the environment, the inspector, and a variety of other possible sources. Thus, most calibration systems focus only on assuring the accuracy of gages vis-à-vis a standard. In the past, control of measurement quality focused almost entirely on gage calibration. Modern quality demands much more. Today gage calibration, while still important, is only one of several methods used to assure meaningful measurements.

Calibration involves three considerations:

1. Identification of variables that might contribute to instrument uncertainty.
2. Measurement of the error contributed by each variable.
3. Determination of the effect of the interaction of the variables on the instrument's measurement capability.

As mentioned above, calibration focuses on variation from standards; it cannot properly account for many of the variables that contribute to uncertainty. However, despite its shortcomings, calibration can still provide valuable information on gage quality. A great deal of the value added by calibration takes place prior to comparing the measurements to the standard value. Figure III.50 lists some questions that should be answered prior to performing the calibration.

IDENTIFICATION

- ❑ Does the information accompanying the instrument agree with the serial number and/or description of the instrument?
- ❑ Do you have the record of past calibrations?
- ❑ Do you have the manual for the instrument?
- ❑ Do you have all applicable specifications?
- ❑ Do you have a report from the user concerning real or imaginary troubles with the instrument?
- ❑ Do you have instructions for disposing of the instrument after calibration?

REQUIREMENTS

- ❑ Is the environment for the calibration suitable for the precision expected? Have the following been checked and minimized: drafts, temperature change, vibration, interference?
- ❑ Are suitable standards on hand and are they in calibration?
- ❑ Are the necessary calibration instruments and accessories on hand and are they in calibration?
- ❑ Is a heat sink of adequate capacity on hand?
- ❑ Are the necessary supplies on hand?
- ❑ Are the necessary packaging materials on hand?
- ❑ Are paper and pencil on hand?

PREPARATION

- ❑ Have the instrument, the standards, and the calibration instruments been normalized?
- ❑ Has the instrument been visually inspected?
- ❑ Have the references and contact surfaces been inspected for damage, wear, and alignment?

Figure III.50. Precalibration check list.

From Harlow, et.al. Used by permission of the publisher.

Control system

A primary concern in calibration is the existence of a documented system which assures that all measuring instruments used to determine quality are properly calibrated. If employee-owned gaging is allowed, they too must be covered by the calibration system. The calibration system involves establishing a system for identifying the calibration status of each measurement instrument and periodically updating the status by recalibration. The calibration system will describe the standards to be used, the calibration interval, calibration procedures to be followed, accuracy required, and how to deal with unacceptable results. A sample gage calibration procedure is shown in Figure III.51.

1.0 Purpose

To establish a system for maintaining the accuracy of gages, tools, and measuring instruments used for the inspection of material purchased or manufactured.

2.0 Scope

Applies to all inspection gages, tools, and measuring instruments used for judging the quality of products or processes.

3.0 Definitions

3.1 The term "gage" is used to describe inspection tools, fixtures, ring gages, thread and cylindrical plugs, dial indicators, and other devices used for determining conformance of the product to dimensional specification.

4.0 Gage storage

All mechanical gages and measuring instruments will be stored in the Calibration Laboratory when not in use, or in other locations as approved by the Quality Engineering Manager.

5.0 Identification

All gages and instruments, whether owned by the company or by an employee, will be identified with a gage-control number.

6.0 Care of gages

The calibration technician will monitor the location, care, and condition of all gages. Responsibility for maintaining the gage in good condition will rest with the persons using the gage. Any gage believed to be damaged or otherwise impaired will be removed from the work area to the calibration laboratory immediately.

7.0 Control and records

All gages and measuring instruments will be identified with a unique record number recorded in the gage control system. The number will be permanently affixed to the gage.

Figure III.51—*Continued on next page . . .*

Figure III.51—*Continued . . .*

8.0 Frequency of calibration

The Calibration Laboratory Manager will establish the recalibration interval for each class of gages.

9.0 Authorization

The Calibration Technician is instructed to inspect and calibrate all gages and instruments on their calibration due dates and is authorized to remove from service any gage that does not meet required standards. No gage will be used beyond its calibration due date.

10.0 Employee-owned measuring instruments

Employees are allowed to own and use their own measuring instruments provided these instruments are approved by Quality Engineering. Employee-owned instruments are covered by the same calibration procedures as company-owned instruments.

Figure III.51. Procedure for control of mechanical gages and measuring instruments.

Control of standard integrity

There are many different classes of standards in common use, including standards for mass, length, flatness, and a wide variety of other properties. As discussed above, a great deal of effort goes into maintaining traceability of the standards to the highest-level national standard. The assumption is that even the lowest-level working standards are valid surrogates of the national standards, which are exact by definition. The integrity of the national standard is maintained by elaborate security systems, extraordinary environmental controls, rigorous handling, and storage requirements, etc. This level of control is appropriate for those standards which will serve as the basis of trade for an entire nation. However, even working standards require a significant effort if their integrity is to be maintained. After all, these standards form the basis for decision-making involving thousands, or even millions, of dollars of production. Each standard requires its own appropriate method of control. Chemical standards require different procedures than dimensional gages. A complete discussion of instrument-specific procedures is beyond the scope of this book. However, for illustration, some considerations for mechanical instruments are discussed.

Most modern mechanical inspection standards have tolerances measured in the millionths of an inch or finer. These standards are highly susceptible to damage from a wide variety of causes, all of which must be carefully controlled:

- Handling damage can be caused from mere contact with human skin, which produces oils, acids, salts and many other contaminants which can effect the standard. Clean hands are an absolute minimum, with clean, white, lint-free gloves being preferred.
- Ambient air contamination must be carefully controlled to prevent damage from airborne pollutants from tobacco smoke, cleaning processes,and so on.
- A temperature-controlled environment may be necessary to prevent thermal-stress damage; this includes storage away from direct sunlight, taking care that the standard is not set near hotplates, etc.
- Oxidation must be carefully monitored, even for such corrosion-resistant materials as aluminum and stainless steel.
- Contact with abrasives must be avoided. This includes contact with ordinary surfaces in the laboratory such as counter surfaces.
- Use of any object that might scratch the standard must be avoided. This includes objects that, to the naked eye, appear smooth, such as spherical contact points. Even a well-rounded spherical surface under one ounce of pressure can cause a dent of a few millionths of an inch in a hardened steel gage block.

Despite all attempts, the standards will eventually wear or become damaged due to all of the causes mentioned above, and many more. The standard should be periodically calibrated using an approved procedure. Check the entire standard to determine if some parts have been affected by wear or damage. However, bear in mind that calibration itself can create problems; don't recalibrate any more than is necessary.

Reliability and Risk Management

IV.A TERMS AND DEFINITIONS

While common usage of the term *reliability* varies, its technical meaning is quite clear: *reliability* is defined as the probability that a product or system will perform a specified function for a specified time without failure. For the reliability figure to be meaningful, the operating conditions must be carefully and completely defined. Although reliability analysis can be applied to just about any product or system, in practice it is normally applied only to complex products. Formal reliability analysis is routinely used for both commercial products, such as automobiles, as well as military products like missiles.

RELIABILITY TERMS

MTBF—Mean time between failures, μ. When applied to repairable products, this is the average time a system will operate until the next failure.

Failure rate—The number of failures per unit of stress. The stress can be time (e.g., machine failures per shift), load cycles (e.g., wing fractures per 100,000 deflections of six inches), impacts (e.g., ceramic cracks per 1,000 shocks of 5 g's each), or a variety of other stresses. The failure rate $\Theta = 1 / \mu$.

MTTF or MTFF—The mean time to first failure. This is the measure applied to systems that can't be repaired during their mission. For example, the MTBF would be irrelevant to the *Voyager* spacecraft.

MTTR—Mean time to repair. The average elapsed time between a unit failing and its being repaired and returned to service.

Availability—The proportion of time a system is operable. Only relevant for systems that can be repaired. Availability is given by the equation

$$\text{Availability} = \frac{\text{MTBF}}{\text{MTBF} + \text{MTTR}} \tag{IV.1}$$

b_{10} life*—The life value at which 10% of the population has failed.

b_{50} life—The life value at which 50% of the population has failed. Also called the median life.

Fault Tree Analysis (FTA)—Fault trees are diagrams used to trace symptoms to their root causes. *Fault tree analysis* is the term used to describe the process involved in constructing a fault tree. (See below for additional discussion.)

Derating—Assigning a product to an application that is at a stress level less than the rated stress level for the product. This is analogous to providing a safety factor.

Censored test—A life test where some units are removed before the end of the test period, even though they have not failed.

Maintainability—A measure of the ability to place a system that has failed back in service. Figures of merit include availability and mean time to repair.

IV.B RELIABILITY LIFE CHARACTERISTIC CONCEPTS

The life cycle of many complex systems can be modeled using a curve called the "bathtub curve," which is shown in Figure IV.1.

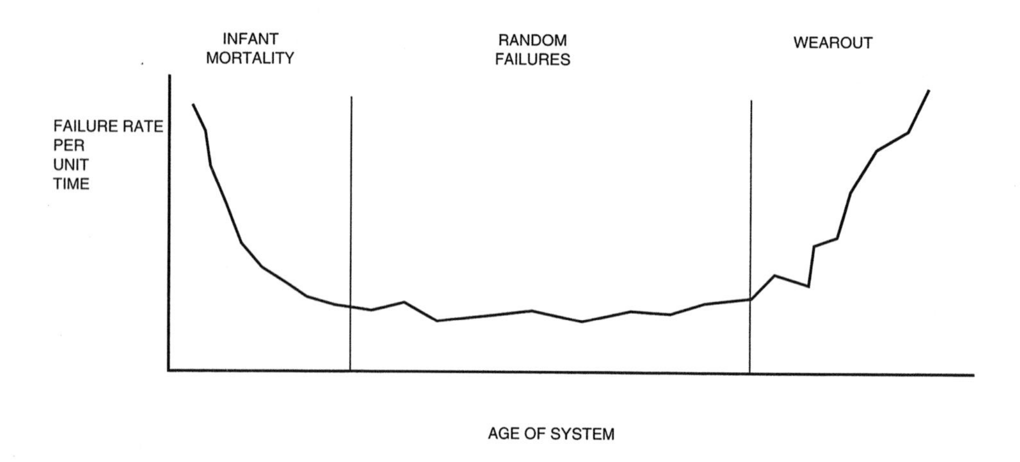

Figure IV.1. The "bathtub" curve of a system's life cycle.

The bathtub curve is typically modeled using the Weibull distribution, which is flexible enough to cover all three phases of the life cycle (Dodson 2002). The pdf of the Weibull distribution is shown in Equation IV.2.

$$f(t)=\frac{\beta(t-\delta)^{\beta-1}}{\theta^{\beta}}\exp\left[-\left(\frac{t-\delta}{\theta}\right)^{\beta}\right],\ t\geq\delta \qquad \text{(IV.2)}$$

*b_{10} life and b_{50} life are terms commonly applied to the reliability of ball bearings.

where β (beta) is the shape parameter, θ (theta) is the scale parameter, and δ (delta) is the location parameter. With three parameters, the Weibull distribution is very flexible. The parameter β provides information as to which phase of the life cycle the system is in. These phases, and the associated value of β, are discussed below.

Infant mortality—The first portion of the bathtub curve, called the infant mortality period, is characterized by a high but rapidly declining rate of failure, called the *infant mortality period.* Systems in the infant mortality phase of their life cycle are deemed unsuitable for routine operation. Typically these early failures are weeded out by *burning in* the system. Burn-in involves running the system under conditions that simulate a high stress operating environment for a period of time sufficient to allow the failure rate to stabilize. If the Weibull distribution is used to characterize reliability, the Weibull parameter β will be less than 1.0 during the infant mortality phase of the life cycle (Dodson 2002).

Random failures—The area of the bathtub curve where the failure rate is relatively constant is the random failure or "useful life" portion of the system. This is the phase of the product's life cycle where it is intended to operate. If the Weibull distribution is used to characterize reliability, the Weibull parameter β will be approximately 1 during the random failure phase of the life-cycle. When β=1 (with reasonable confidence), the Weibull distribution is identical to the exponential distribution.

Wearout—The useful-life phase is followed by increasing failure rates, which is known as the wearout region for the system. If the Weibull distribution is used to characterize reliability, the Weibull parameter β will be greater than 1 during the wearout phase of the life-cycle. To improve reliability, the maintenance plan usually requires replacing units prior to their entering the wearout phase of the life cycle.

While useful, the bathtub curve is not without its critics. In particular, when replacement prior to wearout occurs, the bathtub curve fails to materialize, a situation with obvious advantages. George (1994) describes some alternative techniques to use when planning an optimal replacement schedule.

IV.C DESIGN OF SYSTEMS FOR RELIABILITY

The reliability of a given system is dependent on the reliability of its individual elements combined with how the elements are arranged. For example, a set of Christmas tree lights might be configured so that the entire set will fail if a single light goes out. Or it may be configured so that the failure of a single light will not effect any of the other lights (question: do we define such a set as having failed if only one light goes out? If all but one go out? Or some number in between?).

MATHEMATICAL MODELS

The mathematics of reliability analysis is a subject unto itself. Most systems of practical size require the use of high speed computers for reliability evaluation. However, an introduction to the simpler reliability models is extremely helpful in understanding the concepts involved in reliability analysis.

One statistical distribution that is very useful in reliability analysis is the exponential distribution, which is given by Equation IV.2.

$$R = \exp\left[-\frac{t}{\mu}\right],\ \mathrm{t} \geq 0 \qquad \text{(IV.3)}$$

In Equation IV.3, R is the system reliability, given as a probability; t is the time the system is required to operate without failure; μ is the mean time to failure for the system. The exponential distribution applies to systems operating in the constant failure rate region, which is where most systems are designed to operate.

RELIABILITY APPORTIONMENT

Since reliability analysis is commonly applied to complex systems, it is logical that most of these systems are composed of smaller subsystems. Apportionment is the process involved in allocating reliability objectives among separate elements of a system. The final system must meet the overall reliability goal. Apportionment is something of a hybrid of project management and engineering disciplines.

The process of apportionment can be simplified if we assume that the exponential distribution is a valid model. This is because the exponential distribution has a property that allows the system failure rate to be computed as the reciprocal of the sum of the failure rates of the individual subsystems. Table IV.1 shows the apportionment for a home entertainment center. The complete system is composed of a tape deck, television, compact disk unit, and a phonograph. Assume that the overall objective is a reliability of 95% at 500 hours of operation.

Table IV.1. Reliability apportionment for a home entertainment system.

SUBSYSTEM	RELIABILITY	UNRELIABILITY	FAILURE RATE	OBJECTIVE
Tape deck	0.990	0.010	0.00002	49,750
Television	0.990	0.010	0.00002	49,750
Compact disk	0.985	0.015	0.00003	33,083
Phonograph	0.985	0.015	0.00003	33,083
	0.950	0.050		
TAPE DECK SUBSYSTEM				
SUBSYSTEM	RELIABILITY	UNRELIABILITY	FAILURE RATE	OBJECTIVE
Drive	0.993	0.007	0.000014	71,178
Electronics	0.997	0.003	0.000006	166,417
	0.990	0.010		

The apportionment could continue even further; for example, we could apportion the drive reliability to pulley, engagement head, belt, capstan, etc. The process ends when it has reached a practical limit. The column labeled "objective" gives the minimum acceptable mean time between failures for each subsystem in hours. MTBFs below this will cause the entire system to fail its reliability objective. Note that the required MTBFs are huge compared to the overall objective of 500 hours for the system as a whole. This happens partly because of the fact that the reliability of the system as a whole is the *product* of the subsystem reliabilities which requires the subsystems to have much higher reliabilities than the complete system.

Reliability apportionment is very helpful in identifying design weaknesses. It is also an eye opener for management, vendors, customers, and others to see how the design of an entire system can be dramatically affected by one or two unreliable elements.

SERIES

A system is in series if all of the individual elements must function for the system to function. A series system block diagram is shown in Figure IV.2.

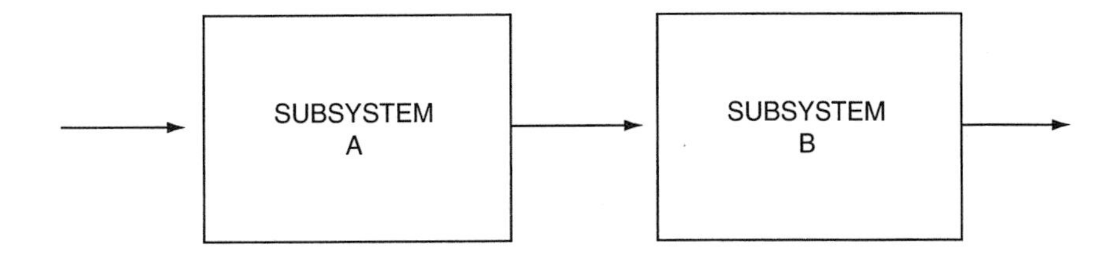

Figure IV.2. A series configuration.

In the above diagram, the system is composed of two subsystems, A and B. Both A and B must function correctly for the system to function correctly. The reliability of this system is equal to the product of the reliabilities of A and B, in other words

$$R_S = R_A \times R_B \tag{IV.4}$$

For example, if the reliability of A is 0.99 and the reliability of B is 0.92, then $R_s = 0.99 \times 0.92 = 0.9108$. Note that with this configuration, R_s is always less than the *minimum* of R_A or R_B. This implies that the best way to improve the reliability of the system is to work on the system component that has the lowest reliability.

PARALLEL

A parallel system block diagram is illustrated in Figure IV.3. The system will function as long as A and B and C have not all failed. The reliability of this type of configuration is computed using Equation IV.5.

$$R_S = 1 - (1 - R_A)(1 - R_B)(1 - R_C) \tag{IV.5}$$

For example, if $R_A = 0.90$, $R_B = 0.95$, and $R_C = 0.93$ then $R_s = 1 - (0.1 \times 0.05 \times 0.07) = 1 - 0.00035 = 0.99965$.

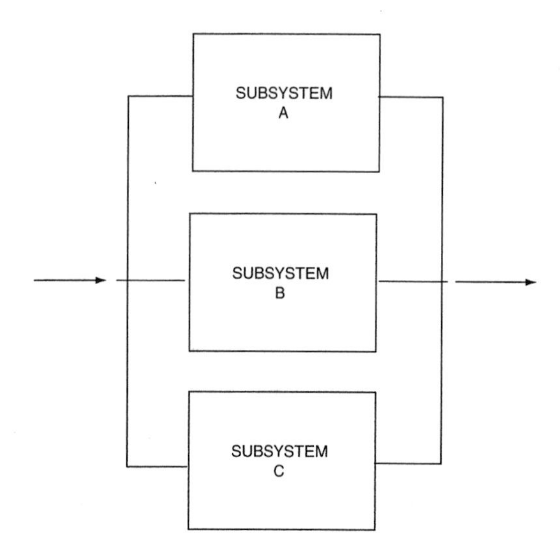

Figure IV.3. A parallel system.

With parallel configurations, the system reliability is always better than the best subsystem reliability. Thus, when trying to improve the reliability of a parallel system you should first try to improve the reliability of the best component. This is precisely opposite of the approach taken to improve the reliability of series configurations.

Series-parallel systems

Systems often combine both series and parallel elements. Figure IV.4 illustrates this approach.

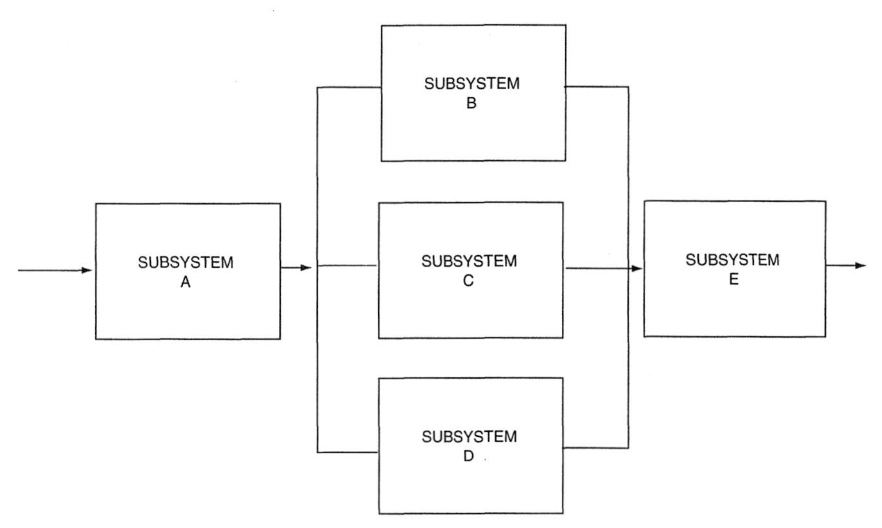

Figure IV.4. A series-parallel system.

The reliability of this system is evaluated in two steps:

1. Convert B-C-D into a single element by solving the parallel configuration.
2. Solve A,B-C-D, E as a series configuration. E.g., Assume RA=0.99, RB=0.90, RC=0.95, RD=0.93 and RE=0.92.

   Then the system reliability Rs is found as follows:

   1. $R_{B,C,D} = 1 - (0.10 \times 0.05 \times 0.07) = 0.99965$.
   2. $R_s = 0.99 \times 0.99965 \times 0.92 = 0.9105$.

Note that this system combines the previously shown series and parallel systems.

REDUNDANT

Redundancy exists when there is more than one component or subsystem that can perform the same function. The parallel configuration shown in Figure IV.4 is an example of a redundant system. Redundancy is a way of dealing with the unreliability of certain subsystems and components. By arranging these elements in parallel configuration, the overall system's reliability can be designed to meet the overall objective.

Another way of dealing with unreliable subsystems and components is to design *standby systems* or units. Standby systems are systems where one or more elements is inactive until some other element or elements in the system fail. In addition to the standby elements, there is also a sensor to detect the conditions required to place the inactive element in service as well as some switching mechanism for engaging the inactive element. Figure IV.5 depicts a standby system.

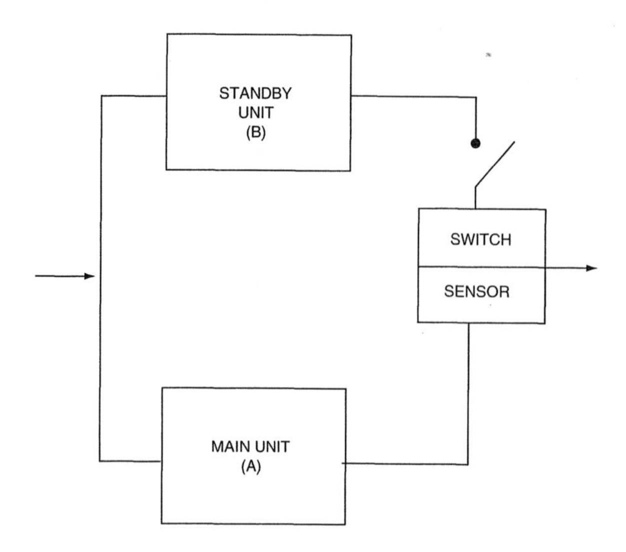

Figure IV.5. Standby system.

The reliability of a standby system is computed with Equation IV.6.

$$R_S = R_A + (1 - R_A)R_{SE}R_{SW}R_B \qquad \text{(IV.6)}$$

In Equation IV.6, R_{SE} is the reliability of the sensor and R_{SW} is the reliability of the switch.

IV.D RELIABILITY AND MAINTAINABILITY

SEVEN STEPS IN PREDICTING DESIGN RELIABILITY

1. Define the product and its functional operation. Use functional block diagrams to describe the systems. Define failure and success in unambiguous terms.
2. Use reliability block diagrams to describe the relationships of the various system elements (e.g., series, parallel, etc.).
3. Develop a reliability model of the system.
4. Collect part and subsystem reliability data. Some of the information may be available from existing data sources. Special tests may be required to acquire other information.
5. Adjust data to fit the special operating conditions of your system. Use care to assure that your "adjustments" have a scientific basis and are not merely reflections of personal opinions.
6. Predict reliability using mathematical model. Computer simulation may also be required.
7. Verify your prediction with field data. Modify your models and predictions accordingly.

SYSTEM EFFECTIVENESS

The effectiveness of a system is a broader measure of performance than simple reliability. There are three elements involved in system effectiveness:

1. Availability.
2. Reliability.
3. Design capability, i.e., assuming the design functions, does it also achieve the desired result?

System effectiveness can be measured with Equation IV.7.

$$P_{SEf} = P_A \times P_R \times P_C \tag{IV.7}$$

In this equation, P_{SEf} is the probability the system will be effective, P_A is the availability as computed with Equation IV.1, P_R is the system reliability, and P_C is the probability that the design will achieve its objective.

SAFETY

Safety and reliability are closely related. A safety problem is created when a critical failure occurs, which reliability theory addresses explicitly with such tools as FMECA and FTA. The modern evaluation of safety/reliability takes into account the probabilistic nature of failures. With the traditional approach a safety factor would be defined using Equation IV.8.

$$SF = \frac{\text{average strength}}{\text{worst expected stress}} \tag{IV.8}$$

The problem with this approach is quite simple: it doesn't account for variation in either stress or strength. The fact of the matter is that both strength and stress will vary over time, and unless this variation is dealt with explicitly we have no idea what the "safety factor" really is. The modern view is that a safety factor is the difference between an improbably high stress (the maximum expected stress, or "reliability boundary") and an improbably low strength (the minimum expected strength). Figure IV.6 illustrates the modern view of safety factors. The figure shows two *distributions*, one for stress and one for strength.

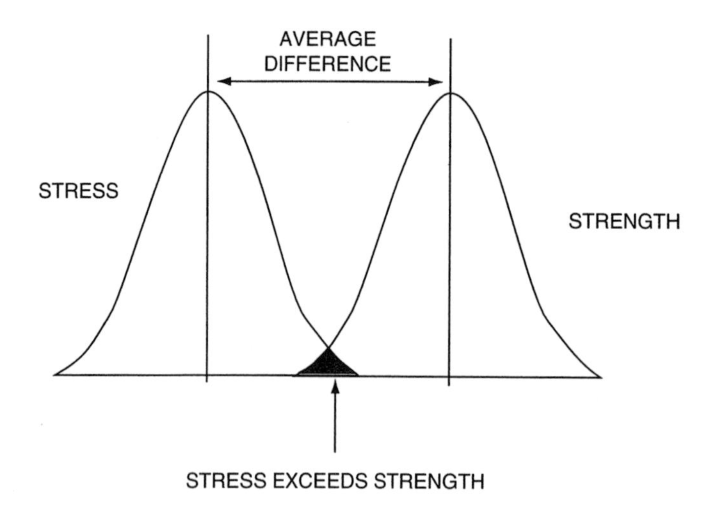

Figure IV.6. Modern view of safety factors.

Since any strength or stress is theoretically possible, the traditional concept of a safety factor becomes vague at best and misleading at worst. To deal intelligently with this situation, we must consider *probabilities* instead of *possibilities.* This is done by computing the probability that a stress/strength combination will occur such that the stress applied exceeds the strength. It is possible to do this since, if we have distributions of stress and strength, then the difference between the two distributions is also a distribution. In particular, if the distributions of stress and strength are normal, the distribution of the difference between stress and strength will also be normal. The average and standard distribution of the difference can be determined using statistical theory, and are shown in Equations IV.9 and IV.10.

$$\sigma_{SF}^2 = \sigma_{STRENGTH}^2 + \sigma_{STRESS}^2 \tag{IV.9}$$

$$\mu_{SF} = \mu_{STRENGTH} - \mu_{STRESS} \tag{IV.10}$$

In Equations IV.9 and IV.10 the SF subscript refers to the safety factor.

EXAMPLE OF COMPUTING PROBABILITY OF FAILURE

Assume that we have normally distributed strength and stress. Then the distribution of strength minus stress is also normally distributed with the mean and variance computed from Equations IV.9 and IV.10. Furthermore, the probability of a failure is the same as the probability that the difference of strength minus stress is less than zero; i.e., a negative difference implies that stress exceeds strength, thus leading to a critical failure.

Assume that the strength of a steel rod is normally distributed with $\mu = 50{,}000^{\#}$ and $\sigma = 5{,}000^{\#}$. The steel rod is to be used as an undertruss on a conveyor system. The stress observed in the actual application was measured by strain gages and it was found to have a normal distribution with $\mu = 30{,}000^{\#}$ and $\sigma = 3{,}000^{\#}$. What is the expected reliability of this system?

Solution

The mean variance and standard deviation of the difference is first computed using Equations IV.9 and IV.10, giving

$$\sigma^2_{DIFFERENCE} = \sigma^2_{STRENGTH} + \sigma^2_{STRESS} = 5{,}000^2 + 3{,}000^2 = 34{,}000{,}000$$

$$\sigma = \sqrt{34{,}000{,}000} = 5{,}831^{\#}$$

$$\mu_{DIFFERENCE} = \mu_{STRENGTH} - \mu_{STRESS} = 50{,}000^{\#} - 30{,}000^{\#} = 20{,}000^{\#}$$

We now compute Z which transforms this normal distribution to a standard normal distribution (see VI.C).

$$Z = \frac{0^{\#} - 20{,}000^{\#}}{5{,}831^{\#}} = -3.43$$

Using a normal table (Appendix Table 4), we now look up the probability associated with this Z value and find it is 0.0003. This is the probability of failure, about 1 chance in 5,000. The reliability is found by subtracting this probability from 1, giving 0.9998. Thus, the reliability of this system (and safety for this particular failure mode) is 99.98%. This example is summarized in Figure IV.7.

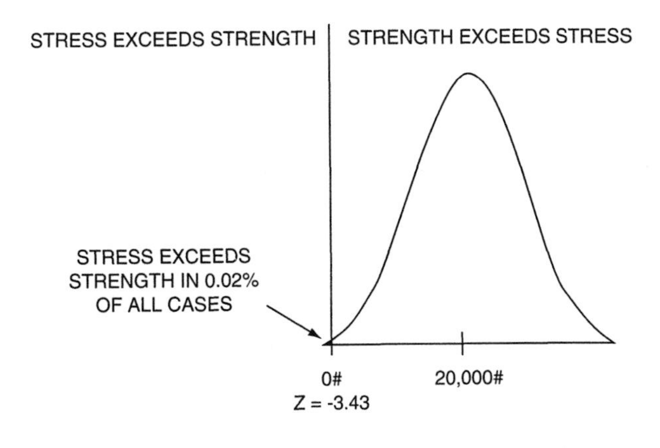

Figure IV.7. Distribution of strength minus stress.

MAINTAINABILITY AND AVAILABILITY*

Many systems are repairable; the system fails—whether it is an automobile, a dishwasher, production equipment, etc.—it is repaired. Maintainability is a measure of the difficulty to repair the system. More specifically, maintainability is:

> The measure of the ability of a system to be retained in, or restored to, a specified condition when maintenance is performed by personnel having specified skill levels, using prescribed procedures and resources, at each prescribed level of maintenance and repair.

Military Handbook 472 (MIL-HDBK-472) defines six components of maintainability. Figure IV.8 shows the relationship of these components.

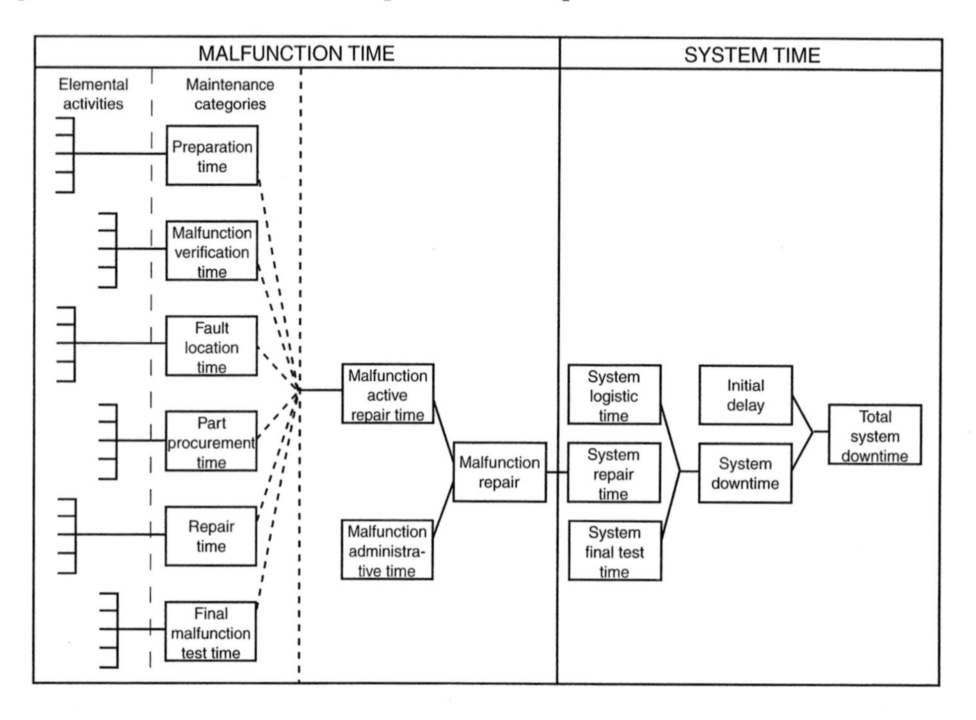

Figure IV.8. Structure of maintainability.

The six components shown in Figure IV.8 are discussed below.

1. **Elemental Activities** are simple maintenance actions of short duration and relatively small variance that do not vary appreciably from one system to another. An example of an elemental activity is the opening and shutting of a door.

*Excerpted from *Reliability Engineering Handbook* by Bryan Dodson and Dennis Nolan. Tucson, Arizona: QA Publishing, LLC.

2. **Malfunction Active Repair Time** consists of the following:
   a. Preparation time
   b. Malfunction verification time
   c. Fault location time
   d. Part procurement time
   e. Repair time
   f. Final malfunction test time

   Items a–f above are composed of elemental activities.
3. **Malfunction Repair Time** consists of the following:
   a. Malfunction active repair time
   b. Malfunction administrative time
4. **System Repair Time** is the product of malfunction repair time and the number of malfunctions.
5. **System Downtime** includes the following:
   a. System logistic time
   b. System repair time
   c. System final test time
6. **Total System Downtime** is a combination of the following distributions:
   a. Initial delay
   b. System downtime

MIL-HDBK-472 provides a procedure for predicting maintainability based on the structure described above. The philosophy of the procedure is based on the principles of synthesis and transferability. The synthesis principle involves a buildup of downtimes, step-by-step, progressing from the distribution of downtimes of elemental activities through various stages culminating finally with the distribution of system downtime. The transferability principle embodies the concept that data applicable to one type of system can be applied to similar systems, under like conditions of use and environment, to predict system maintainability.

Other useful maintainability references are Military Standard 470, which describes a maintainability program for systems and equipment, and Military Standard 471, which provides procedures for maintainability verification, demonstration, and evaluation.

Availability is a measure of the readiness of a system. More specifically, availability is:

> A measure of the degree to which a system is in an operable and comitable state at the start of a mission when the mission is called for at a random time.

There are three categories of availability.

1. Inherent Availability is the ideal state for analyzing availability. It is a function only of the mean time to fail, MTBF, and the mean time to repair, MTTR; preventive maintenance is not considered. Inherent availability is defined as

$$A_I = \frac{MTBF}{MTBF + MTTR} \tag{IV.11}$$

2. Achieved Availability includes preventive maintenance as well as corrective maintenance. It is a function of the mean time between maintenance actions, MTMA, and the mean maintenance time, MMT. Achieved availability is defined as

$$A_A = \frac{MTMA}{MTMA + MMT} \tag{IV.12}$$

3. Operational Availability includes preventive maintenance, corrective maintenance, and delay time before maintenance begins, such as waiting for parts or personnel. It is a function of the mean time between maintenance actions and the mean down time, MDT, and is defined as

$$A_O = \frac{MTMA}{MTMA + MDT} \tag{IV.13}$$

It is important to note that the type of availability being described is often not distinguished. Many authors simply refer to "availability;" MTTR may be the equivalent of MMT or MDT, and MTBF may be the equivalent MTMA.

QUANTIFYING MAINTAINABILITY AND AVAILABILITY

Maintainability is defined as

$$M(t) = P(T < t) = \int_0^t f(x)dx \tag{IV.14}$$

where: $P(T<t)$ is the probability of completing the repairs in time < T, and $f(x)$ is the repair time probability density function.

It has been shown that the lognormal distribution can often be used to model repair times. There are times when repairs are completed quickly, but it is rather uncommon to achieve repair times much shorter than normal. However, in some instances, problems arise causing repairs to take much longer than normal. The skewness of the lognormal distribution models this situation well.

Perhaps the most widely used measure of maintainability is the mean-time-to-repair (MTTR).

$$MTTR = \frac{\sum_{i=1}^{n} \lambda_i \tau_i}{\sum_{i=1}^{n} \lambda_i} \tag{IV.15}$$

where: n is the number of sub-systems,
λ_i is the failure rate of the ith sub-system, and
τ_i is the time to repair of the ith unit.

Another useful measure is the median corrective maintenance time (MCMT). This is the time in which half of the maintenance actions can be accomplished: the median of the time to repair distribution. For normally distributed repair times, MTTR = MCMT.

Maximum Corrective Maintenance Time (CMAX%) is the maximum time required to complete a given percentile of maintenance tasks. Unless otherwise specified, CMAX% refers to 95% of all maintenance tasks.

EXAMPLE IV.1

The time to repair for a system follows a lognormal distribution with a scale parameter of 1.2 and a location parameter of 2.2. Plot the maintainability function for this system, and determine the probability of completing a system repair in less than time = 50.

Solution

The maintainability function is

$$M(t) = \int_0^t f(x)dx$$

The lognormal probability density function is given as

$$f(x) = \frac{1}{\sigma x \sqrt{2\pi}} \exp\left[-\frac{1}{2}\left(\frac{\ln x - \mu}{\sigma}\right)^2\right], x > 0$$

where: μ is the location parameter, and
σ is the scale parameter.

The integral of this function from zero to t can be expressed as

$$M(t) = \Phi\left(\frac{\ln t - \mu}{\sigma}\right)$$

where $\Phi(z)$ is the area under the standard normal probability density function to the left of z. This value is tabulated in Appendix Table 20.

The maintainability function is shown in Figure IV.9.

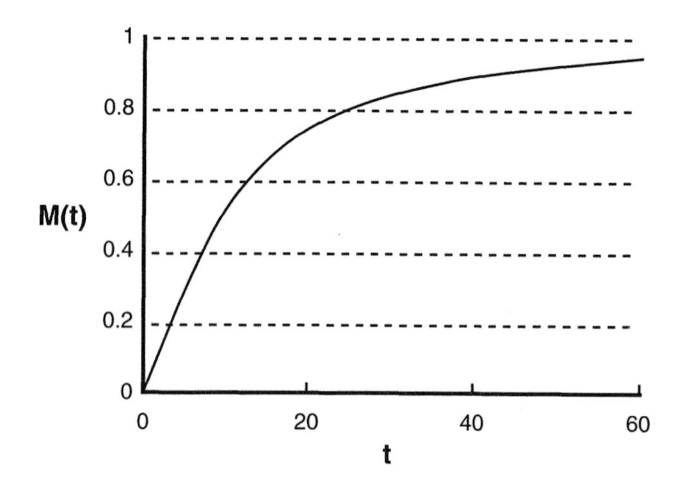

Figure IV.9. Maintainability example.

The probability of completing a repair in less than time = 50 is

$$M(50) = \Phi\left(\frac{\ln 50 - 2.2}{1.2}\right) = \Phi(1.4267) = 0.9232$$

In the special case where the time to fail and the time to repair follow exponential distributions, availability is

$$A(t) = \frac{\mu}{\mu + \lambda} + \frac{\lambda}{\mu + \lambda} e^{-(\mu+\lambda)t} \qquad \text{(IV.16)}$$

where: μ is the mean repair rate, and
λ is the mean failure rate.

It can be seen from Equation IV.16 that as time increases, availability reaches a steady state of

$$A = \frac{\mu}{\mu + \lambda} \qquad \text{(IV.17)}$$

This expression is equivalent to

$$A = \frac{\theta}{\theta + \alpha} \quad \text{(IV.18)}$$

where: θ is the mean time to fail, and
α is the mean time to repair.

The steady-state availability is independent of the time to fail and time to repair distributions.

The average availability for the period from time = 0 to time = T is equal to the average uptime, which is

$$U(t) = \frac{1}{t}\int_0^t A(s)ds \quad \text{(IV.19)}$$

For the special case of exponentially distributed times to fail and repair, this reduces to

$$U(t) = \frac{\mu}{\lambda + \mu} + \frac{\mu}{(\lambda + \mu)^2 t} - \frac{\mu}{(\lambda + \mu)^2 t} e^{-(\lambda + \mu)t} \quad \text{(IV.20)}$$

Note that the steady-state uptime is equal to the steady-state availability.

EXAMPLE IV.2

The time to fail for a component is exponentially distributed with a mean of 25. The time to repair for this component is also exponentially distributed with a mean of 40. Plot availability as a function of time. What is the steady-state availability?

Solution

The mean failure rate is

$$\lambda = \frac{1}{25} = 0.04$$

The mean repair rate is

$$\mu = \frac{1}{40} = 0.025$$

Availability is

$$A(t) = \frac{0.025}{0.025 + 0.04} + \frac{0.04}{0.025 + 0.04} e^{-(0.025+0.04)t}$$

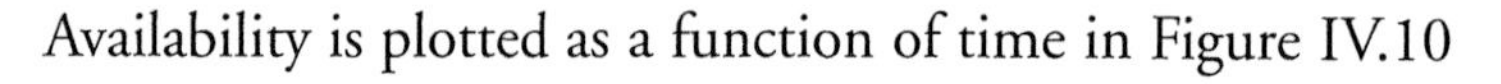
Availability is plotted as a function of time in Figure IV.10

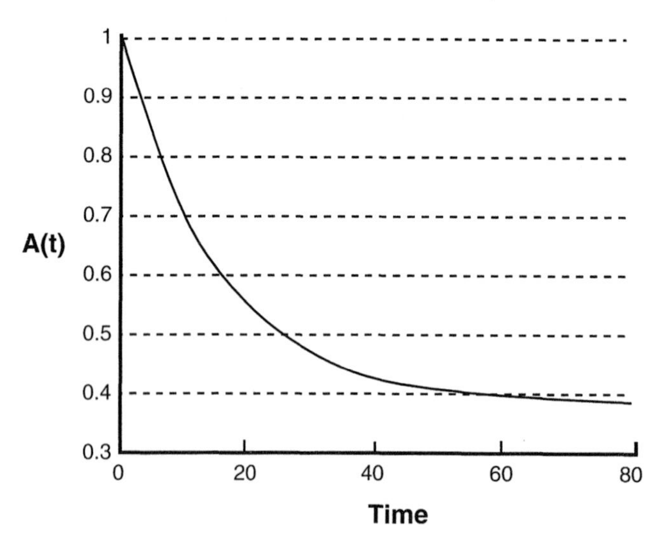

Figure IV.10. Availability as a function of time.

The steady-state availability is

$$A = \frac{25}{25 + 40} = 0.3846$$

This is verified by Figure IV.10.

To determine system availability, it is important to specify the operating assumptions. With the trivial assumption that all components in the system remain operating while failed components are repaired, system availability for a series system consisting of n components is

$$A_s(t) = \prod_{i=1}^{n} A_i(t) \tag{IV.21}$$

This assumption is obviously not valid in many circumstances. Consider a six-stand rolling mill. It is impossible to operate five stands while repairing the other stand. Equation IV.21 also requires that n repairmen be available. If the n–1 operating components continue to run while the failed component is being repaired, it is possible to have all n components in a failed state.

Assuming all components remain running while the failed component is being repaired makes more sense for a parallel system. The purpose of redundant components is to prevent system failure or downtime. Using this assumption, and assuming n repairmen are available, the availability of a system with n systems in parallel is

$$A_s(t) = 1 - \prod_{i=1}^{n}[1 - A_i(t)] \quad \text{(IV.22)}$$

EXAMPLE IV.3

Five components in series each have a mean time to fail that is normally distributed, with a mean of 60 hours, and a standard deviation of 10. The mean time to repair of each of these components is exponentially distributed with a mean of 30 hours. There are five repairmen and components are only removed from service during repair time (if the system is down because component number 1 failed, components 2–5 continue to operate). What is the system steady-state availability?

Solution

The steady-state availability of each component is

$$A = \frac{60}{60+30} = 0.6667$$

The system steady-state availability is

$$A_s = (0.6667)^5 = 0.1317$$

EXAMPLE IV.4

Rework Example IV.3 assuming the five components are in parallel.

Solution

The system steady-state availability is

$$A_S=1-(1-0.6667)5=0.9959$$

As stated earlier, the assumption that all components in a series system continue to operate even though the failure of one component causes a system failure does not often apply to series systems. More often, all components are shut down until the failed item is repaired. Then, the entire system is brought on-line. In this case, system availability can be computed from the system mean time to fail and the system mean time to repair.

EXAMPLE IV.5

Assuming exponential distributions for the mean time between failures and the mean time to repair for all components, compute the steady-state system availability of the series system shown below. All components are idled when the system fails.

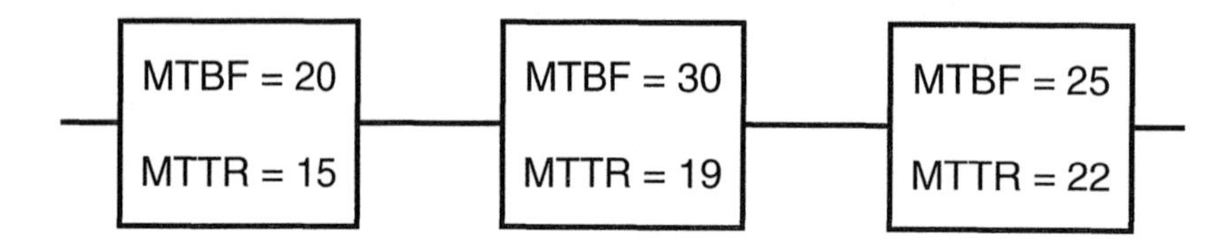

Solution

The system mean time between failures is

$$MTBF_s = \frac{1}{\frac{1}{20}+\frac{1}{30}+\frac{1}{25}} = 8.1081$$

The system mean time to repair is found by computing the weighted average of each component's mean repair time and the percentage of system failures caused by that component.

COMPONENT	MTBF	FAILURE RATE	MTTR	WEIGHTED AVERAGE
1	20	0.05	15	0.75
2	30	0.0333	19	0.6333
3	25	0.04	22	0.88
TOTAL		0.1233		2.2633

The system mean time to repair is

$$MTTR_S = \frac{2.2633}{0.1233} = 18.3560$$

The steady-state system availability is

$$A_S = \frac{8.1081}{8.1081 + 18.356} = 0.3061$$

Assuming all components continue to operate when the system fails, the system availability would be found by computing the product of the component availabilities.

COMPONENT	MTBF	MTTR	AVAILABILITY
1	20	15	0.5714
2	30	19	0.6122
3	25	22	0.5319

System availability in this case is

$$A_s = (0.5714)(0.6122)(0.5319) = 0.1861$$

which is substantially less than the availability achieved by shutting the remaining components down during a system failure.

PREVENTIVE MAINTENANCE

In some cases it is possible to prevent failures with preventive maintenance. The question is to determine if preventive maintenance is applicable, and, if so, how often it should be scheduled. Referring to Figure IV.11, failures can be grouped into three categories based on the behavior of the failure rate. Infant mortality failures are characterized by a decreasing failure rate. The hazard function (failure rate) of the Weibull distribution is decreasing if the shape parameter, b, is less than one. Random* failures exhibit a constant failure rate; the shape parameter of the Weibull distribution is equal to one. Wear-out failures have an increasing failure rate; the shape parameter of the Weibull distribution is greater than one.

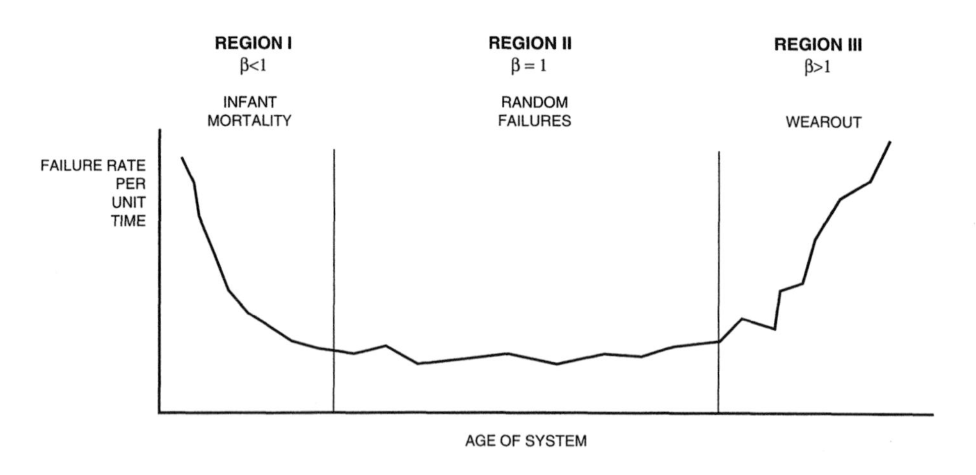

Figure IV.11. The bathtub curve.

Infant mortality failures are premature failures that often can be prevented by management. If infant mortality failures cannot be prevented, a burn-in procedure can be implemented to eliminate the failures before the product is shipped. Preventive maintenance is not applicable for an item with a decreasing failure rate. Performing preventive maintenance restores the system to its initial state which has a higher failure rate; preventive maintenance increases the number of failures.

Some causes of infant mortality failures include the following:

- Improper use
- Inadequate materials
- Over-stressed components
- Improper setup
- Improper installation
- Poor quality control
- Power surges
- Handling damage

Random failures cannot be prevented with preventive maintenance. The failure rate is constant, so preventive maintenance has no effect on failures. Reliability can be increased by redesigning the item, or in some cases, by implementing an inspection program.

Wear-out failures can be prevented with preventive maintenance. The failure rate is increasing with time, so preventive maintenance restores the system to a state with a lower failure rate. The question is how often preventive maintenance should be scheduled.

---

*This term can be confusing, since all failures occur randomly. This portion of the bathtub curve is also referred to as the "useful life". This term is also confusing since many products exhibit a decreasing or increasing failure fate for their entire life.

The time to fail for an item is variable, and can be represented by a probability distribution, $f(x)$. Referring to Figure IV.12, the cost of failures per unit time decreases as preventive maintenance is performed more often, but the cost of preventive maintenance per unit time increases. There exists a point where the total cost of failures and preventive maintenance per unit time is at a minimum; this is the optimum schedule for preventive maintenance.

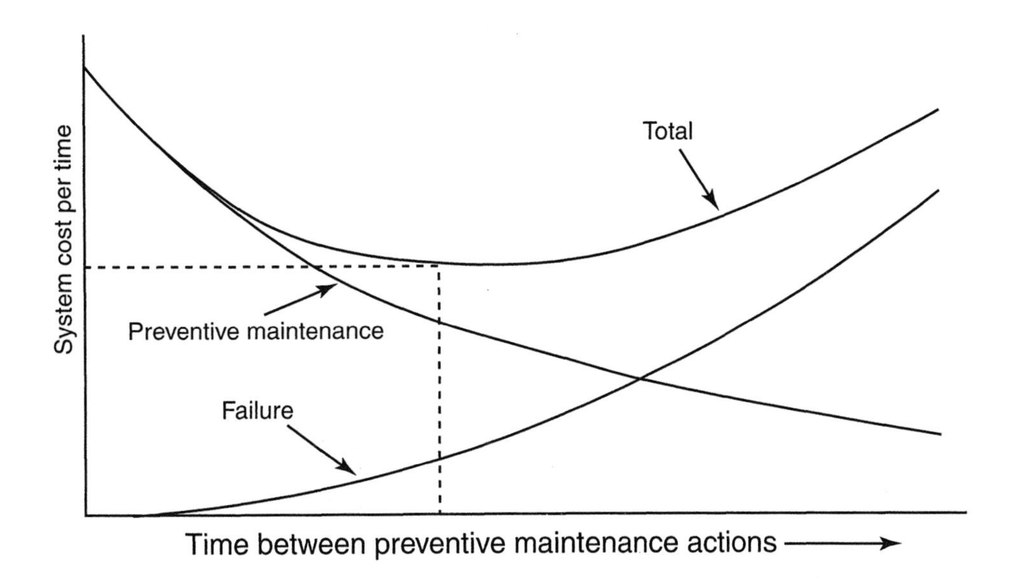

Figure IV.12. Optimum schedule for preventive maintenance.

The optimum time between maintenance actions is found by minimizing the total cost per unit time.

$$C_T = \frac{C_p \int_T^{\infty} f(t)dt + C_f \int_0^T f(t)dt}{T \int_T^{\infty} f(t)dt + \int_0^T t f(t)dt} \qquad \text{(IV.23)}$$

where: C_p is the cost of preventive maintenance,
C_f is the cost of a failure, and
T is the time between preventive maintenance actions.

Minimizing Equation IV.23 is tedious, and numerical routines are usually required. Dodson (2002) developed a tabular solution for this problem given the following assumptions:

1. The time to fail follows a Weibull distribution.
2. Preventive maintenance is performed on an item at time T, at a cost of C_p.
3. If the item fails before time = T, a failure cost of C_f is incurred.
4. Each time preventive maintenance is performed, the item is returned to its initial state; that is, the item is "as good as new."

The optimum time between preventive maintenance actions is

$$T = m\theta + \delta \qquad \text{(IV.24)}$$

where: m is a function of the ratio of the failure cost to the preventive maintenance cost, and the value of the shape parameter, and is given in Table IV.2;
θ is the scale parameter of the Weibull distribution; and
δ is the location parameter of the Weibull distribution.

Table IV.2. Values of m.

C_f/C_p	β							
	1.5	2.0	2.5	3.0	4.0	5.0	7.0	10.0
2.0	2.229	1.091	0.883	0.810	0.766	0.761	0.775	0.803
2.2	1.830	0.981	0.816	0.760	0.731	0.733	0.755	0.788
2.4	1.579	0.899	0.764	0.720	0.702	0.711	0.738	0.777
2.6	1.401	0.834	0.722	0.688	0.679	0.692	0.725	0.766
2.8	1.265	0.782	0.687	0.660	0.659	0.675	0.713	0.758
3.0	1.158	0.738	0.657	0.637	0.642	0.661	0.702	0.749
3.3	1.033	0.684	0.620	0.607	0.619	0.642	0.687	0.739
3.6	0.937	0.641	0.589	0.582	0.600	0.627	0.676	0.730
4.0	0.839	0.594	0.555	0.554	0.579	0.609	0.662	0.719
4.5	0.746	0.547	0.521	0.526	0.557	0.591	0.648	0.708
5	0.676	0.511	0.493	0.503	0.538	0.575	0.635	0.699
6	0.574	0.455	0.450	0.466	0.509	0.550	0.615	0.683
7	0.503	0.414	0.418	0.438	0.486	0.530	0.600	0.671
8	0.451	0.382	0.392	0.416	0.468	0.514	0.587	0.661
9	0.411	0.358	0.372	0.398	0.452	0.500	0.575	0.652
10	0.378	0.337	0.355	0.382	0.439	0.488	0.566	0.645
12	0.329	0.304	0.327	0.357	0.417	0.469	0.550	0.632
14	0.293	0.279	0.306	0.338	0.400	0.454	0.537	0.621
16	0.266	0.260	0.288	0.323	0.386	0.441	0.526	0.613
18	0.244	0.244	0.274	0.309	0.374	0.430	0.517	0.605
20	0.226	0.230	0.263	0.298	0.364	0.421	0.508	0.598

Continued on next page...

Table IV.2—*continued...*

Cf/Cp	β 1.5	2.0	2.5	3.0	4.0	5.0	7.0	10.0
25	0.193	0.205	0.239	0.275	0.343	0.402	0.492	0.584
30	0.170	0.186	0.222	0.258	0.328	0.387	0.478	0.573
35	0.152	0.172	0.207	0.245	0.315	0.374	0.468	0.564
40	0.139	0.160	0.197	0.234	0.304	0.364	0.459	0.557
45	0.128	0.151	0.187	0.225	0.295	0.356	0.451	0.550
50	0.119	0.143	0.179	0.217	0.288	0.348	0.444	0.544
60	0.105	0.130	0.167	0.204	0.274	0.335	0.432	0.534
70	0.095	0.120	0.157	0.193	0.264	0.325	0.422	0.526
80	0.087	0.112	0.148	0.185	0.255	0.316	0.415	0.518
90	0.080	0.106	0.141	0.177	0.248	0.309	0.407	0.513
100	0.074	0.101	0.135	0.172	0.241	0.303	0.402	0.507
150	0.057	0.082	0.115	0.150	0.217	0.278	0.379	0.487
200	0.047	0.071	0.103	0.136	0.203	0.263	0.363	0.472
300	0.035	0.058	0.087	0.119	0.182	0.243	0.343	0.454
500	0.025	0.045	0.071	0.100	0.161	0.219	0.319	0.431
1000	0.016	0.032	0.054	0.079	0.135	0.190	0.288	0.403

EXAMPLE IV.6

The cost of failure for an item is $1,000. The cost of preventive maintenance for this item is $25. The following Weibull distribution parameters were determined from time-to-fail data: $\beta = 2.5$, $\theta = 181$ days, $\delta = 0$. How often should preventive maintenance be done?

Solution

The ratio of failure cost to PM cost is

$$\frac{C_f}{C_p} = \frac{1000}{25} = 40$$

Entering Table IV.2 with this ratio and a shape parameter of 2.5, give 0.197 for the value of m.

A PM should be done every

$$T = (0.197)(181) + 0 = 35.657 \text{ days}$$

This is shown graphically in Figure IV.13.

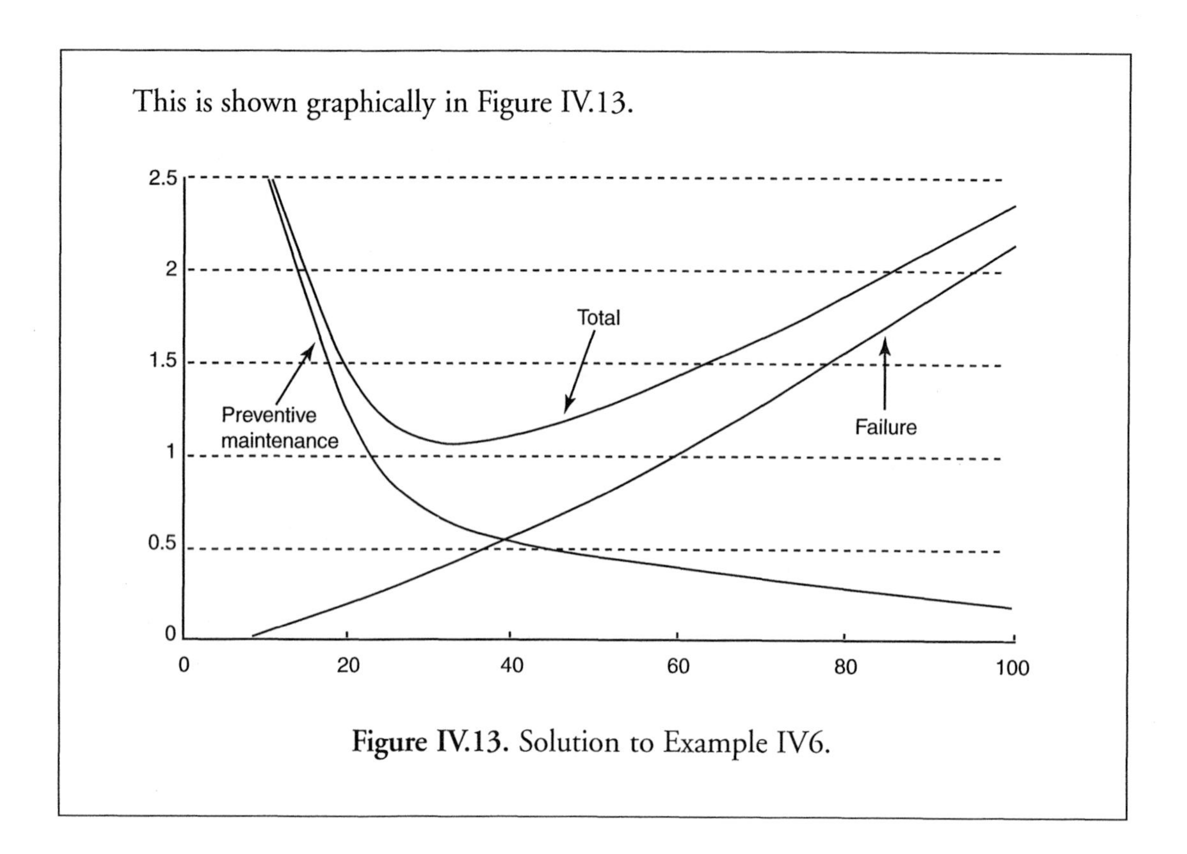

Figure IV.13. Solution to Example IV6.

EXAMPLE IV.7

Repeat Example IV.6 with the following Weibull distribution parameters: β =0.8, θ = 181 days, δ = 0.

Solution

When the shape parameter of the Weibull distribution is less than or equal to 1.0, the optimum solution is to do no preventive maintenance.

DESIGNING FOR MAINTAINABILITY

Maintenance may be divided into the following tasks:

1. **Disassembly**—Equipment disassembly to the extent necessary, to gain access to the item that is to be replaced. How difficult is it to access the failed part? How much time does it take?
2. **Interchange**—Removing the defective item and installing the replacement. How difficult is it to exchange the defective part? How much time does it take?
3. **Reassembly**—Closing and reassembly of the equipment after the replacement has been made. Again difficulty and time should be a consideration.

Consider that the ease of maintenance may be measured in terms of man-hours, material, and money. Savings are attributable to the fact that designing for ease of maintenance is considered to be a design enhancement because it provides for the recognition and elimination of areas of poor maintainability during the early stages of the design life cycle. Otherwise, this area of poor maintainability would only become apparent during demonstration testing or actual use. After which time, correction of design deficiencies would be costly and would unduly delay schedules and missions. The ease with which corrective action may be taken is the key to designing for maintainability.

Moss (1985) gives eight guidelines that should be considered when designing for maintainability. Following these guidelines will help produce a design that is easily repaired, thus increasing maintainability and availability.

1. **Standardization**—Ensure compatibility among mating parts, and minimize the number of different parts in the system. Instead of using a 4mm bolt in one area and a 5mm bolt in another area, use a 5mm bolt in both areas. This will reduce spare parts requirements.
2. **Modularization**—Enforce conformance of assembly configurations to standards based on modular units with standard sizes, shapes, and interface locations. This allows use of standardized assembly and disassembly procedures.
3. **Functional packaging**—Expedite repair of faulty systems by containing all components of an item performing a function in a package. Replacing the faulty package repairs the unit.
4. **Interchangeability**—Control dimensional and functional tolerances to assure that items can be replaced in the field without physical rework.
5. **Accessibility**—Control spatial arrangements to allow accessibility. Items should be arranged to allow room for a worker to be comfortable when making repairs. All items should be accessible. Unfailed items should not have to be removed to gain access to a failed item.
6. **Malfunction annunciation**—Provide a means of notifying the operator when a malfunction is present. The oil pressure and temperature gauges on an automobile are examples.

7. **Fault isolation**—Assure that a malfunction can be traced. A simple method is to provide terminals that can be connected to external test equipment to determine if a component is operating properly. Often built-in diagnostics are provided with a system.
8. **Identification**—Provide a method for unique identification of all components as well as a method for recording preventive and corrective maintenance.

SYSTEM EFFECTIVENESS

The Weapons System Effectiveness Industry Advisory Committee (WSEIAC) developed a performance measurement for repairable systems that incorporates availability, capability, and reliability. The measurement is *system effectiveness*. In order for a system to be effective, the system must be available when called on, the system must have the capability to complete the stated task, and the system must achieve the mission (i.e. be reliable).

The ARINC Research Corporation has also developed a measure of system effectiveness based on mission reliability, operational readiness, and design adequacy. This relationship is shown in Figure IV.14.

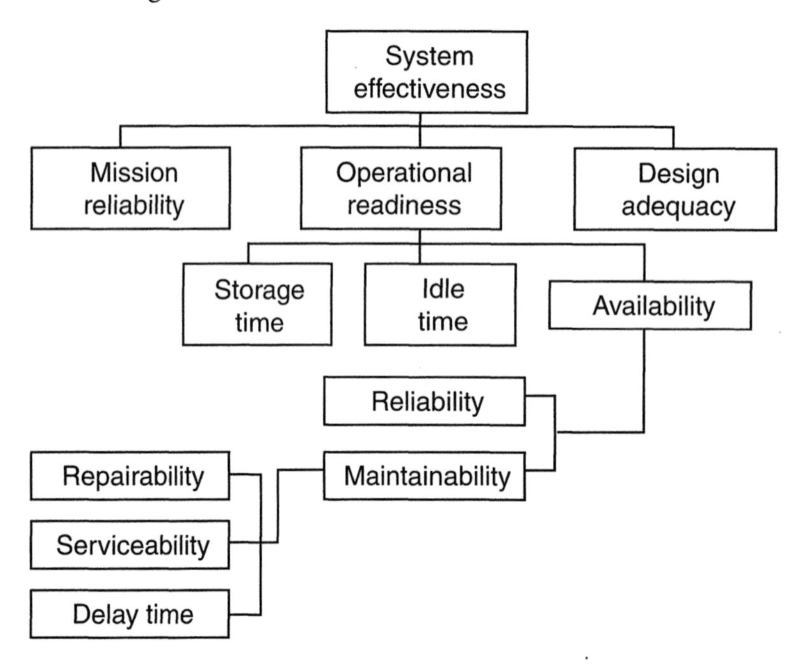

Figure IV.14. System effectiveness.

System effectiveness can be quantified as

$$SE = (MR)(OR)(DA) \quad \text{(IV.25)}$$

where: MR is mission reliability,
OR is operational readiness, and
DA is design adequacy.

EXAMPLE IV.8

A missile system not originally designed to intercept Scud missiles has been employed to do so. Assuming this missile's design adequacy for intercepting Scud missiles is 0.9. determine the system effectiveness if operational readiness is 0.98 and mission reliability is 0.87.

Solution

System effectiveness is

$$SE = (0.87)(0.98)(0.9) = 0.7673$$

PRODUCT TRACEABILITY SYSTEMS AND RECALL PROCEDURES

While section III.B dealt with traceability and control of product in-house, in-transit, or at a supplier's facility, this section deals with traceability of product after delivery into the stream of commerce. Responsible producers have long been concerned with minimizing customer exposure to hazardous products that have been discovered after delivery. With modern product liability laws and consumer attitudes, the need for rapid removal of unsafe products from the market is more important than ever before. The recall can be greatly facilitated if the procedures described for internal traceability and control are used. The scope of the recall can then be precisely determined, allowing the manufacturer to avoid unnecessary expense as well as minimizing customer fear.

ASQ (__ 1981) lists the following steps for implementing and evaluating a recall:

1. Initial activities to implement a recall
    a. Establishment of a management control center
    b. Description of the product, features, problem
    c. Identification of products
    d. Definition of product groups to be recalled

- e. Data on population of products
- f. Planning and designing reporting forms
- g. Estimating, ordering, and distributing replacement parts

2. Release of recall information
   - a. Preparation of notification information
   - b. Special news releases
   - c. Notification to customers and users
   - d. Internal employee information
   - e. Timing
3. Field response
   - a. Response from the field
   - b. Collection of internal and field cost information
   - c. Parts usage traceability
   - d. Field data reporting
     1. Periodic reporting
     2. Summary reports
4. Monitoring activity
   - a. Audit of field corrective action
   - b. Data verification and correlation
5. Management evaluation, program completion
   - a. Project review on periodic basis
   - b. Success-level evaluation

IV.E RELIABILITY FAILURE ANALYSIS AND REPORTING*

Data should be collected and analyzed prior to making decisions that affect the quality of a product, process, or operation. Data provides the foundation for forming and testing hypotheses. The outcome of decisions derived from collected data is positively correlated to the quality of data on which the decision is made. Therefore, it is essential to develop a system for retaining and retrieving accurate information in order to maintain or improve product reliability.

Before starting a good data collecting program, the objectives for collecting data should be understood. The basic objectives are to:

1. Identify the problem
2. Report the problem
3. Verify the problem
4. Analyze the problem
5. Correct the problem

*Excerpted from *Reliability Engineering Handbook* by Bryan Dodson and Dennis Nolan. Tucson, Arizona: QA Publishing, LLC.

According to Ireson (1995) the benefits of any good data collecting, analysis, and reporting system are the following:

1. Reduction of research or redundant test time
2. Assurance that proven parts are specified
3. Assistance for quality assurance and purchasing in obtaining parts
4. Assurance that products are capable of meeting specifications
5. Reduction in time and expense of testing
6. Improvement of capability to estimate reliability

There are other important questions to ask: what level of accuracy is needed? What methods should be used? Who will do the collecting? Before formulating questions and attempting to answer them, it is important to distinguish between the basic types of data: continuous and discrete. Anything that can be measured—weight, height, length, time, etc.—is an example of *continuous* data. Anything that can be counted—number of accidents, number of failures, or defects—is an example of *discrete* data.

According to MIL-STD-785, the purpose of a failure reporting, analysis, and corrective-action system (FRACAS) is to determine the root cause of the problem, document the problem, and record the corrective-action taken. Considerations for establishing such an information gathering program are outlined in this chapter.

RECORDING, REPORTING, AND PROCESSING

Before a quality analysis can be made, a closed loop system that collects and records failures that occur at specified levels must be in place. A "closed loop" system is a feedback loop that determines what failed, how it failed, why it failed, and tracks corrective-action. A program should be developed first to initiate the recording, reporting, and processing of all failures into the design, test, and manufacturing processes. The first thing to consider in the recording process is the recording form. It should be clear, simple, and easy to understand by the user. As a minimum the form should include the following information:

- Equipment identification
- Failed part identification
- Type or category of failure
- Time of failure
- Who reported the failure
- Possible causes
- Action taken
- Who initiated action if any

Training should be provided for the more complex forms as necessary. Always provide valid reasons for collecting data. If the data collectors understand why they are taking data they will ultimately understand the need for accuracy when filling in the forms.

A Failure Review Board (FRB) should be established to monitor failure trends, significant failures and corrective-actions. The mission of the FRB is to review functional failure-data from appropriate inspections and testing including qualification, reliability, and acceptance test failures. All failure occurrence information should be made available to the FRB, including a description of conditions at the time of failure, symptoms of failure, failure isolation procedures, and known or suspected causes of the failure. Reports should include data on successes as well as failures. Details of any action taken must also be reported to enable the review board to make exceptions for items that may have been wrongfully replaced during test or service. Open items should then be followed up by the FRB until failure mechanisms have been satisfactorily identified and corrective-action initiated. FRB members should include appropriate representatives from design, reliability, safety, maintainability, manufacturing, and quality assurance. The leader of this group is usually a reliability engineer.

It is important that the term "failure" be defined. For example, what does failure mean to the organization or process? Failure is defined by MIL-STD-721 as, "The event, or inoperable state, in which any item or part of an item does not, or would not, perform as previously specified."

Failures should also be defined and sorted into categories: relevant, nonrelevant, chargeable, nonchargeable, and alternate failures. All failures occurring during reliability qualification and production reliability acceptance tests should be classified as relevant or nonrelevant. Relevant failures should be further classified as chargeable or nonchargeable. Rules for classification of failures should be in agreement with approved failure definitions or other criteria. Flow diagrams or other modeling techniques depicting failed hardware and data flow should be used where possible. Figure IV.15 is a flow diagram for failure categories.

Costs should be included in the recording process, where management approval is needed. Dollars and cents are sometimes the most important factors in management decisions. For example, the costs of redesigning or replacing a certain item could be compared to the cost of not replacing the item. The cost of not replacing the item, thus allowing a failure, should include warranty, liability, customer satisfaction, etc. These costs are usually only qualified estimates since they are difficult or impossible to quantify precisely. In any case, include as many cost comparisons as possible for the decision-making process.

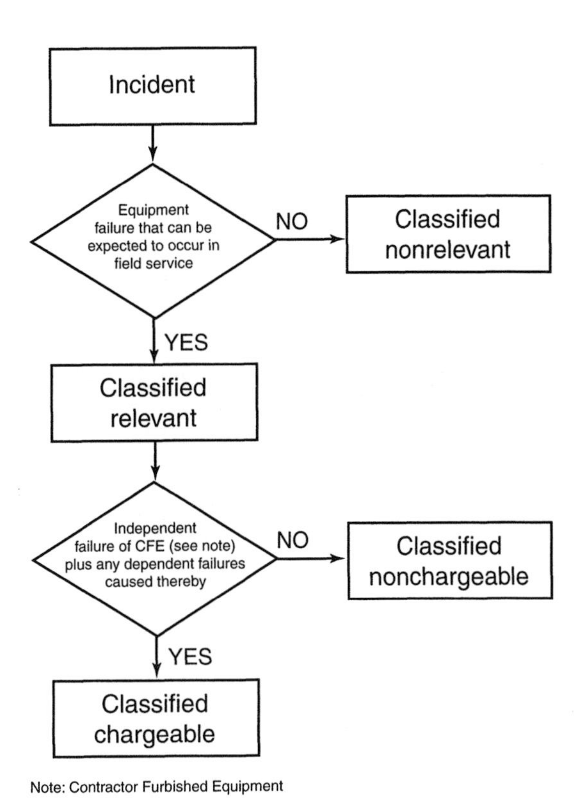

Figure IV.15. Flow diagram of failure categories (MIL-STD-721).

Failure terms and definitions

Catastrophic failure—A failure that can cause item loss.

Critical failure—A failure, or combination of failures, that prevents an item from performing a specified mission.

Dependent failure—Failure caused by the failure of an associated item(s).

Independent failure—Failure that occurs without being caused by the failure of any other item.

Equipment design failure—Any failure that can be traced directly to the design of the equipment; that is, the design of the equipment caused the part in question to degrade or fail, resulting in an equipment failure.

Equipment manufacturing failure—A failure caused by poor workmanship or inadequate manufacturing process control during equipment construction, testing, or repair prior to the start of testing. For example, the failure due to cold solder joints.

Intermittent failure—Failure for a limited period of time, followed by the item's recovery or its ability to perform within limits without any remedial action.

Mechanism failure—The physical, chemical, electrical, thermal, or other process that results in failure.

Multiple failures—The simultaneous occurrence of two or more independent failures. When two or more failed parts are found during trouble-shooting and failures cannot be shown to be dependent, multiple failures are presumed to have occurred.

Pattern failures—The occurrence of two or more failures of the same part in identical or equivalent applications when the failures are caused by the same basic failure mechanism and the failures occur at a rate that is inconsistent with the part's predicted failure rate.

Part design failure—Failures that can be traced directly to inadequate design.

Part manufacturing failure—These failures are the result of poor workmanship or inadequate manufacturing process control during part assembly, inadequate inspection, or improper testing.

Primary failure—This type of failure is the result of a component deficiency. These failures result while operating within design limits.

Secondary failure—This type of failure produces operating conditions outside of design limits thereby causing primary failures. The primary failure would not have occurred if the secondary failure had not caused stress outside the design limits.

Software error failure—A failure caused by an error in the computer program.

Unverified failure—The opposite of a verified failure. Lack of failure verification, by itself, is not sufficient rationale to conclude the absence of a failure.

Verified failure—A failure that is determined either by repeating the failure mode on the reported item or by physical or electrical evidence of failure.

Failure categories

Relevant failures—Relevant failures are specified in 1–4:

1. Intermittent failures
2. Unverified failures (failures that cannot be duplicated, which are still under investigation, or for which no cause could be determined)
3. Verified failures not otherwise excluded under approved failure categories
4. Pattern failures

Nonrelevant failures—Nonrelevant failures are specified in 1–7:

1. Installation damage
2. Accident or mishandling
3. Failures of the test facility or test-peculiar instrumentation
4. Equipment failures caused by an externally applied overstress condition in excess of the approved test requirements

5. Normal operating adjustments (nonfailures) specified in the approved equipment operating instructions
6. Secondary failures within the equipment, which are directly caused by non-relevant or relevant primary failures (the secondary failures must be proved to be dependent on the primary failure)
7. Failures caused by human errors

Chargeable failures—A relevant, independent failure of equipment under test and any dependent failures caused thereby which are classified as one failure and used to determine contractual compliance with acceptance and rejection criteria. Chargeable failures are specified in 1–4:

1. Intermittent failures
2. Unverified or verified failures
3. Independent failures
   a. Equipment design
   b. Equipment manufacturing
   c. Part design
   d. Part manufacturing
   e. Software errors identified, corrected, and verified during the pretest, the system verification, and the test, should not be chargeable as equipment failures
   f. Contractor-furnished equipment (CFE)
4. Relevant failures

Nonchargeable failures—Nonchargeable failures are specified in 1–3:

1. Nonrelevant failures
2. Failures induced by operating, maintenance, or repair procedures
3. Failures of items having a specified life expectancy and operated beyond the specified replacement time of the item

Alternate failure categories—The alternate failure categories specified in 1–3 can be used to categorize both hardware and software failures for systems.

1. *Critical failure* is a failure which prevents the system from performing its mission.
2. *Major failures* are specified below:
   a. Any failure that reduces performance of on-line hardware or software so that the system mission capability is degraded but not eliminated
   b. Any failure that prevents built-in test equipment (BITE) from detecting major failures in on-line hardware or software
   c. Any failure which causes BITE to continuously indicate a major failure in on-line hardware or software when such a failure does not exist
   d. Any failure that prevents BITE from isolating and localizing on-line hardware or software major failures to the lowest group

   e. Any failure that causes BITE to continuously isolate and localize a major failure in on-line hardware or software when such failure does not exist
   f. Any failure that causes BITE false alarm rate to exceed the specified level
3. *Minor failures* are specified as:
   a. Any failure that impairs on-line hardware or software performance, but permits the system to fully perform its objective
   b. Any failure that impairs BITE performance (creates false alarms) but permits BITE, when recycled, to isolate and localize to the required level, except when false alarm rate exceeds specified levels

Determining what information to collect

Deciding what data to collect will depend on the phase of the project—the conceptual, design, production, or maintenance phase. In any case, data should include failures due to equipment failure and human error. The conceptual phase requires the use of data from similar products. The design phase requires research or actual test data for the specific product. The production phase requires the use of a more historical type data, sometimes derived from the design stages. The maintenance phase requires the use of actual failure data that may have been acquired with various failure analysis techniques. In short, all failures must be included from development to acceptance.

Five basic steps are outlined below that will help determine what data to collect:

1. Find out what happened, and be as specific as possible. At what level in the overall system, product, or process was the event discovered?
2. Method of detection. Internally? Externally?
3. Find out when the event happened. During testing? During production run?
4. Find out if there is a similar event in historical records. If the answer is "yes," it could save time by eliminating some data collection.
5. Find out if there have been any recent changes. Check vendor materials, test conditions, etc.

How will data be collected and reported

Data may be collected by either manual or automatic means. Most test results or observations are recorded manually on forms customized to collect specific information, then input into a computer database. An example of a typical manual data collecting form is shown in Figure IV.16. Data is sometimes taken automatically through the use of electronic devices that send information directly to a database.

The automatic data information gathering technique is usually desirable where continuous monitoring is necessary.

There are no standards for how to record or store data. When data is input into a computer, manually or automatically, both retrieval and use become obviously enhanced. There are many software packages on the market that can be readily tailored to fit specific needs.

Who will collect the information

Deciding who will collect the information depends on who will use the data, the accuracy needed, and time and cost constraints. Keep in mind that the person who collects data is not necessarily the one who will use it.

1. Report No.:	2. Report Date:	3. Time of Failure:	4. Equip Type:	5. Equip S/N:

7. Failed Part Description:	8. Failed Part ID:	9. Failed Part S/N:	10. Failed Part Mod No.:

ITEM USAGE:	11. Hrs:	12. Mins:	13. Secs:	14. Cycle Time.	15. Cal. Time:	16. Miles:

17. Failure Occurred During:	18. Action Taken:	19. Replacement:	20. Critical Cause:	21. Date of Failure:
❑ 1. Inspection	❑ 1. Adjustment	❑ 1. Complete	❑ 1. Priority 1	________
❑ 2. Production	❑ 2. Replaced	❑ 2. Partial	Name: ________	
❑ 3. Maintenance	❑ 3. Repaired	❑ 3. Tested		22. Time of Failure:
❑ 4. Shipping	❑ 4. Retested	❑ 4. Not Tested	❑ 2. Priority 2	
❑ 5. Field Use	❑ 5. Other	❑ 5. None	Name:	________

REASON FOR REPORT (check one) ➠

23. Removal or Maintenance Action Required as a Result of:
❑ 1. Suspected failure or malfunction ❑ 2. Due to improper maintenance ❑ 3. Damaged accidentally

24. Symptom(s):

❑ A. Excessive vibration
❑ B. High fuel consumption
❑ C. High oil consumption
❑ D. Inoperative
❑ E. Interference
❑ F. Leakage
❑ G. Noisy
❑ H. Out of balance
❑ I. Overheating
❑ J. Out of limits
❑ K. Unstable op

25. Condition(s):

❑ 110. Arced
❑ 990. Bent
❑ 880. Cracked
❑ 440. Galled
❑ 505. Chafed
❑ 606. Peeled
❑ 707. Eroded
❑ 009. Frayed
❑ 008. Loose
❑ 005. Open
❑ 555. Plugged
❑ 444. Ruptured
❑ 333. Split
❑ 222. Sheared
❑ 120. Shorted
❑ 130. Stripped
❑ 140. Worn
❑ 150. Corroded
❑ 160. Dented
❑ 170. Chipped
❑ 101. Burned
❑ 202. Other

26. Cause of Trouble:

❑ A. Design deficiency
❑ B. Faulty maintenance
❑ C. Faulty manufacturing
❑ C. Faulty overhaul
❑ D. Faulty packaging
❑ E. Foreign object
❑ F. Fluid contamination
❑ G. Faulty installation
❑ H. Faulty replacement
❑ I. Operator adjustment
❑ J. Undetermined
❑ K. Weather conditions
❑ M. Wrong part used
❑ N. Unknown contaminates
❑ O. Other (amplify)

27. Disposition or Corrective-action (select appropriate code(s) from list below and check boxes at left to indicate action taken in respect to failed item):

REASON FOR: NO ACTION
❑ Hold for 90 days
❑ Lack of facilities
❑ Lack of repair parts
❑ Lack of personnel
❑ Other (explain below): ________

ACTION
❑ Adjusted
❑ Removed and replaced
❑ Repaired and reinstalled
❑ Retest and hold
❑ Other (explain below): ________

28. Maintainability Information:

	HOURS	MINUTES
Time to locate trouble	____	____
Time to repair/replace	____	____
Total time non-operable	____	____

29. Component/Subcomponent Replaced With:
Description: ________
Part no.: ________
Serial no.: ________

30. Remarks (Furnish additional information concerning failure or corrective-action not covered above—do not repeat information checked above.):

Location:	Department:	Signature:	Date:

Figure IV.16. Failure report form.

What level of accuracy is needed

Accuracy will depend on the product and its intended use. For example, a cook may only need to take time and temperature data at minutes and degrees, while a race car designer may want time and temperature in tenths of seconds and degrees. For another example, if someone is asked their age 10 days before their 40th birthday, they may reply 39 or 40. Which is more accurate? Which is accurate enough? It could be important enough to require an answer like: 39 years, 355 days, 12 hours, and 15 minutes. Of course, asking a person's age usually will not require that much detail, but when asking how long a verified equipment problem has persisted, details do become important.

The program outlined below will help assure that accurate and complete data is collected which meets the objectives for data collecting—identifying, reporting, verifying, analyzing, and correcting problems.

1. **Identify and control failed items.**

   A tag should be affixed to the failed item immediately upon the detection of any failure or suspected failure. The failure tag should provide space for the failure report serial number and for other pertinent entries from the item failure record. All failed parts should be marked conspicuously and then controlled to ensure disposal in accordance with all laws, rules or regulations. Failed parts should not be handled in any manner that may obliterate facts which might be pertinent to the analysis; they should be stored pending disposition of the failure analysis agent.

2. **Reporting of problems or failures.**

   A failure report should be initiated at the occurrence of each problem or failure of hardware, software, or equipment. The report should contain the information required to permit determination of the origin and correction of failures. The following information should be included in the report:

   a. Descriptions of failure symptoms, conditions surrounding the failure, failed hardware identification, and operating time (or cycles) at time of failure
   b. Information on each independent and dependent failure and the extent of confirmation of the failure symptoms, the identification of failure modes, and a description of all repair action taken to return the item to operational readiness
   c. Information describing the results of the investigation, the analysis of all part failures, an analysis of the item design, and the corrective-action taken to prevent failure recurrence (if no corrective-action is taken, the rationale for this decision should be recorded)

3. **Verify failures.**

   Reported failures should be verified as actual failures or an acceptable explanation provided for lack of failure verification. Failure verification is determined either by repeating the failure mode on the reported item or by physical or electrical evidence of failure (leakage residue, damaged hardware, etc.) Lack of failure verification, by itself, is not sufficient rationale to conclude the absence of a failure.

4. **Investigation and analysis of problems or failures.**

   An investigation and analysis of each reported failure should be performed. Investigation and analysis should be conducted to the level of hardware or software necessary to identify causes, mechanisms, and potential effects of the failure. Any applicable method (test, microscopic analysis, applications study, dissection, x-ray analysis, spectrographic analysis, etc.) of investigation and analysis that may be needed to determine failure cause shall be used. When the removed item is not defective or the cause of failure is external to the item, the analysis should be extended to include the circuit, higher hardware assembly, test procedures, and subsystem if necessary. Investigation and analysis of supplier failures should be limited to verifying that the supplier failure was not the result of the supplier's hardware, software, or procedures. This determination should be documented for notification of the procuring activities.

5. **Corrective-action and follow-up.**

   When the cause of a failure has been determined, a corrective-action shall be developed to eliminate or reduce the recurrence of the failure. The procuring activity should review the corrective-actions at scheduled status reviews prior to implementation. In all cases, the failure analysis and the resulting corrective-actions should be documented. The effectiveness of the corrective-action should be demonstrated by restarting the test at the beginning of the cycle in which the original failure occurred.

Who will use the information

Deciding who will use the data is probably of less concern than what data to use. Usually everyone involved in the project will use some portion of the information. Assuring that collected information is available to all, as needed, and making it easily accessible is the key.

CORRECTIVE ACTION AND FOLLOW-UP

The purpose of a corrective-action plan is to eliminate or reduce failures. Prior to implementation, the corrective action plan should be submitted to management for approval. The system for such action should include provisions to assure that effective corrective actions are taken on a timely basis by a follow-up audit that reviews and reports all open failure delinquencies to management. The failure cause and classification for each failure must be clearly stated. It is not necessary to include the cost of failure in the corrective-action status report if it was included in the initial report to management for corrective-action approval. This also describes the duties of the FRB.

The purpose for corrective-action status or follow-up should be obvious—to determine the effectiveness of the corrective-action taken. There are four questions that must be answered during the follow-up phase:

1. Was the failure or problem completely eliminated?
2. Was the failure or problem partially eliminated?
3. Did the corrective-action cause an insignificant change or no change?
4. Did the corrective-action cause a worse condition?

In essence, effective follow-up closes the loop in the "closed loop" system. Each failure must be monitored by continued reporting and analysis until effective results are obtained.

Summary

Data should be collected, reported, and investigated on all failures. It is important that all failures be categorized as relevant or nonrelevant while at the same time resisting the temptation to classify a failure as nonrelevant due to cost or time constraints.

As changes are made to correct a failure, follow-up is essential to assure the effectiveness of the corrective-action. For example, the corrective-action taken may cause overstressing of another item that would cause a secondary failure, etc. The effectiveness of any corrective-action should never be assumed without 100% confidence that the decision or change will prevent a like failure. Always keep in mind that the outcome of decisions derived from collected data is positively correlated to the quality of data on which the decision is made.

IV.F RELIABILITY, SAFETY, HAZARD ASSESSMENT TOOLS

While reliability prediction is a valuable activity, it is even more important to design reliable systems in the first place. Proposed designs must be evaluated to detect potential failures prior to building the system. Some failures are more important than others, and the assessment should highlight those failures most deserving of attention and scarce resources. Once failures have been identified and prioritized, a system can be designed that is robust; i.e., it is insensitive to most conditions that might lead to problems.

DESIGN REVIEW

Design reviews are conducted by specialists, usually working on teams. Designs are, of course, reviewed on an ongoing basis as part of the routine work of a great number of people. However, the term as used here refers to the formal design review process. The purposes of formal design review are threefold:

1. Determine if the product will actually work as desired and meet the customer's requirements.
2. Determine if the new design is producible and inspectable.
3. Determine if the new design is maintainable and repairable.

Design review should be conducted at various points in the design and production process. Review should take place on preliminary design sketches, after prototypes have been designed, and after prototypes have been built and tested, as developmental designs are released, etc. Designs subject to review should include parts, subassemblies, and assemblies.

FAULT TREES*

Fault tree analysis provides a formal method of determining the combinations of primary events that result in the occurrence of a specified system level event. H. R. Watson conceived fault-tree analysis while working on safety evaluation for the Minuteman Launch Control System. Watson realized that the logic flow in data processing equipment could be used for analyzing the logic of system failures resulting from component failures.

Fault-trees provide a graphical representation of the system being studied, and their use often results in the discovery of failure combinations that would not have been obvious if other failure identification methods were used. Fault trees are useful for both qualitative and quantitative analyses. Some common uses for fault trees are:

- Root cause failure analysis
- Failure modes and effects analysis
- Preliminary hazard analysis
- Probability calculations

Fault-tree construction

Fault trees are built from gates and events. A list of fault-tree gates is provided in Table IV.3; a list of fault tree events is provided in Table IV.4.

Consider how an automobile might fail to complete a trip. The event under consideration is the failure of the automobile to complete the trip. In the fault tree, this will be the *Top* event, and will be represented by a conditional event.

Simplifying the problem, an automobile fails to complete a trip if

1. two of the five tires go flat (including the spare),
2. the engine fails, or
3. the transmission fails.

For this example, it is assumed all other systems work. This is depicted with a house event. The fault tree for this example is shown in Figure IV.17.

The *Top* event in this fault tree is the conditional event—automobile fails to complete the trip. Following the *Top* event is an *OR* gate. The *Top* event will take place if any of the four events below the *OR* gate take place. Two of these four events are primary events; engine fails and transmission fails. The conditional event, 4 tires not available, is dependent on the outcome of the *r-out-of-n* gate. The number "2" in the *r-out-of-n* gate signifies that two of the five events below the gate must transpire for the conditional event above

*Excerpted from *Reliability Engineering Handbook* by Bryan Dodson and Dennis Nolan. Tucson, Arizona: QA Publishing, LLC.

the gate to occur. The house event is used to indicate if a component not detailed in the fault tree failed. If this house event is turned off, all of the automobile's components not detailed in the fault tree function properly. If turned on, the top event occurs. A logical analysis for this example would be to turn the house event off.

Table IV.3. Fault tree gates.

SYMBOL	NAME	DESCRIPTION
	OR Gate	The event above this gate occurs if *any* of the events below this gate occur. An *OR* gate signifies a union of events.
	AND Gate	The event above this gate occurs if *all* the events below this gate occur. An *AND* gate signifies an intersection of events.
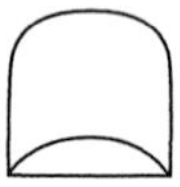	Exclusive OR Gate	The event above this gate occurs if one of the events below the gate occur and the others do not. The triggering event is specified within the gate.
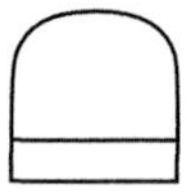	Priority AND Gate	The event above this gate occurs if all the events below the gate occur in the order specified. The order may be listed in the gate or in an attached oval. If no order is specified, the default order is left to right.
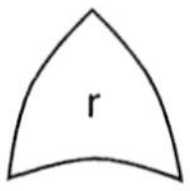 r	r-out-of-n Gate	The event above this gate occurs if r of the n events below the gate take place.
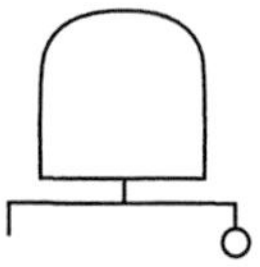	AND NOT Gate	The event above this gate occurs if all the events on a leg without a circle occur, and all the events on a leg with a circle *do not* occur.
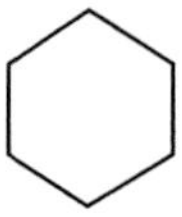	Inhibit Gate	A special type of *AND* gate. The event above this gate transpires if the event listed within the Inhibit gate is satisfied *and* the event below the gate takes place.

Table IV.4. Fault tree events.

SYMBOL	NAME	DESCRIPTION
	Primary event. Also known as a Basic event.	The lowest level of failure possible. Primary events are individual component failures, human errors and software errors.
	Conditional Event	Conditional events are dependent on Primary events. They are located above gates in the fault tree.
	House Event	An event that can be turned on or off. If turned on, the probability of occurrence is 1; if turned off, the probability of occurrence is 0. Often used for scenario analysis.
	Undeveloped Event	This event has not been defined. Some reasons for not defining an event are; lack of interest, lack of information, or further development will not provide additional understanding.
	Triangle In	Use to repeat a portion of the tree or to continue the tree on another page.
	Triangle Out	Used in conjunction with the Triangle In block.

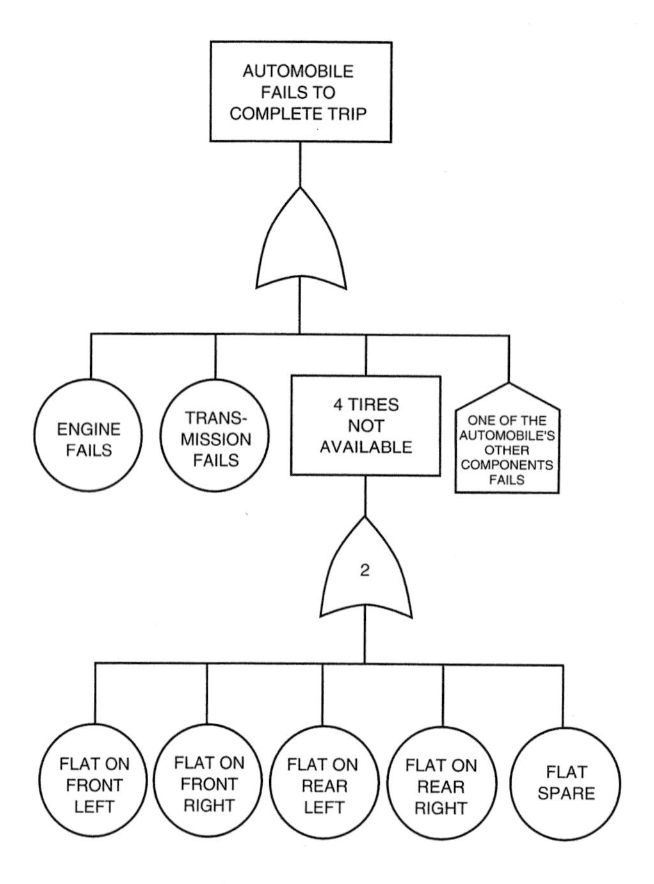

Figure IV.17. Automobile fault tree.

Cut sets and path sets

A *cut set* is a set of primary events that cause the *Top* event to transpire if they *all* transpire. A cut set may be a subset of another cut set. For the automobile trip example: if the engine fails, the transmission fails, there is a flat on the front left tire, there is a flat on the front right, and a flat spare tire, then the automobile will fail to make the trip. Thus, the occurrence of all of these events is a cut set. However, if there is a flat on the front left tire and the front right tire, the automobile also fails to make the trip. The set: flat on front left and flat on front right, is also a cut set.

A *minimum cut set* is a cut set that is no longer a cut set if any events are removed from the cut set. The first cut set mentioned above is not a minimum cut set. Several events were removed (engine fails, transmission fails, left, flat spare) and the set remained a cut set. The cut set: flat on front left and flat on front right, is a minimum cut set. If either of the two events: flat on front left or flat on front right, is removed from the cut set, the remaining event in the set is not sufficient to cause the Top event to occur. Thus: the set, flat on front left and flat on front right, is a minimum cut set. The remaining minimum cut sets for this problem are:

1. Engine fails
2. Transmission fails
3. House event turned on
4. Flat on front left and flat on rear left
5. Flat on front left and flat on rear right
6. Flat on front left and flat spare
7. Flat on front right and flat on rear left
8. Flat on front right and flat on right
9. Flat on front right and flat spare
10. Flat on rear left and flat on rear right
11. Flat on rear left and flat spare
12. Flat on rear right and flat spare

Minimum cut sets can be determined intuitively in small fault trees, however, for large trees, quantitative methods are needed. The four most common techniques for determining minimum cut sets are:

1. Use of Boolean algebra
2. The prime number method
3. The binary string bit method
4. The combination testing method

For more information concerning these methods consult Sundararajan (1991).

In some cases the interest is in what prevents the *Top* event from occurring, rather than what causes it to occur. This is done by identifying path sets; the opposite of cut sets. If *none* of the events in a *path set* occur, the *Top* event is prevented from occurring.

A path set, that is no longer a path set if any events are removed from the set, is a *minimum path set*. The minimum path sets for the previous example are (assuming the house event is turned off):

1. a. engine fails and
   b. transmission fails and
   c. flat on front left and
   d. flat on front right and
   e. flat on rear left and
   f. flat on rear right

2. a. engine fails and
   b. transmission fails and
   c. flat on front left and
   d. flat on front right and
   e. flat on rear left and
   f. flat spare
3. a. engine fails and
   b. transmission fails and
   c. flat on front left and
   d. flat on front right and
   e. flat spare rear left and
   f. flat on rear right
4. a. engine fails and
   b. transmission fails and
   c. flat on front left and
   d. flat spare and
   e. flat on rear left
   f. flat on rear right
5. a. engine fails and
   b. transmission fails and
   c. flat spare and
   d. flat on front right and
   e. flat on rear left and
   f. flat on rear right

The fault tree displayed in Figure IV.18 demonstrates the *AND* gate, the *AND NOT* gate, and the Inhibit gate. The *Top* Event is followed by an *AND* gate and takes place only if Event A and Event B take place. Event A is followed by an *AND NOT* gate and transpires if Event 1 occurs and Event 2 does not occur. Event B is followed by an Inhibit gate and occurs if Events 3 and 4 take place.

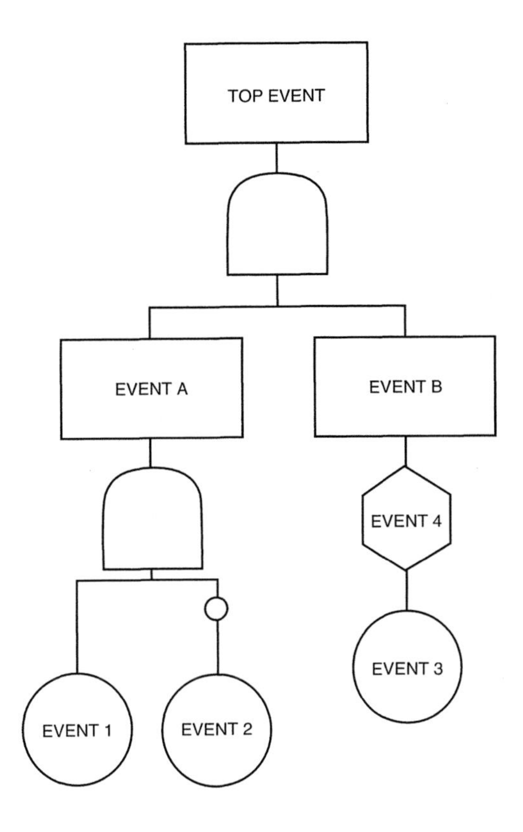

Figure IV.18. Example fault tree.

The fault tree presented in Figure IV.19 illustrates the Exclusive *OR* gate and the Priority *AND* gate. Being followed by an Exclusive *OR* gate, the *Top* Event transpires if either Event A or Event 3 occur, but does not transpire if both Event A and Event 3 occur. Event A takes place if Event 1 and Event 2 take place, but only if Event 1 occurs before Event 2.

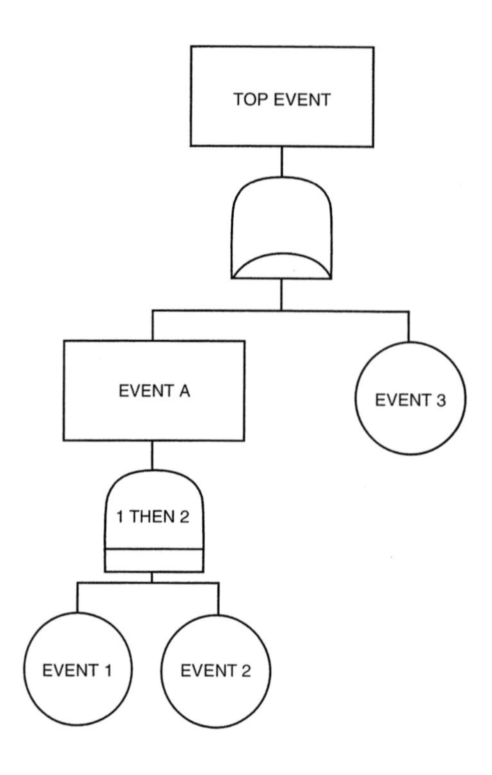

Figure IV.19. Example fault tree.

Quantitative fault tree analysis
Example

If an event does not occur in more than one place in a fault tree, the probability of the Top event occurring can be determined by starting at the bottom of the tree and evaluating the probabilities of each branch of the tree. For example, compute the probability of the Top Event occurring in the fault tree in Figure IV.20, given the following probabilities for occurrence

1. probability of Event 1 = 0.3,
2. probability of Event 2 = 0.06,
3. probability of Event 3 = 0.12,
4. probability of Event 4 = 0.22, and
5. probability of Event 5 = 0.08.

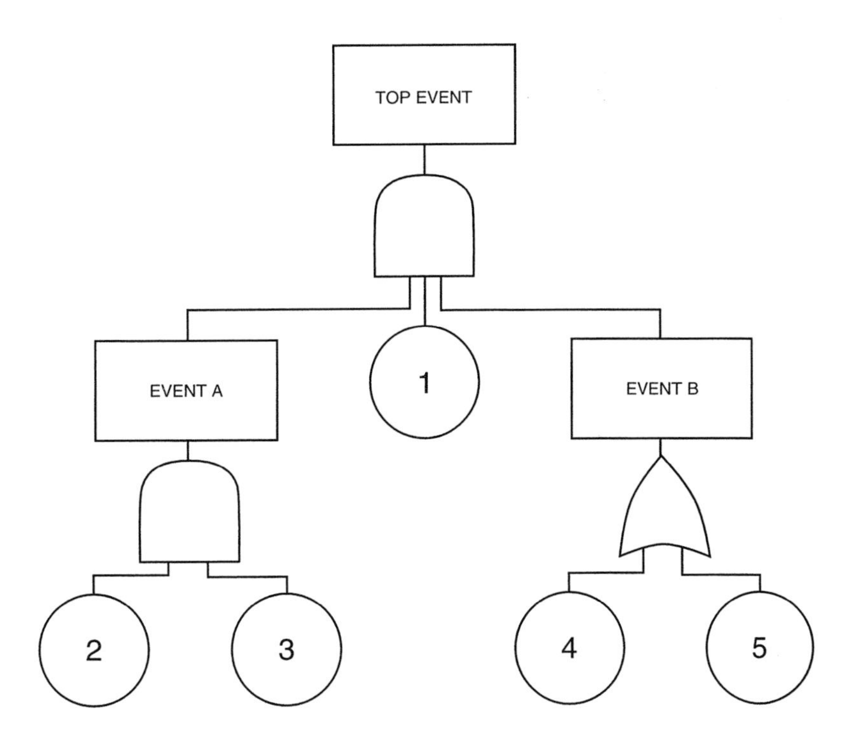

Figure IV.20. Quantitative fault tree example.

Solution

Event A occurs if Events 2 and 3 both occur; the probability of this happening is (0.06)(0.12) = 0.0072. Event B occurs if either event 4 or event 5 occur; this probability is [1–(1–0.22)(1–0.08)] = 0.2824. The Top Event occurs if events A, B and 1 all occur; this probability is (0.0072)(0.2824)(0.3) = 0.00061.

Example

Now consider the fault tree in Figure IV.21. This tree is identical to the tree depicted in Figure IV.20, except for two changes; Event 5 has been replaced by event 1, and the gate immediately below the *Top* Event has been changed from an *AND* gate to an *OR* gate. The probability of Events 1–4 occurring is the same as in the previous example.

Because Event 1 appears in more than one location in the fault tree, the branches of the tree are no longer independent. Thus, the method used to solve the previous problem cannot be used.

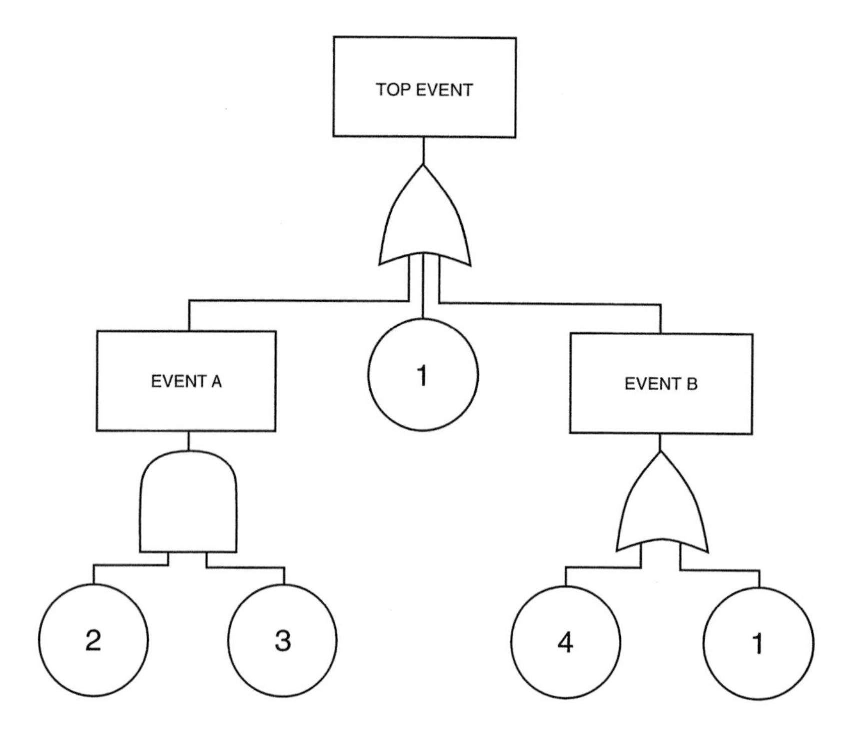

Figure IV.21. Quantitative fault tree example.

The probability of the *Top* Event occurring in a fault tree can be determined if the minimum cut sets for the tree are known. The expression for the probability of the *Top* Event occurring is

$$P_T = \sum_{i=1}^{N} P(C_i) - \sum_{i=1}^{N}\sum_{j=1}^{N} P(C_i \cap C_j) + \sum_{i=1}^{N}\sum_{j=1}^{N}\sum_{k=1}^{N} P(C_i \cap C_j \cap C_k) \ldots + (-1)^{N-1} P\left(\bigcap_{i=1}^{N} C_i\right) \qquad \text{(IV.26)}$$

where N is the number of minimum cut sets and $P(C_i)$ represents the probability of the minimum cut set i occurring.

Solution

The cut sets for this tree are; (1), (4), (1,2), (1,3), (1,4), (2,3), (2,4), (3,4), (1,2,3), (1,2,4), (1,3,4), (2,3,4), and (1,2,3,4). The minimum cut sets are (1), (4), and (2,3). The probability of each minimum cut set occurring is

Minimum Cut Set	*Probability of Occurrence*
Event 1	0.3
Event 4	0.22
Event 2 and Event 3	(0.06)(0.12)=0.0072

Using Equation IV.26, the probability of the Top Event occurring is:

$$PT = 0.3 + 0.22 + 0.0072 - (0.3)(0.22) - (0.3)(0.0072) - (0.22)(0.0072) + (0.3)(0.22)(0.0072)$$
$$= 0.4579$$

FAILURE MODE, EFFECTS AND CRITICALITY ANALYSIS*

Failure Mode, Effects and Criticality Analysis (FMECA) is a powerful design analysis tool that is used to increase system reliability. It can be applied during the initial design phase or to existing equipment. To be more effective, the FMECA should relate to the nature of the design process itself. In either case, it considers overall design, operating and service problems, while at the same time addressing process and safety problems.

If used as a design tool, the benefit of FMECA depends upon the timeliness in which information is communicated in the early design phase. Timeliness is probably the most important factor in differentiating between effective and ineffective implementation of the FMECA. The efforts and sophistication of the approach used much depends on the requirements of each individual program. In any case, the FMECA should contribute to the overall program decision.

CRITICALITY ANALYSIS

The objective of an FMECA is to identify all failure modes in a system design. Its purpose is to identify all catastrophic and critical failure probabilities so they can be minimized as early as possible. Therefore, the FMECA should be started as soon as preliminary design information is available and extended as more information becomes available in suspected problem areas.

The effects of all failure modes are not equal with respect to the total impact on the system concerning safety and overall system performance. The designer, faced with this dilemma, needed a tool that would rank the significance of each potential failure for each component in the system's design alternatives. Because of the need for such justification, the criticality analysis function was added to the Failure Mode and Effects Analysis (FMEA) process, thus creating Failure Mode, Effects and Criticality Analysis (FMECA).

This tool has been used extensively by the military in the last three decades. In recent years, more commercial industries have been requiring the FMECA to evaluate new designs and even more recently to improve the reliability of existing equipment.

*Excerpted from *Reliability Engineering Handbook* by Bryan Dodson and Dennis Nolan. Tucson, Arizona: QA Publishing, LLC.

The qualitative approach to FMECA

This approach should be used when specific failure rate data is not available. Failure modes identified by the FMECA process are assessed by their probability of occurrence. To establish qualitative measures of occurrence, severity and detection, criteria must be established that subjectively relates to the overall effect on the process. Examples are offered in Tables IV.5, IV.6, and IV.7, to serve as guides in establishing qualitative measures. The product of the measures of occurrence, severity and detection is called the Risk Priority Number (RPN). Tables IV.5, IV.6, and IV.7 are for example only. The numbers or criteria assigned to any particular ranking system are at the discretion of the user. Detailed instructions on how to use this criterion on the FMECA form is explained later in this section.

Table IV.5. Occurrence probabilities.

Rank	Occurrence Criteria	Occurrence Rates (cycles, hrs, etc.)
1	Unlikely. Unreasonable to expect this failure mode to occur.	—
2	Isolated. Based on similar designs having a low number of failures.	1/10,000
3	Sporadic. Based on similar designs that have experienced occasional failures.	1/1,000
4	Conceivable. Based on similar designs that have caused problems.	1/100
5	Recurrent. Certain that failures will ensue.	1/10

NOTE: The ranking criteria selected must be consistent throughout the FMECA.

Table IV.6. Severity Probabilities.

Rank	Severity Criteria
1	Minor. No noticeable effect. Unable to realize that a failure has occurred.
2	Marginal. Annoying. No system degradation.
3	Moderate. Causing dissatisfaction. Some system degradation.
4	Critical. Causing a high degree of dissatisfaction. Loss of system function.
5	Catastrophic. A failure which may cause death or injury. Extended repair outages.

NOTE: The ranking criteria selected must be consistent throughout the FMECA.

Table IV.7. Detection probabilities.

Rank	Detection Criteria	Probability
1	Very High probability of detecting the failure before it occurs. Almost always preceded by a warning.	80%–100%
2	High probability of detecting the failure before it occurs. Preceded by a warning most of the time.	60%–80%
3	Moderate probability of detecting the failure before it occurs. About a 50% chance of getting a warning.	40%–60%
4	Low probability of detecting the failure before it occurs. Always comes with little or no warning.	20%–40%
5	Remote probability of detecting the failure before it occurs. Always without a warning.	0%–20%

NOTE: The ranking criteria selected must be consistent throughout the FMECA.

FMECA quantitative approach

Figure IV.22 provides the worksheet for the quantitative FMECA process. The failure mode and criticality number (C_m) is the portion of the criticality number for the item due to a particular failure mode. This criticality number replaces the RPN number used in the qualitative method described in the previous section. The C_m for a failure mode is determined by the expression

$$C_m = \beta\alpha\lambda_p t \qquad \text{(IV.27)}$$

where: β = conditional probability of loss of function,
α = failure mode ratio,
λ_p = part failure rate, and
t = duration or operating time.

System ______ Date ______

Indenture Level ______ Sheet ______ of ______

Reference Drawing ______ Compiled by ______

Mission ______ Approved by ______

ID#	Item Identification	Function	Failure Modes & Causes	Mission Operational Mode	Severity Class	Failure Probability & Data Source	Failure Effect Probability	Failure Mode Ratio	Failure Rate	Operating Time	Failure Mode Criticality	Item Criticality	Remarks

Figure IV.22. MIL-STD-1629 worksheet for method 102.

The β values represent the analyst's judgment as to the conditional probability that the loss will occur and should be quantified in general accordance with the following:

FAILURE EFFECT	PROBABILITY OF LOSS OF FUNCTION
Actual loss	$\beta=1$
Probable loss	$0.10<\beta<1.00$
Possible loss	$0<\beta<0.10$
No effect	$\beta=0$

The failure mode ratio, α, is the probability that the part or item will fail. If all potential failure modes of a particular part or item are listed, the sum of the α values for that part or item will equal one. Individual failure mode multipliers may be derived from failure rate source data or from test and operational data. If failure mode data are not available, the α values should represent the analyst's judgment based upon an analysis of the item's functions.

Part failure rates, λ_p, are derived from appropriate reliability prediction methods using mean-time-between-failure (MTBF) data or possibly other data obtained from handbooks or reference material. Manufacturers often supply failure data; however, it is important that the environment the item will be subjected to is similar to the environment the manufacturer used when obtaining the failure data.

The operating time, t, is usually expressed in hours or the number of operating cycles of the item being analyzed.

The α and β values are often subjective, thus making the supposed quantitative method somewhat qualitative. All things considered, it is generally understood that the FMECA process is a qualitative method of analysis.

FMECA process

The process outlined here demonstrates a method, often used throughout industry today, that incorporates the basic FMECA and a qualitative Criticality Analysis on the same form. Exact forms, as defined by various standards, are usually modified by different users to fit individual conditions.

The FMECA process is a combination of group discussions, team meetings, and individual studies. The meetings are typically led by the design engineer or an expert in Failure Mode, Effects and Criticality Analysis.

Entire systems are much too complex to be covered by one FMECA. The system should be defined and separated into sub-systems, also known as the infrastructure, early in the process. It should be a consensus as to how and why the system is separated, since addressing a single part is usually not an efficient use of resources. The FMECA should be done at some major level. For example, for an automobile you might list the *drive train, frame, body,* and *support equipment* separately.

The team is led into discussion about the individual functions of each sub-system and how they interact with the overall system. At this time, team members should be supplied with pertinent technical data including layouts, functional descriptions, prints, etc., of the system being analyzed. After the initial discussions and examination, the infrastructure is separated into infrastructure sub-systems. For example, an automobile drive train might be divided into the engine, the transmission, the drive gearing, and support equipment. This should be done for each infrastructure.

At this time, the engineer or team leader prompts the team to identify the process or machine functions. The functions of each infrastructure sub-system is then discussed, and the data is input on the FMECA form. After all functions are established, *FAILURE MODES, EFFECTS, CAUSES, PROBABILITIES,* and *RECOMMENDED CORRECTIVE ACTIONS* are then input on the form.

SYSTEM:									INFRASTRUCTURE:
COMPANY / LOCATION:									PROJECT ENGINEER:
OUTSIDE SUPPLIER(S):									DATE:
INFRASTRUCTURE SUB-SYSTEM	**SUB-SYSTEM FUNCTION**	**FAILURE MODE**	**EFFECTS OF FAILURE**	**CAUSE(S) OF FAILURE**	**OCC**	**SEV**	**DET**	**RPN**	**RECOMMENDED CORRECTIVE ACTION & STATUS**
WHAT DOES IT DO?	WHAT FUNCTION CAN BE LOST?	HOW **COULD** IT FAIL TO PERFORM THIS FUNCTION? 1. 2. 3.	WHAT HAPPENS IF IT FAILS?	WHAT CAUSES IT TO FAIL?	LIKELIHOOD	CONSEQUENCES	PREDICTABILITY	OXSXD	HOW CAN THIS FAILURE BE PREVENTED?

Figure IV.23. Sample FMECA form for existing equipment.

Steps for completing the form

Refer to Figure IV.23 for the remainder of this section.

First Column—Infrastructure sub-system

What does this sub-system do? Describe briefly what the equipment is designed to do. Be as concise in the description as possible. For example, hold temperature below 200° F during continuous operation.

Second Column—Sub-system function

What function is lost if the sub-system fails? For example, temperature regulation is lost.

Third Column—Failure mode

Anticipation of failures is the goal. The question here is how could it fail, not will it fail. Try to determine all possibilities for the system failing to meet objectives. Some distinctive failure modes are:

Over/undersize	Cracked	Breaks
Torn or Split	Open	Mis-aligned
Rough	Out-of-balance	Pitting
Eccentric	Leaking	Deformed

NOTE: Try to be specific. For example, say, "gas line leaking and deformed" instead of "gas line fails." Always find a word to replace "fails or failed."

Fourth Column—Effects of failure

Describe what will happen as a result of each failure mode. Will the equipment be totally inoperative, inefficient, inaccurate, etc.?

Fifth Column—Cause of failure

Again, try to anticipate the cause of the failure mode described. Ask what might cause this particular failure mode. Fault-tree analysis is sometimes used at this point to determine root causes, as discussed earlier in this section. If time does not permit the use of a fault tree, list all conceivable causes for each failure mode. Examples of causes might be:

Worn or broken tooling	Improper part location
Worn bearings	Overload
Poor handling	Improper weld
Contamination	Interference
Inefficient cooling	Over pressure
Heat treat shrinkage	Inadequate lubrication

NOTE: In columns 7, 8, and 9, it is important to remember that the ratings apply to the failure mode and not to the cause of the failure.

Sixth Column—Occurrence

Estimate the likelihood that the failure will occur using a previously set evaluation scale. Table IV.5 gives an example of an evaluation scale for the likelihood of a failure. The lowest number of the scale must indicate the lowest probability of occurrence, and the highest number must indicate the highest probability of occurrence.

Seventh Column—Severity

Estimate the consequences of the failure using a predetermined scale with the lowest number of the scale indicating minor concerns and the highest number indicating major consequences. Table IV.6 gives an example of an evaluation scale for the consequences of a failure.

Eighth Column—Detection

Estimate the ability to predict the failure. Table IV.7 gives an example of an evaluation scale for the ability to predict a failure. Use a predetermined scale with the lowest number indicating a very high probability that the failure mode will be detected before reaching the customer.

Ninth Column—RPN

Estimate the Risk Priority Number (RPN). The product of the occurrence, severity, and detection provides a relative priority to each failure mode. From this number critical summaries can be developed for prioritizing follow-up actions.

Tenth Column—Recommended corrective action and status

Corrective actions are most usually found to be design changes, either to the equipment or the process. Changes should be focused on reducing failure occurrences, reducing the consequences of failures, and increasing the ability to predict failures. Remember the idea is to reduce failures, not just detect them.

FMECA follow-up report

Follow-up is critical to the FMECA's success. The basic recommendations that the FMECA and the follow-up provide should set a solid foundation for corrective action.

A good tool to use for the follow-up report is a matrix. The first thing to do is establish the matrix criteria (see Table IV.8). Keep in mind that this information will be used to present recommendations to management for approval. The end result should provide answers to all major concerns, such as, why are these problems being worked on, how much will this project cost, what is the time frame, etc.?

Table IV.8. Sample matrix report criteria.

Equipment	Difficulty	% Solved	Cost ($)	Time Required	Actions
Failure mode	1= Very easy	Percent	a = 0	1 = 1–5 units	Recom-
and rank #	2 = easy	of total	b = very low	2 = 6–10 units	mended
	3 = medium	ranked	c = low	3 = 11–15 units	actions
	4 = hard	order	d = medium	4 = 15–20 units	
	5 = Very hard		e = high	5 = 21–30 units	
				6 = 31–50 units	

The next step is to compile the RPNs from the highest number to the lowest (see Table IV.9). This example shows that the number one problem identified by the FMECA is the conveyor system's guide rollers breaking. Also note that the redesigning of these rollers will eliminate three other failure modes; guide rollers seizing, tracking bars bending and bent track rips belt.

Table IV.9. Ranked order chart.

Rank	RPN	Component	Failure mode
1	125	Conveyor system	Guide rollers break
2	100	Drying chamber	Plugged drains
3	90	Air compressor valves	Corrosion
4	80	Ventilation	Louvers stick
5	60	Conveyor system	Guide rollers seize
6	40	Conveyor system	Tracking bars bend
7	20	Conveyor system	Bent track rips belt
8	5	Drying chamber	High humidity

Now total the RPN numbers for a number representing 100% of the problems in the sub-system. In the example in Table IV.9, this number is 520.

If the RPNs for just the CONVEYOR SYSTEM are totaled, the result is 245. Accepting that 245 is 47% of 520, with 520 representing at, or near, 100% of the total population of failure modes, it can be reasonably said that by taking actions to eliminate the cause of the top failure mode, 47% of the overall problem can be solved. This is shown on the matrix in Table IV.10. The cost, complexity, time, etc., of any potential changes should be discussed as a team and a consensus drawn. Table IV.11 is an example of how a safety concern of the FMECA process could be handled on a matrix.

Table IV.10. Completed sample matrix for equipment.

Equipment	Difficulty	% Solved	Cost ($)	Time Required	Actions
Conveyor System; Rank = 1; RPN = 125; Other ranked items effected 5, 6, and 7	Hard	47%	low	Category 6, 45 days	Installed self-lubricating double-guide rollers. Redesigned track bars. Redesigned retainers. Redesigned pins

NOTE: This table is based on Tables IV.5, IV.6, and IV.7, where 5 is the highest possible value in ranking criteria. Thus, 125 would be the highest possible RPN (5x5x5).

Table IV.11. Sample matrix report criteria.

Equipment	Difficulty	% Solved	Cost ($)	Time Required	Actions
Inline plugs	Hard	50%	Very low	5 weeks	Installed inline plugs in discharge

What it takes to do a successful FMECA

The keys to a successful FMECA are the same when applied to the initial design or to existing designs. Management support and the diversity of the team are the most important considerations, followed by training for the facilitator and the team members.

Steps must be taken to assure management's support for the project from beginning to end. First, management must understand the benefits of the FMECA. After completing the FMECA, a quantifying report reflecting the total impact of the FMECA's suggested corrective actions and implementation plans should be presented to management.

A good FMECA team will include personnel that are directly involved with the process to be analyzed. Examples include, but certainly not limited to: electrical and mechanical maintenance technicians, machine operators, technical specialists, design engineers, and even a representative from the equipment supplier if possible.

The FMECA facilitator should be trained to ask appropriate key questions and be prepared to probe the team for answers to describe the equipment's functions, failure modes and their effects. A good facilitator should also be able to assist the team with establishing criteria for the equipment's ranking indexes and organize information as it comes from the team members.

Some time should be allotted for new FMECA team members to become familiar with the FMECA process prior to beginning the actual FMECA. The team should have a good understanding of the purpose and benefits of the FMECA as well as an understanding of the team's objectives. If time allows, training should include an actual FMECA on some less complex piece of equipment such as a flashlight or cigarette lighter. On the basis of past experiences, this training can be accomplished in about four hours.

The use of a computer with the ability to project the screen during the FMECA team meetings is highly recommended. Software containing a standard FMECA blank form should be filled in as data is received from the team members. A copy of the form should be distributed after each meeting to enable the team to review the data and recommend any changes at the next meeting.

Although the use of the computer and software are not an absolute necessity for the successful FMECA, this practice is not only a time-saver, but helps keep the team's ideas organized and on track.

ENVIRONMENTAL REQUIREMENTS AND INFLUENCE

When the FMECA team members are properly selected, the FMECA process will address all problems concerning design, manufacturing, process, safety and environment. The consequences of these problems are addressed during the study of the Failure Mode during the FMECA. Preventive measures for failures where an individual's well-being in the workplace is concerned should always be given top priority. According to John Moubray (2001), "A failure mode has safety consequences if it causes a loss of function or other damage which could hurt or kill someone."

The overall design must consider how the environment influences the system during testing, storage, installing and operating. When considering the failure mode effects, think about the impact on society in general if the failure happened. Society may, in some cases, be considered as a customer because if the product is not accepted due to environmental effects, it may not be successful. "A failure mode has environmental consequences if it causes a loss of function or other damage which could lead to the breach of any known environmental standard or regulation," (Moubray, 2001).

DIFFERENCES BETWEEN PART FUNCTION, PART FAILURE MECHANISM, FAILURE EFFECT, FAILURE DETECTION, AND FAILURE MODE

The definitions which follow form the fundamentals of FMECAs. These definitions will enable the reader to understand the differences between Part Function, Part Failure Mechanism, Failure Effect, Failure Detection, and Failure Modes.

1. **Failure Mode and Effects Analysis (FMEA)**—A procedure by which each potential failure mode in a system is analyzed to determine the results, or effects thereof, on the system and to classify each potential failure mode according to its severity.
2. **Failure mode**—The manner by which a failure is observed. Generally describes the way the failure occurs and its impact on equipment operation.
3. **Failure effect**—The consequence(s) a failure mode has on the operation, function, or status of an item. Failure effects are usually classified according to how the entire system is impacted.
4. **Failure cause**—The physical or chemical process, design defects, part misapplication, quality defects, or other processes that are the basic reason for failure or which initiate the physical process by which deterioration proceeds to failure.
5. **Environments**—The conditions, circumstances, influences, stresses and combinations thereof, surrounding and affecting systems or equipment during storage, handling, transportation, testing, installation, and use in standby status and operations.
6. **Detection mechanism**—The means or methods by which a failure can be discovered by an operator under normal system operation or can be discovered by the maintenance crew by some diagnostic action.
7. **Severity**—The consequences of a failure as a result of a particular failure mode. Severity considers the worst potential consequence of a failure, determined by the degree of injury, property damage, or system damage that could ultimately occur.
8. **Criticality**—A relative measure of the consequences of a failure mode and its frequency of occurrence.
9. **Criticality analysis (CA)**—A procedure by which each potential failure mode is ranked according to the combined influence of severity and probability of occurrence.
10. **FMECA-Maintainability information**—A procedure by which each potential failure is analyzed to determine how the failure is detected and the actions to be taken to repair the failure.
11. **Single failure point**—The failure of an item that would result in failure of the system and is not compensated for by redundancy or alternative operational procedure.
12. **Undetectable failure**—A postulated failure mode in the FMEA for which there is no failure detection method by which the operator is made aware of the failure.

13. **FMECA planning**—Planning the FMECA work involves the contractor's procedures for implementing their specified requirements. Planning should include updating to reflect design changes and analysis results. Worksheet formats, ground rules, assumptions, identification of the level of analysis, failure definitions, and identification of coincident use of the FMECA by the contractor and other organizational elements should also be considered.
14. **Ground rules and assumptions**—The ground rules identify the FMECA approach (e.g., hardware, functional or combination), the lowest level to be analyzed, and include statements of what might constitute a failure in terms of performance criteria. Every effort should be made to identify and record all ground rules and analysis assumptions prior to initiation of the analysis; however, ground rules and analysis assumptions may be adjusted as requirements change.
15. **Analysis approach**—Variations in design complexity and available data will generally dictate the analysis approach to be used. There are two primary approaches for the FMECA. One is the hardware approach that lists individual hardware items and analyzes their possible failure modes. The other is the functional approach that recognizes that every item is designed to perform a number of outputs. The outputs are listed and their failures analyzed. For more complex systems, a combination of the functional and hardware approaches may be considered.
16. **Hardware approach**—The hardware approach is normally used when hardware items can be uniquely identified from schematics, drawings, and other engineering and design data. This approach is recommended for use in a part level up approach often referred to as the bottom-up approach.
17. **Functional approach**—The functional approach is normally used when hardware items cannot be uniquely identified or when system complexity requires analysis from the top down.
18. **Level of analysis**—The level of analysis applies to the system hardware or functional level at which failures are postulated. In other words, how the system being analyzed is segregated. (e.g., a section of the system, component, sub-component, etc.)
19. **Failure definition**—This is a general statement of what constitutes a failure of the item in terms of performance parameters and allowable limits for each specified output.
20. **Trade-off study reports**—These reports should identify areas of marginal and state-of-the-art design and explain any design compromises and operating restraints agreed upon. This information will aid in determining the possible and most probable failure modes and causes in the system.

21. **Design data and drawings**—Design data and drawings identify each item and the item configuration that perform each of the system functions. System design data and drawings will usually describe the system's internal and interface functions beginning at system level and progressing to the lowest indenture level of the system. Design data will usually include either functional block diagrams or schematics that will facilitate construction of reliability block diagrams.
22. **Block diagrams**—Block diagrams that illustrate the operation, inter-relationships, and interdependencies of the functions of a system are required to show the sequence and the series dependence or independence of functions and operations. Block diagrams may be constructed in conjunction with, or after defining the system and shall present the system breakdown of its major functions. More than one block diagram is sometimes required to represent alternative modes of operation, depending upon the definition established for the system.
23. **Functional block diagrams**—Functional block diagrams illustrate the operation and interrelationships between functional entities of a system as defined in engineering data and schematics. An example of a functional block diagram is shown in Figure IV.24.

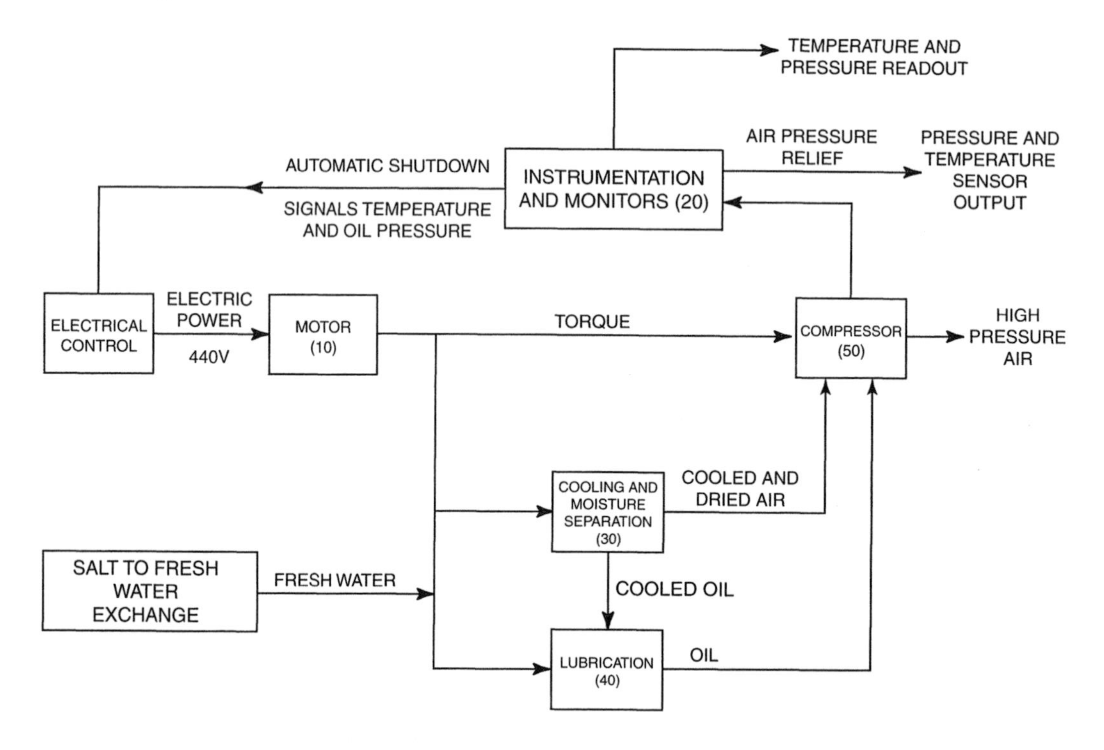

Figure IV.24. Functional block diagram.

24. **Reliability block diagrams**—Reliability block diagrams define the series dependence, or independence, of all functions of a system or functional group for each life-cycle event. An example of a reliability block diagram is shown in Figure IV.25.

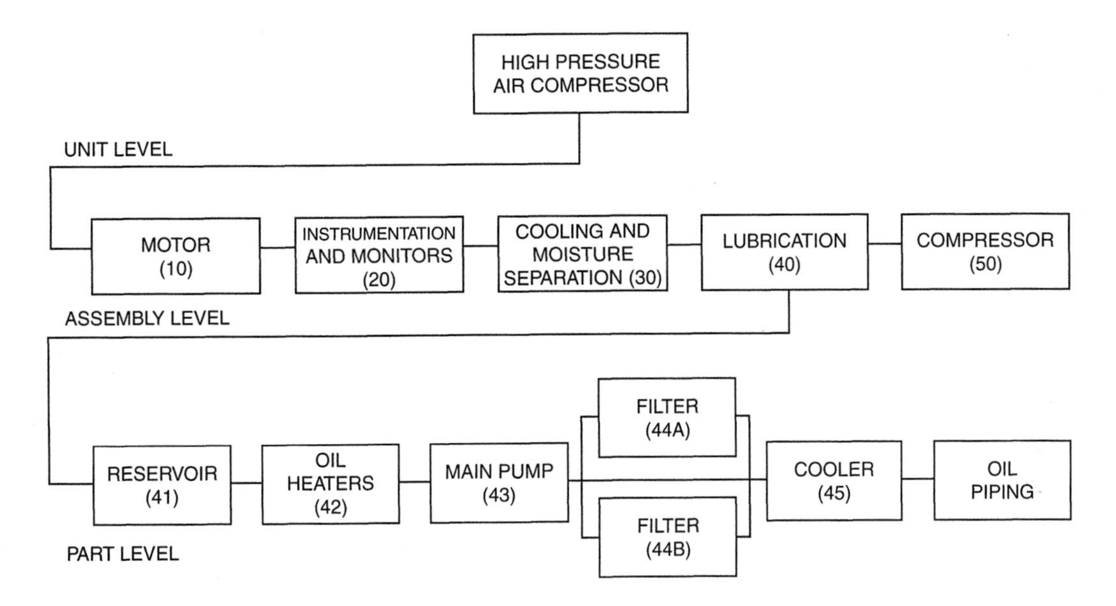

Figure IV.25. Reliability block diagram.

Problem Solving and Quality Improvement

V.A APPROACHES

CONTINUOUS PROCESS IMPROVEMENT

In this section we discuss the fundamental principles of continuous process improvement. There are many other aspects of quality improvement discussed elsewhere in this book. Process improvement is often accomplished via projects, which are discussed in section I.B. Chapters V and VI cover process improvement and process-control tools. Linking process results to customer needs is discussed in chapter I, section H.

There are three premises that underlie the intense interest in continuous process improvement by the quality profession:

- All work is a process
- All processes vary
- Process variation equates to reduced quality

This implies, in turn, that meeting engineering requirements is not good enough. After all, engineering requirements are an allowance for variation. A better approach is that advocated by Genichi Taguchi: minimize process variability around a target value.

Process elements

The dictionary defines process as a series of actions, changes, or functions bringing about a result. Another definition is that a process is a system of related activities that act upon inputs to produce an output. There are three basic parts to a process: inputs, actions, and outputs. It has a well-defined beginning and end. The actions essentially transform the inputs into outputs by some repetitive series of activities. The transformation can be either simple or complex. The transformation can take the form of a physical modification of the inputs (e.g., converting ore to steel, assembling an appliance), it may involve relocating the inputs (e.g., storage in a warehouse, delivery of a package), it may involve a transaction that changes the relationship between the parties involved (e.g., signing a contract, a loan agreement), or it may add value to data or information (e.g., creating a report from raw data, creating a chart from a report). Processes add value to the customer, the one who receives the output from the process. "Added value" means that the outputs are valued more highly by the customer than would the unprocessed inputs. The essence of processing is adding value.

Process control

For process control to occur, there must be information on the process state as it operates, combined with information on the effect of the process state on the output. For example, one might monitor raw material chemistry and relate it to the result of a material hardening process. Of course, there are many potential variables to monitor for any given process. Process control points are selected to maximize the amount of control obtained at minimum cost. The linking of information on process conditions to information on process results is known as *feedback*. Without a feedback loop, process improvement will occur only by chance, and it is just as likely that the process will degrade by chance.

To be useful, feedback must be accompanied by the means to control the process. Control is the ability to modify the process based on feedback, which implies that there are standards of performance to compare with process measurements. Deviations beyond a certain limit are cause for action designed to return the process to its controlled state. To prevent tampering, basic knowledge of the theory of variation is required (see sections I.G and V.A). If control is achieved by human intervention, the person responsible must know what action to take, have the authority to take the necessary action, and know that the action taken will produce a predictable result, at least within reasonable limits.

For processes of interest to quality managers, process activities are repetitive. Process should not be confused with product in this regard. Repetitive processes often produce unique *products*. For example, a communication satellite may well be "one of a kind," custom-made to a particular customer's specifications. But the *process* of producing one satellite has much in common with the process used to produce any other satellite: circuit boards are assembled, terminals are staked, joints are soldered, etc. Were this not so, each product would be approached as a totally new experience, and the customer might just as well choose a shoe cobbler as a manufacturer with a long history of successful satellite manufacturing. Obviously, this is absurd. Yet, the author has lost count of the number of times he has heard "We can't use process control here, every product we produce is unique."

Naturally, when the quality professional hears "repetitive" he immediately thinks "standardize." To the extent that it is justified by sound principles of economics and management, repetitive process operations should be standardized and the standard approach documented. This records the lessons that have been learned in the past and provides newcomers a starting point for operating the process. Of course, provision should be made for experimentation and modifying the process based on new lessons learned. This is a fundamental paradox in quality improvement: the need to standardize combined with the need to experiment. Too much or too little of either is harmful.

Process improvement

Although desirable, process control merely maintains the process in its current state. Since all outputs are the result of processes, continuously improving customer satisfaction requires the continuous improvement of processes. Improvement is accomplished by enhancing the ability of the process to deliver value to the customer. There are two popular and effective strategies for reaching this goal: gradual improvement and radical improvement. Both strategies have their place.

KAIZEN*

KAIZEN is a philosophy of continuous improvement, a belief that all aspects of life should be constantly improved. In Japan, where the concept originated, KAIZEN applies to all aspects of life, not just the workplace. In America the term is usually applied to work processes.

The KAIZEN approach focuses attention on ongoing improvement that involves everyone. Its domain is that of small improvements from ongoing efforts. Over time these small improvements produce changes every bit as dramatic as the "big project" approach. KAIZEN does not concern itself with changing fundamental systems. Rather, it works to optimize the existing systems.

All jobs in any given organization have two components: process improvement and process control. Control, as described above, involves taking action on deviations to maintain a given process state. In the absence of signals indicating that the process has gone astray, control is achieved by adhering to established standard operating procedures (SOPs). In contrast, improvement requires experimentally modifying the process to produce better results. When an improvement has been identified, the SOPs are changed to reflect the new way of doing things. Imai (1986) describes the perception of job responsibilities (improvement or maintenance) based on job function. (See Figure V.1.)

---

*KAIZEN™ is a trademark of the KAIZEN Institute, Ltd.

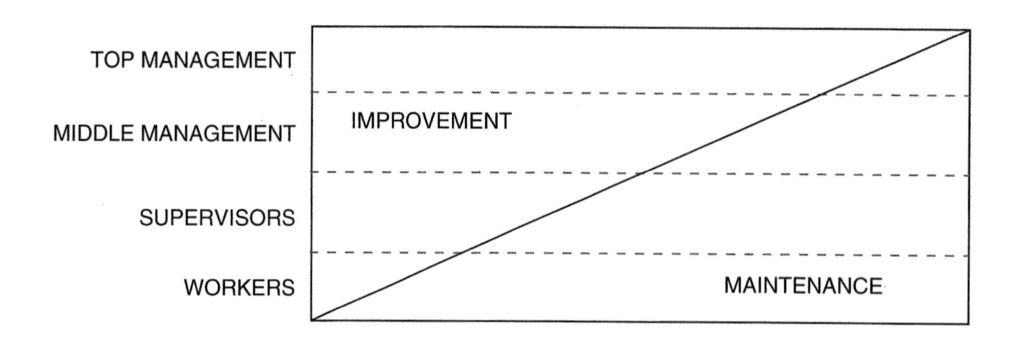

Figure V.1. Responsibilities and job functions.

The lower portion involves maintaining the process at its current level by following the SOPs. KAIZEN fits into the upper portion of this picture. However, the upper portion goes beyond KAIZEN. Imai illustrates the relationship as shown in Figure V.2.

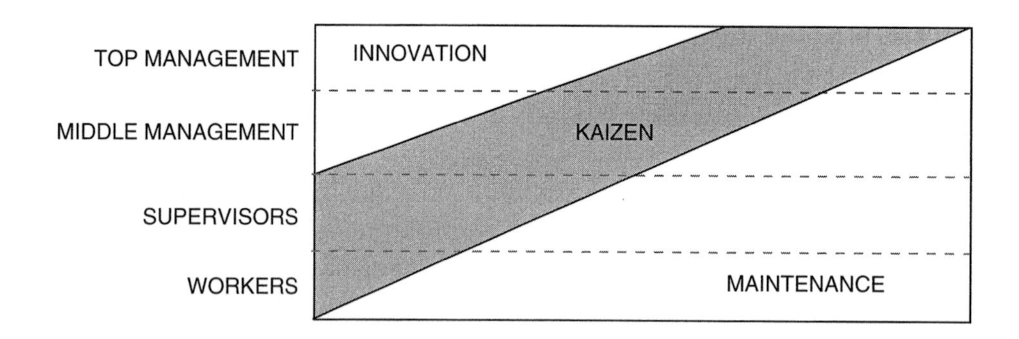

Figure V.2. Responsibility for KAIZEN and KAIZEN's role in process improvement.

The figure illustrates that, as mentioned earlier, KAIZEN does not cover radical innovations. It can be seen that all levels of management share responsibility for KAIZEN. Since work is always done according to standards, the standards must be viewed as dynamic documents. The fundamental idea behind KAIZEN comes straight from the Deming/Shewhart PDCA cycle:

- Someone has an idea for doing the job better (Plan)
- Experiments will be conducted to investigate the idea (Do)
- The results evaluated to determine if the idea produced the desired result (Check)
- If so, the SOP will be changed (Act)

Thus, this "Japanese" approach has its roots well-established in the scientific method. The Japanese contribution was to integrate the approach into its management systems to assure that it was done routinely, at all levels of the organization. Also, the Japanese apply KAIZEN to every process and to the entire production cycle, while non-Japanese companies tend to restrict improvement to Research and Development or new-process start-ups.

Imai lists the following KAIZEN duties for the different levels in management.

Table V.1. Hierarchy of KAIZEN involvement.

From Imai, M. *KAIZEN: The Key to Japan's Competitive Success*, page 8.

Copyright © 1986 by the KAIZEN Institute, Ltd.

POSITION	RESPONSIBILITIES
Top management	• Be determined to introduce KAIZEN as a corporate strategy • Provide support and direction for KAIZEN by allocating resources • Establish policy for KAIZEN and cross-functional goals • Realize KAIZEN goals through policy deployment and audits • Build systems, procedures, and structures conducive to KAIZEN
Middle management and staff	• Deploy and implement KAIZEN goals as directed by top management through policy deployment and cross-functional management • Use KAIZEN in functional capabilities • Establish, maintain, and upgrade standards • Make employees KAIZEN-conscious through intensive training programs • Help employees develop skills and tools for problem solving
Supervisors	• Use KAIZEN in functional roles • Formulate plans for KAIZEN and provide guidance to workers • improve communication with workers and sustain high morale • Support small group activities (such as quality circles) and the individual suggestion system • Introduce discipline in the workshop • Provide KAIZEN suggestions
Workers	• Engage in KAIZEN through the suggestion system and small group activities • Practice discipline in the workshop • Engage in continuous self-development to become better problem solvers • Enhance skills and job performance expertise with cross-education

Generally, KAIZEN is implemented via quality improvement teams at various levels of management. Team formation, training, and management is discussed in section I.C and I.D. The tools used in KAIZEN are discussed in section V.C. Individual KAIZEN projects are generally not done on a scale that warrants full-blown project management tools such as PERT-CPM, but the "7M" tools described in section V.B are often used to manage KAIZEN projects.

Reengineering

Reengineering is the complement of KAIZEN, it applies to the innovation portion of Figure V.2 Unlike the small, cumulative changes made by KAIZEN, reengineering seeks innovation-based dramatic improvements. The goal of reengineering is large scale improvements from radical redesign of systems and business processes. Reengineering projects begin with a "blank sheet of paper." That is, the process is redesigned from scratch for maximum value-added potential. Reengineering change is discontinuous change, rather than gradual change. It breaks from the past completely.

Information technology

Information technology is treated as an enabler in reengineering. Recent and rapid improvements in this area make possible dramatically different organizational relationships. Past assumptions regarding limits on communication and coordination of activities should be reexamined in the light of new and future information technologies. Organization structures created to communicate the wishes of the leadership throughout the organization may no longer be needed. Traditional communications between customer and supplier via gatekeeper activities such as purchasing departments may be obsolete. Many paper-based systems can be replaced with far more responsive, and less expensive, electronic systems.

Resynthesis of disaggregated work

Section I.G traces the history of management up to the present time. It makes clear that old management structures are now outdated and unable to deal effectively with current and future realities. It's not that the old structures were "wrong." Rather, these structures were designed to deal with conditions that no longer exist. One key area where this is so is in the design of work itself. The approach to work design used by many organizations still treats employees as uneducated and unskilled cogs-in-the-wheel.

Most jobs are still designed using an assembly-line mentality. Jobs are done step-by-step with a different person responsible for each step or small groups of steps. This type of job design creates a number of problems:

- More work in process exists;
- The many "handoffs" between people performing the various steps creates inefficiencies, delays, and errors;

- The work is boring and uninteresting, increasing the likelihood of error;
- It is difficult for people to see the whole picture, which makes it harder to conceive of process improvements.

Because of their process-wide perspective, reengineering projects often focus on non-sequential tasks. Activities are viewed from the perspective of the value added to customers or to the business, i.e., they are process-specific rather than job-specific. Rather than having all work design done by management and engineering specialists, workers decide how they will organize the processes. Rather than one large, specialized, difficult-to-change process, work is organized into small, flexible processes. When work is reintegrated into value-added delivery processes, there is a reduced need for checking and control. Not only does this require fewer quality control and quality assurance personnel, it also requires fewer supervisory personnel. Typically, reengineered processes result in flatter and more horizontal organizational structures.

New worker requirements

These new approaches to work require a different set of skills and abilities from employees. The term "self-management" implies that employees have key management skills, such as communication skills, time management, organizational skills, etc. When work is redesigned by combining tasks, each employee must be more highly skilled than before. Thus, reengineering usually requires significant additional investment in education and training, not just during the project but on an ongoing basis. Flexible systems require flexible people, and this is achieved by an investment in education and training. Longer training periods are required to provide the additional knowledge, skills, and abilities necessary for the new assignments. Also, employees must possess greater problem-solving ability than in the past. Self-directed teams must have planning skills as well. Finally, the team-based approach and cooperative relationships involved with reintegrated work require people with greater interpersonal skills than in the past.

CYCLE-TIME REDUCTION

Cycle-time measurements

All of the products and services we see around us began as ideas in someone's mind. There is a virtual consensus that the term "cycle-time" refers to the time elapsed from the initial idea to some other event. However, there is no consensus as to which event to use to terminate the measurement of cycle-time. Various authors have measured cycle-time as the time from the initial concept to

- The completion of the design subprocess
- Functional prototype completion and test
- First production unit
- Start of full scale manufacturing

- First delivery to customer
- Break even on initial investment
- Full contribution to margin

In this book the term cycle-time will refer to the time elapsed from the initial written record of an idea to the time acceptable product is first delivered to a paying customer. The list above makes it clear that this cycle consists of a number of subprocesses. The reader may choose to define cycle-time according to other criteria.

The development process

The development process consists of the following steps:

1. Think of an idea (e.g., "Controlled flight in heavier-than-air machines")
2. Clarify the ideas (fixed wing craft, control surfaces, propeller driven)
3. Product development (detailed engineering requirements)
4. Prototype (prove the concept is feasible, e.g., Kitty Hawk)
5. Production requirements definition (detailed engineering specifications)
6. Production process design
7. Pilot production (no pun intended!)
8. Full-scale production
9. Sales and distribution
10. Support

(Note that Orville and Wilbur Wright's success faltered badly after item 4!)

Identifying the need to improve cycle-time

There are certain signals that indicate a need to place a priority on reducing cycle-time. Donovan (1995) lists the following "symptoms" of poor cycle-time:

- Poor quality
- Low throughput
- Falling sales
- Excessive inventory
- Excessive nonvalue-added activity
- Poor delivery
- Dissatisfied customers

An interrelationship digraph could be used to show how these different symptoms share the common underlying cause of excessive cycle-time. For example, the previous discussion of engineering changes indicated that long cycle-times are often caused by large number of design changes near or after launch. These changes disrupt production, leading to quality problems, reduced throughput and poor delivery performance. Customers become dissatisfied and some leave. When planned sales fail to materialize, inventories build up. Employees spend a great deal of time responding to customer complaints, processing inventory, and dealing with quality problems (nonvalue-added activities).

Causes of long cycle-times

The major causes of long cycle-times are (Donovan, 1995):

- Bottlenecks
- Poor supplier relations
- Improper performance metrics
- Overly complicated business processes
- Inadequate production scheduling

We will discuss each of these causes in turn.

Bottlenecks—Work flow through a series of operations must be balanced. It often happens that one of the process steps takes longer than the others, leading to a bottleneck. When bottlenecks exist they produce excessive work-in-progress inventory (WIP) and delays. Bottlenecks "pace" the entire process. Work cannot be completed any faster than it can be processed by the bottleneck. Thus, dealing with bottlenecks demands a different strategy than dealing with an ordinary process. Management of bottlenecks should, in general, *oversupply the bottleneck with resources.* The quality of resources going into bottlenecks should be very high. Bottlenecks should be staffed every possible moment.

Poor supplier relations—Supplier quality and lead times are critical factors in cycle-time management. Obviously, production cannot begin until a supplier delivers acceptable quality raw materials or component parts for the production process. Since the typical level of purchased material varies from 50% to 80% of the total cost, suppliers have a tremendous influence on overall cycle-time. Section I.I discusses supplier relations in greater detail.

Improper performance metrics—The prevailing management world view tends to focus on what systems engineers call "local optimization." A local optimum, when viewed from a global perspective, often leads to decidedly *suboptimal* performance for the entire system. Metrics play an important role in reinforcing this distorted view of reality. For example, a manufacturing department manager may focus her attention on efficiency reports. Supervision can improve efficiency in a variety of ways. For example, they may run non-bottleneck processes at a rate faster than the bottleneck can use. The result will be high efficiency numbers for the non-bottlenecks, but at the price of excessive work-in-progress and other costs. Overall throughput will be completely unaffected.

Overly complicated business processes—It is common for business processes to increase in complexity as time goes by. The complexity is often added in response to unusual events that are unlikely to recur or situations that no longer exist. For example, a company might add additional approvals for non-budgeted expenses during a cashflow crunch. Over a period of years, processes can become burdened with a great many steps that are not worth the expense. One way of dealing with this is to reevaluate the process completely using an approach such as business process reengineering, which was discussed earlier.

Inadequate production scheduling—Production scheduling is both an art and a science. Improper scheduling creates many of the problems described above. If scheduling problems are chronic, it may be worthwhile to enlist outside consulting assistance to deal with the problem.

The check sheet shown in Figure V.3 can be used to evaluate cycle-times for your organization.

		YES	NO	POINTS
1.	Have key personnel been thoroughly trained in all aspects of cycle-time reduction and quick response?	❑	❑	______
2.	Has the business impact of cycle-time reduction and quick response been quantified?	❑	❑	______
3.	Have we organized and trained multifunctional teams to shorten cycle-time for all manufacturing, nonmanufacturing and administrative processes?	❑	❑	______
4.	Have we mapped all supply-chain processes, clearly identifying value-added and nonvalue-added activities, bottlenecks, queues, cycle-times, etc.?	❑	❑	______
5.	Have we specifically defined the barriers that prevent us from achieving reductions in cycle-time, with an action plan to remove the barriers?	❑	❑	______
6.	Are we actively working with key vendors to achieve mutually agreed upon goals for cycle-time reduction?	❑	❑	______
7.	Have we produced substantial improvements by reducing lot sizes, changeover times and queues in manufacturing and nonmanufacturing areas?	❑	❑	______
8.	Are our primary performance measurements and reward system heavily weighted toward fast, on-time customer response?	❑	❑	______
9.	Can we *precisely* predict our delivery lead time for customers and/or to replenish inventories?	❑	❑	______
10.	Are our lead times equal to or less than what our customers need and want?	❑	❑	______

Figure V.3. Cycle-time self-assessment checklist.

From Donovan, R.M. (1995). "Improving manufacturing cycle-time," *Quality Digest*, June, 51–53.

To evaluate and score your organization's effectiveness with cycle-time reduction efforts, have your entire management team candidly answer the question on the checklist. A "yes" answer receives 10 points and a "no" answer receives "0" points. The performance criteria contained within the questions are, individually and collectively, very important to cycle-time performance, making a score of nothing less than 100 % really acceptable. More important than the quantitative score is the focus each question and the various responses can bring to management team discussions. This should then result in initiating the action required to achieve significant cycle-time and delivery performance improvements.

Implementing cycle-time reduction

One of the contributions of the quality profession to the field of management is a general approach to implementing process improvements. This approach can be applied to improving the development process as well. The basic steps are as follows:

1. Identify the process owner
2. Identify the process boundaries, e.g.,
   a. Concept-to-manufacturing
   b. Design subprocess of development process
3. Define the process
   a. Flowchart and/or process map
   b. Identify subprocesses
4. Establish control points
   a. E.g., design review, model evaluation, etc.
5. Measure cycle-time at control points
   a. Determine baselines
   b. Determine benchmarks
6. Apply Kaizen or reengineer design process
   (E.g., use modular designs, use more off-the-shelf components, use advanced statistics in qualification testing.)

These steps are part of a general approach to managing major quality projects. See section I.B for additional discussion of this important topic.

Results

The approach described above was implemented for an important subprocess at a computer manufacturing company (Melan, 1993, p. 213). The company reported the following results:

- Cycle-time reduced from 39 months to 19 months.
- $2 million cost savings.
- ROI 10:1 in first year alone.

Results similar to these have been reported by companies in other industries that undertook cycle-time reduction projects. For a chemical manufacturing example, see Wason and Bhalla (1994).

PDCA (PLAN-DO-CHECK-ACT)

The PDCA cycle, which Deming refers to as the PDSA cycle (Deming, 2000), is a flow chart for learning and process improvement. The basic idea began with Shewhart's attempt to understand the nature of knowledge. Shewhart believed that knowledge begins and ends in experimental data but that it does not end in the data in which it begins. He felt there were three important components of knowledge (Shewhart, 1939, 1986): a) the data of experience in which the process of knowing *begins*, b) the prediction in terms of data that one would expect to get if one were to perform certain experiments in the *future*, and c) the degree of belief in the prediction based on the original data or some summary thereof as evidence. Shewhart arranged these three components schematically as shown in Figure V.4.

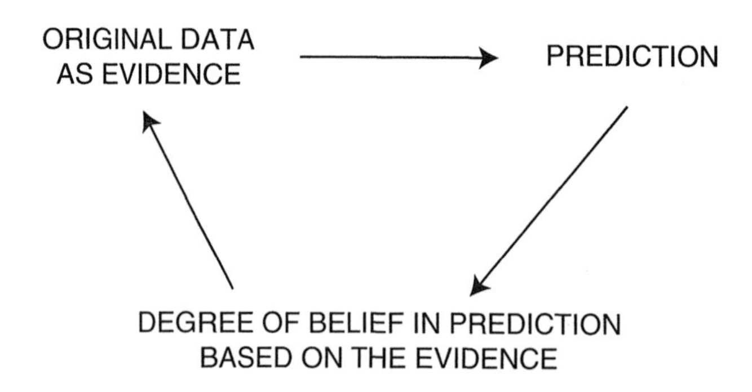

Figure V.4. The three components of knowledge.

Since knowledge begins with the original data and ends in new data, these future data constitute the operationally verifiable meaning of the original data. However, since inferences or predictions based upon experimental data can never be certain, the knowledge based upon the original data can inhere in these data only to the extent of some degree of rational belief. In other words, according to Shewhart, knowledge can only be *probable*. Also, the data are not "facts" in and of themselves, they are merely measurements that allow us to draw inferences about something. In other words, *we can not have facts without some theory.*

Shewhart applied these principles in many practical ways. For example, he identified the three steps of quality control in manufacturing as specification, production, and judgment of quality (inspection). He noted that, in practice, specifications could not be set without first having some information from inspection to help establish process capability, and that this information could not be obtained until some units had been produced.

In short, Shewhart modified the sequence of specification-production-inspection as shown in Figure V.5. He also observed that the specification-production-inspection sequence corresponded respectively to making a hypothesis, carrying out an experiment, and testing the hypothesis. Together the three steps constitute a dynamic scientific process of acquiring knowledge.

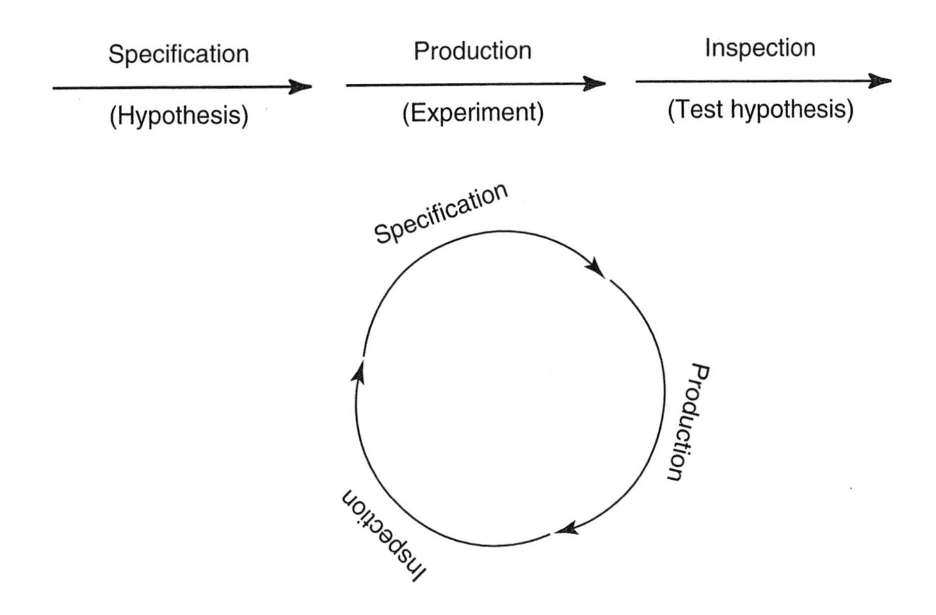

Figure V.5. Scientific process of acquiring knowledge (Figure II.1a repeated).

Note that Shewhart's model of knowledge forms a circle. Shewhart followed the teachings of philosopher C.I. Lewis, who believed that all good logics are circular. The essence of this view is to see knowledge as dynamic. It changes as new evidence comes in. As Shewhart put it (Shewhart, 1939, 1986 p. 104):

> Knowing in this sense is somewhat a continuing process, or method, and differs fundamentally in this respect from what it would be if it were possible to attain certainty in the making of predictions.

Shewhart and Deming revised the above model for application to the improvement of products and processes. The new model was first called the PDCA cycle, later revised by Deming to the Plan-Do-Study-Act, or PDSA cycle (Deming, 2000). The Shewhart-Deming PDSA cycle is shown in Figure V.6.

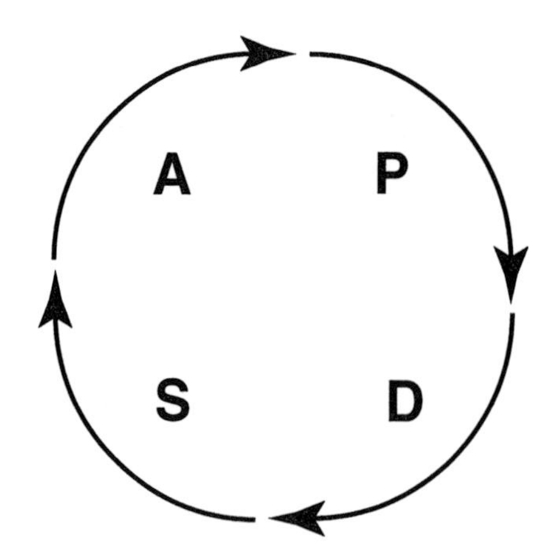

Figure V.6. The Shewhart-Deming PDSA cycle for learning and improvement.

Plan a change or a test, aimed at improvement. This is the foundation for the entire PDCA-PDSA cycle. The term "plan" need not be limited to large-scale planning on an organization-wide scale, it may simply refer to a small process change one is interested in exploring.

Do. Carry out the change or the test (preferably on a small scale). It is important that the DO step carefully follow the plan, otherwise learning will not be possible.

Study the results. What did we learn? What went wrong?

Act. Adopt the change, or abandon it, or run through the cycle again.

The PDCA approach is essentially a management-oriented version of the original Shewhart cycle, which focused on engineering and production. A number of other variations have been developed, two of Deming's variations are shown in Figure V.7.

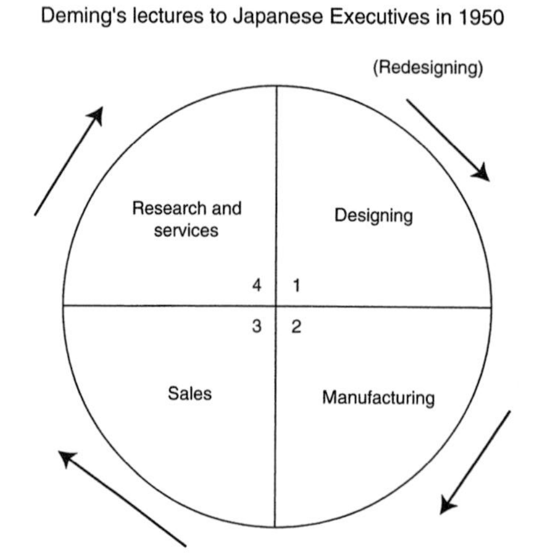

Figure V.7. Some variations of the PDCA-PDSA cycle.

Continued on next page . . .

Figure V.7—*Continued . . .*

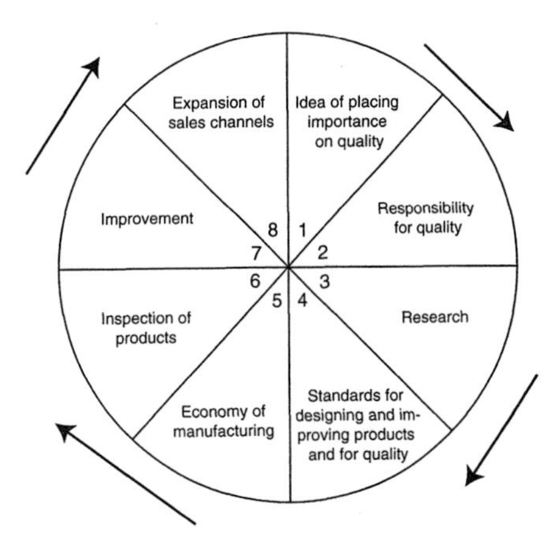

Figure V.7. Some variations of the PDCA-PDSA cycle.

Juran depicts quality as a "spiral," as shown in Figure V.8.

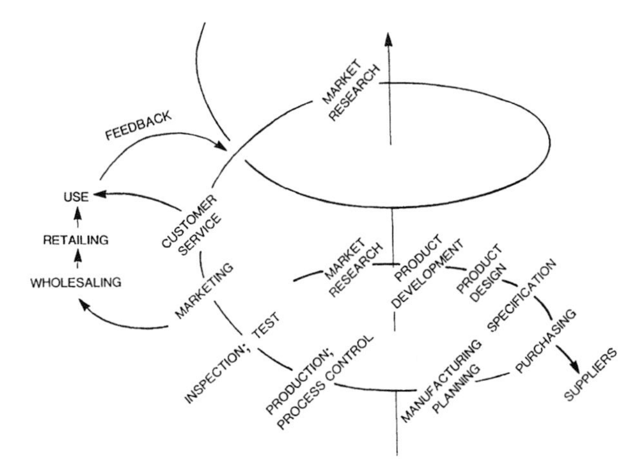

Figure V.8. Juran's spiral of progress in quality (Figure II.1c repeated).

Because of their historical origins and logical appeal, circular diagrams are ubiquitous in the quality field. In quality management, the circle represents continuous improvement of quality by continuous acquisition of knowledge.

V.B MANAGEMENT AND PLANNING TOOLS

Since Dr. Shewhart launched modern quality control practice in 1931, the pace of change in recent years has been accelerating. The 7M tools are an example of the rapidly changing face of quality technology. While the traditional QC tools (Pareto analysis, control charts, etc.) are used in the analysis of quantitative data analysis, the 7M tools apply to qualitative data as well. The "M" stands for Management, and the tools are focused on managing and planning quality improvement activities. In recognition of the planning emphasis, these tools are often referred to as the "7 MP" tools. This section will provide definitions of the 7M tools. The reader is referred to Mizuno (1988) for additional information on each of these techniques.

AFFINITY DIAGRAMS

The word *affinity* means a "natural attraction" or kinship. The affinity diagram is a means of organizing ideas into meaningful categories by recognizing their underlying similarity. It is a means of *data reduction* in that it organizes a large number of qualitative inputs into a smaller number of major dimensions, constructs, or categories. The basic idea is that, while there are many *variables*, the variables are measuring a smaller number of important factors. For example, if patients are interviewed about their hospital experience they may say "the doctor was friendly," "the doctor knew what she was doing," and "the doctor kept me informed." Each of these statements relates to a single thing, the doctor. Many times affinity diagrams are constructed using existing data, such as memos, drawings, surveys, letters, and so on. Ideas are sometimes generated in brainstorming sessions by teams. The technique works as follows:

1. Write the ideas on small pieces of paper (Post-its™ or 3x5 cards work very well).
2. The team works in silence to arrange the ideas into separate categories. Silence is believed to help because the task involves pattern recognition and some research shows that for some people, particularly males, language processing involves the left side of the brain. Research also shows that left-brain thinking tends to be more linear, which is thought to inhibit creativity and pattern-recognition. Thus, by working silently, the right brain is more involved in the task. To put an idea into a category a person simply picks up the Post-it™ and moves it.
3. The final groupings are then reviewed and discussed by the team. Usually, the grouping of ideas helps the team to develop a coherent plan.

Affinity diagrams are useful for analysis of quality problems, defect data, customer complaints, survey results, etc. They can be used in conjunction with other techniques such as cause and effect diagrams or interrelationship digraphs (see below). Figure V.9 is an example of an affinity diagram.

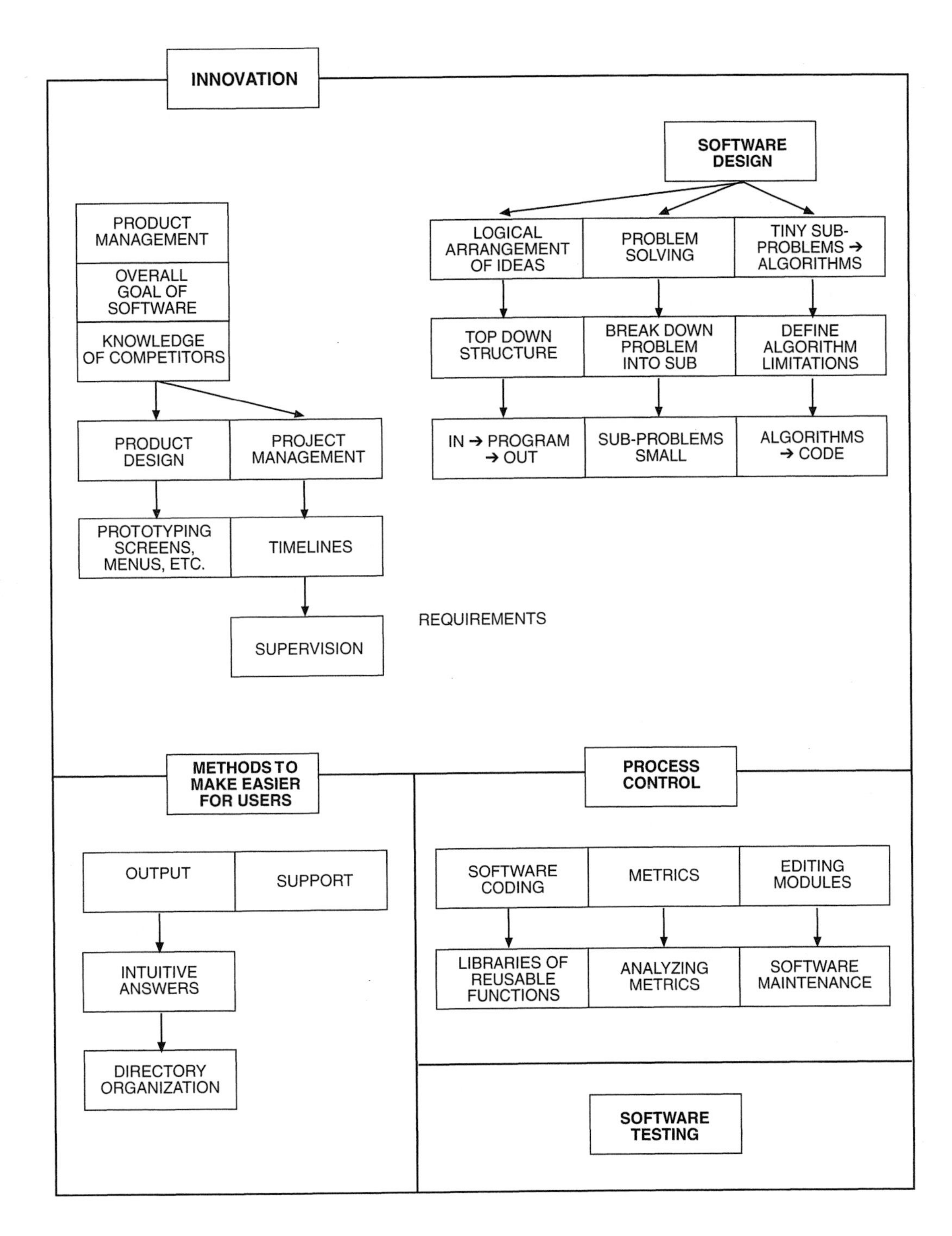

Figure V.9. Software development process affinity diagram.

From "Modern approaches to software quality improvement," figure 3. *Australian Organization for Quality: Qualcon 90*. Copyright © 1990 by Thomas Pyzdek.

TREE DIAGRAMS

Tree diagrams are used to break down or stratify ideas in progressively greater detail. The objective is to partition a big idea or problem into its smaller components. By doing this you will make the idea easier to understand, or the problem easier to solve. The basic idea behind this is that, at some level, a problem's solution becomes relatively easy to find. Figure V.10 shows an example of a tree diagram. Quality improvement would progress from the rightmost portion of the tree diagram to the leftmost. Another common usage of tree diagrams is to show the goal or objective on the left side and the means of accomplishing the goal to the right.

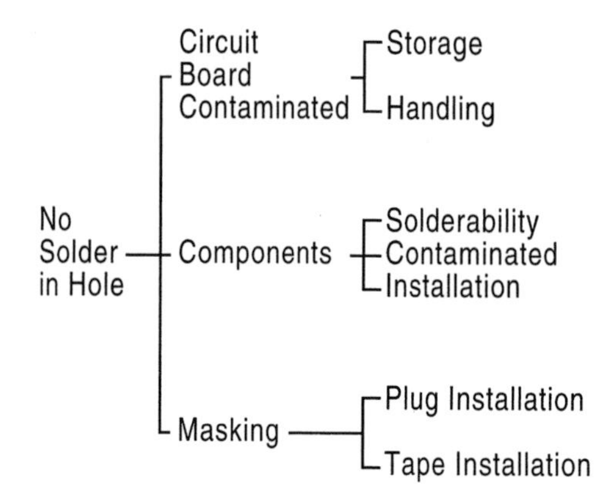

Figure V.10. An example of a tree diagram.

PROCESS DECISION PROGRAM CHARTS

The process decision program chart (PDPC) is a technique designed to help prepare contingency plans. It is modeled after reliability engineering methods of Failure Mode, Effects, and Criticality Analysis (FMECA) and Fault Tree Analysis (see section IV.F). The emphasis of PDPC is the impact of the "failures" (problems) on project schedules. Also, PDPC seeks to describe specific actions to be taken to prevent the problems from occurring in the first place, and to mitigate the impact of the problems if they do occur. An enhancement to classical PDPC is to assign subjective probabilities to the various problems and to use these to help assign priorities. Figure V.11 shows a PDPC.

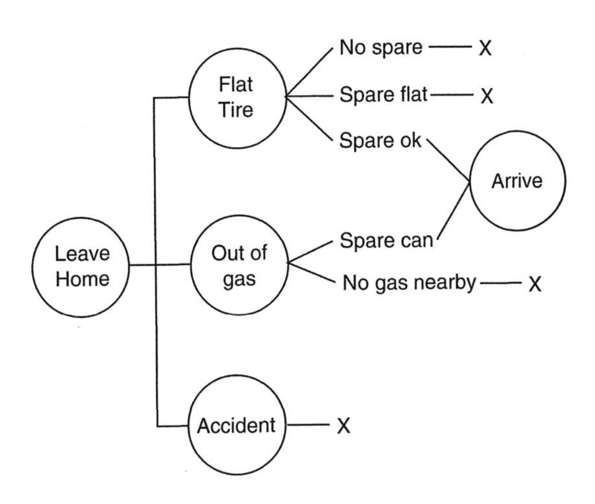

Figure V.11. Process decision program chart.

MATRIX DIAGRAMS

A matrix diagram is constructed to analyze the correlations between two groups of ideas. Actually, quality function deployment (QFD) is an enhanced matrix diagram (see I.H for a discussion of QFD). The major advantage of constructing matrix diagrams is that it forces you to *systematically* analyze correlations. Matrix diagrams can be used in conjunction with decision trees. To do this, simply use the most detailed level of two decision trees as the contents of rows and columns of a matrix diagram. An example of a matrix diagram is shown in Figure V.12.

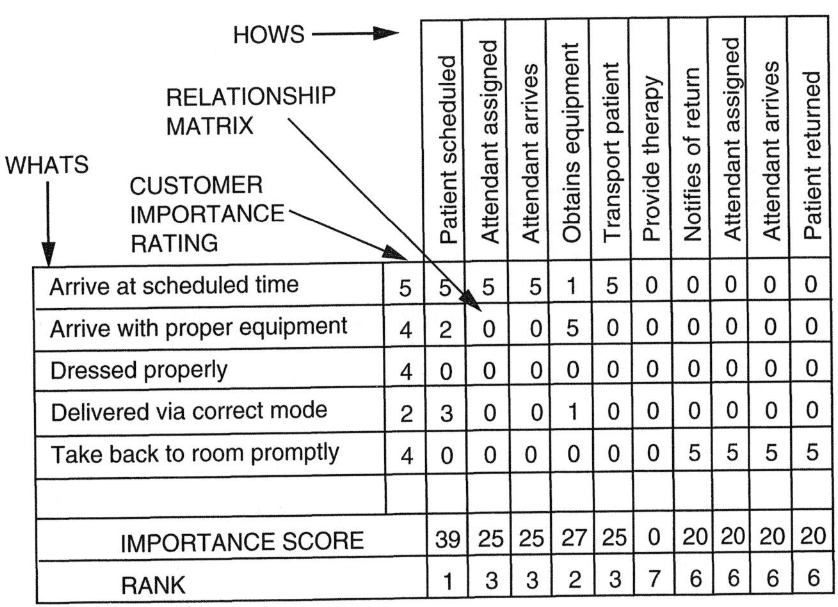

		Patient scheduled	Attendant assigned	Attendant arrives	Obtains equipment	Transport patient	Provide therapy	Notifies of return	Attendant assigned	Attendant arrives	Patient returned
Arrive at scheduled time	5	5	5	5	1	5	0	0	0	0	0
Arrive with proper equipment	4	2	0	0	5	0	0	0	0	0	0
Dressed properly	4	0	0	0	0	0	0	0	0	0	0
Delivered via correct mode	2	3	0	0	1	0	0	0	0	0	0
Take back to room promptly	4	0	0	0	0	0	0	5	5	5	5
IMPORTANCE SCORE		39	25	25	27	25	0	20	20	20	20
RANK		1	3	3	2	3	7	6	6	6	6

5 = high importance, 3 = average importance, 1 = low importance

Figure V.12. An example of a matrix diagram.

INTERRELATIONSHIP DIGRAPHS

Like affinity diagrams, interrelationship digraphs are designed as a means of organizing disparate ideas, usually (but not always) ideas generated in brainstorming sessions. However, while affinity diagrams seek to simply arrange related ideas into groups, interrelationship digraphs attempt to define the ways in which ideas influence one another. It is best to use both affinity diagrams and interrelationship digraphs.

The interrelationship digraph begins by writing down the ideas on small pieces of paper, such as Post-its™. The pieces of paper are then placed on a large sheet of paper, such as a flip-chart sheet or a piece of large-sized blueprint paper. Arrows are drawn between related ideas. An idea that has arrows leaving it but none entering is a "root idea." By evaluating the relationships between ideas you will get a better picture of the way things happen. The root ideas are often keys to improving the system. Figure V.13 illustrates a simple interrelationship digraph.

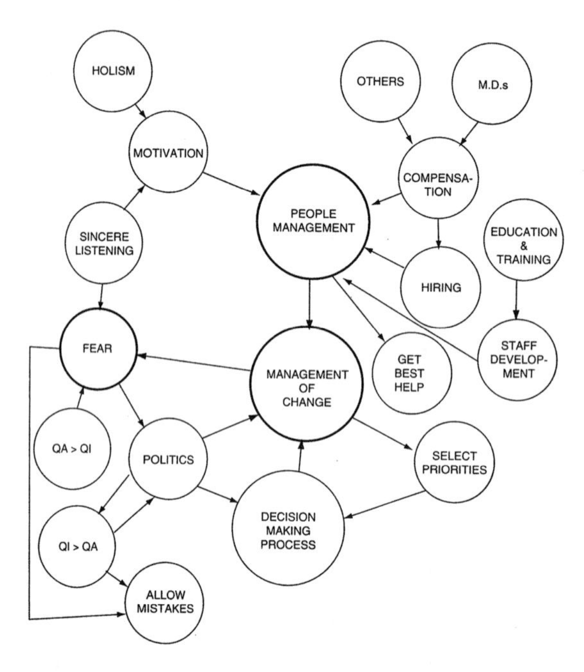

Figure V.13. How does "people management" impact change?

PRIORITIZATION MATRICES*

To prioritize is to arrange or deal with in order of importance. A prioritization matrix is a combination of a tree diagram and a matrix chart and it is used to help decision makers determine the order of importance of the activities or goals being considered. Prioritization matrices are designed to rationally narrow the focus of the team to those key issues and options which are most important to the organization. Brassard (1996) presents three methods for developing prioritization matrices: the full analytical criteria method, the combination interrelationship digraph (ID)/matrix method, and the consensus criteria method. We will discuss the three different methods.

Full analytical criteria method

The full analytical criteria method is based upon work done by Saaty (1988). Saaty's approach is called the analytic hierarchy process (AHP). While analytically rigorous, AHP is cumbersome in both data collection procedures and the analysis. This author recommends that this approach be reserved for truly "heavy-duty" decisions of major strategic importance to the organization. In those cases, you may wish to obtain consulting assistance to assure that the approach is properly applied. In addition, you may want to acquire software to assist in the analysis**. Brassard (1996) and Saaty (1988) provide detailed examples of the application of the full analytical criteria approach.

Combination ID/matrix method

The interrelationship digraph (ID) is a method used to uncover patterns in cause and effect relationships (see above). This approach to creating a prioritization matrix begins with a tree diagram (see above). Items at the rightmost level of the tree diagram (the most detailed level) are placed in a matrix (i.e., both the rows and columns of the matrix are the items from the rightmost position of the tree diagram) and their impact on one another evaluated. The ID matrix is developed by starting with a single item, then adding items one-by-one. As each item is added, the team answers the question "is this item caused by X? where X is another item." The process is repeated item-by-item until the relationship between each item and every other item has been determined. If the answer is "yes," then an arrow is drawn between the "cause" item and the "effect" item. The strength of the relationship is determined by consensus. The final result is an estimate of the relative strength of each item and its effect on other items.

*This chart replaces the matrix data analysis chart, formerly one of the 7M tools. The matrix data analysis chart was based on factor analysis or principal components analysis. This dependence on heavy-duty statistical methods made it unacceptable as a tool for use by nonstatisticians on a routine basis.

**At the time of this writing, software was available from Quality America, Inc. in Tucson, Arizona.

◎ (9) = Strong influence ○ (3) = Some influence △ (1) = Weak/possible influence ↑ Means row leads to column item ← Means column leads to row item	Add features to existing products	Make existing product faster	Make existing product easier to use	Leave as-is and lower price	Devote resources to new products	Increase technical support budget	Out arrows	In arrows	Total arrows	Strength
Add features to existing products	■	↑◎	↑◎	↑◎	←◎	↑◎	4	1	5	45
Make existing product faster	←◎	■	↑◎	↑◎			2	1	3	27
Make existing product easier to use	←◎	←◎	■	↑○			1	2	3	21
Leave as-is and lower price	←◎	←◎	←○	■			0	3	3	21
Devote resources to new products	←◎				■	↑◎	1	1	2	18
Increase technical support budget	←◎				←◎	■	0	2	2	18

Figure V.14. Combination I.D./matrix method.
Use the best mix of marketing medium.

In Figure V.14, an "in" arrow points left and indicates that the row item leads to the column item. On the ID, this would be indicated by an arrow *from* the row item *to* the column item. An "out" arrow points upward and indicates the opposite of an "in" arrow. To maintain symmetry, if an in arrow appears in a row/column cell, an out arrow must appear in the corresponding column/row cell, and vice versa.

Once the final matrix has been created, priorities are set by evaluating the strength column, the total arrows column, and the relationship between the number of in and out arrows. An item with a high strength and a large number of out arrows would be a strong candidate because it is important (high strength) and it influences a large number of other options (many arrows, predominately out arrows). Items with high strength and a large number of in arrows are candidates for outcome measures of success.

Consensus criteria method

The consensus criteria method is a simplified approach to selecting from several options according to some criteria. It begins with a matrix where the different options under consideration are placed in rows and the criteria to be used are shown in columns. The criteria are given weights by the team using the consensus decision rule. For example, if criterion #1 were given a weight of 3 and the group agreed that criterion #2 was twice as important, then criterion #2 would receive a weight of 6. Another way to do the weighting is to give the team $1 in nickels and have them "spend" the dollar on the various criteria. The resulting value allocated to each criterion is its weight. The group then rank orders the options based on each criterion. Ranks are labeled such that the option that best meets the criterion gets the highest rank; e.g., if there are five options being considered for a given criterion, the option that best meets the criterion is given a rank of 5.

The options are then prioritized by adding up the option's rank for each criterion multiplied by the criterion weight.

Example of consensus criteria method

A team had to choose which of four projects to pursue first. To help them decide, they identified four criteria for selection and their weights as follows: high impact on bottom line (weight=0.25), easy to implement (0.15), low cost to implement (0.20) and high impact on customer satisfaction (0.40). The four projects were then ranked according to each criterion; the results are shown in the table below.

	CRITERIA AND WEIGHTS				
WEIGHT →	0.25	0.15	0.2	0.4	
	Bottom line	Easy	Low cost	Customer satisfaction	Total
Project 1	1	2	2	1	1.35
Project 2	3	4	4	3	3.35
Project 3	2	1	3	4	2.85
Project 4	4	3	1	2	2.45

In the above example, the team would begin with project #2 because it has the highest score. If the team had difficulty reaching consensus on the weights or ranks, they could use totals or a method such as the nominal group technique described below.

ACTIVITY NETWORK DIAGRAMS

Activity network diagrams, sometimes called arrow diagrams, have their roots in well-established methods used in operations research. The arrow diagram is directly analogous to the Critical Path Method (CPM) and the Program Evaluation and Review Technique (PERT) discussed in I.B. These two project management tools have been used for many years to determine which activities must be performed, when they must be performed, and in what order. Unlike CPM and PERT, which require training in project management or systems engineering, arrow diagrams are greatly simplified so that they can be used with a minimum of training. Figure V.15, an illustration of an arrow (PERT) diagram, is reproduced here.

ACTIVITIES LIST

1. Excavate
2. Foundation
3. Rough wall
4. Roof
5. Rough exterior plumbing
6. Exterior siding
7. Rough interior plumbing
8. Exterior painting
9. Wall boards
10. Exterior fixtures
11. Flooring
12. Interior painting
13. Interior fixtures

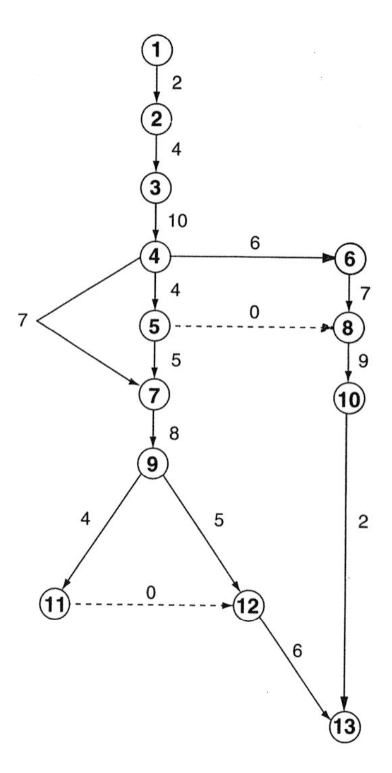

Figure V.15. PERT network for constructing a house (Figure I.9 repeated).

Other continuous improvement tools

Over the years, the tools of quality improvement have proliferated. By some estimates there are now over 400 tools in the "TQM Toolbox." This author believes that it is possible to make dramatic improvements with the tools already described, combined with the powerful statistical techniques described in other parts of this book. However, in addition to the tools already discussed, there are two more simple tools that the author believes deserve mention: the nominal group technique and force field analysis. These tools are commonly used to help teams move forward by obtaining input from all interested parties and identifying the obstacles they face.

Nominal group technique

The nominal group technique (NGT) is a method for generating a "short list" of items to be acted upon. NGT uses a highly structured approach designed to reduce the usual give-and-take among group members. Usage of NGT is indicated 1) when the group is new or has several new members, 2) when the topic under consideration is controversial, or 3) when the team is unable to resolve a disagreement. Scholtes (1996) describes the steps involved in the NGT. A summary of the approach is shown below.

Part I—A formalized brainstorm

1. Define the task in the form of a question.
2. Describe the purpose of this discussion and the rules and procedures of the NGT.
3. Introduce and clarify the question.
4. Generate ideas. Do this by having the team write down their ideas in silence.
5. List the ideas obtained.
6. Clarify and discuss the ideas.

Part II—Making the selection

1. Choose the top 50 ideas. Note: members can remove their ideas from consideration if they wish, but no member can remove another's idea.
2. Pass out index cards to each member, using the following table as a guide:

IDEAS	INDEX CARDS
less than 20	4 cards
20–35	6 cards
36–50	8 cards

3. Each member writes down their choices from the list, one choice per card.
4. Each member rank-orders their choices and writes the rank on the cards.
5. Record the group's choices and ranks.
6. Group reviews and discusses the results. Consider: How often was an item selected? What is the total of the ranks for each item?

If the team can agree on the importance of the item(s) that got the highest score(s) (sum of ranks), then the team moves on to preparing an action plan to deal with the item or items selected.

Force-field analysis

Force-field analysis (FFA) is a method borrowed from the mechanical engineering discipline known as free-body diagrams. Free-body diagrams are drawn to help the engineer identify all the forces surrounding and acting on a body. The objective is to ascertain the forces leading to an *equilibrium state* for the body.

In FFA the "equilibrium" is the status quo. FFA helps the team understand the forces that keep things the way they are. Some of the forces are "drivers" that move the system towards a desired goal. Other forces are "restrainers" that prevent the desired movement and may even cause movement away from the goal. Once the drivers and restrainers are known, the team can design an action plan which will 1) reduce the forces restraining progress and 2) increase the forces which lead to movement in the desired direction.

FFA is useful in the early stages of planning. Once restrainers are explicitly identified, a strategic plan can be prepared to develop the drivers necessary to overcome them. FFA is also useful when progress has stalled. By performing FFA, people are brought together and guided toward consensus, an activity that, by itself, often overcomes a number of obstacles. Pyzdek (1994) lists the following steps for conducting FFA.

1. Determine the goal
2. Create a team of individuals with the authority, expertise, and interest needed to accomplish the goal.
3. Have the team use brainstorming or the NGT to identify restrainers and drivers.
4. Create a force-field diagram or table which lists the restrainers and drivers.
5. Prepare a plan for removing restrainers and increasing drivers.

An example of a force-field diagram is shown in Figure V.16.

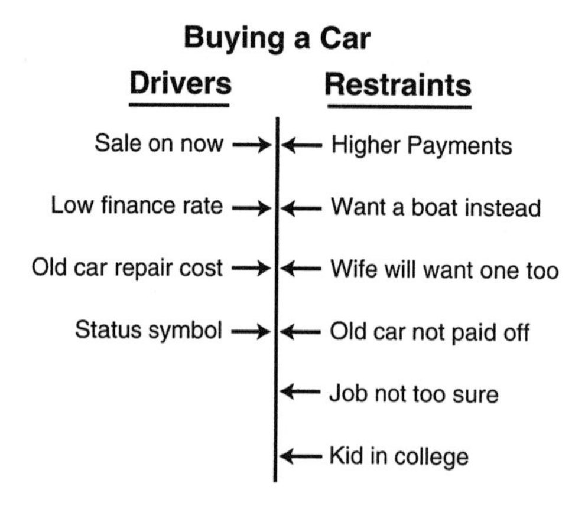

Figure V.16. Example of a force-field diagram.
From *Pocket Guide to Quality Tools*, p. 10. Copyright © 1994 by Thomas Pyzdek.

It may be helpful to assign "strength weights" to the drivers and restrainers (e.g., weak, moderate, strong).

V.C. Quality tools

The 7 quality tools have a long history. All of these tools have been in use for over twenty years and some date prior to 1920. These are the "old standby" tools that have been used successfully by quality engineers for decades.

PARETO CHARTS

Definition—Pareto analysis is the process of ranking opportunities to determine which of many potential opportunities should be pursued first. It is also known as "separating the vital few from the trivial many."

Usage—Pareto analysis should be used at various stages in a quality improvement program to determine which step to take next. Pareto analysis is used to answer such questions as "What department should have the next SPC team?" or "On what type of defect should we concentrate our efforts?"

How to perform a Pareto analysis

1. Determine the classifications (Pareto categories) for the graph. If the desired information does not exist, obtain it by designing check sheets and logsheets.
2. Select a time interval for analysis. The interval should be long enough to be representative of typical performance.
3. Determine the total occurrences (i.e., cost, defect counts, etc.) for each category. Also determine the grand total. If there are several categories which account for only a small part of the total, group these into a category called "other."
4. Compute the percentage for each category by dividing the category total by the grand total and multiplying by 100.
5. Rank-order the categories from the largest total occurrences to the smallest.
6. Compute the "cumulative percentage" by adding the percentage for each category to that of any preceding categories.
7. Construct a chart with the left vertical axis scaled from 0 to at least the grand total. Put an appropriate label on the axis. Scale the right vertical axis from 0 to 100%, with 100% on the right side being the same height as the grand total on the left side.
8. Label the horizontal axis with the category names. The leftmost category should be the largest, second largest next, and so on.
9. Draw in bars representing the amount of each category. The height of the bar is determined by the left vertical axis.
10. Draw a line that shows the cumulative percentage column of the Pareto analysis table. The cumulative percentage line is determined by the right vertical axis.

Example of Pareto analysis

The data in Table V.2 has been recorded for peaches arriving at Super Duper Market during August.

Table V.2. Raw data for Pareto analysis.

PROBLEM	PEACHES LOST
Bruised	100
Undersized	87
Rotten	235
Underripe	9
Wrong variety	7
Wormy	3

The Pareto table for the data in Table V.2 is shown in Table V.3.

Table V.3. Data organized for Pareto analysis.

RANK	CATEGORY	COUNT	PERCENTAGE	CUM %
1	Rotten	235	53.29	53.29
2	Bruised	100	22.68	75.97
3	Undersized	87	19.73	95.70
4	Other	19	4.31	100.01

Note that, as often happens, the final percentage is slightly different than 100%. This is due to round-off error and is nothing to worry about. The finished diagram is shown in Figure V.17.

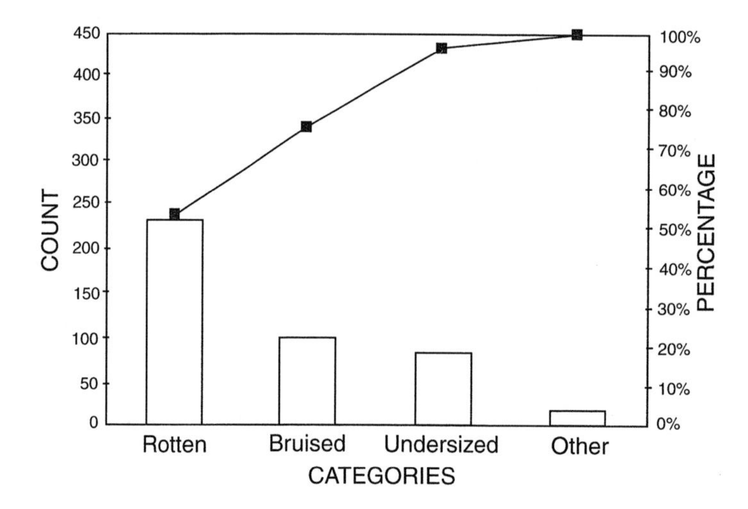

Figure V.17. The completed Pareto diagram.

CAUSE AND EFFECT DIAGRAMS

Process improvement involves taking action on the causes of variation. With most practical applications, the number of possible causes for any given problem can be huge. Dr. Kaoru Ishikawa developed a simple method of graphically displaying the causes of any given quality problem. His method is called by several names, the Ishikawa diagram, the fishbone diagram, and the cause and effect diagram.

Cause and effect diagrams are tools that are used to organize and graphically display all of the knowledge a group has relating to a particular problem. Usually, the steps are the following:

1. Develop a flow chart of the area to be improved.
2. Define the problem to be solved.
3. Brainstorm to find all possible causes of the problem.
4. Organize the brainstorming results in rational categories.
5. Construct a cause and effect diagram that accurately displays the relationships of all the data in each category.

Once these steps are complete, constructing the cause and effect diagram is very simple.

1. Draw a box on the far right-hand side of a large sheet of paper and draw a horizontal arrow that points to the box. Inside of the box, write the description of the problem you are trying to solve.
2. Write the names of the categories above and below the horizontal line. Think of these as branches from the main trunk of the tree.
3. Draw in the detailed cause data for each category. Think of these as limbs and twigs on the branches.

A good cause and effect diagram will have many "twigs," as shown in Figure V.18. If your cause and effect diagram doesn't have a lot of smaller branches and twigs, it shows that the understanding of the problem is superficial. Chances are you need the help of someone outside of your group to aid in the understanding, perhaps someone more closely associated with the problem.

Cause and effect diagrams come in several basic types. The dispersion analysis type is created by repeatedly asking "why does this dispersion occur?" For example, we might want to know why all of our fresh peaches don't have the same color.

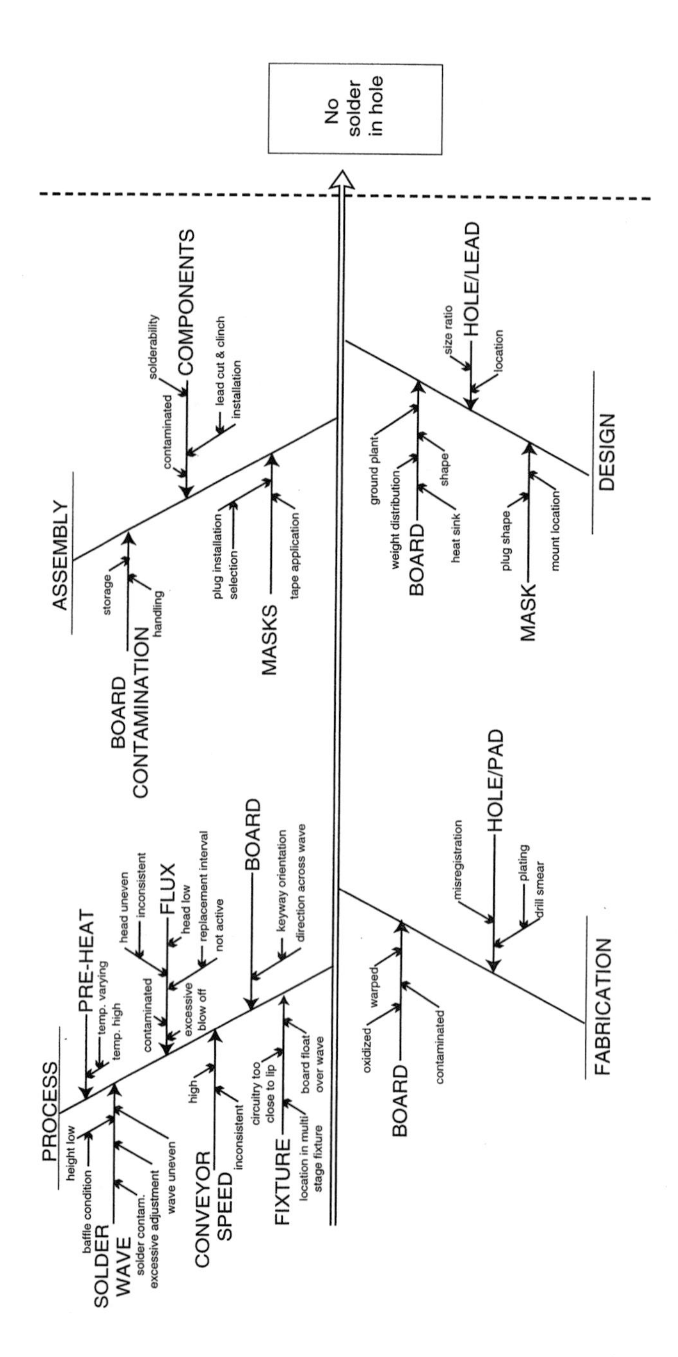

Figure V.18. Cause and effect diagram.

The production process class cause and effect diagram uses production processes as the main categories, or branches, of the diagram. The processes are shown joined by the horizontal line. Figure V.19 is an example of this type of diagram.

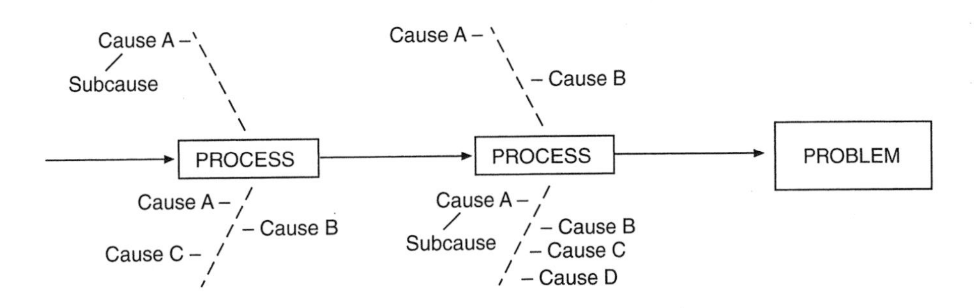

Figure V.19. Process production class cause and effect diagram.

The cause enumeration cause and effect diagram simply displays all possible causes of a given problem grouped according to rational categories. This type of cause and effect diagram lends itself readily to the brainstorming approach we are using.

Cause and effect diagrams have a number of uses. Creating the diagram is an education in itself. Organizing the knowledge of the group serves as a guide for discussion and frequently inspires more ideas. The cause and effect diagram, once created, acts as a record of your research. Simply record your tests and results as you proceed. If the true cause is found to be something that wasn't on the original diagram, write it in. Finally, the cause and effect diagram is a display of your current level of understanding. It shows the existing level of technology as understood by the team. It is a good idea to post the cause and effect diagram in a prominent location for all to see.

A variation of the basic cause and effect diagram, developed by Dr. Ryuji Fukuda of Japan, is cause and effect diagrams with the addition of cards, or CEDAC. The CEDAC differs since the group gathers ideas outside of the meeting room on small cards, as well as in group meetings. The cards also serve as a vehicle for gathering input from people who are not in the group; they can be distributed to anyone involved with the process. Often the cards provide more information than the brief entries on a standard cause and effect diagram. The cause and effect diagram is built by actually placing the cards on the branches.

FLOW CHARTS

A process flow chart is simply a tool that graphically shows the inputs, actions, and outputs of a given system. These terms are defined as follows:

Inputs—the factors of production: land, materials, labor, equipment, and management.

Actions—the way in which the inputs are combined and manipulated in order to add value. Actions include procedures, handling, storage, transportation, and processing.

Outputs—the products or services created by acting on the inputs. Outputs are delivered to the customer or other user. Outputs also include *unplanned* and *undesirable* results, such as scrap, rework, pollution, etc. Flow charts should contain these outputs as well.

Flow charting is such a useful activity that the symbols have been standardized by various ANSI standards. There are special symbols for special processes, such as electronics or information systems (see Figure I.16). However, in most cases one can use the symbols shown in Figure V.20.

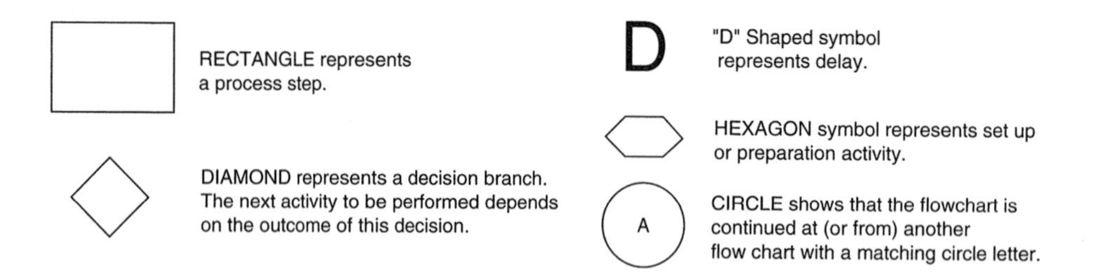

Figure V.20. Selected flow chart symbols.

The flow chart in Figure V.21 shows a high-level view of a process capability analysis. The flow chart can be made either more complex or less complex. As a rule of thumb, to paraphrase Albert Einstein, "Flow charts should be as simple as possible, but not simpler." The purpose of the flow chart is to help people understand the process and this is not accomplished with flow charts that are either too simple or too complex.

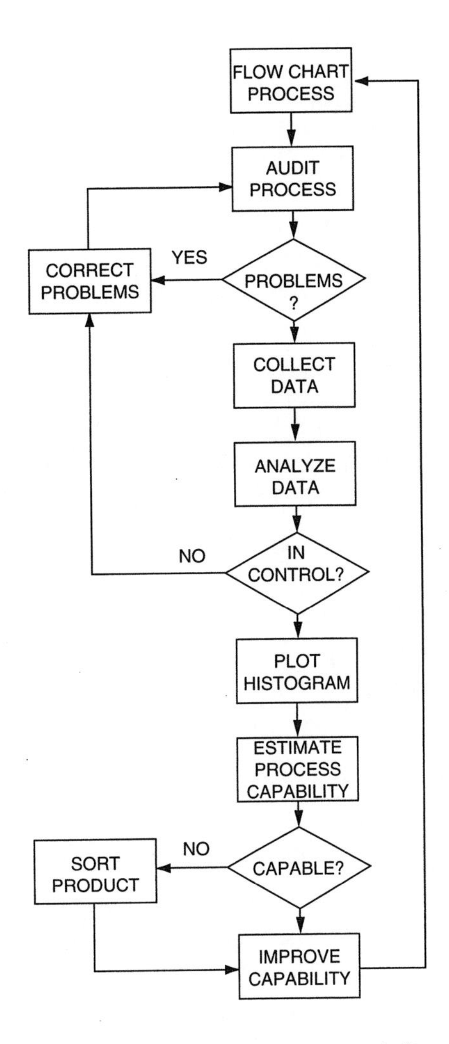

Figure V.21. Flow chart of process capability analysis.

CONTROL CHARTS

In this section, we discuss the broad principles that underlie all Shewhart control charts. Additional discussion, including specific control chart calculations are discussed in VI.G.

Statistical process control principles

In his landmark book, *Economic Control of Quality of Manufactured Product*, Shewhart described the following:

> Write the letter "a" on a piece of paper. Now make another "a" just like the first one; then another and another until you have a series of a's (a, a, a...). You try to make all the a's alike but you don't; you can't. You are willing to accept this as an empirically established fact. But what of it? Let us see just what this means in respect to control. Why can we not do a simple thing like making all the a's exactly alike? Your answer leads to a generalization which all of us are perhaps willing to accept. It is that there are many causes of variability among the a's: the paper was not smooth, the lead in the pencil was not uniform, and the unavoidable variability in your external surroundings reacted upon you to introduce variation in the a's. But are these the only causes of variability in the a's? Probably not.
>
> We accept our human limitations and say that likely there are many other factors. If we could but name all the reasons why we cannot make the a's alike, we would most assuredly have a better understanding of a certain part of nature than we now have. Of course, this conception of what it means to be able to do what we want to do is not new; it does not belong exclusively to any one field of human thought; it is commonly accepted.
>
> The point to be made in this simple illustration is that we are limited in doing what we want to do; that to do what we set out to do, even in so simple a thing as making a's that are alike, requires almost infinite knowledge compared with that which we now possess. It follows, therefore, since we are thus willing to accept as axiomatic that we cannot do what we want to do and cannot hope to understand why we cannot, that we must also accept as axiomatic that a controlled quality will not be a constant quality. Instead, a controlled quality must be a variable quality. This is the first characteristic.
>
> But let us go back to the results of the experiment on the a's and we shall find out something more about control. Your a's are different from my a's; there is something about your a's that makes them yours and something about my a's that makes them mine. True, not all of your a's are alike. Neither are all of my a's alike. Each group of a's varies within a certain range and yet each group is distinguishable from the others. This distinguishable and, as it were, constant variability within limits is the second characteristic of control.

Shewhart goes on to define control:

> A phenomenon will be said to be controlled when, through the use of past experience, we can predict, at least within limits, how the phenomenon may be expected to vary in the future. Here it is understood that prediction within limits means that we can state, at least approximately, the probability that the observed phenomenon will fall within the given limits.

The critical point in this definition is that control is not defined as the complete absence of variation. Control is simply a state where all variation is predictable variation. In all forms of prediction there is an element of chance. Any unknown cause of variation is called a chance cause. If the influence of any particular chance cause is very small, and if the number of chance causes of variation are very large and relatively constant, we have a situation where the variation is predictable within limits. You can see from our definition above that a system such as this qualifies as a controlled system. Deming uses the term common cause rather chance cause, and we will use Deming's term common cause in this book.

An example of such a controlled system might be the production and distribution of peaches. If you went into an orchard to a particular peach tree at the right time of the year, you would find a tree laden with peaches (with any luck at all). The weights of the peaches will vary. However, if you weighed every single peach on the tree you would probably notice that there was a distinct pattern to the weights. In fact, if you drew a small random sample of, say, 25 peaches you could probably predict the weights of those peaches remaining on the tree. This predictability is the essence of a controlled phenomenon. The number of common causes that account for the variation in peach weights is astronomical, but relatively constant. A constant system of common causes results in a controlled phenomenon.

Needless to say, not all phenomena arise from constant systems of common causes. At times, the variation is caused by a source of variation that is not part of the constant system. These sources of variation were called assignable causes by Shewhart; Deming calls them "special causes" of variation. Experience indicates that special causes of variation can usually be found and eliminated.

Statistical tools are needed to help us effectively identify the effects of special causes of variation. This leads us to another definition:

Statistical process control (SPC)—the use of statistical methods to identify the existence of special causes of variation in a process.

The basic rule of statistical process control is:

> Variation from common cause systems should be left to chance, but special causes of variation should be identified and eliminated.

The charts in Figure V.22 illustrate the need for statistical methods to determine the type of variation. The answer to the question "should these variations be left to chance?" can be obtained through the use of statistical theory. Figure V.23 illustrates the basic concept. Variation between the control limits designated by the two lines will be considered to be variation from the common cause system. Any variability beyond these limits will be treated as having come from special causes of variation. We will call any system exhibiting only common cause variation *statistically controlled.* It must be noted that the control limits are not simply pulled out of the air, they are calculated from the data using statistical theory. A control chart is a practical tool that provides an operational definition of a special cause.

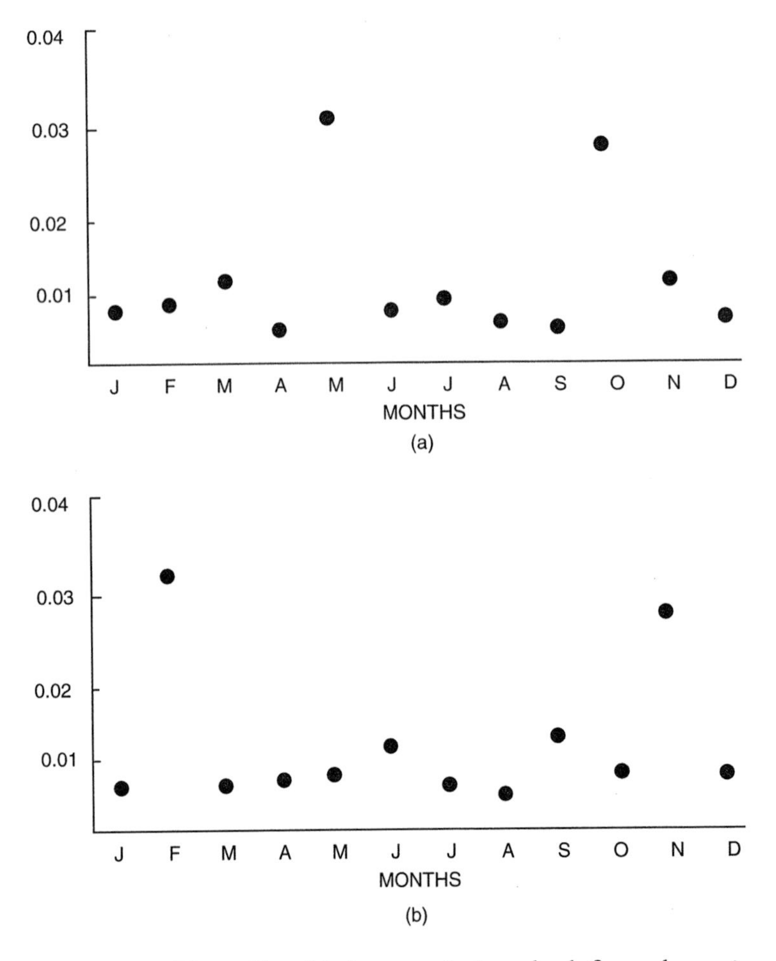

Figure V.22. Should these variations be left to chance?

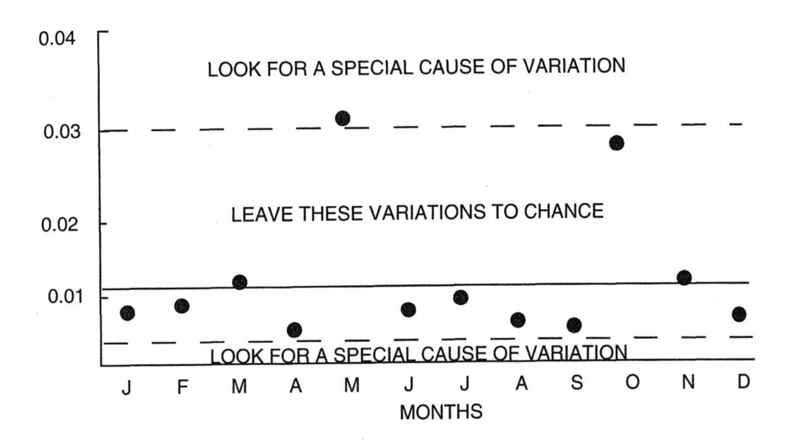

Figure V.23. Special and common causes of variation.

Distributions

A fundamental concept of statistical process control is that almost every measurable phenomenon is a statistical distribution. In other words, an observed set of data constitutes a sample of the effects of unknown common causes. It follows that, after we have done everything to eliminate special causes of variations, there will still remain a certain amount of variability exhibiting the state of control. Figure V.24 illustrates the relationships between common causes, special causes, and distributions.

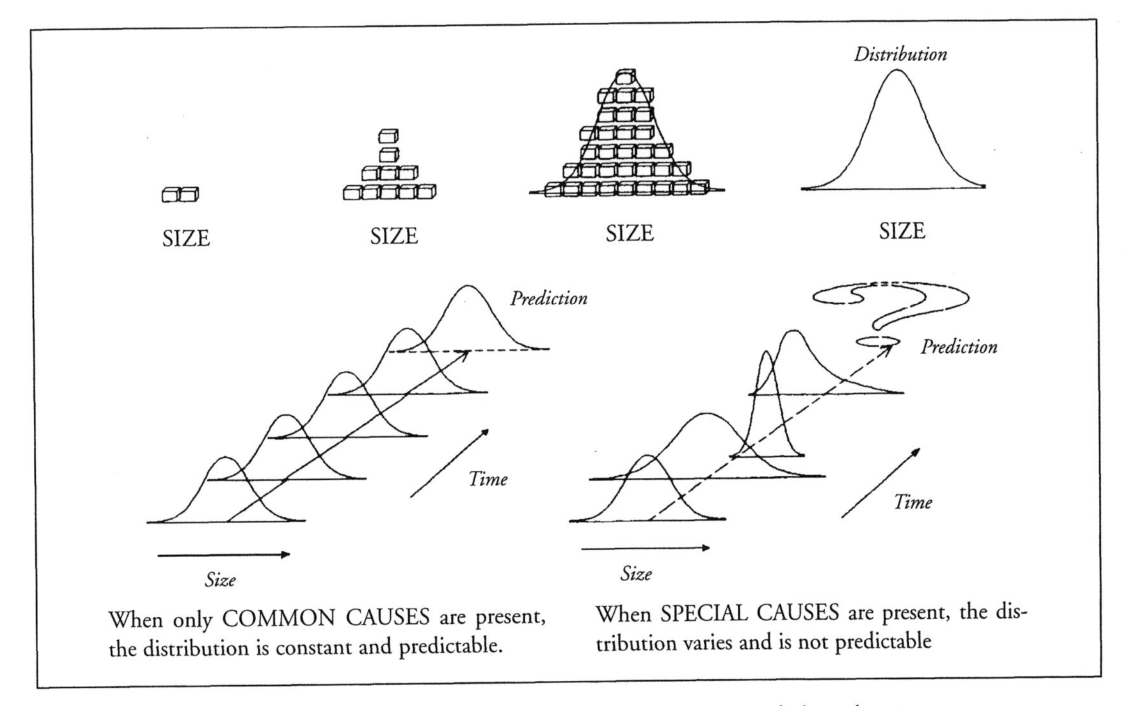

Figure V.24. Common causes, special causes, and distributions.
From *Pyzdek's Guide to SPC-Volume One: Fundamentals*, p. 43. Copyright © 1998 by Thomas Pyzdek.

There are three basic properties of a distribution: location, spread, and shape. The distribution can be characterized by these three parameters. Figure V.25 illustrates these three properties. The location refers to the typical value of the distribution. The spread of the distribution is the amount by which smaller values differ from larger ones. And the shape of a distribution is its pattern—peakedness, symmetry, etc. Note that a given phenomenon may have any of a number of distributions, i.e., the distribution may be bell-shaped, rectangular shaped, etc. In this book we will discuss methods which facilitate the analysis and control of distributions.

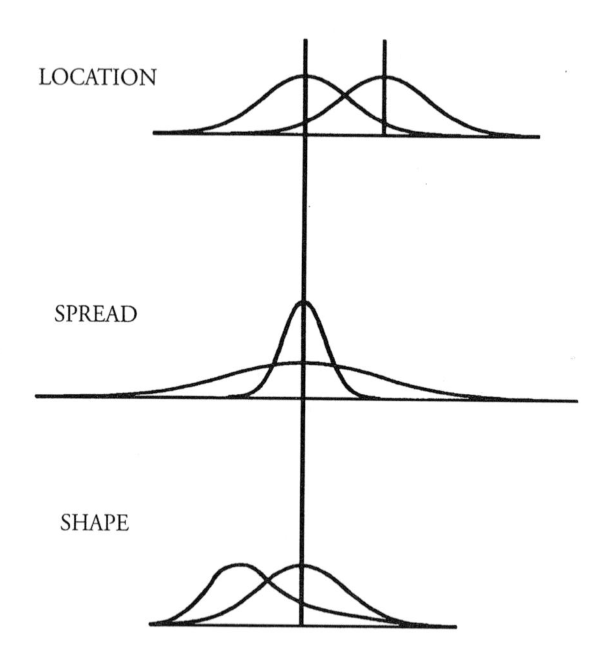

Figure V.25. How distributions vary from one another.

From *Pyzdek's Guide to SPC-Volume One: Fundamentals*, p. 44. Copyright © 1998 by Thomas Pyzdek.

Process control versus meeting requirements

In the past, the emphasis was on meeting requirements. Such programs as zero defects received a great deal of attention because firms were only concerned with producing output that met the minimum requirements. Anything that met requirements was acceptable, anything that failed requirements was unacceptable. This point of view is illustrated in Figure V.26.

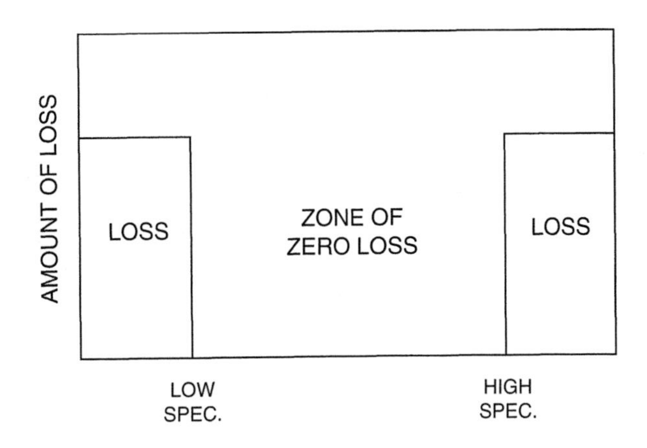

Figure V.26. Model of requirement-based "Goal Post" control system.
From *Pyzdek's Guide to SPC-Volume One: Fundamentals*, p. 52. Copyright © 1998 by Thomas Pyzdek.

If one accepts this model, there is no incentive whatever to improve a process that meets requirements. Since there is no loss, there is no benefit. Any cost/benefit analysis will necessarily result in rejecting any investment in reducing variation because there is always cost, but no benefit. It is not hard to show that this view of loss is naive.

In most real world applications, there is a "preferred" level of performance. This might be the largest value possible (e.g., the strongest material), the smallest value possible (e.g., a contamination measurement) or a nominal target value (e.g., a hole location). When a process is performing at this preferred level, or target level, it runs smoother, costs less to maintain, and produces better quality output. This situation has been frequently demonstrated empirically. A more realistic model based on this mental model was developed by Dr. Genichi Taguchi. Figure V.27 illustrates Taguchi's loss function.

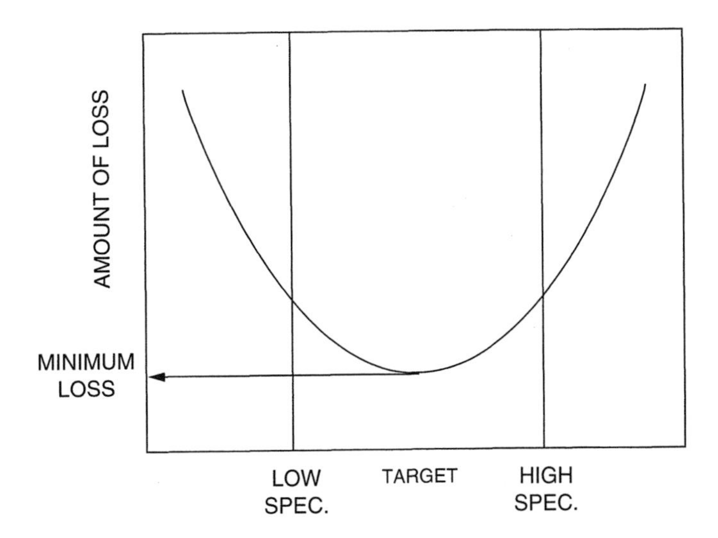

Figure V.27. Model of Taguchi loss function.
From *Pyzdek's Guide to SPC-Volume One: Fundamentals*, p. 53. Copyright © 1998 by Thomas Pyzdek.

Figure V.27 shows that an optimum exists at some target value. The target may or may not be at the center of the specifications. Often the location of the target depends on the amount of variability in the process. For example, if very tight control of a bolt hole diameter can be maintained it may be wise to run the diameter to the small end of the specification to obtain a better fit. Furthermore, *the smaller the variation around the target, the smaller the total loss.* Thus, according to the Taguchi school of thought, it is not enough to merely meet the requirements. Continuous improvement is reasonable with this model. Although Dr. Taguchi's methods go far beyond SPC (e.g., see VI.F), the concept of never-ending, continuous improvement by reducing variation is also at the heart of the SPC approach. Meeting requirements is a side benefit. In fact, specifications are viewed simply as the point where it becomes more economical to scrap or rework than to ship. Specifications become points on the loss function curve. With SPC and Taguchi methods, quality levels will improve far beyond the marginal quality represented by meeting specifications.

CHECK SHEETS

Check sheets are devices which consist of lists of items and some indicator of how often each item on the list occurs. In their simplest form, checklists are tools that make the data collection process easier by providing pre-written descriptions of events likely to occur. A well-designed check sheet will answer the questions posed by the investigator. Some examples of questions are the following: "Has everything been done?" "Have all inspections been performed?" "How often does a particular problem occur?" "Are problems more common with part X than with part Y?" They also serve as reminders that direct the attention of the data collector to items of interest and importance. Such simple check sheets are called *confirmation check sheets.* Check sheets have been improved by adding a number of enhancements, a few of which are described below.

Although they are simple, check sheets are extremely useful process-improvement and problem-solving tools. Their power is greatly enhanced when they are used in conjunction with other simple tools, such as histograms and Pareto analysis. Ishikawa estimated that 80% to 90% of all workplace problems could be solved using only the simple quality improvement tools.

Process check sheets

These check sheets are used to create frequency distribution tally sheets that are, in turn, used to construct histograms (see below). A process check sheet is constructed by listing several ranges of measurement values and recording a mark for the actual observations. An example is shown in Figure V.28. Notice that if reasonable care is taken in recording tick marks, the check sheet gives a graphical picture similar to a histogram.

RANGE OF MEASUREMENTS	FREQUENCY
0.990-0.995 INCHES	////
0.996-1.000 INCHES	~~////~~
1.001-1.005 INCHES	~~////~~ ////
1.006-1.010 INCHES	~~////~~ ~~////~~ //
1.011-1.015 INCHES	////
1.016-1.020 INCHES	//

Figure V.28. Process check sheet.

Defect check sheets

Here the different types of defects are listed and the observed frequencies observed. An example of a defect check sheet is shown in Figure V.29. If reasonable care is taken in recording tick marks, the check sheet resembles a bar chart.

DEFECT	FREQUENCY
COLD SOLDER	////
NO SOLDER IN HOLE	~~////~~ ~~////~~ ////
GRAINY SOLDER	~~////~~ ////
HOLE NOT PLATED THROUGH	~~////~~ ~~////~~ ///
MASK NOT PROPERLY INSTALLED	~~////~~ ////
PAD LIFTED	/

Figure V.29. Defect check sheet.

Stratified defects check sheets

These check sheets stratify a particular defect type according to logical criteria. This is helpful when the defect check sheet fails to provide adequate information regarding the root cause or causes of a problem. An example is shown in Figure V.30.

SAMPLES OF 1,000 SOLDER JOINTS	PART NUMBER X-1011	PART NUMBER X-2011	PART NUMBER X-3011	PART NUMBER X-4011	PART NUMBER X-5011
COLD SOLDER	////			卌	
NO SOLDER IN HOLE	卌		//	//	
GRAINY SOLDER	卌	/		///	
HOLE NOT PLATED THROUGH	卌			///	
MASK NOT PROPERLY INSTALLED	卌		////	卌	
PAD LIFTED	/				

Figure V.30. Stratified defect check sheet.

Defect location check sheet

These "check sheets" are actually drawings, photographs, layout diagrams or maps which show where a particular problem occurs. The spatial location is valuable in identifying root causes and planning corrective action. In Figure V.31, the location of complaints from customers about lamination problems on a running shoe are shown with an "X." The diagram makes it easy to identify a problem area that would be difficult to depict otherwise. In this case, a picture is truly worth a thousand words of explanation.

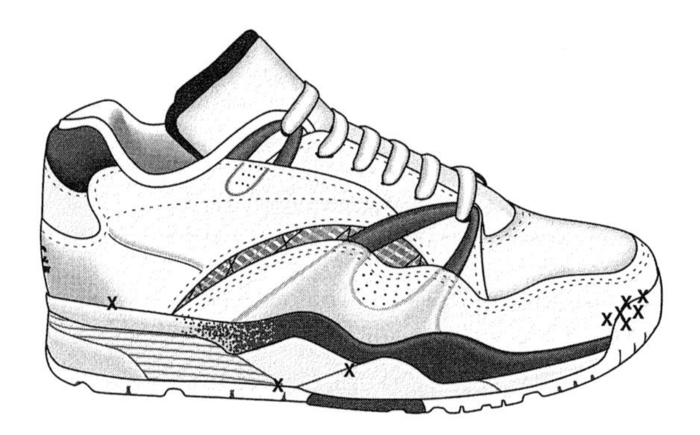

Figure V.31. Defect location check sheet lamination complaints.

Cause and effect diagram check sheet

Cause and effect diagrams can also serve as check sheets. Once the diagram has been prepared, it is posted in the work area and the appropriate arrow is marked whenever that particular cause or situation occurs. Teams can also use this approach for historic data, when such data is available.

SCATTER DIAGRAMS

Definition—A scatter diagram is a plot of one variable versus another. One variable is called the *independent variable* and it is usually shown on the horizontal (bottom) axis. The other variable is called the *dependent variable* and it is shown on the vertical (side) axis.

Usage—Scatter diagrams are used to evaluate cause and effect relationships. The assumption is that the independent variable is causing a change in the dependent variable. Scatter plots are used to answer such questions as "Does vendor A's material machine better than vendor B's?" "Does the length of training have anything to do with the amount of scrap an operator makes?" and so on.

How to construct a scatter diagram

1. Gather several paired sets of observations, preferably 20 or more. A paired set is one where the dependent variable can be directly tied to the independent variable.
2. Find the largest and smallest independent variable and the largest and smallest dependent variable.
3. Construct the vertical and horizontal axes so that the smallest and largest values can be plotted. Figure V.32 shows the basic structure of a scatter diagram.

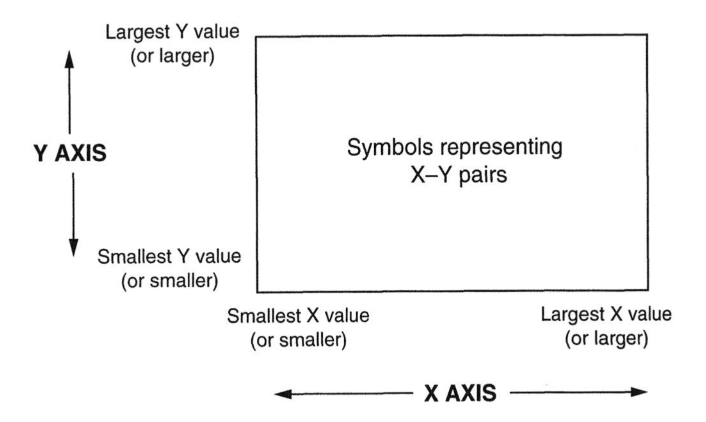

Figure V.32. Layout of a scatter diagram.

4. Plot the data by placing a mark at the point corresponding to each X-Y pair, as illustrated by Figure V.33 If more than one classification is used, you may use different symbols to represent each group.

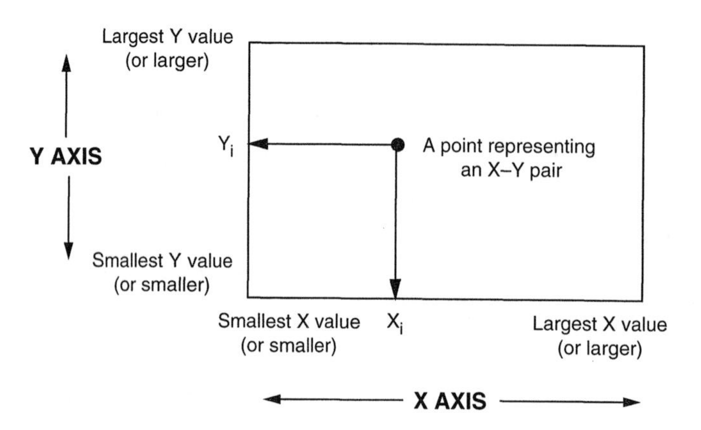

Figure V.33. Plotting points on a scatter diagram.

From *Pyzdek's Guide to SPC-Volume One: Fundamentals*, p. 66. Copyright © 1998 by Thomas Pyzdek.

Example of a scatter diagram

The orchard manager has been keeping track of the weight of peaches on a day by day basis. The data are provided in Table V.4.

Table V.4. Raw data for scatter diagram.

From *Pyzdek's Guide to SPC-Volume One: Fundamentals*, p. 67. Copyright © 1998 by Thomas Pyzdek.

NUMBER	DAYS ON TREE	WEIGHT (OUNCES)
1	75	4.5
2	76	4.5
3	77	4.4
4	78	4.6
5	79	5.0
6	80	4.8
7	80	4.9
8	81	5.1
9	82	5.2
10	82	5.2

Continued on next page . . .

Table V.4—*Continued . . .*

NUMBER	DAYS ON TREE	WEIGHT (OUNCES)
11	83	5.5
12	84	5.4
13	85	5.5
14	85	5.5
15	86	5.6
16	87	5.7
17	88	5.8
18	89	5.8
19	90	6.0
20	90	6.1

1. Organize the data into X-Y pairs, as shown in Table V.4. The independent variable, X, is the number of days the fruit has been on the tree. The dependent variable, Y, is the weight of the peach.
2. Find the largest and smallest values for each data set. The largest and smallest values from Table V.4 are shown in Table V.5.

Table V.5. Smallest and largest values.

From *Pyzdek's Guide to SPC-Volume One: Fundamentals*, p. 68. Copyright © 1998 by Thomas Pyzdek.

VARIABLE	SMALLEST	LARGEST
Days on tree (X)	75	90
Weight of peach (Y)	4.4	6.1

3. Construct the axes. In this case, we need a horizontal axis that allows us to cover the range from 75 to 90 days. The vertical axis must cover the smallest of the small weights (4.4 ounces) to the largest of the weights (6.1 ounces). We will select values beyond these minimum requirements, because we want to estimate how long it will take for a peach to reach 6.5 ounces.
4. Plot the data. The completed scatter diagram is shown in Figure V.34.

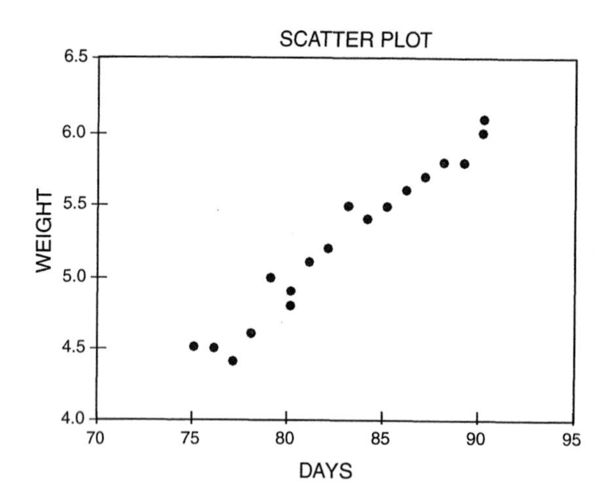

Figure V.34. Completed scatter diagram.
From *Pyzdek's Guide to SPC-Volume One: Fundamentals*, p. 68. Copyright © 1998 by Thomas Pyzdek.

Pointers for using scatter diagrams

- Scatter diagrams display different patterns that must be interpreted;Figure V.35 provides a scatter diagram interpretation guide.

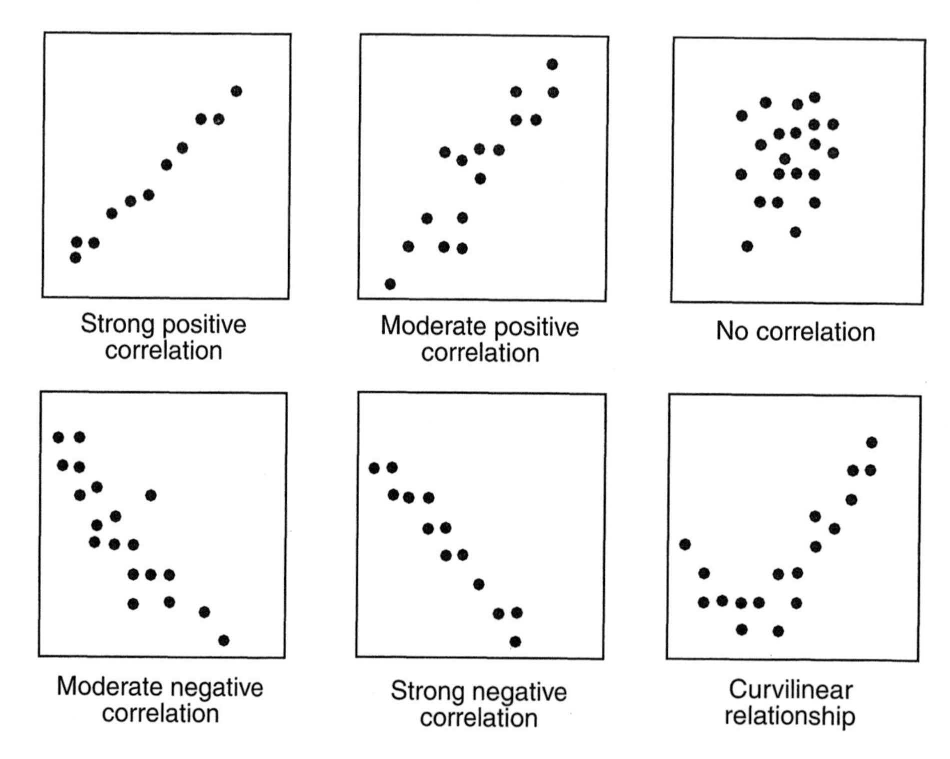

Figure V.35. Scatter diagram interpretation guide.
From *Pyzdek's Guide to SPC-Volume One: Fundamentals*, p. 69. Copyright © 1998 by Thomas Pyzdek.

- Be sure that the independent variable, X, is varied over a sufficiently large range. When X is changed only a small amount, you may not see a correlation with Y, even though the correlation really does exist.
- If you make a prediction for Y, for an X value that lies outside of the range you tested, be advised that the prediction is highly questionable and should be tested thoroughly. Predicting a Y value beyond the X range actually tested is called extrapolation.
- Watch for the effect of variables you didn't evaluate. Often, an uncontrolled variable will wipe out the effect of your X variable. It is also possible that an uncontrolled variable will be causing the effect and you will mistake the X variable you are controlling as the true cause. This problem is much less likely to occur if you choose X levels at random. An example of this is our peaches. It is possible that any number of variables changed steadily over the time period investigated. It is possible that these variables, and not the independent variable, are responsible for the weight gain (e.g., was fertilizer added periodically during the time period investigated?).
- Beware of "happenstance" data! Happenstance data is data that was collected in the past for a purpose different than constructing a scatter diagram. Since little or no control was exercised over important variables, you may find nearly anything. Happenstance data should be used only to get ideas for further investigation, never for reaching final conclusions. One common problem with happenstance data is that the variable that is truly important is not recorded. For example, records might show a correlation between the defect rate and the shift. However, perhaps the real cause of defects is the ambient temperature, which also changes with the shift.
- If there is more than one possible source for the dependent variable, try using different plotting symbols for each source. For example, if the orchard manager knew that some peaches were taken from trees near a busy highway, he could use a different symbol for those peaches. He might find an interaction, that is, perhaps the peaches from trees near the highway have a different growth rate than those from trees deep within the orchard.

Although it is possible to do advanced analysis without plotting the scatter diagram, this is generally bad practice. This misses the enormous learning opportunity provided by the graphical analysis of the data.

HISTOGRAMS

A histogram is a pictorial representation of a set of data. It is created by grouping the measurements into "cells." Histograms are used to determine the shape of a data set. Also, a histogram displays the numbers in a way that makes it easy to see the dispersion and central tendency and to compare the distribution to requirements. Histograms can be valuable troubleshooting aids. Comparisons between histograms from different machines, operators, vendors, etc., often reveal important differences.

How to construct a histogram

1. Find the largest and the smallest value in the data.
2. Compute the range by subtracting the smallest value from the largest value.
3. Select a number of cells for the histogram. Table V.6 provides some useful guidelines. The final histogram may not have exactly the number of cells you choose here, as explained below.

Table V.6. Histogram cell determination guidelines.

SAMPLE SIZE	NUMBER OF CELLS
100 or less	7 to 10
101–200	11 to 15
201 or more	13 to 20

As an alternative, the number of cells can be found as the square root of the number in the sample. For example, if n=100, then the histogram would have 10 cells. Round to the nearest integer.

4. Determine the width of each cell. We will use the letter W to stand for the cell width. W is computed from Equation V.1.

$$W = \frac{\text{Range}}{\text{Number of Cells}} \tag{V.1}$$

The number W is a starting point. Round W to a convenient number. Rounding W will affect the number of cells in your histogram.

5. Compute "cell boundaries." A cell is a range of values and cell boundaries define the start and end of each cell. Cell boundaries should have one more decimal place than the raw data values in the data set; for example, if the data are integers, the cell boundaries would have one decimal place. The low boundary of the first cell must be less than the smallest value in the data set. Other cell boundaries are found by adding W to the previous boundary. Continue until the upper boundary is larger than the largest value in the data set.
6. Go through the raw data and determine into which cell each value falls. Mark a tick in the appropriate cell.
7. Count the ticks in each cell and record the count, also called the frequency, to the right of the tick marks.
8. Construct a graph from the table. The vertical axis of the graph will show the frequency in each cell. The horizontal axis will show the cell boundaries. Figure V.36 illustrates the layout of a histogram.

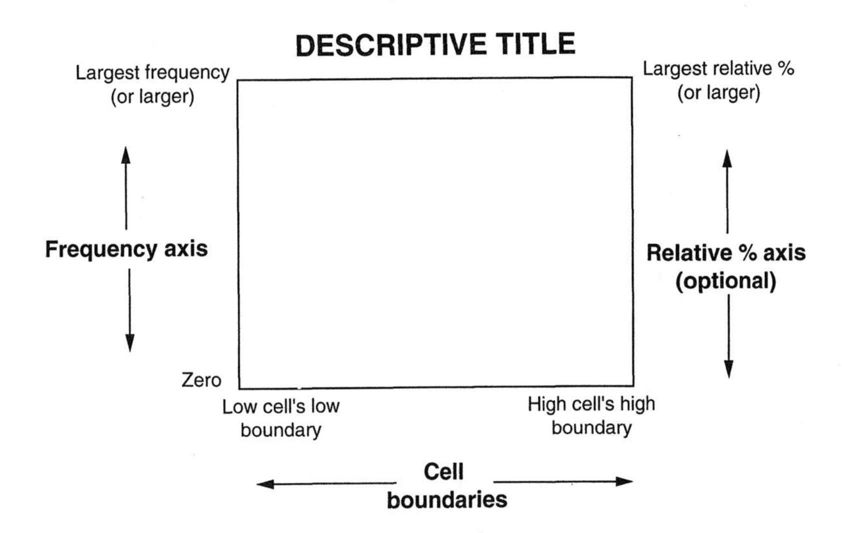

Figure V.36. Layout of a histogram.

From *Pyzdek's Guide to SPC-Volume One: Fundamentals*, p. 61. Copyright © 1998 by Thomas Pyzdek.

9. Draw bars representing the cell frequencies. The bars should all be the same width, the height of the bars should equal the frequency in the cell.

Histogram example

Assume you have the data in Table V.7 on the size of a metal rod. The rods were sampled every hour for 20 consecutive hours and 5 consecutive rods were checked each time (20 subgroups of 5 values per group).

Table V.7. Data for histogram.

From *Pyzdek's Guide to SPC-Volume One: Fundamentals*, p. 62. Copyright © 1998 by Thomas Pyzdek.

ROW	SAMPLE 1	SAMPLE 2	SAMPLE 3	SAMPLE 4	SAMPLE 5
1	1.002	0.995	1.000	1.002	1.005
2	1.000	0.997	1.007	0.992	0.995
3	0.997	1.013	1.001	0.985	1.002
4	0.990	1.008	1.005	0.994	1.012
5	0.992	1.012	1.005	0.985	1.006
6	1.000	1.002	1.006	1.007	0.993
7	0.984	0.994	0.998	1.006	1.002
8	0.987	0.994	1.002	0.997	1.008
9	0.992	0.988	1.015	0.987	1.006
10	0.994	0.990	0.991	1.002	0.988

Continued on next page . . .

Table V.7—*Continued . . .*

ROW	SAMPLE 1	SAMPLE 2	SAMPLE 3	SAMPLE 4	SAMPLE 5
11	1.007	1.008	0.990	1.001	0.999
12	0.995	0.989	0.982*	0.995	1.002
13	0.987	1.004	0.992	1.002	0.992
14	0.991	1.001	0.996	0.997	0.984
15	1.004	0.993	1.003	0.992	1.010
16	1.004	1.010	0.984	0.997	1.008
17	0.990	1.021*	0.995	0.987	0.989
18	1.003	0.992	0.992	0.990	1.014
19	1.000	0.985	1.019	1.002	0.986
20	0.996	0.984	1.005	1.016	1.012

1. Find the largest and the smallest value in the data set. The smallest value is 0.982 and the largest is 1.021. Both values are marked with an (*) in Table V.7.
2. Compute the range, R, by subtracting the smallest value from the largest value. R = 1.021 – 0.982 = 0.039.
3. Select a number of cells for the histogram. Since we have 100 values, 7 to 10 cells are recommended. We will use 10 cells.
4. Determine the width of each cell, W. Using Equation V.1, we compute W=0.039 / 10 = 0.0039. We will round this to 0.004 for convenience. Thus, W= 0.004.
5. Compute the cell boundaries. The low boundary of the first cell must be below our smallest value of 0.982, and our cell boundaries should have one decimal place more than our raw data. Thus, the lower cell boundary for the first cell will be 0.9815. Other cell boundaries are found by adding W = 0.004 to the previous cell boundary until the upper boundary is greater than our largest value of 1.021. This gives us the cell boundaries in Table V.8.

Table V.8. Histogram cell boundaries.

From *Pyzdek's Guide to SPC-Volume One: Fundamentals*, p. 63. Copyright © 1998 by Thomas Pyzdek.

CELL NUMBER	LOWER CELL BOUNDARY	UPPER CELL BOUNDARY
1	0.9815	0.9855
2	0.9855	0.9895
3	0.9895	0.9935
4	0.9935	0.9975
5	0.9975	1.0015
6	1.0015	1.0055
7	1.0055	1.0095
8	1.0095	1.0135
9	1.0135	1.0175
10	1.0175	1.0215

6. Go through the raw data and mark a tick in the appropriate cell for each data point.
7. Count the tick marks in each cell and record the frequency to the right of each cell. The results of all we have done so far are shown in Table V.9. Table V.9 is often referred to as a "frequency table" or "frequency tally sheet."

Table V.9. Frequency tally sheet.

From *Pyzdek's Guide to SPC-Volume One: Fundamentals*, p. 64. Copyright © 1998 by Thomas Pyzdek.

CELL NUMBER	CELL START	CELL END	TALLY	FREQUENCY
1	0.9815	0.9855	𝍸III	8
2	0.9855	0.9895	𝍸IIII	9
3	0.9895	0.9935	𝍸𝍸𝍸II	17
4	0.9935	0.9975	𝍸𝍸𝍸I	16
5	0.9975	1.0015	𝍸IIII	9
6	1.0015	1.0055	𝍸𝍸𝍸IIII	19
7	1.0055	1.0095	𝍸𝍸I	11
8	1.0095	1.0135	𝍸I	6
9	1.0135	1.0175	III	3
10	1.0175	1.0215	II	2

Construct a graph from the table in step 7. The frequency column will be plotted on the vertical axis, and the cell boundaries will be shown on the horizontal (bottom) axis. The resulting histogram is shown in Figure V.37.

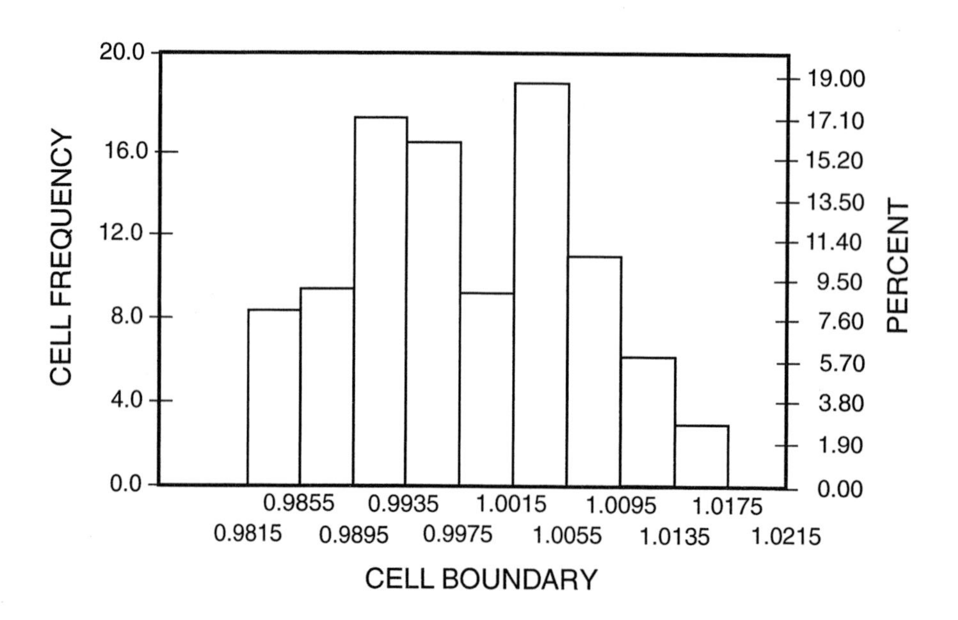

Figure V.37. Completed histogram.

From *Pyzdek's Guide to SPC-Volume One: Fundamentals*, p. 64. Copyright © 1998 by Thomas Pyzdek.

Pointers for using histograms

- Histograms can be used to compare a process to requirements if you draw the specification lines on the histogram. If you do this, be sure to scale the histogram accordingly.
- Histograms should not be used alone. Always construct a run chart or a control chart before constructing a histogram. They are needed because histograms will often conceal out of control conditions since they don't show the time sequence of the data.
- Evaluate the pattern of the histogram to determine if you can detect changes of any kind. The changes will usually be indicated by multiple modes or "peaks" on the histogram. Most real-world processes produce histograms with a single peak. However, histograms from small samples often have multiple peaks that merely represent sampling variation. Also, multiple peaks are sometimes caused by an unfortunate choice of the number of cells. Processes heavily influenced by behavior patterns are often multi-modal. For example, traffic patterns have distinct "rush-hours," and prime time is prime time precisely because more people tend to watch television at that time.

- Compare histograms from different periods of time. Changes in histogram patterns from one time period to the next can be very useful in finding ways to improve the process.
- Stratify the data by plotting separate histograms for different sources of data. For example, with the rod diameter histogram we might want to plot separate histograms for shafts made from different vendors' materials or made by different operators or machines. This can sometimes reveal things that even control charts don't detect.

V.D CORRECTIVE ACTION

Corrective Action is defined as "an action taken to eliminate the causes of an existing non-conformity, defect or other undesirable situation(s) in order to prevent recurrence." (ISO/DIS 8402 [1992]). The first element of Corrective Action is, of course, identification that a problem exists. Problem detection may occur through many methods discussed in this book, including inspection, quality audits, customer feedback, and statistical process control, amongst others.

Once the problem has been identified, solving the problem requires locating the root cause of the problem. This is not a trivial matter. Identifying the root cause(s) of problems is challenging work. Many times the problem is treated as if *it* were a cause, i.e., action is taken to "manage the problem" rather than addressing its cause. Examples of this are inspection to remove defects or testing software to catch bugs. Wilson, Dell and Anderson (1993, 9) define *root cause* as that most basic reason for an undesirable condition or problem which, if eliminated or corrected, would have prevented it from existing or occurring. Root causes are usually expressed in terms of specific or systematic factors. A root cause usually is expressed in terms of the least common organizational, personal, or activity denominator.

In many cases, process personnel lack the resources to identify the root cause of a problem. The Quality Engineer is expected to perform the necessary analysis and specify the corrective action necessary to address the cause(s) of the problem. In the past, corrective actions often involved little more than a cursory statement, e.g. "Cause: operator. Corrective Action: Operator will be instructed to take greater care in processing." Although such activities still occur, the modern expectations require much more rigor. Control charts are used to determine if the problem is due to special or common causes of variation; statistical evidence is required to show a link between the problem and its cause, often using Designed Experiments. Finally, corrective action plans are reviewed in light of the statistical evidence.

V.E PREVENTIVE ACTION

Preventive Action refers to activities designed to prevent the instance of poor quality in products or services. Examples include design review, robust design practices, quality planning, supplier capability surveys, process capability analysis, quality improvement projects, and training (where it is done to improve quality). Each of these activities is discussed in other sections of this book.

A primary task of the quality function is to assure that specifications are met. This is accomplished by allocating resources to product and process appraisal and prevention activities, and by designing systems to control and assure conformance to the requirements. There has been an historic trend towards emphasis on prevention and systems design rather than on appraisal of product. Recent emphasis on continuous improvement begins with the specification as a minimum requirement, but does not stop when the specifications are met.

A process control system is essentially a feedback system. There are four main elements involved: the process itself, information about the process, action taken on the process, and action taken on the output from the process. The way these elements fit together is shown in Figure V.38.

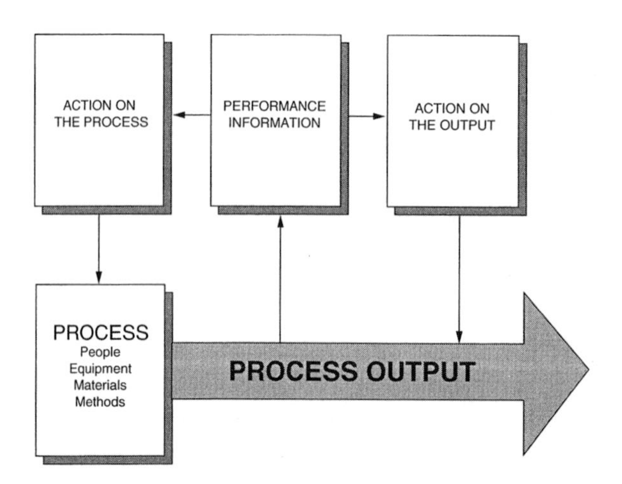

Figure V.38. Process control system.

From *Pyzdek's Guide to SPC-Volume One: Fundamentals*, p. 51. Copyright © 1998 by Thomas Pyzdek.

By the process, we mean the whole combination of people, equipment, materials, methods, and environment that work together to create added value. The performance information is obtained, in part, from evaluation of the process output. The output of a process includes more than product; output also includes information about the operating state of the process such as temperature, cycle times, etc. Action taken on a process is *future-oriented* in the sense that it will affect output yet to come. Action on the output is *past-oriented* because it involves detecting non-conformances in output that has already been produced.

Past-oriented strategies are problem *management systems.* Future-oriented strategies are problem *prevention systems.*

Historically, industry has concentrated its attention on the past-oriented, problem management strategy of inspection. With this approach we wait until output has been produced, then the output is inspected and either accepted or rejected. This does nothing to prevent problems from occurring.

Prevention of poor quality is an investment in the discovery, incorporation, and maintenance of defect prevention disciplines for all operations affecting the quality of product or service (Campenella, 1990). As such, prevention needs to be applied correctly and not evenly across the board. Much improvement has been demonstrated through reallocation of prevention efforts from areas having little effect to areas where it really pays off; once again, the Pareto principle in action.

CATEGORIES OF HUMAN ERRORS

When trying to eliminate or reduce errors, it is often helpful to "divide and conquer." By carefully dividing errors into different categories we can sometimes better see what type of action is appropriate.

Inadvertent errors

Many human errors occur due to lack of attention. People are notorious for their propensity to commit this type of error. Inadvertent errors have certain hallmarks:

- There is usually no advance knowledge that an error is imminent.
- The incidence of error is relatively small. That is, the task is normally performed without error.
- The occurrence of errors is random, in a statistical sense.

Examples of inadvertent errors are not hard to find. This is the type of error we all make ourselves in everyday life when we find a mistake balancing the checkbook, miss a turn on a frequently traveled route, dial a wrong number on the phone, or forget to pay a bill. At home, these things can be overlooked, but not in quality control.

Preventing inadvertent errors may seem an impossible task. Indeed, these errors are among the most difficult of all to eliminate. And the closer the error rate gets to zero, the more difficult improvement becomes. Still, in most cases it is possible to make substantial improvements economically. At times it is even possible to eliminate the errors completely.

One way of dealing with inadvertent errors is fool-proofing. *Fool-proofing* involves changing the design of a process or product to make the commission of a particular human error impossible. For example, a company was experiencing a sporadic problem (note: the words "sporadic problem" should raise a flag in your mind that inadvertent human error is likely!) with circuit board defects. Occasionally entire orders were lost because the circuit boards were drilled wrong. A study revealed that the problem occurred because the circuit boards could be mounted backwards on an automatic drill unless the manufacturing procedure was followed carefully. Most of the time there was no problem, but as people got more experience with the drills they sometimes got careless. The problem was solved by adding an extra hole in an unused area of the circuit board panel, and then adding a pin to the drill fixture. If the board was mounted wrong, the pin wouldn't go through the hole. Result: no more orders lost.

Another method of reducing human errors is automation. People tend to commit more errors when working on dull, repetitive tasks. Also, people tend to make more errors when working in unpleasant environments, where the unpleasantness may arise from heat, odors, noise, or a variety of other factors. It happens that machines are very well suited to exactly this type of work. As time passes, more progress is made with robotics. Robots redefine the word "repetitive." A highly complicated task for a normal machine becomes a simple repetitive task for a robot. Elimination of errors is one item in the justification of an investment in robots. On a more mundane level, simpler types of automation, such as numerically controlled machining centers, often produce a reduction in human errors.

Another approach to the human error problem is ergonomics, or human factors engineering. Many errors can be prevented through the application of engineering principles to design of products, processes, and workplaces. By evaluating such things as seating, lighting, sound levels, temperature change, workstation layout, etc., the environment can often be improved and errors reduced. Sometimes human factors engineering can be combined with automation to reduce errors. This involves automatic inspection and the use of alarms (lights, buzzers, etc.) that warn the operator when he's made an error. This approach is often considerably less expensive than full automation.

Technique errors

As an example of technique errors, consider the following real-life problem with gearbox housings. The housings were gray iron castings and the problem was cracks. The supplier was made aware of the problem and the metallurgist and engineering staff had worked long and hard on the problem, but to no avail. Finally, in desperation, the customer sat down with the supplier to put together a "last-gasp" plan. If the plan failed, the customer would be forced to try an alternate source for the casting.

As might be expected, the plan was grand. The team identified many important variables in the product, process, and raw materials. Each variable was classified as either a "control variable," which would be held constant, or an "experimental variable" which would be changed in a prescribed way. The results of the experiment were to be analyzed using all the muscle of a major mainframe statistical analysis package. All of the members of the team were confident that no stone had been left unturned.

Shortly after the program began, the customer quality engineering supervisor received a call from his quality engineering representative at the supplier's foundry. "We can continue with the experiment if you really want to," he said, "but I think we've identified the problem and it isn't on our list of variables." It seems that the engineer was in the inspection room inspecting castings for our project and he noticed a loud "clanging sound" in the next room. The clanging occurred only a few times each day, but the engineer soon noticed that the cracked castings came shortly after the clanging began. Finally he investigated and found the clanging sound was a relief operator pounding the casting with a hammer to remove the sand core. Sure enough, the cracked castings had all received the "hammer treatment!"

This example illustrates a category of human error different than the inadvertent errors described earlier. Technique errors share certain common features:

- They are unintentional.
- They are usually confined to a single characteristic (e.g., cracks) or class of characteristics.
- They are often isolated to a few workers who consistently fail.

Solution of technique errors involve the same basic approaches as the solution of inadvertent errors, namely automation, fool-proofing, and human factors engineering. In the meantime, unlike inadvertent errors, technique errors may be caused by a simple lack of understanding that can be corrected by developing better instructions and training.

Willful errors (sabotage)

This category of error is unlike either of the two previous categories. Willful errors are often very difficult to detect; however they do bear certain trademarks:

- They are not random.
- They don't "make sense" from an engineering point of view.
- They are difficult to detect.
- Usually only a single worker is involved.
- They begin at once.
- They do not occur when an observer is present.

Another real-life example may be helpful. An electromechanical assembly suddenly began to fail on some farm equipment. An examination of the failures revealed that the wire had been broken *inside of the insulation*. However, the assemblies were checked 100% for continuity after the wire was installed and the open circuit should have been easily discovered. After a long and difficult investigation no solution had been found. However, the problem had gone away and never come back.

About a year later, the quality engineer was at a company party when a worker approached him. The worker said he knew the answer to the now-infamous "broken wire mystery," as it had come to be known. The problem was caused, he said, when a newly hired probationary employee was given his two-weeks notice. The employee decided to get even by sabotaging the product. He did this by carefully breaking the wire, but not the insulation, and then pushing the broken sections together so the assembly would pass the test. However, in the field, the break would eventually separate, resulting in failure. Later, the quality engineer checked the manufacturing dates and found that every failed assembly had been made during the two weeks prior to the saboteur's termination date.

In most cases, the security specialist is far better equipped and trained to deal with this type of error than quality control or engineering personnel. In serious cases criminal charges may be brought as a result of the sabotage. If the product is being made on a government contract, federal agencies may be called in. Fortunately, willful errors are extremely rare. They should be considered a possibility only after all other explanations have been investigated and ruled out.

V.F OVERCOMING BARRIERS TO QUALITY IMPROVEMENT

The structure of modern organizations is based on the principle of division of labor. The idea is simple: divide the work to be done into small pieces and let people become expert at doing a given piece. The principle was first suggested by Adam Smith in 1776 and it worked quite well when applied to industrial work. In the early part of the twentieth century, Frederick Taylor applied the idea to management and scientific management was born. Most organizations today consist of a number of departments, each devoted to their own specialty.

Considering the bounty produced by today's organizations when compared to pre-industrial times, we are forced to admit that the current system works quite well. However, that is not to say that there isn't room for improvement. Modern organizations are faced with a new reality with which traditional organizational forms are ill-equipped to deal. A fundamental problem is that the separate functional departments tend to optimize their own operations, often to the detriment of the organization as a whole. Traditional organizations, in effect, create barriers between departments. Departmental managers are often forced to compete for shares of limited budgets; in other words, they are playing a "zero sum game" where another manager's gain is viewed as their department's loss. Behavioral research has shown that people engaged in zero sum games exhibit self-destructive, cut-throat behavior. Overcoming this tendency requires improved communication and cooperation between departments. Deming's point #9 addresses this issue. (See I.G.)

Destructive competition is not the result of character flaws in the people in charge; it is the result of the system. Likewise, cooperative behavior will only result from transformation of the system. It will do no good to exhort people to cooperate with one another. Nor will punitive measures do the trick. The system itself must be redesigned to produce the desired results. The most common system changes involve simultaneous process redesign and the use of interdepartmental teams.

Interdepartmental teams are groups of people with the skills needed to deliver the value desired. Processes are designed by the team to create the value in an effective and efficient manner. Management must see to it that the needed skills exist in the organization (if not, they must be developed or acquired from outside the organization). It is also management's job to see that they remove barriers to cooperation, such as poorly designed incentive and reward systems, rigid procedures, and so on. Some organizations restructure themselves around the team approach. These self-directed teams are given the authority to hire team members, determine the compensation of team members, control certain assets, and so on. The amount of self-direction allowed varies enormously from one firm to the next, but the practice seems to be growing.

The objective of all of this is to create self-directed, self-sustaining systems, or what is sometimes called *spontaneous order*. People working in such systems will pursue the organization's mission by pursuing what is best for them. The need for oversight and control is greatly diminished, thus eliminating layers of management and supervision.

MANAGING CHANGE

Experts agree: change is difficult, disruptive, expensive, and a major cause of error. Given these problems, it's reasonable to ask: Why change? Here are the most common reasons organizations choose to face the difficulties involved with change:

- **Leadership**—Some organizations choose to maintain product or service leadership as a matter of policy. Change is a routine.
- **Competition**—When competitors improve their products or services such that their offering provides greater value than yours, you are forced to change. Refusal to do so will result in the loss of customers and revenues and can even lead to complete failure.
- **Technological advances**—Effectively and quickly integrating new technology into an organization can improve quality and efficiency and provide a competitive advantage. Of course, doing so involves changing management systems.
- **Training requirements**—Many companies adopt training programs without realizing that many such programs implicitly involve change. For example, a company that provides employees with SPC training should be prepared to implement a process control system. Failure to do so leads to morale problems and wastes training dollars.
- **Rules and regulations**—Change can be forced on an organization from internal regulators via policy changes and changes in operating procedures. Government and other external regulators and rule-makers (e.g., ISO for manufacturing, JCAHO for hospitals) can also mandate change.

- **Customer demands**—Customers, large and small, have the annoying habit of refusing to be bound by your policies. The nice customers will demand that you change your policy and procedures. The really nasty customers don't say anything at all, they simply go somewhere else to do business.

Johnson (1993, p. 233) gives the following summary of change management:

1. Change will meet resistance for many different reasons.
2. Change is a balance between the stable environment and the need to implement TQM. Change can be painful while it provides many improvements.
3. There are four times change can most readily be made by the leader: when the leader is new on the job, receives new training, has new technology, or when outside pressures demand change.
4. Leaders must learn to implement change they deem necessary, change suggested from above their level, and change demanded from above their level.
5. There are all kinds of reaction to change. Some individuals will resist, some will accept, and others will have mixed reactions.
6. There is a standard process that supports the implementation of change. Some of the key requirements for change are leadership, empathy, and solid communications.
7. It is important that each leader become a change leader. This requires self-analysis and the will to change those things requiring change.

ROLES

Change requires new behaviors from everyone involved. However, four specific roles commonly appear during most successful change processes (Hutton 1994, pp. 2–4):

- Official change agent. An officially designated person who has primary responsibility for helping management plan and manage the change process.
- Sponsors. Sponsors are senior leaders with the formal authority to legitimize the change. The sponsor makes the change a goal for the organization and ensures that resources are assigned to accomplish it. No major change is possible without committed and suitably placed sponsors.
- Advocate. An advocate for change is someone who sees a need for change and sets out to initiate the process by convincing suitable sponsors. This is a selling role. Advocates often provide the sponsor with guidance and advice. Advocates may or may not hold powerful positions in the organization.
- Informal change agent. Persons other than the official change agent who voluntarily help plan and manage the change process. While the contribution of these people is extremely important, it is generally not sufficient to cause truly significant, organization-wide change.

The position of these roles within a typical organizational hierarchy is illustrated graphically in Figure V.39.

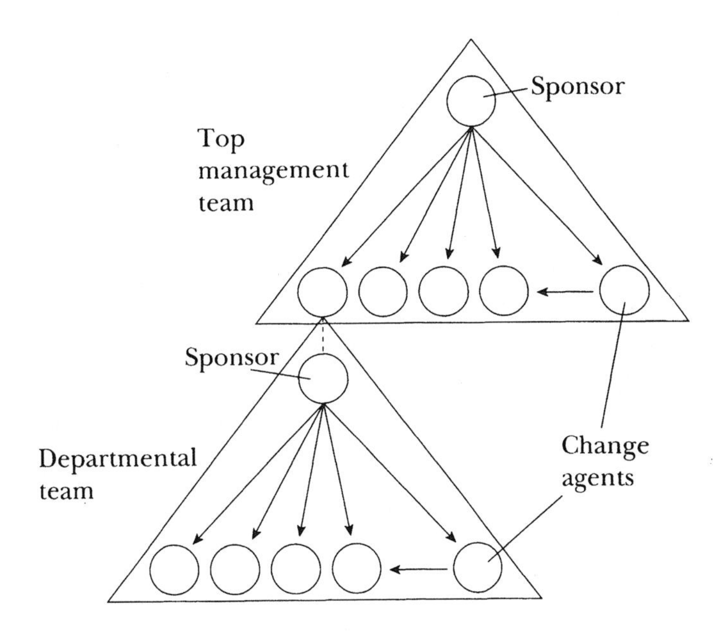

Figure V.39. Cascading of sponsorship.

From Hutton, D.W. (1994). *The Change Agent's Handbook: A Survival Guide for Quality Improvement Champions.* Copyright © 1994 by David W. Hutton. Reprinted with permission.

THE JOB OF CHANGE AGENT

Goals

There are three goals of change:

1. Change the way people in the organization think. Helping people change the way they think is a primary activity of the change agent. All change begins with the individual, at a personal level. Unless the individual is willing to change his behavior, no real change is possible. Changing behavior requires a change in thinking. In an organization where people are expected to use their minds, people's actions are guided by their thoughts and conclusions. The change agent's job starts here.
2. Change the norms. Norms consist of standards, models, or patterns which guide behavior in a group. All organizations have norms or expectations of their members. Change cannot occur until the organization's norms change.
3. Changing the organization's systems or processes. This is the "meat" of the change. Ultimately, all work is a process and quality improvement requires change at the process and system level. However, this cannot occur on a sustained basis until individuals change their behavior and organizational norms are changed.

Mechanisms used by change agents

The change agents help accomplish the above goals in a variety of ways. Education and training are important means of changing individual perceptions and behaviors. In this discussion, a distinction is made between training and education. Training refers to instruction and practice designed to teach a person how to perform some task. Training focuses on concretes that need to be done. Education refers to instruction in how to think. Education focuses on integrating abstract concepts into one's knowledge of the world. An educated person will view the world differently after being educated than they did before. This is an essential part of the process of change.

Change agents help organize an assessment of the organization to identify its strengths and weaknesses. Change is usually undertaken to either reduce areas of weakness, or exploit areas of strength. The assessment is part of the education process. Knowing one's specific strengths and weaknesses is useful in mapping the process for change.

Change agents play an important role in quality improvement (remember, "improvement" implies change). As shown in Figure V.39 change agents are in strategic positions throughout the organization. This makes it possible for them to assist in the coordination of the development and implementation of quality improvement plans. Quality improvement of any significance nearly always involves multiple departments and levels in the organization.

In the final analysis, all we humans really have to spend is our time. Change agents see to it that senior management spends sufficient time on the transformation process. Senior managers' time is in great demand from a large number of people inside and outside of the organization. It is all too easy to schedule a weekly meeting to discuss "TQM" for an hour, then think you've done your part. In fact, transforming an organization, large or small, requires a prodigious commitment of the time of senior leadership. At times the executive will not understand what he or she is contributing by, say, attending team meetings. The change agent must constantly assure the leader that time spent on transformation activities is time well spent.

One way of maximizing the value of an executive's time investment is for the executive to understand the tremendous power of certain symbolic events. Some events generate stories that capture the essence of management's commitment (or lack of it) to the change being undertaken. People repeat stories and remember them far better than proclamations and statements. For example, there's a story told by employees of a large U.S. automotive firm that goes as follows:

In the early 1980s the company was just starting their quality improvement effort. At a meeting between upper management and a famous quality consultant, someone casually mentioned that quality levels were seasonal—quality was worse in the summer months. The consultant asked why this should be so. Were different designs used? Were the machines different? How about the suppliers of raw materials? The answer to each of these questions was "No." An investigation revealed that the problem was vacations. When one worker went on vacation, someone else did her job, but not quite as well. And that "someone" also vacated a job, which was done by a replacement, etc. It turned out that the one person going on vacation lead to six people doing jobs they did not do routinely. The solution was to have a vacation shutdown of two weeks. This greatly reduced the number of people on new jobs and brought summer quality levels up to the quality levels experienced the rest of the year.

This worked fine for a couple of years since there was a recession in the auto industry and there was plenty of excess capacity. However, one summer the senior executives were asked by the finance department to reconsider their shutdown policy. Demand had picked up and the company could sell every car it could produce. The accountants pointed out that the shutdown would cost $100 million per day in lost sales.

The vice president of the truck division asked if anything had been done to address the cause of the quality slippage in the summer. No, nothing had been done. The president asked the staff "If we go back to the old policy, would quality levels fall like they did before?" Yes, he was told, they would. "Then we stay with our current policy and shut down the plants for vacations," the president announced.

The President was challenged by the vice president of finance. "I know we're committed to quality, but are you sure you want to lose $1.4 billion in sales just to demonstrate our commitment?" The President replied, "Frank, I'm not doing this to 'demonstrate' anything. We almost lost our company a few years back because our quality levels didn't match our overseas competition. Looking at this as a $1.4 billion loss is just the kind of short-term thinking that got us in trouble back then. I'm making this decision to *save* money."

This story had tremendous impact on the managers who heard it, and it spread like wildfire throughout the organization. It demonstrated many things simultaneously: senior leadership's commitment to quality, political parity between operations and finance, how seemingly harmless policies can have devastating effects, an illustration of how short-term thinking had damaged the company in the past, and how long-term thinking worked in a specific instance, etc. It is a story worth 100 speeches and mission statements.

Leadership support activities

The change agent provides technical guidance to the leadership team. This guidance takes the form of presenting management with alternative strategies for pursuing the transformation, education on methods that can be used to implement the strategies, and selection of key personnel for key transformation jobs.

Change agents help to monitor the status of quality teams and quality projects relating to the transformation (see I.B for a complete discussion of project management). In addition to being a vehicle for local quality improvement, projects can be used as one of the mechanisms for actually implementing the transformation. If used in this way, it is important that projects be properly chartered to align the project activities with the goals of the transformation. All teams, chartered or not, must avoid projects and methods that conflict with the goals of the transformation. Project team membership must be carefully planned to assure that both task and group maintenance roles are properly filled (see I.C and I.D). Project charters must clearly identify the scope of the project to prevent the confusion between teams that results from overlapping charters.

Change agents also serve as coaches to senior leaders. Culture involves innumerable subtle characteristics and behaviors that become unconsciously "absorbed" into one's being. At times, it is nearly impossible for the individual executive to see how his or her behavior or relationships are interpreted by others. The change agent must quietly advise leadership on these issues.

The press of day-to-day business, combined with the inherent difficulties of change, make it easy to let time slip by without significant progress. Keeping operations going is a full-time job, and current problems present themselves with an urgency that meeting a future goal can't match. Without the constant reminders from change agents that goals aren't being met, the leadership can simply forget about the transformation. It is the change agent's job to become the "conscience" of the leadership and to challenge them when progress falls short of goals.

Change networks

Change agents should work to develop an internal support network. The network provides resources to support the change process by disseminating education and guidance. The network's tasks will eventually be subsumed by normal operations, but in the early stages of the transformation it is vital that the network exist since the control of resources is determined by the existing infrastructure and may be difficult to divert to the change process. Usually, the members of the network are formal and informal change agents in various areas of the organization.

Once the network has been established, it is the change agent's job to assure that the activities in the network are consistent with and in support of the organization's vision. For example, if a hospital has a vision where physicians obtain real-time clinical information and decision support at the patient's bedside, then a financially based and centralized information system is inconsistent with that vision. The change agent, especially the formal change agent, provides leadership and moral support to network members, who may otherwise feel isolated and vulnerable. Change agents ensure that members of the network receive the education and guidance they need. Finally, the change agent acts as a conduit and a stimulant to maintain regular communication in the network. This may take the form of setting up an email list, scheduling lunches for network members, etc.

Transforming staff functions

Table V.10 illustrates the contrast between the way that staff functions used to operate under the traditional system of management, and the way they can operate more effectively.

Table V.10. How staff functions are changing.

From Hutton, D.W. (1994). *The Change Agent's Handbook: A Survival Guide for Quality Improvement Champions.* Page 220. Copyright © 1994 by David W. Hutton. Reprinted with permission.

	FROM	TO
Role	Customer—for information, evidence, and reports from others	Supplier—of information, expertise, and other services
Strategy	Control—by imposition of policies and procedures, and by audit and inspection	Support—by gearing efforts to the needs of others Self-control by client
Goal	Departmental achievement of departmental objectives	Collective achievement of the organization's objectives
Style of working with others	Competitive, adversarial	Integrating, collaborative
Focus of attention	Some aspects of outcomes; for example, product quality, financial results Some pieces of the process; for example, adherence to policy and procedure	The relationship between the entire underlying process and the achievement of all the desired outcomes
Image	Regulator, inspector, policeman	Educator, helper, guide

There are several ways in which change agents can assist staff functions in transforming their roles:

- Collaborate with staff functions.
- Encourage staff functions to take a proactive approach to change.
- Make support functions partners in the support network.
- Encourage staff members to become role models.
- Help staff functions develop transition plans that are aligned and integrated with the overall transformation plan.
- Encourage staff members to share their concerns.

Quantitative Methods

VI.A CONCEPTS OF PROBABILITY AND STATISTICS

DEFINITIONS OF BASIC STATISTICAL TERMS

Many basic statistical terms essential to quality engineering are defined in the glossary at Appendix 3.

ENUMERATIVE AND ANALYTICAL STUDIES

How would you respond to the following question?

> A sample of 100 bottles taken from a filling process has an average of 11.95 ounces and a standard deviation of 0.1 ounce. The specifications are 11.9–12.1 ounces. Based on these results, should you
>
> a. Do nothing?
> b. Adjust the average to precisely 12 ounces?
> c. Compute a confidence interval about the mean and adjust the process if the nominal fill level is not within the confidence interval?
> d. None of the above?

The correct answer is *d*, none of the above. The other choices all make the mistake of applying enumerative statistical concepts to an analytic statistical situation. In short, *the questions themselves are wrong!* For example, based on the data, there is no way to determine if doing nothing is appropriate. "Doing something" implies that there is a known cause and effect mechanism which can be employed to reach a known objective. There is nothing to suggest that this situation exists. Thus, we can't simply adjust the process average to the nominal value of 12 ounces, even though the process appears to be 5 standard errors below this value. This might have happened because the first 50 were 10 standard errors below the nominal and the last 50 were exactly equal to the nominal (or any of a nearly infinite number of other possible scenarios). The confidence interval calculation fails for the same reason. Figure VI.1 illustrates some processes that could produce the statistics provided above.

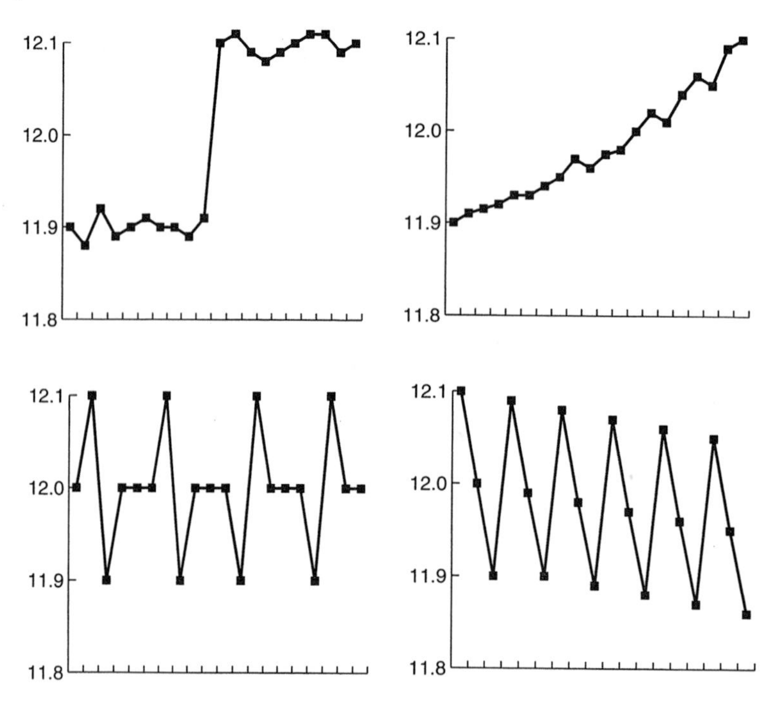

Figure VI.1. Possible processes with similar means and sigmas.

Some appropriate analytic statistics questions might be:

- Is the process central tendency stable over time?
- Is the process dispersion stable over time?
- Is the process distribution stable over time?

If any of the above are answered "no," then what is the cause of the instability? To help answer this question, ask "what is the nature of the variation as revealed by the patterns?" when plotted in time-sequence and stratified in various ways.

If none of the above are answered "no," then, and only then, we can ask such questions as the following:

- Is the process meeting the requirements?
- *Can* the process meet the requirements?
- Can the process be improved by recentering it?
- How can we reduce variation in the process?

What are enumerative and analytic studies?

Deming (1975) defines enumerative and analytic studies as follows:

Enumerative study—a study in which action will be taken on the universe.

Analytic study—a study in which action will be taken on a process to improve performance in the future.

The term "universe" is defined in the usual way: the entire group of interest, e.g., people, material, units of product, which possess certain properties of interest. An example of an enumerative study would be sampling an isolated lot to determine the quality of the lot.

In an analytic study the focus is on a *process* and how to improve it. The focus is the *future*. Thus, unlike enumerative studies which make inferences about the universe actually studied, analytic studies are interested in a universe which has yet to be produced. Table VI.1 compares analytic studies with enumerative studies (Provost 1988).

Table VI.1. Important aspects of analytic studies.

ITEM	ENUMERATIVE STUDY	ANALYTIC STUDY
Aim	Parameter estimation	Prediction
Focus	Universe	Process
Method of access	Counts, statistics	Models of the process (e.g., flow charts, cause and effects, mathematical models)
Major source of uncertainty	Sampling variation	Extrapolation into the future
Uncertainty quantifiable?	Yes	No
Environment for the study	Static	Dynamic

Deming (1986) points out that "Analysis of variance, t-tests, confidence intervals, and other statistical techniques taught in the books, however interesting, are inappropriate because they provide no basis for prediction and because they bury the information contained in the order of production." These traditional statistical methods have their place, but they are widely abused in the real world. When this is the case, statistics do more to cloud the issue than to enlighten.

Analytic study methods provide information for *inductive thinking*, rather than the largely *deductive* approach of enumerative statistics. Analytic methods are primarily graphical devices such as run charts, control charts, histograms, interrelationship digraphs, etc.

Analytic statistics provide operational guidelines, rather than precise calculations of probability. Thus, such statements as "There is a 0.13% probability of a type I error when acting on a point outside a three-sigma control limit" are false (the author admits to having made this error in the past). The future cannot be predicted with a known level of confidence. Instead, based on knowledge obtained from every source, including analytic studies, one can state that they have a certain degree of belief (e.g., high, low) that such and such will result from such and such action on a process.

Another difference between the two types of studies is that enumerative statistics proceed from predetermined hypotheses while analytic studies try to help the analyst generate new hypotheses. In the past, this extremely worthwhile approach has been criticized by some statisticians as "fishing" or "rationalizing." However, this author believes that using data to develop plausible explanations retrospectively is a perfectly legitimate way of creating new theories to be tested. To refuse to explore possibilities suggested by data is to take a very limited view of the scope of statistics in quality improvement and control.

CENTRAL LIMIT THEOREM

The central limit theorem can be stated as follows:

> Irrespective of the shape of the distribution of the population or universe, the distribution of average values of samples drawn from that universe will tend toward a normal distribution as the sample size grows without bound.

It can also be shown that the average of sample averages will equal the average of the universe and that the standard deviation of the averages equals the standard deviation of the universe divided by the square root of the sample size. Shewhart performed experiments that showed that small sample sizes were needed to get approximately normal distributions from even wildly non-normal universes. Figure VI.2 was created by Shewhart using samples of 4 units.

The practical implications of the central limit theorem are immense. Consider that without the central limit theorem effects we would have to develop a separate statistical model for every non-normal distribution encountered in practice. This would be the only way to determine if the system were exhibiting chance variation. Because of the central limit theorem, we can use *averages* of small samples to evaluate *any* process using the normal distribution. The central limit theorem is the basis for the most powerful of statistical process control tools: Shewhart control charts.

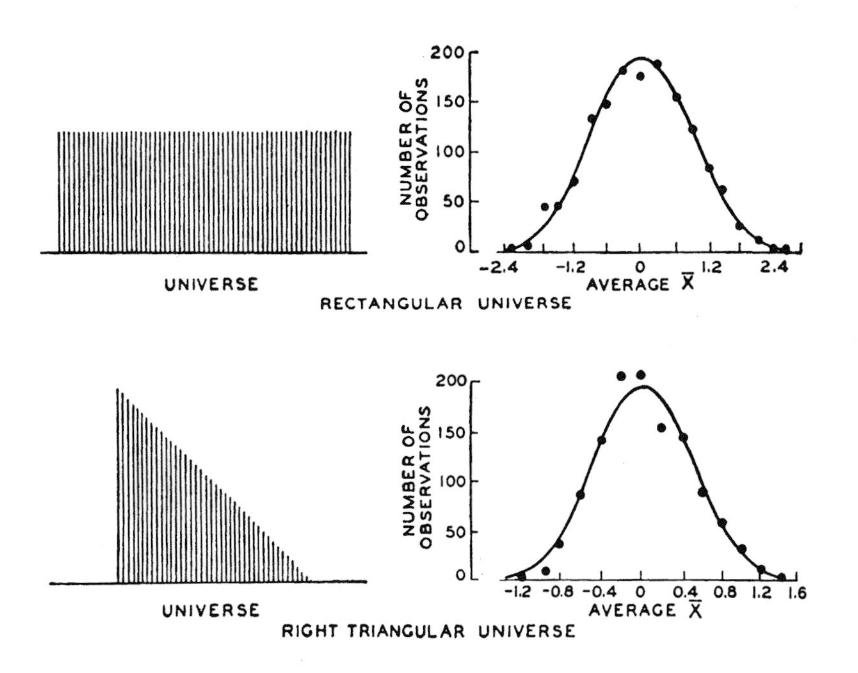

Figure VI.2. Illustration of the central limit theorem.
From Shewhart, Figure 59. Used by permission of the publisher.

BASIC PROBABILITY CONCEPTS

In quality engineering, most decisions involve uncertainty. Can a process meet the requirements? Is a gage accurate and repeatable enough to be used for process control? Does the quality of the lot on the receiving dock meet the contract quality requirements? We use statistical methods to answer these questions. Probability theory forms the basis of statistical decision-making. This section discusses those probability concepts that underlie the statistical methods commonly used in quality engineering.

Random variables

Every time an observation is made and recorded, a random experiment has occurred. The experiment may not have even involved any deliberate change on the part of the observer, but it is an experiment nonetheless. The observation is an experimental outcome which cannot be determined in advance with certainty. However, probability theory makes it possible to define the set of all possible outcomes and to assign unique numbers to every possible outcome in the set. For example, if a die is rolled it must result in either

1, 2, 3, 4, 5, or 6. We cannot determine in advance which number will come up, but we know it will be one of these. This set of numbers is known as the *sample space* and the variable to which we assign the outcome is known as a *random variable.* If we assign the outcome of the roll of a die to the variable X, then X is a random variable and the space of X is {1, 2, 3, 4, 5, 6}.

Sets

Probability theory makes heavy use of parts of set theory. In addition to presenting important concepts, the notation used in set theory provides a useful shorthand for the discussion of probability concepts.

The *universal set* is the totality of objects under consideration and it is denoted *S*. In probability theory, S is called the *sample space*. Each object in S is an *element* of S. When x is an element of the set A, we write $x \in A$. If a set includes some, but not all, of the elements in S, it is termed a *subset* of S. If A is a subset of S we write $A \subset S$. The part of S that is not in the set A is called the *complement* of A, denoted A'. A set that contains no elements of S is called the *null* or *empty* set and is denoted by $\varnothing$ (pronounced "fee").

Let A and B be two sets. The set of elements in either A or B or both A and B is called the *union* of A and B and is denoted by $A \cup B$. The set of elements in both A and B is called the *intersection* of A and B and is denoted by $A \cap B$. If A and B have no elements in common, then A and B are said to be *mutually exclusive*, denoted $A \cap B = \varnothing$. If A and B cover the entire sample space, then they are said to be *exhaustive*, denoted $A \cup B = S$.

A simple graphical method that is useful for illustrating many probability relationships, and other relationships involving sets, is the Venn diagram. Figure VI.3 is a Venn diagram of the flip of a coin.

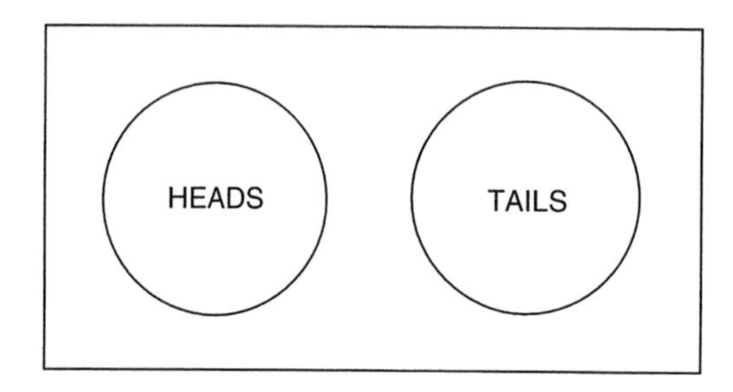

Figure VI.3. Venn diagram of two mutually exclusive events.

We say the events A and B are *independent* if A does not depend on B, and vice versa. When this is the case, the probability of two events occurring simultaneously is simply the product of them occurring individually. In fact, by definition, events A and B are independent if and only if $P(A \cap B) = P(A)P(B)$. For example, if a coin is flipped twice the probability of getting a head followed by a tail is $P(H \cap T) = P(H)P(T) = (0.5)(0.5) = 0.25$.

Figure VI.3 shows two circles that do not overlap, which means that they are mutually exclusive. If it is possible to get both events simultaneously, we have a different situation. Let's say, for example, that we have a group of 1000 parts (the sample space). Let event A be that a part is defective, and event B be that the part is from Vendor A. Since it is possible to find a defective part from Vendor A, the two events can occur simultaneously. The Venn diagram in Figure VI.4 illustrates this.

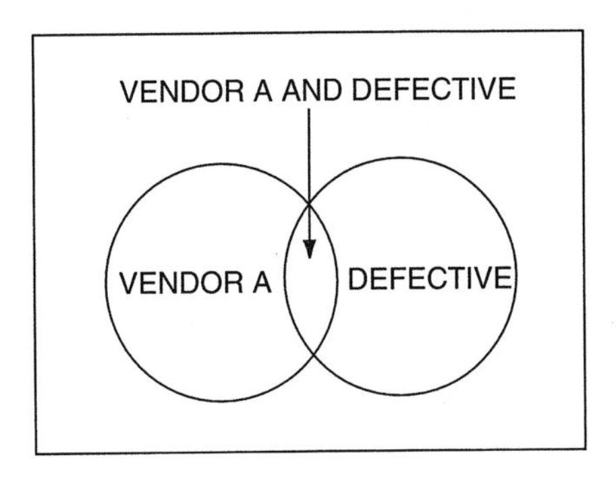

Figure VI.4. Venn diagram of two events that can occur simultaneously.

If we have the situation illustrated by Figure VI.4, then we can no longer find the probability of either event occurring by just adding the individual probabilities. Instead we must use the equation

$$P(A \cup B) = P(A)+P(B)-P(A \cap B) \qquad \text{(VI.1)}$$

For example, let A be the event that the part is defective (P=0.05), and B be the event that the part is from Vendor A (P=0.50). Then, assuming the events are independent, the probability that the part is either defective or from Vendor A is $.05 + .50 - (.05 \times .50) = .55 - .025 = .525$. The same rules can be applied to more than two events. For example, if A, B, and C are three events then

$$P(A \cup B \cup C) = P(A)+P(B)+P(C) - P(A \cap B) - P(A \cap C) - P(B \cap C) + P(A \cap B \cap C) \qquad \text{(VI.2)}$$

Probability functions

With this background it is now possible to define probability formally. Probability is a set function P that assigns to each event A in the sample space S a number P(A), called the probability of the event A, such that

1. $P(A) \geq 0$,
2. $P(S)=1$,
3. If A, B, C, . . . are mutually exclusive events then $P(A \cup B \cup C \cup \ldots) = P(A)+P(B)+P(C)+ \ldots$

Methods of enumeration

Enumeration involves counting techniques for very large numbers of possible outcomes. This occurs for even surprisingly small sample sizes. In quality engineering, these methods are commonly used in a wide variety of statistical procedures.

The basis for all of the enumerative methods described here is the multiplication principle. The multiplication principle states that the number of possible outcomes of a series of experiments is equal to the product of the number of outcomes of each experiment. For example, consider flipping a coin twice. On the first flip there are two possible outcomes (heads/tails) and on the second flip there are also two possible outcomes. Thus, the series of two flips can result in 2x2=4 outcomes. Figure VI.5 illustrates this example.

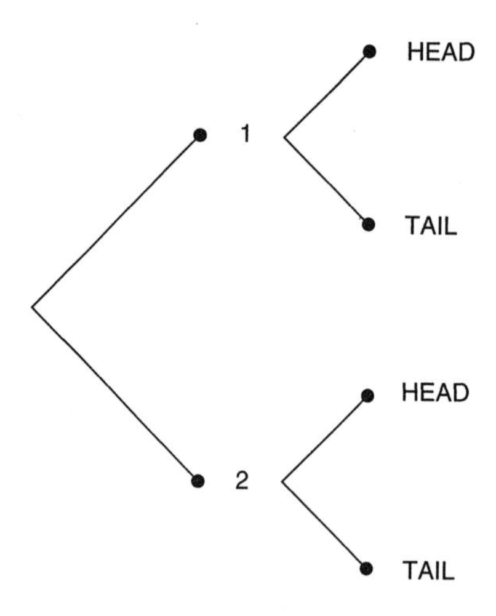

Figure VI.5. Multiplication principle applied to coin flips.

An ordered arrangement of elements is called a *permutation*. Suppose that you have four objects and four empty boxes, one for each object. Consider how many different ways the objects can be placed into the boxes. The first object can be placed in any of the four boxes. Once this is done there are three boxes to choose from for the second object, then two boxes for the third object and finally one box left for the last object. Using the multiplication principle you find that the total number of arrangements of the four objects into the four boxes is 4x3x2x1=24. In general, if there are n positions to be filled with n objects there are

$$n(n-1)\ldots(2)(1) = n! \qquad \text{(VI.3)}$$

possible arrangements. The symbol n! is read n factorial. By definition, 0!=1.

In applying probability theory to discrete variables in quality control we frequently encounter the need for efficient methods of counting. One counting technique that is especially useful is combinations. The combination formula is shown in Equation VI.4.

$$C_r^n = \frac{n!}{r!(n-r)!} \qquad \text{(VI.4)}$$

Combinations tell how many unique ways you can arrange *n* objects taking them in groups of *r* objects at a time, where *r* is a positive integer less than or equal to *n*. For example, to determine the number of combinations we can make with the letters X,Y, and Z in groups of 2 letters at a time, we note that $n = 3$ letters, $r = 2$ letters at a time and use the above formula to find

$$C_2^3 = \frac{3!}{2!(3-2)!} = \frac{3\times 2\times 1}{(2\times 1)(1)} = \frac{6}{2} = 3$$

The 3 combinations are XY, XZ, and YZ. Notice that this method does not count reversing the letters as separate combinations, i.e., XY and YX are considered to be the same.

THEORETICAL EXPECTED VALUE

The theoretical expected value, or mathematical expectation, or simply expectation, of a random variable is a theoretically defined quantity intended to be an analog of the arithmetic mean. It can be regarded as corresponding to a limit of the arithmetic mean as the sample size increases indefinitely.

The concept of mathematical expectation is useful in summarizing important characteristics of probability distributions. There are two expected values of particular importance in quality engineering: the mean and the variance; these are discussed below. An example of the application of the expected value concept is provided below.

Example of expectation

A contractor must reserve an expensive piece of construction equipment from a rental company at least three days in advance. The rental costs \$500. If it rains, the contractor will still pay the \$500, but he will not get to use the equipment. If it does not rain, the contractor will earn \$1000 for the work done. If the probability of rain is 20%, what is the expected value of renting the equipment?

The solution involves solving the equation

$$E = \tfrac{1}{5}\times(-\$500) + \tfrac{4}{5}\times \$1{,}000 = \$700$$

In other words, the contractor can expect to earn \$700. Note that, in fact, the number \$700 cannot actually occur; either the contractor will lose \$500 or he will gain \$1,000.

The expectation applies *on the average*. The probabilities are "weights" applied to the different outcomes. Thus, mathematical expectation can be thought of as a weighted average of the various outcomes.

Expectation for discrete random variables

There are two different aspects to expectation: the outcome and the probability of the outcome. In the example, the outcomes were a loss of $500 or a gain of $1,000. The probabilities were 1/5 and 4/5. It can be seen that the expectation is just a sum of the outcome multiplied by its probability. If we designate the outcomes as $u(x)$ and the probability function as $f(x)$ then expectation can be expressed mathematically.

$$E[u(X)] = \sum_{R} u(x) f(x) \quad \text{(VI.5)}$$

The subscript R indicates that the sum is taken over the entire region of x values. For our example,

$$f(x) = \begin{cases} \frac{1}{5}, & \text{rain} \\ \frac{4}{5}, & \text{shine} \end{cases}$$

$$u(x) = \begin{cases} -\$500, & \text{x = rain} \\ +\$1{,}000, & \text{x = shine} \end{cases}$$

Properties of expectation

When it exists, expectation satisfies the following properties:

1. If c is a constant, E[c]=c.
2. If c is a constant and u is a function, E[cu(x)]=cE[u(X)].
3. If c_1 and c_2 are constants and u_1 and u_2 are functions, then $E[c_1u_1(x)+c_2u_2(x)] = c_1E[u_1(x)] + c_2E[u_2(x)]$.

The mean, variance, and standard deviation as expectations

As mentioned above, expectation can be thought of as a weighted average. In fact, the mean itself is an expectation. In general, if X is a random variable with the discrete pdf $f(x)$ defined on a region R, then

$$\mu = E(X) = \sum_{R} x f(x) \quad \text{(VI.6)}$$

Where μ is the mean of the distribution. μ is often called the population mean. The sample mean is discussed in section C of this chapter.

If X is a random variable with the continuous pdf f(x) defined on a region R, then integration is used instead of summation,

$$\mu = E(X) = \int_R xf(x)dx \tag{VI.7}$$

This is merely the continuous analog of the previous equation.

Example: finding the mean for a fair die

For a fair die, the probability of getting x=1, 2, 3, 4, 5, or 6 = 1/6; thus *f(x)*=1/6. Substituting x and *f(x)* into Equation VI.6,

$$\mu = \sum_{x=1}^{6} xf(x) = 1\left(\frac{1}{6}\right) + 2\left(\frac{1}{6}\right) + 3\left(\frac{1}{6}\right) + 4\left(\frac{1}{6}\right) + 5\left(\frac{1}{6}\right) + 6\left(\frac{1}{6}\right) = \frac{1+2+3+4+5+6}{6} = 3.5$$

The mean provides a measure of the center of a distribution. In mechanics, the mean is the centroid.

The *variance* of a distribution is also an expectation. Where the mean provides information about the location of a distribution, the variance provides information about the dispersion or spread of a distribution. Mathematically, for a discrete distribution,

$$\sigma^2 = E\left[(X-\mu)^2\right] = \sum_R (X-\mu)^2 f(x) \tag{VI.8}$$

When x is a continuous variable, the variance equation is found by integration,

$$\sigma^2 = E\left[(X-\mu)^2\right] = \int_R (X-\mu)^2 f(x)dx \tag{VI.9}$$

As with the mean, this is simply the continuous variable equivalent to the previous equation.

Example: finding the variance for a fair die

We already know that the mean for a fair die is 3.5. Thus, utilizing the equation for the variance,

$$\sigma^2 = E\left[(X-3.5)^2\right] = \sum_{x=1}^{6} (X-3.5)^2 \frac{1}{6} = \left[(1-3.5)^2 + (2-3.5)^2 + \cdots + (6-3.5)^2\right]\frac{1}{6} = \frac{35}{12}$$

The *standard deviation* is the positive square root of the variance and is denoted by

$$\sigma = \sqrt{Var(X)} = \sqrt{\sigma^2} \tag{VI.10}$$

σ is often called the *population standard deviation*. The sample standard deviation is discussed in section C of this chapter. For the example,

$$\sigma = \sqrt{35/12} = 1.708,$$

approximately.

VI.B COLLECTING AND SUMMARIZING DATA

DISCRETE AND CONTINUOUS DATA

Data are said to be *discrete* when they take on only a finite number of points that can be represented by the non-negative integers. An example of discrete data is the number of defects in a sample. Data are said to be *continuous* when they exist on an interval, or on several intervals. An example of continuous data is the measurement of pH. Quality methods exist based on probability functions for both discrete and continuous data.

LEVELS OF MEASUREMENT

A *measurement* is simply a numerical assignment to something, usually a non-numerical element. Measurements convey certain information about the relationship between the element and other elements. Measurement involves a theoretical domain, an area of substantive concern represented as an empirical relational system, and a domain represented by a particular selected numerical relational system. There is a mapping function that carries us from the empirical system into the numerical system. The numerical system is manipulated and the results of the manipulation are studied to help the engineer better understand the empirical system.

In reality, measurement is problematic; the engineer can never know the "true" value of the element being measured. Yet, measurement is a sine qua non of all science and engineering. Fundamentally, any item measure should meet two tests:

1. The item measures what it is intended to measure (i.e., it is valid).
2. A remeasurement would order individual responses in the same way (i.e., it is reliable).

The numbers provide information on a certain scale and they represent measurements of some unobservable variable of interest. Some measurements are richer than others, i.e., some measurements provide more information than other measurements. The information content of a number is dependent on the scale of measurement used. This scale determines the types of statistical analysis that can be properly employed in studying the numbers. Until one has determined the scale of measurement, one cannot know if a given method of analysis is valid.

There are four measurement scales: nominal, ordinal, interval, and ratio. Harrington (1992) summarizes the properties of each scale as shown in Table VI.2.

Table VI.2. Types of measurement scales and permissible statistics.
From *Harrington.* Used by permission of the publisher.

SCALE	DEFINITION	EXAMPLE	STATISTICS
Nominal	Only the presence/absence of an attribute; can only count items	go/no go; success/fail; accept/reject	percent; proportion; chi-square tests
Ordinal	Can say that one item has more or less of an attribute than another item; can order a set of items	taste; attractiveness	rank-order correlation
Interval	Difference between any two successive points is equal; often treated as a ratio scale even if assumption of equal intervals is incorrect; can add, subtract, order objects	calendar time; temperature	correlations; t-tests; F-tests; multiple regression
Ratio	True zero point indicates absence of an attribute; can add, subtract, multiply, and divide	elapsed time; distance; weight	t-test; F-test; correlations; multiple regression

Numbers on a nominal scale aren't measurements at all, they are merely *category labels* in numerical form. Nominal measurements might indicate membership in a group (1=male, 2=female) or simply represent a designation (John Doe is #43 on the team). Nominal scales represent the simplest and weakest form of measurement. Nominal variables are perhaps best viewed as a form of classification rather than as a measurement scale. Ideally, categories on the nominal scale are constructed in such a way that all objects in the universe are members of one and only one class. Data collected on a nominal scale are called *attribute data.* The only mathematical operations permitted on nominal scales are = (which shows that an object possesses the attribute of concern) or ≠.

An ordinal variable is one that has a natural ordering of its possible values, but for which the distances between the values are undefined. An example is product preference rankings such as good, better, best. Ordinal data can be analyzed with the mathematical operators, = (equality), ≠ (inequality), > (greater than), and < (less than). There is a wide variety of statistical techniques which can be applied to ordinal data including the Pearson correlation. Other ordinal models include odds-ratio measures, log-linear models and logit models, all of which are used to analyze cross-classifications of ordinal data presented in contingency tables. In quality engineering, ordinal data is commonly converted into nominal data and analyzed using binomial or Poisson models. For example, if parts were classified using a poor-good-excellent ordering the quality engineer might plot a p-chart of the proportion of items in the poor category.

Interval scales consist of measurements where the ratios of differences are invariant. For example, 90°C = 194°F, 180°C = 356°F, 270°C = 518°F, 360°C = 680°F. Now 194°F/90°C ≠356°F/180°C but

$$\frac{356°F - 194°F}{680°F - 518°F} = \frac{180°C - 90°C}{360°C - 270°C}$$

Conversion between two interval scales is accomplished by the transformation y=ax+b, a>0. For example, °F=32+9/5(°C), where a=9/5, b=32. As with ratio scales, when permissible transformations are made, statistical results are unaffected by the interval scale used. Also, 0° (on either scale) is arbitrary in that it does not indicate an absence of heat.

Ratio scale measurements are so-called because measurement of an object in two different metrics are related to one another by an invariant ratio. For example, if an object's mass was measured in pounds (x) and kilograms (y) then x/y=2.2 for all values of x and y. This implies that a change from one ratio measurement scale to another is performed by a transformation of the form y=ax, a>0, e.g., pounds=(2.2)(kilograms). When permissible transformations are used, statistical results based on the data are identical regardless of the ratio scale used. Zero has an inherent meaning; in this example it signifies an absence of mass.

DESCRIPTIVE STATISTICS

Some descriptive statistical methods are covered in previous chapters, e.g., scatter diagrams and histograms (section V.C). Typically, descriptive statistics are computed to describe properties of empirical distributions, that is, distributions of data from samples. There are three areas of interest: the distribution's location or central tendency, its dispersion, and its shape. The analyst may also want some idea of the magnitude of possible error in the statistical estimates. Table VI.3 describes some of the more common descriptive statistical measures.

Table VI.3. Common descriptive statistics.

SAMPLE STATISTIC	DISCUSSION	EQUATION/SYMBOL
Measures of location		
Population mean	The center of gravity or centroid of the distribution.	$\mu = \frac{1}{N}\sum_{i=1}^{N} x_i$, where x is an observation, N is the population size.
Sample mean	The center of gravity or centroid of a sample from a distribution.	$\overline{X} = \frac{1}{n}\sum_{i=1}^{n} x_i$, where x is an observation, n is the sample size.
Median	The 50%/50% split point. Precisely half of the data set will be above the median, and half below it.	$\tilde{X}$
Mode	The value that occurs most often. If the data are grouped, the mode is the group with the highest frequency.	None
Measures of dispersion		
Range	The distance between the sample extreme values.	R=Largest–Smallest
Population variance	A measure of the variation around the mean, units are the square of the units used for the original data.	$\sigma^2 = \sum_{i=1}^{N} \frac{(x_i - \mu)^2}{N}$
Population standard deviation	A measure of the variation around the mean, in the same units as the original data.	$\sigma = \sqrt{\sigma^2}$
Sample variance	A measure of the variation around the mean, units are the square of the units used for the original data.	$s^2 = \sum_{i=1}^{n} \frac{(x_i - \overline{X})^2}{n-1}$
Sample standard deviation	A measure of the variation around the mean, in the same units as the original data.	$s = \sqrt{s^2}$

Continued on next page . . .

Table VI.3—*Continued . . .*

SAMPLE STATISTIC	DISCUSSION	EQUATION/SYMBOL
Measures of shape		
Skewness	A measure of asymmetry. Zero indicates perfect symmetry; the normal distribution has a skewness of zero. Positive skewness indicates that the "tail" of the distribution is more stretched on the side above the mean. Negative skewness indicates that the tail of the distribution is more stretched on the side below the mean. The normal distribution has a skewness of 0.	$k = \dfrac{\dfrac{\sum_{i=1}^{n} x_i^3}{n} - \dfrac{3\overline{X}\sum_{i=1}^{n} x_i^2}{n} + 2\overline{X}^3}{s^3}$
Kurtosis	Kurtosis is a measure of flatness of the distribution. Heavier tailed distribution have larger kurtosis measures. The normal distribution has a kurtosis of 3.	
	$\beta_2 = \dfrac{\dfrac{\sum_{i=1}^{n} x_i^4}{n} - 4\overline{X}\dfrac{\sum_{i=1}^{n} x_i^3}{n} + 6\overline{X}^2\dfrac{\sum_{i=1}^{n} x_i^2}{n} - 3\overline{X}^4}{s^4}$	

The figures below illustrate distributions with different descriptive statistics.

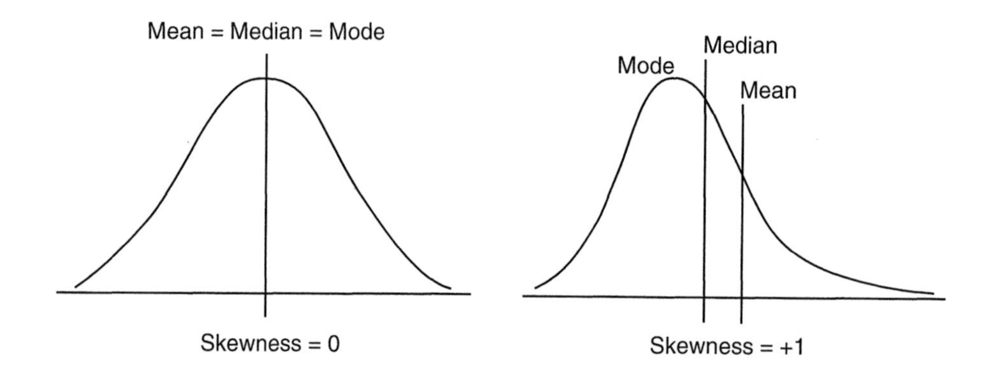

Figure VI.6. Illustration of mean, median, and mode.

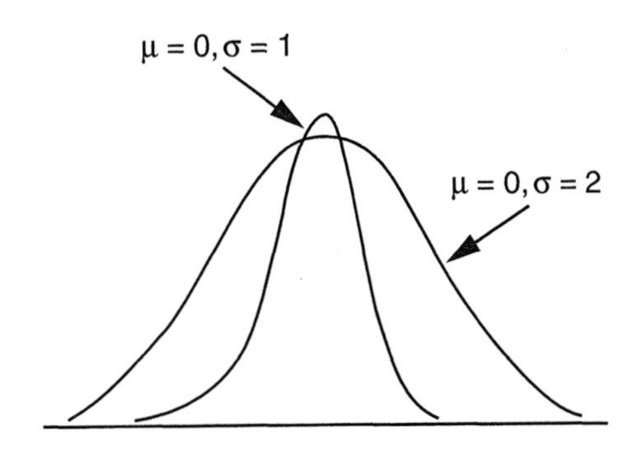

Figure VI.7. Illustration of sigma.

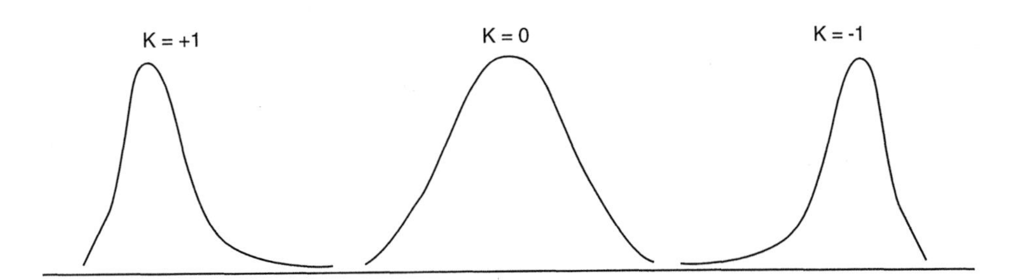

Figure VI.8. Illustration of skewness.

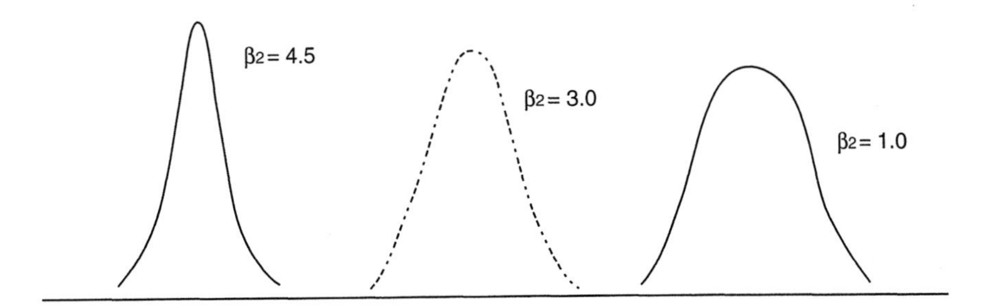

Figure VI.9. Illustration of kurtosis.

Exploratory data analysis

Data analysis can be divided into two broad phases: an exploratory phase and a confirmatory phase. Data analysis can be thought of as detective work. Before the "trial" one must collect evidence and examine it thoroughly. One must have a basis for developing a theory of cause and effect. Is there a gap in the data? Are there patterns that suggest some mechanism? Or, are there patterns that are simply mysterious (e.g., are all of the numbers even or odd)? Do outliers occur? Are there patterns in the variation of the data? What are the shapes of the distributions? This activity is known as exploratory data analysis (EDA). Tukey's 1977 book with this title elevated this task to acceptability among "serious" devotees of statistics. [See also Hoaglin, Mosteller and Tukey (2000).]

Four themes appear repeatedly throughout EDA: resistance, residuals, re-expression, and visual display. *Resistance* refers to the insensitivity of a method to a small change in the data. If a small amount of the data is contaminated, the method shouldn't produce dramatically different results. *Residuals* are what remain after removing the effect of a model or a summary. For example, one might subtract the mean from each value, or look at deviations about a regression line. *Re-expression* involves examination of different scales on which the data are displayed. Tukey focused most of his attention on simple power transformations such as $y = \sqrt{x}$, $y = x^2$, $y = 1/x$. *Visual display* helps the analyst examine the data graphically to grasp regularities and peculiarities in the data.

EDA is based on a simple basic premise: it is important to understand what you can do before you learn to measure how well you seem to have done it (Tukey, 1977). The objective is to investigate the appearance of the data, not to confirm some prior hypothesis. While there are a large number of EDA methods and techniques, there are two which are commonly encountered in quality engineering work: stem-and-leaf plots and boxplots. These techniques are commonly included in most statistics packages. (SPSS was used to create the figures used in this book.) However, the graphics of EDA are simple enough to be done easily by hand.

Stem-and-leaf plots

Stem-and-leaf plots are a variation of histograms and are especially useful for smaller data sets ($n<200$). A major advantage of stem-and-leaf plots over the histogram is that the raw data values are preserved, sometimes completely and sometimes only partially. There is a loss of information in the histogram because the histogram reduces the data by grouping several values into a single cell.

Figure VI.10 is a stem-and-leaf plot of diastolic blood pressures. As in a histogram, the length of each row corresponds to the number of cases that fall into a particular interval. However, a stem-and-leaf plot represents each case with a numeric value that corresponds to the actual observed value. This is done by dividing observed values into two components—the leading digit or digits, called the *stem*, and the trailing digit, called the *leaf*. For example, the value 75 has a stem of 7 and a leaf of 5.

```
FREQUENCY      STEM & LEAF
      .00         6 *
     7.00         6 .  5558889
    13.00         7 *  0000111223344
    32.00         7 .  55555555667777777777788888889999
    44.00         8 *  00000000000000000000001111222222333333334444
    45.00         8 .  555555555555666666677777777777778888899999999
    31.00         9 *  0000000001111111222222222333334
    27.00         9 .  556666677777788888888899999
    13.00        10 *  0000122233333
    11.00        10 .  55555577899
     5.00        11 *  00003
     5.00        11 .  55789
     2.00        12 *  01
     4.00 Extremes     (125), (133), (160)

Stem width:    10
Each leaf:      1 case (s)
```

Figure VI.10. Stem-and-leaf plot of diastolic blood pressures.

From *SPSS for Windows Base System User's Guide*, p. 183. Copyright © 1993.

Used by permission of the publisher, SPSS, Inc., Chicago, IL.

In this example, each stem is divided into two rows. The first row of each pair has cases with leaves of 0 through 4, while the second row has cases with leaves of 5 through 9. Consider the two rows that correspond to the stem of 11. In the first row, we can see that there are four cases with diastolic blood pressure of 110 and one case with a reading of 113. In the second row, there are two cases with a value of 115 and one case each with a value of 117, 118, and 119.

The last row of the stem-and-leaf plot is for cases with extreme values (values far removed from the rest). In this row, the actual values are displayed in parentheses. In the frequency column, we see that there are four extreme cases. Their values are 125, 133, and 160. Only distinct values are listed.

When there are few stems, it is sometimes useful to subdivide each stem even further. Consider Figure VI.11 a stem-and-leaf plot of cholesterol levels. In this figure, stems 2 and 3 are divided into five parts, each representing two leaf values. The first row, designated by an asterisk, is for leaves of 0 and 1; the next, designated by t, is for leaves of 2's and 3's; the third, designated by f, is for leaves of 4's and 5's; the fourth, designated by s is for leaves of 6's and 7's; and the fifth, designated by a period, is for leaves of 8's and 9's. Rows without cases are not represented in the plot. For example, in Figure VI.11, the first two rows for stem 1 (corresponding to 0-1 and 2-3) are omitted.

```
FREQUENCY     STEM & LEAF
   1.00    Extremes    (106)
   2.00        1  f   55
   6.00        1  s   677777
  12.00        1  .   888889999999
  23.00        2  *   0000000000000111111111 1
  36.00        2  t   222222222222222223333333333333333333
  35.00        2  f   44444444444444444455555555555555555
  42.00        2  s   666666666666666666666777777777777777777777
  28.00        2  .   8888888888888899999999999999
  18.00        3  *   000000011111111111
  17.00        3  t   22222222222233333
   9.00        3  f   444445555
   6.00        3  s   666777
   1.00        3  .   8
   3.00    Extremes    (393), (425), (515)
 Stem width:       100
 Each leaf:     1 case (s)
```

Figure VI.11. Stem-and-leaf plot of cholesterol levels.

This stem-and-leaf plot differs from the previous one in another way. Since cholesterol values have a wide range—from 105 to 515 in this example—using the first two digits for the stem would result in an unnecessarily detailed plot. Therefore, we will use only the hundreds digit as the stem, rather than the first two digits. The stem setting of 100 appears in the column labeled Stem width. The leaf is then the tens digit. The last digit is ignored. Thus, from this particular stem-and-leaf plot, it is not possible to determine the exact cholesterol level for a case. Instead, each is classified by only its first two digits.

Boxplots

A display that further summarizes information about the distribution of the values is the boxplot. Instead of plotting the actual values, a *boxplot* displays summary statistics for the distribution. It is a plot of the 25th, 50th, and 75th percentiles, as well as values far removed from the rest.

Figure VI.12 shows an annotated sketch of a boxplot. The lower boundary of the box is the 25th percentile. Tukey refers to the 25th and 75th percentiles "hinges." Note that the 50th percentile is the median of the overall data set, the 25th percentile is the median of those values below the median, and the 75th percentile is the median of those values above the median. The horizontal line inside the box represents the median. 50% of the cases are included within the box. The box length corresponds to the interquartile range, which is the difference between the 25th and 75th percentiles.

The boxplot includes two categories of cases with outlying values. Cases with values that are more than 3 box-lengths from the upper or lower edge of the box are called *extreme values*. On the boxplot, these are designated with an asterisk (*). Cases with values that are between 1.5 and 3 box-lengths from the upper or lower edge of the box are called *outliers* and are designated with a circle. The largest and smallest observed values that aren't outliers are also shown. Lines are drawn from the ends of the box to these values. (These lines are sometimes called *whiskers* and the plot is then called a *box-and-whiskers plot.*)

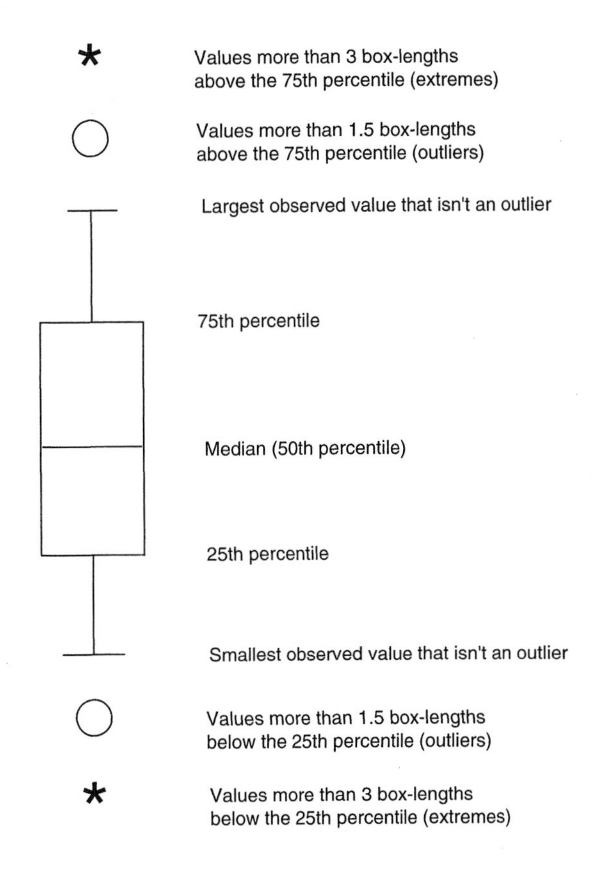

Figure VI.12. Annotated boxplot.

Despite its simplicity, the boxplot contains an impressive amount of information. From the median you can determine the central tendency, or location. From the length of the box, you can determine the spread, or variability, of your observations. If the median is not in the center of the box, you know that the observed values are skewed. If the median is closer to the bottom of the box than to the top, the data are positively skewed. If the median is closer to the top of the box than to the bottom, the opposite is true: the distribution is negatively skewed. The length of the tail is shown by the whiskers and the outlying and extreme points.

Boxplots are particularly useful for comparing the distribution of values in several groups. Figure VI.13 shows boxplots for the salaries for several different job titles.

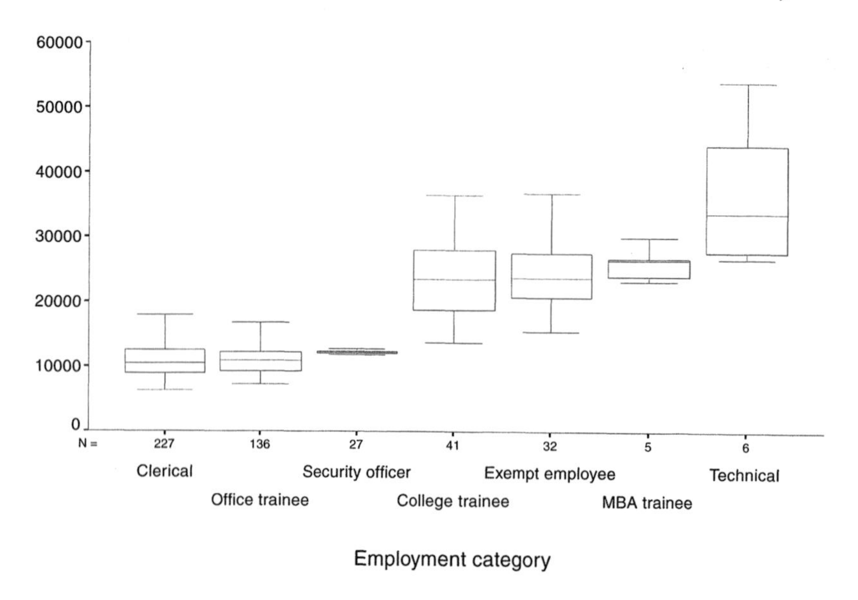

Figure VI.13. Boxplots of salary by job category.

The boxplot makes it easy to see the different properties of the distributions. The location, variability, and shapes of the distributions are obvious at a glance. This ease of interpretation is something that statistics alone cannot provide.

VI.C PROPERTIES AND APPLICATIONS OF PROBABILITY DISTRIBUTIONS

Distributions are a set of numbers collected from a well-defined universe of possible measurements arising from a property or relationship under study. Distributions show the way in which the probabilities are associated with the numbers being studied. Assuming a state of statistical control, by consulting the appropriate distribution one can determine the answer to such questions as the following:

- What is the probability that x will occur?
- What is the probability that a value less than x will occur?
- What is the probability that a value greater than x will occur?
- What is the probability that a value will occur that is between x and y?

By examining plots of the distribution shape, one can determine how rapidly, or slowly, probabilities change over a given range of values. In short, distributions provide a great deal of information.

FREQUENCY AND CUMULATIVE DISTRIBUTIONS

A frequency distribution is an empirical presentation of a set of observations. If the frequency distribution is *ungrouped*, it simply shows the observations and the frequency of each number. If the frequency distribution is *grouped*, then the data are assembled into cells, each cell representing a subset of the total range of the data. The frequency in each cell completes the grouped frequency distribution. Frequency distributions are often graphically displayed in histograms or stem-and-leaf plots (see above).

While histograms and stem-and-leaf plots show the frequency of specific values or groups of values, analysts often wish to examine the *cumulative frequency* of the data. The cumulative frequency refers to the total up to and including a particular value. In the case of grouped data, the cumulative frequency is computed as the total number of observations up to and including a cell boundary. Cumulative frequency distributions are often displayed on an *ogive*, as depicted in Figure VI.14.

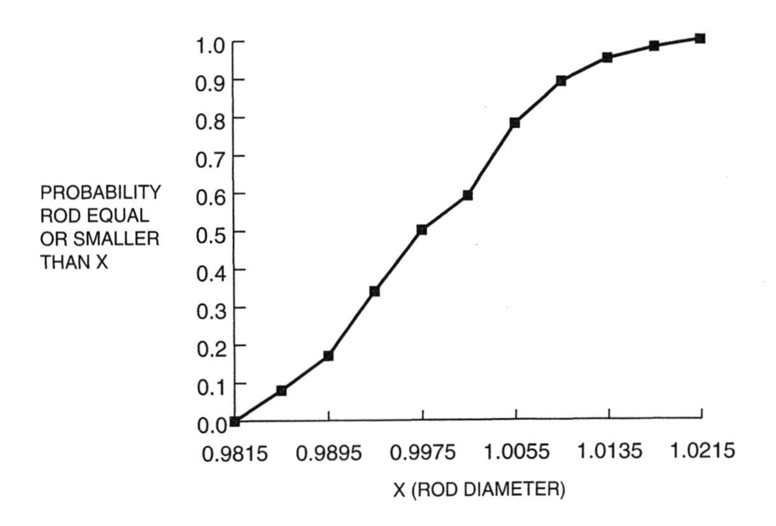

Figure VI.14. Ogive of rod diameter data.

SAMPLING DISTRIBUTIONS

In most engineering situations involving enumerative statistics, we deal with samples, not populations. In the previous section, sample data were used to construct an ogive and, earlier in this book, sample data were used to construct histograms, stem-and-leaf plots, boxplots, and to compute various statistics. We now consider the estimation of certain characteristics or parameters of the distribution from the data.

The empirical distribution assigns the probability $1/n$ to each X_i in the sample, thus the mean of this distribution is

$$\overline{X} = \sum_{i=1}^{n} X_i \frac{1}{n} \qquad \text{(VI.11)}$$

The symbol $\overline{X}$ is called "x bar." Since the empirical distribution is determined by a sample, $\overline{X}$ is simply called the *sample mean.*

Earlier in this chapter, the population variance was shown to be

$$\sigma^2 = E\left[(X-\mu)^2\right] = \sum_R (X-\mu)^2 f(x)$$

The equivalent variance of the empirical distribtion is given by

$$S^2 = \frac{1}{n}\sum_{i=1}^{n}\left(X_i-\overline{X}\right)^2 \qquad \text{(VI.12)}$$

However, it can be shown that $E[S^2] \neq \sigma^2$. In other words, this statistic is a *biased estimator* of the population parameter σ^2. In fact,

$$E(S^2) = \frac{n-1}{n}\sigma^2$$

Thus, to correct for the bias, it is necessary to multiply the value found in Equation VI.12 by the factor

$$\frac{n}{n-1}$$

This gives

$$S^2 = \frac{1}{n-1}\sum_{i=1}^{n}\left(X_i-\overline{X}\right)^2 \qquad \text{(VI.13)}$$

This equation for S^2 is commonly referred to as the *sample variance.* The unbiased *sample standard deviation* is given by

$$S = \sqrt{S^2} = \sqrt{\frac{\sum_{i=1}^{n}\left(X_i-\overline{X}\right)^2}{n-1}} \qquad \text{(VI.14)}$$

Another sampling statistic of special interest in quality engineering is the standard deviation of the sample average, also referred to as the *standard error of the mean* or simply the standard error. This statistic is given by

$$S_{\overline{X}} = \frac{S}{\sqrt{n}} \qquad \text{(VI.15)}$$

As can be seen, the standard error of the mean is inversely proportional to the square root of the sample size; i.e., the larger the sample size, the smaller is the standard deviation of the sample average. This relationship is shown in Figure VI.15. It can be seen that averages of $n=4$ have a distribution half as variable as the population from which the samples are drawn.

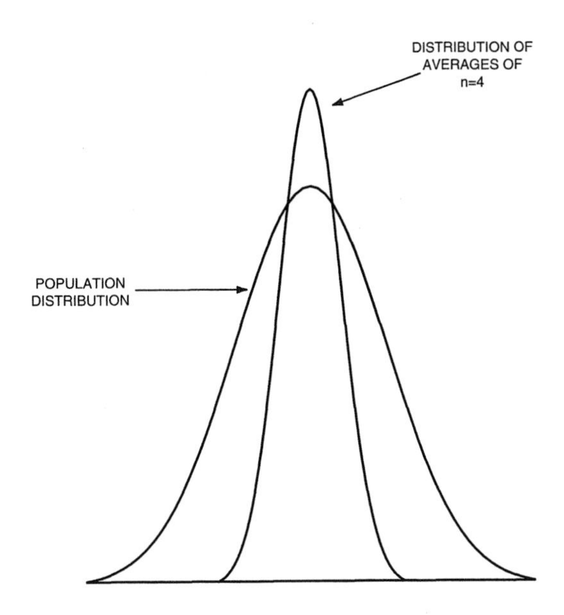

Figure VI.15. Effect of sample size on the standard error.

PROBABILITY DENSITY FUNCTIONS

Although probability density functions (pdfs) have been introduced and used previously, they have not been formally discussed. This will be done, briefly, in this section.

Discrete pdfs and cdfs

Let $f(x)$ be the pdf of a random variable of the discrete type defined on the space R of X. The pdf $f(x)$ is a real-valued function and satisfies the following properties.

1. $f(x)>0,\ x\in R$;

2. $\sum_{x\in R} f(x)=1$;

3. $P(X\in A)=\sum_{x\in A} f(x),$ where $A\subset R$.

For discrete data, $f(x)$ gives the probability of obtaining an observation exactly equal to x, i.e., $P(X=x)$. The cumulative distribution function (cdf) is denoted F(x) and it gives the probability of obtaining an observation equal to or less than x, i.e., $F(x)=P(X\leq x)$.

Example of discrete pdf

Let the random variable X of the discrete type have the pdf $f(x)$=x/10, x = 1, 2, 3, 4. Then, for example

$$\sum_{x \in R} f(x) = \frac{1+2+3+4}{10} = 1$$

$$F(X=1) = 1/10,$$

and

$$F(X \le 3) = \frac{1+2+3}{10} = \frac{3}{5}.$$

The cdf and the pdf of the example are plotted in Figure VI.16.

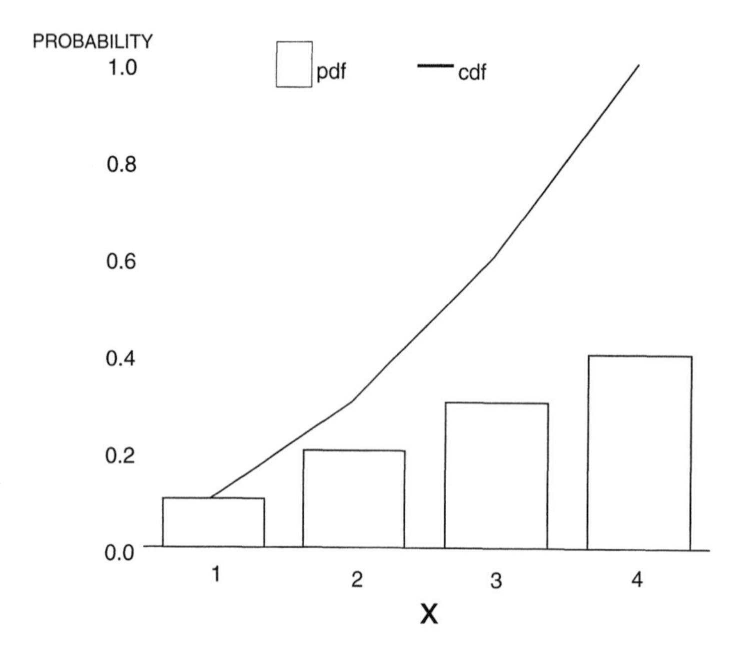

Figure VI.16. Plots of $f(x)$=x/10 and F(X).

Continuous pdfs and cdfs

Basically, the pdf and cdf concepts are the same for both continuous and discrete variables. The major difference is that calculus is required to obtain the cdf for continuous variables. One cannot merely add up a pdf for continuous data; instead, integration must be used. However, the equations for most continuous distributions commonly used in quality engineering are presented later in this section or in tabular form in the Appendix. This section presents a brief conceptual overview of pdfs and cdfs for continuous data. Although calculus is required for complete understanding, it is possible to grasp the ideas involved without a calculus background.

Continuous pdfs

Let $f(x)$ be the pdf of a random variable of the continuous type defined on the space R of X. The pdf $f(x)$ is a real-valued function and satisfies the following properties.

1. $f(x)>0$, $x \in R$;

2. $\int_{:\in R} f(x)dx = 1$;

3. $P(X \in A) = \int_A f(x)\mathrm{dx}$.

Discussion of continuous pdf properties

Property 1 states that the pdf can never be zero or negative for the region (space) over which it applies. For example, the normal distribution (see below) covers the region from negative infinity to positive infinity, thus, theoretically at least, it is possible to obtain *any* value for *any* normally distributed random variable. Of course, extreme values have a probability that is vanishingly small and, in practice, they never appear.

Property 2 states that the total area under the pdf curve is 1. For continuous random variables probability is measured as the area under the pdf curve. Since the pdf is a probability function, and P(S)=1 is a property of probability functions (see section A above), the area under the pdf curve must equal 1.

Property 3 states that the probability of some value (X) occurring within a subset (A) of the space covered by the pdf is found by integrating the pdf over the space covered by the subset. For example, if the amount of medication in a capsule is normally distributed with a mean of 40mg and a standard deviation of 0.1mg, then the probability of obtaining a capsule which has between 40.1mg and 40.2mg is found by integrating the normal pdf between 40.1mg and 40.2mg. This is equivalent to finding the area under the curve between these two values. In practice, these probabilities are usually obtained from tables.

Continuous cdfs

The cumulative distribution function (cdf) is denoted F(x) and it gives the probability of obtaining an observation equal to or less than x, i.e., $F(x)=P(X \leq x)$. For continuous data,

$$F(X \leq x) = \int_{-\infty}^{x} f(t)dt$$

In other words, the cdf for any value is found by integrating the pdf up to that value. This is equivalent to finding the area under the pdf curve up to the value of interest. In practice, these probabilities are found by using tables.

The probability that x lies between two values a and b can be found using

$$F(a \leq X \leq b) = \int_a^b f(x)dx$$

This merely states that the probability of obtaining a result between two values is the area under the pdf that lies between the two values. Since the cdf gives the area *less than* a particular value, the probability of an observation being between two values, *a* and *b,* can also be calculated as

$$F(a \leq X \leq b) = F(b) - F(a)$$

Several examples of finding probabilities for continuous distributions commonly encountered in quality engineering are shown later in this chapter.

Probability distributions for quality engineering

This section discusses the following probability distributions often used in quality engineering:

- Hypergeometric distribution
- Binomial distribution
- Poisson distribution
- Normal distribution
- Exponential distribution
- Chi-square distribution
- Student's t distribution
- F distribution

BINOMIAL, POISSON, HYPERGEOMETRIC, NORMAL, EXPONENTIAL DISTRIBUTIONS

Binomial distribution

Assume that a process is producing some proportion of non-conforming units, which we will call *p*. If we are basing p on a sample we find p by dividing the number of non-conforming units in the sample by the number of items sampled. The equation that will tell us the probability of getting x defectives in a sample of n units is shown by Equation VI.16.

$$P(x) = C_x^n p^x (1-p)^{n-x} \qquad \text{(VI.16)}$$

This equation is known as the *binomial probability distribution.* In addition to being useful as the exact distribution of non-conforming units for processes in continuous production, it is also an excellent approximation to the cumbersome hypergeometric probability distribution when the sample size is less than 10% of the lot size.

Example of applying the binomial probability distribution

A process is producing glass bottles on a continuous basis. Past history shows that 1% of the bottles have one or more flaws. If we draw a sample of 10 units from the process, what is the probability that there will be 0 non-conforming bottles?

Using the above information, n = 10, p = .01, and x = 0. Substituting these values into Equation VI.16 gives us

$$P(0) = C_0^{10} 0.01^0 (1-0.01)^{10-0} = 1 \times 1 \times 0.99^{10} = 0.904 = 90.4\%$$

Another way of interpreting the above example is that a sampling plan "inspect 10 units, accept the process if no non-conformances are found" has a 90.4% probability of accepting a process that is averaging 1% non-conforming units.

Poisson distribution

Another situation encountered often in quality control is that we are not just concerned with *units* that don't conform to requirements; instead we are concerned with the number of non-conformances themselves. For example, let's say we are trying to control the quality of a computer. A complete audit of the finished computer would almost certainly reveal some non-conformances, even though these non-conformances might be of minor importance (for example, a decal on the back panel might not be perfectly straight). If we tried to use the hypergeometric or binomial probability distributions to evaluate sampling plans for this situation, we would find they didn't work because our lot or process would be composed of 100% non-conforming units. Obviously, we are interested not in the units per se, but in the non-conformances themselves. In other cases, it isn't even possible to count sample units *per se.* For example, the number of accidents must be counted as occurrences. The correct probability distribution for evaluating counts of non-conformances is the *Poisson distribution.* The pdf is given in Equation VI.17.

$$P(x) = \frac{\mu^x e^{-\mu}}{x!} \qquad \text{(VI.17)}$$

In Equation VI.17, μ is the average number of non-conformances per unit, x is the number of non-conformances in the sample, and e is the constant approximately equal to 2.7182818. P(x) gives the probability of exactly x occurrences in the sample.

Example of applying the Poisson distribution

A production line is producing guided missiles. When each missile is completed, an audit is conducted by an Air Force representative and every non-conformance to requirements is noted. Even though any major non-conformance is cause for rejection, the prime contractor wants to control minor non-conformances as well. Such minor problems as blurred stencils, small burrs, etc., are recorded during the audit. Past history shows that on the average each missile has 3 minor non-conformances. What is the probability that the next missile will have 0 non-conformances?

We have $\mu = 3$, $x = 0$. Substituting these value into Equation VI.17 gives us

$$P(0) = \frac{3^0 e^{-3}}{0!} = \frac{1 \times 0.05}{1} = 0.05 = 5\%$$

In other words, 100% – 5% = 95% of the missiles will have at least one non-conformance.

The Poisson distribution, in addition to being the exact distribution for the number of non-conformances, is also a good approximation to the binomial distribution in certain cases. To use the Poisson approximation, you simply let $\mu = np$ in Equation VI.17. Juran (1980) recommends considering the Poisson approximation if the sample size is at least 16, the population size is at least 10 times the sample size, and the probability of occurrence p on each trial is less than 0.1. The major advantage of this approach is that it allows you to use the tables of the Poisson distribution, such as Table 9 in the Appendix. Also, the approach is useful for designing sampling plans. The subject of sampling plans is taken up in detail in section III.C.

Hypergeometric distribution

Assume we have received a lot of 12 parts from a distributor. We need the parts badly and are willing to accept the lot if it has 3 or fewer non-conforming parts. We decide to inspect only 4 parts since we can't spare the time to check every part. Checking the sample, we find 1 part that doesn't conform to the requirements. Should we reject the remainder of the lot?

This situation involves sampling without replacement. We draw a unit from the lot, inspect it, and draw another unit from the lot. Furthermore, the lot is quite small; the sample is 25% of the entire lot. The formula needed to compute probabilities for this procedure is known as the hypergeometric probability distribution; it is shown in Equation VI.18.

$$P(x) = \frac{C_{n-x}^{N-m} C_x^m}{C_n^N} \qquad \text{(VI.18)}$$

In the above equation, N is the lot size, m is the number of defectives in the lot, n is the sample size, x is the number of defectives in the sample, and P(x) is the probability of getting exactly x defectives in the sample. Note that the numerator term C_{n-x}^{N-m} gives the number of combinations of non-defectives while C_x^m is the number of combinations of defectives. Thus the numerator gives the total number of arrangements of samples from lots of size N with m defectives where the sample n contains exactly x defectives. The term C_n^N in the denominator is the total number of combinations of samples of size n from lots of size N, regardless of the number of defectives. Thus, the probability is a ratio of the likelihood of getting the result under the assumed conditions.

For our example, we must solve the above equation for x = 0 as well as x = 1, since we would also accept the lot if we had no defectives. The solution is shown as follows.

$$P(0) = \frac{C_{4-0}^{12-3} C_0^3}{C_4^{12}} = \frac{126 \times 1}{495} = 0.255$$

$$P(1) = \frac{C_{4-1}^{12-3} C_1^3}{C_4^{12}} = \frac{84 \times 3}{495} = \frac{252}{495} = 0.509$$

$$P(1 \text{ or less}) = \text{P}(0) + \text{P}(1).$$

Adding the two probabilities tells us the probability that our sampling plan will accept lots of 12 with 3 non-conforming units. The plan of inspecting 4 parts and accepting the lot if we have 0 or 1 non-conforming has a probability of .255 + .509 = .764, or 76.4%, of accepting this "bad" quality lot. This is the "consumer's risk" for this sampling plan. Such a high sampling risk would be unacceptable to most people. The evaluation of sampling plans is taken up in more detail in III.C

Normal distribution

The most common continuous distribution encountered in quality control work is, by far, the normal distribution. Sometimes the process itself produces an approximately normal distribution; other times a normal distribution can be obtained by performing a mathematical transformation on the data or by using averages. The probability density function for the normal distribution is given by Equation VI.19.

$$f(x) = \frac{1}{\sigma\sqrt{2\pi}} e^{-(x-\mu)^2/2\sigma^2} \qquad \text{(VI.19)}$$

If $f(x)$ is plotted versus x, the well-known "bell curve" results. The normal distribution is also known as the Gaussian distribution. An example is shown in Figure VI.17.

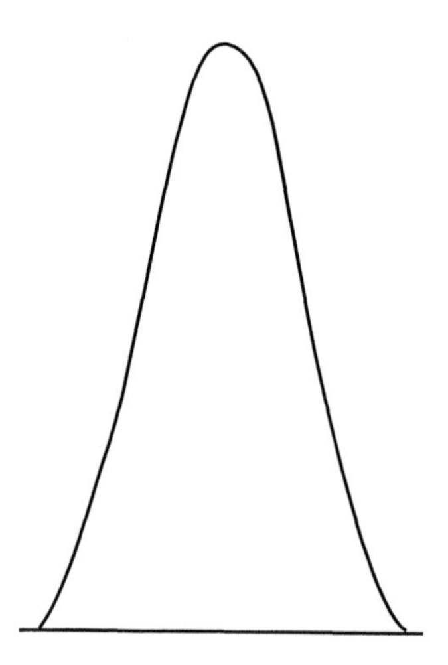

Figure VI.17. The normal/Gaussian curve.

In Equation VI.19, μ is the population average or mean and σ is the population standard deviation. These parameters have been discussed in section A of this chapter.

Example of calculating μ, σ^2 and σ

Find μ, σ^2 and σ for the following data:

i	x_i
1	17
2	23
3	5

Table VI.3 gives the equation for the population mean as:

$$\mu = \frac{1}{N}\sum_{i=1}^{N} x_i$$

To find the mean for our data we compute

$$\mu = \frac{1}{3}(17 + 23 + 5) = 15$$

The variance and standard deviation are both measures of dispersion or spread. The equations for the population variance σ^2 and standard deviation σ are given in Table VI.3.

$$\sigma^2 = \sum_{i=1}^{N} \frac{(x_i - \mu)^2}{N}$$

$$\sigma = \sqrt{\sigma^2}$$

Referring to the data above with a mean μ of 15, we compute σ^2 and σ as follows:

i	x_i	$x_i - \mu$	$(x_i - \mu)^2$
1	17	2	4
2	23	8	64
3	5	–10	100
			SUM 168

$$\sigma^2 = 168 / 3 = 56$$

$$\sigma = \sqrt{\sigma^2} = \sqrt{56} \approx 7.483$$

Usually we have only a sample and not the entire population. A population is the entire set of observations from which the sample, a subset, is drawn. Calculations for the sample mean, variance, and standard deviation were shown earlier in this chapter.

The areas under the normal curve can be found by integrating Equation VI.19, but, more commonly, tables are used. Table 4 in the Appendix gives areas under the normal curve. The table is indexed by using the Z transformation, which is

$$Z = \frac{x_{i-\mu}}{\sigma} \qquad \text{(VI.20)}$$

for population data, or

$$Z = \frac{x_{i-\bar{x}}}{s} \qquad \text{(VI.21)}$$

for sample data.

By using the Z transformation, we can convert any normal distribution into a normal distribution with a mean of 0 and a standard deviation of 1. Thus, we can use a single normal table to find probabilities.

Example of calculating Z

The normal distribution is very useful in predicting long-term process yields. Assume we have checked the breaking strength of a gold wire bonding process used in microcircuit production and we have found that the process average strength is 9# and the standard deviation is 4#. The process distribution is normal. If the engineering specification is 3# minimum, what percentage of the process will be below the low specification?

Since our data are a sample, we must compute Z using Equation VI.21.

$$Z = \frac{3-9}{4} = \frac{-6}{4} = -1.5$$

Figure VI.18 illustrates this situation.

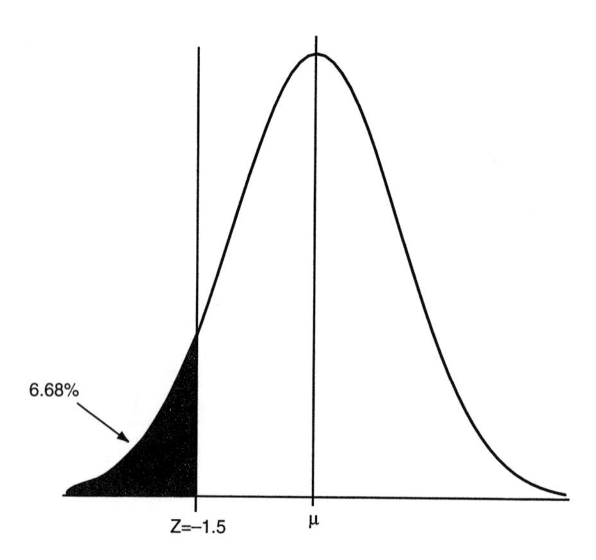

Figure VI.18. Illustration of using Z tables for normal areas.

Entering Table 4 in the Appendix for Z = –1.5, we find that 6.68% of the area is below this Z value. Thus 6.68% of our breaking strengths will be below our low specification limit of 3#. In quality control applications, we usually try to have the average at least 3 standard deviations away from the specification. To accomplish this, we would have to improve the process by either raising the average breaking strength or reducing the process standard deviation, or both.

Exponential distribution

Another distribution encountered often in quality control work is the exponential distribution. The exponential distribution is especially useful in analyzing reliability (see chapter IV). The equation for the probability density function of the exponential distribution is

$$f(x) = \frac{1}{\mu} e^{-x/\mu}, \ x \geq 0 \tag{VI.22}$$

Unlike the normal distribution, the shape of the exponential distribution is highly skewed and there is a much greater area below the mean than above it. In fact, over 63% of the exponential distribution falls below the mean. Figure VI.19 shows an exponential pdf.

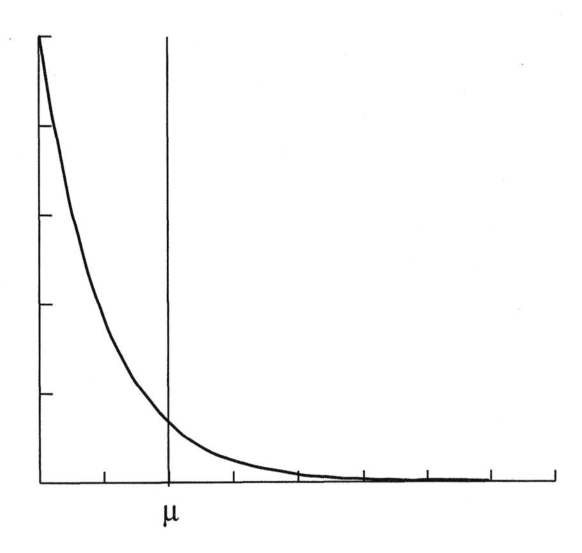

Figure VI.19. Exponential pdf curve.

Unlike the normal distribution, the exponential distribution has a closed form cumulative density function, i.e., there is an equation which gives the cumulative probabilities directly. Since the probabilities can be determined directly from the equation, no tables are necessary.

$$P(X \leq x) = 1 - e^{-x/\mu} \tag{VI.23}$$

Example of using the exponential cdf

A city water company averages 500 system leaks per year. What is the probability that the weekend crew, which works from 6 p.m. Friday to 6 a.m. Monday, will get no calls?

We have μ = 500 leaks per year, which we must convert to leaks per hour. There are 365 days of 24 hours each in a year, or 8760 hours. Thus, mean time between failures (MTBF) is 8760/500 = 17.52 hours. There are 60 hours between 6 p.m. Friday and 6 a.m. Monday. Thus x_i = 60. Using Equation VI.23 gives

$$P(X \leq 60) = 1 - e^{-60/17.52} = 0.967 = 96.7\%$$

Thus, the crew will get to loaf away 3.3% of the weekends.

CHI-SQUARE, STUDENT'S *t*, AND F DISTRIBUTIONS

These three distributions are used in quality engineering to test hypotheses, construct confidence intervals, and to compute control limits.

Chi-Square

Many characteristics encountered in quality engineering have normal or approximately normal distributions. It can be shown that in these instances the distribution of sample variances has the form (except for a constant) of a chi-square distribution, symbolized χ^2. Tables have been constructed giving abscissa values for selected ordinates of the cumulative χ^2 distribution. One such table is Table 6 in the Appendix.

The χ^2 distribution varies with the quantity v, which for our purposes is equal to the sample size minus 1. For each value of v there is a different χ^2 distribution. Equation VI.24 gives the pdf for the χ^2.

$$f(\chi^2)=\frac{e^{-\chi^2/2}\left(\chi^2\right)^{(v-2)/2}}{2^{v/2}\left(\frac{v-2}{2}\right)!} \qquad \text{(VI.24)}$$

Figure VI.20 shows the pdf for v=4.

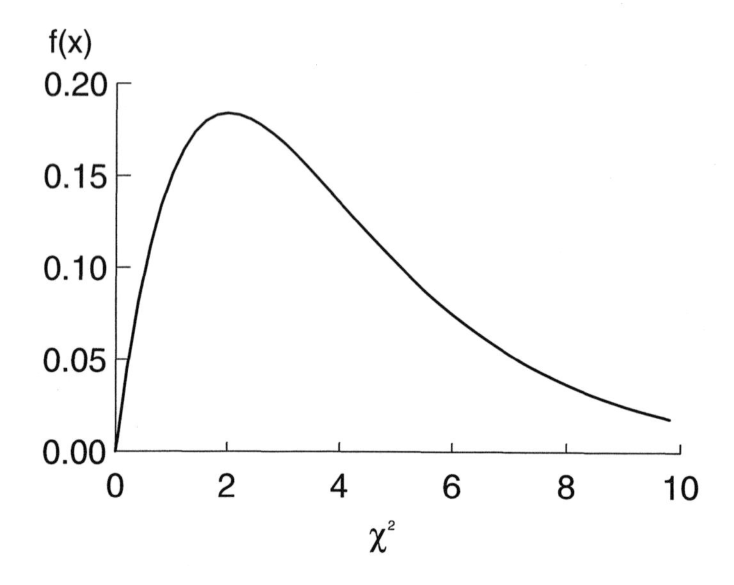

Figure VI.20. χ^2 pdf for v=4.

Example of using Chi-Square

The use of χ^2 is illustrated in this example to find the probability that the variance of a sample of n items from a specified normal universe will equal or exceed a given value s^2; we compute $\chi^2 = (n-1)s^2/\sigma^2$. Now, let's suppose that we sample n=10 items from a process with σ^2=25 and wish to determine the probability that the sample variance will exceed 50. Then

$$\frac{(n-1)s^2}{\sigma^2} = \frac{9(50)}{25} = 18$$

We enter the χ^2 Appendix Table 6 at the line for v=10–1=9 and note that 18 falls between the columns for the percentage points of 0.025 and 0.05. Thus, the probability of getting a sample variance in excess of 50 is about 3%.

It is also possible to determine the sample variance that would be exceeded only a stated percentage of the time. For example, we might want to be alerted when the sample variance exceeded a value that should occur only once in 100 times. Then we set up the χ^2 equation, find the critical value from Table 6 in the Appendix, and solve for the sample variance. Using the same values as above, the value of s^2 that would be exceeded only one in 100 times is found as follows:

$$\frac{9s^2}{\sigma^2} = \frac{9s^2}{25} = 21.7 \Rightarrow s^2 = \frac{21.7 \times 25}{9} = 60.278$$

In other words, the variance of samples of size 10, taken from this process, should be less than 60.278, 99% of the time.

Student's t distribution

The t statistic is commonly used to test hypotheses regarding means, regression coefficients and a wide variety of other statistics used in quality engineering. "Student" was the pseudonym of W.S. Gosset, whose need to quantify the results of small scale experiments motivated him to develop and tabulate the probability integral of the ratio which is now known as the t statistic and is shown in Equation VI.25.

$$t = \frac{\mu - \overline{X}}{s/\sqrt{n}} \qquad \text{(VI.25)}$$

In EquationVI.25, the denominator is the standard deviation of the sample mean. Percentage points of the corresponding distribution function of t may be found in Table 5 in the Appendix. There is a t distribution for each sample size of n>1. As the sample size increases, the t distribution approaches the shape of the normal distribution, as shown in Figure VI.21.

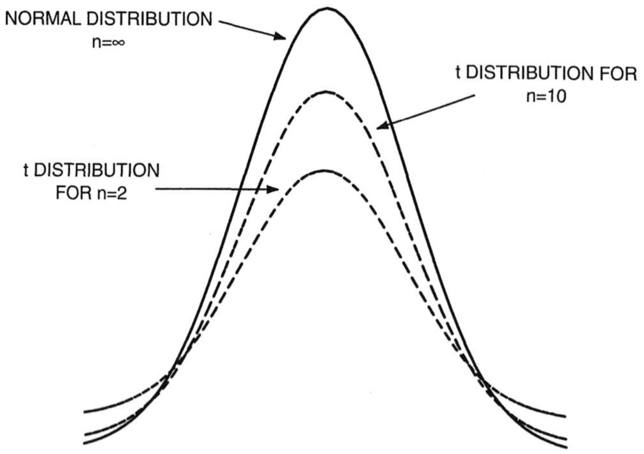

NORMAL DISTRIBUTION
n=∞
t DISTRIBUTION FOR
n=10
t DISTRIBUTION
FOR n=2

Figure VI.21. Student's *t* distributions.

One of the simplest (and most common) applications of the Student's *t* test involves using a sample from a normal population with mean μ and variance σ^2. This is demonstrated in VI.D below.

F distribution

Suppose we have two random samples drawn from a normal population. Let s_1^2 be the variance of the first sample and s_2^2 be the variance of the second sample. The two samples need not have the same sample size. The statistic F given by

$$F = \frac{s_1^2}{s_2^2} \tag{VI.26}$$

has a sampling distribution called the *F distribution*. There are two sample variances involved and two sets of degrees of freedom, n_1-1 in the numerator and n_2-1 in the denominator. The Appendix includes tables for 1% and 5% percentage points for the F distribution. The percentages refer to the areas to the right of the values given in the tables. Figure VI.22 illustrates two *F* distributions.

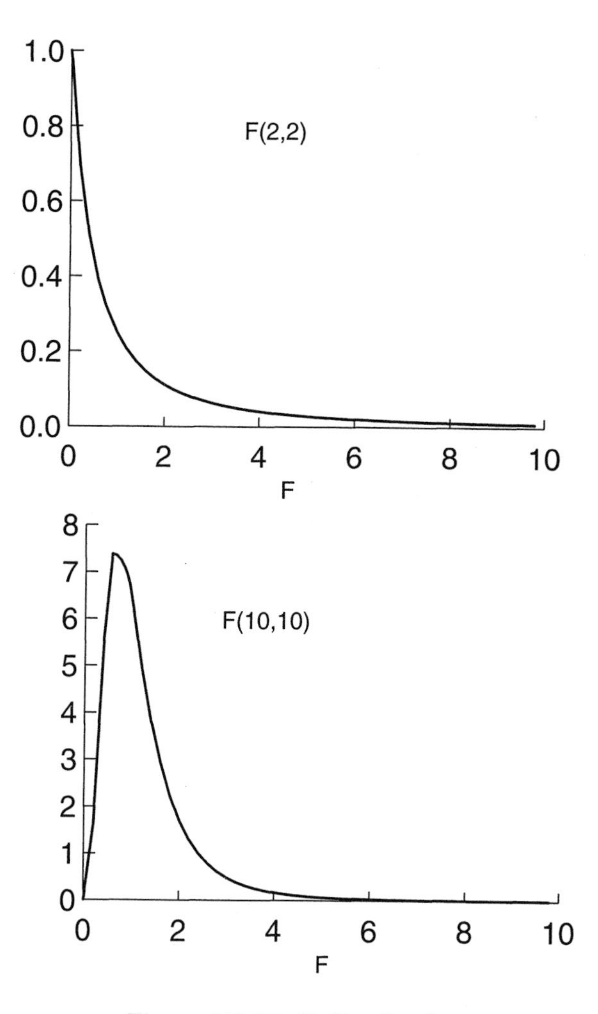

Figure VI.22. F distributions.

VI.D STATISTICAL DECISION-MAKING

All statements made in this section are valid only for stable processes, i.e., processes in statistical control. The statistical methods described in this section are enumerative. Although most applications of quality engineering are analytic, there are times when enumerative statistics prove useful. In reading this material, the engineer should keep in mind the fact that analytic methods should also be used to identify the underlying process dynamics and to control and improve the processes involved. The subject of statistical inference is large and it is covered in many different books on introductory statistics. In this book we review that part of the subject matter of particular interest in quality engineering.

POINT AND INTERVAL ESTIMATION

So far, we have introduced a number of important statistics including the sample mean, the sample standard deviation, and the sample variance. These sample statistics are called *point estimators* because they are single values used to represent population parameters. It is also possible to construct an interval about the statistics that has a predetermined probability of including the true population parameter. This interval is called a *confidence interval.* Interval estimation is an alternative to point estimation that gives us a better idea of the magnitude of the sampling error. Confidence intervals can be either one-sided or two-sided. A one-sided confidence interval places an upper or lower bound on the value of a parameter with a specified level of confidence. A two-sided confidence interval places both upper and lower bounds.

In almost all practical applications of enumerative statistics, including quality control applications, we make *inferences* about *populations* based on data from *samples.* In this chapter, we have talked about sample averages and standard deviations; we have even used these numbers to make statements about future performance, such as long term yields or potential failures. A problem arises that is of considerable practical importance: any estimate that is based on a sample has some amount of sampling error. This is true even though the sample estimates are the "best estimates" in the sense that they are (usually) unbiased estimators of the population parameters.

Estimates of the mean

For random samples with replacement, the sampling distribution of $\overline{X}$ has a mean μ and a standard deviation equal to $\sigma/\sqrt{n}$. For large samples the sampling distribution of $\overline{X}$ is approximately normal and normal tables can be used to find the probability that a sample mean will be within a given distance of μ. For example, in 95% of the samples we will observe a mean within $\pm 1.96\,\sigma/\sqrt{n}$ of μ. In other words, in 95% of the samples the interval from $\overline{X} - 1.96 \sigma/\sqrt{n}$ to $\overline{X} + 1.96 \sigma/\sqrt{n}$ will include μ. This interval is called a "95% confidence interval for estimating μ." It is usually shown using inequality symbols:

$$\overline{X} - 1.96 \sigma/\sqrt{n} < \mu < \overline{X} + 1.96 \sigma/\sqrt{n}$$

The factor 1.96 is the Z value obtained from the normal Table 4 in the Appendix. It corresponds to the Z value beyond which 2.5% of the population lie. Since the normal distribution is symmetric, 2.5% of the distribution lies above Z and 2.5% below –Z. The notation commonly used to denote Z values for confidence interval construction or hypothesis testing is $Z_{\alpha/2}$ where $100(1-\alpha)$ is the desired confidence level in per cent; e.g., if we want 95% confidence, $\alpha=0.05$, $100(1-\alpha)=95\%$, and $Z_{0.025}=1.96$. In hypothesis testing the value of α is known as the *significance level.*

Example : estimating μ when σ is known

Suppose that σ is known to be 2.8. Assume that we collect a sample of n=16 and compute $\bar{X} = 15.7$. Using the above equation we find the 95% confidence interval for μ as follows:

$$\bar{X} - 1.96\sigma/\sqrt{n} < \mu < \bar{X} + 1.96\sigma/\sqrt{n}$$

$$15.7 - 1.96\left(2.8/\sqrt{16}\right) < \mu < 15.7 + 1.96\left(2.8/\sqrt{16}\right)$$

$$14.33 < \mu < 17.07$$

There is a 95% *level of confidence* associated with this interval. The numbers 14.33 and 17.07 are sometimes referred to as the *confidence limits.*

Note that this is a two-sided confidence interval. There is a 2.5% probability that 17.07 is lower than μ and a 2.5% probability that 14.33 is greater than μ. If we were only interested in, say, the probability that μ were greater than 14.33, then the one-sided confidence interval would be $\mu > 14.33$ and the one-sided confidence level would be 97.5%.

Example: estimating μ when σ is unknown

When σ is not known and we wish to replace σ with s in calculating confidence intervals for μ, we must replace $Z_{\alpha/2}$ with $t_{\alpha/2}$ and obtain the percentiles from tables for Student's t distribution instead of the normal tables. Let's revisit the example above and assume that instead of knowing σ, it was estimated from the sample; that is, based on the sample of n=16, we computed s=2.8 and $\bar{X} = 15.7$. Then the 95% confidence interval becomes:

$$\bar{X} - 2.131\, s/\sqrt{n} < \mu < \bar{X} + 2.131\, s/\sqrt{n}$$

$$15.7 - 2.131\left(2.8/\sqrt{16}\right) < \mu < 15.7 + 2.131\left(2.8/\sqrt{16}\right)$$

$$14.21 < \mu < 17.19$$

It can be seen that this interval is wider than the one obtained for known σ. The $t_{\alpha/2}$ value found for 15 df is 2.131 (see Table 5 in the Appendix), which is greater than $Z_{\alpha/2}$ = 1.96 above.

TOLERANCE AND CONFIDENCE INTERVALS, SIGNIFICANCE LEVEL

Tolerance intervals

We have found that confidence limits may be determined so that the interval between these limits will cover a population parameter with a certain confidence, that is, a certain proportion of the time. Sometimes it is desirable to obtain an interval which will cover a fixed portion of the population distribution with a specified confidence. These intervals are called *tolerance intervals*, and the end points of such intervals are called *tolerance limits*. For example, a manufacturer may wish to estimate what proportion of product will have dimensions that meet the engineering requirement. In quality engineering, tolerance intervals are typically of the form $\overline{X} \pm Ks$, where K is determined, so that the interval will cover a proportion P of the population with confidence γ. Confidence limits for μ are also of the form $\overline{X} \pm Ks$. However, we determine k so that the confidence interval would cover the population mean μ a certain proportion of the time. It is obvious that the interval must be longer to cover a large portion of the distribution than to cover just the single value μ. Table 10 in the Appendix gives K for P = 0.75, 0.90, 0.95, 0.99, 0.999 and γ = 0.75, 0.90, 0.95, 0.99 and for many different sample sizes n.

Example of calculating a tolerance interval

Assume that a sample of n=20 from a stable process produced the following results: $\overline{X} = 20$, $s = 1.5$. We can estimate that the interval $\overline{X} \pm Ks = 20 \pm 3.615(1.5) = 20 \pm 5.4225$, or the interval from 14.5775 to 25.4225 will contain 99% of the population with confidence 95%. The K values in the table assume normally distributed populations.

HYPOTHESIS TESTING/TYPE I AND TYPE II ERRORS

Hypothesis testing

Statistical inference generally involves four steps:

1. Formulating a hypothesis about the population or "state of nature,"
2. Collecting a sample of observations from the population,
3. Calculating statistics based on the sample,
4. Either accepting or rejecting the hypothesis based on a pre-determined acceptance criterion.

There are two types of error associated with statistical inference:

Type I error (α error)—The probability that a hypothesis that is actually true will be rejected. The value of α is known as the significance level of the test.

Type II error (ß error)—The probability that a hypothesis that is actually false will be accepted.

Type II errors are often plotted in what is known as an operating characteristics curve. Operating characteristics curves will be used extensively in subsequent chapters of this book in evaluating the properties of various statistical quality control techniques.

Confidence intervals are usually constructed as part of a *statistical test of hypotheses.* The hypothesis test is designed to help us make an inference about the true population value at a desired level of confidence. We will look at a few examples of how hypothesis testing can be used in quality control applications.

Example: hypothesis test of sample mean

Experiment: The nominal specification for filling a bottle with a test chemical is 30 cc's. The plan is to draw a sample of n=25 units from a stable process and, using the sample mean and standard deviation, construct a two-sided confidence interval (an interval that extends on either side of the sample average) that has a 95% probability of including the true population mean. If the interval includes 30, conclude that the lot mean is 30, otherwise conclude that the lot mean is not 30.

Result: A sample of 25 bottles was measured and the following statistics computed

$$\overline{X} = 28\ cc$$
$$s = 6\ cc$$

The appropriate test statistic is t, given by the formula

$$t = \frac{\overline{X} - \mu}{s/\sqrt{n}} = \frac{28 - 30}{6/\sqrt{25}} = -1.67$$

Table 5 in the Appendix gives values for the t statistic at various degrees of freedom. There are n–1 degrees of freedom. For our example we need the $t_{.975}$ column and the row for 24 df. This gives a t value of 2.064. Since the absolute value of this t value is greater than our test statistic, we fail to reject the hypothesis that the lot mean is 30 cc's. Using statistical notation this is shown as:

H_0: μ = 30 cc's (the *null hypothesis*)

H_1: μ is not equal to 30 cc's (the *alternate hypothesis*)

α = .05 (*type I error or level of significance*)

Critical region: $-2.064 \leq t_0 \leq +2.064$

Test statistic: t = –1.67.

Since t lies inside the critical region, fail to reject H_0, and accept the hypothesis that the lot mean is 30cc for the data at hand.

Example: hypothesis test of two sample variances

The variance of machine X's output, based on a sample of n = 25 taken from a stable process, is 100. Machine Y's variance, based on a sample of 10, is 50. The manufacturing representative from the supplier of machine X contends that the result is a mere "statistical fluke." Assuming that a "statistical fluke" is something that has less than 1 chance in 100, test the hypothesis that both variances are actually equal.

The test statistic used to test for equality of two sample variances is the F statistic, which, for this example, is given by the equation

$$F = \frac{s_1^2}{s_2^2} = \frac{100}{50} = 2\text{, numerator df} = 24\text{, denominator df} = 9$$

Using Table 7 in the Appendix for $F_{.99}$ we find that for 24 df in the numerator and 9 df in the denominator F = 4.73. Based on this we conclude that the manufacturer of machine X could be right, the result could be a statistical fluke. This example demonstrates the volatile nature of the sampling error of sample variances and standard deviations.

Example: hypothesis test of a standard deviation compared to a standard value

A machine is supposed to produce parts in the range of 0.500 inches plus or minus 0.006 inches. Based on this, your statistician computes that the absolute worst standard deviation tolerable is 0.002 inches. In looking over your capability charts you find that the best machine in the shop has a standard deviation of 0.0022, based on a sample of 25 units. In discussing the situation with the statistician and management, it is agreed that the machine will be used if a one-sided 95% confidence interval on sigma includes 0.002.

The correct statistic for comparing a sample standard deviation with a standard value is the chi-square statistic. For our data we have s=0.0022, n=25, and σ_0=0.002. The χ^2 statistic has n–1 = 24 degrees of freedom. Thus,

$$\chi^2 = \frac{(n-1)s^2}{\sigma^2} = \frac{24 \times (0.0022)^2}{(0.002)^2} = 29.04$$

Appendix Table 6 gives, in the 0.05 column (since we are constructing a one-sided confidence interval) and the df = 24 row, the critical value χ^2 = 36.42. Since our computed value of χ^2 is less than 36.42, we use the machine. The reader should recognize that all of these exercises involved a number of assumptions, e.g., that we "know" that the best machine has a standard deviation of 0.0022. In reality, this knowledge must be confirmed by a stable control chart.

Resampling (bootstrapping)

A number of criticisms have been raised regarding the methods used for estimation and hypothesis testing:

- They are not intuitive.
- They are based on strong assumptions (e.g., normality) that are often not met in practice.
- They are difficult to learn and to apply.
- They are error-prone.

In recent years a new method of performing these analyses has been developed. It is known as resampling or bootstrapping. The new methods are conceptually quite simple: using the data from a sample, calculate the statistic of interest repeatedly and examine the distribution of the statistic. For example, say you obtained a sample of n=25 measurements from a lot and you wished to determine a confidence interval on the statistic C_{pk}. Using resampling, you would tell the computer to select a sample of n=25 *from the sample results*, compute C_{pk}, and repeat the process many times, say 10,000 times. You would then determine whatever percentage point value you wished by simply looking at the results. The samples would be taken "with replacement;" i.e., a particular value from the original sample might appear several times (or not at all) in a resample.

Resampling has many advantages, especially in the era of easily available, low-cost computer power. Spreadsheets can be programmed to resample and calculate the statistics of interest. Compared with traditional statistical methods, resampling is easier for most people to understand. It works without strong assumptions and it is simple. Resampling doesn't impose as much baggage between the engineering problem and the statistical result as conventional methods. It can also be used for more advanced problems, such as modeling, design of experiments, etc.

For a discussion of the theory behind resampling, see Efron (1982), Chernick (1999), or Good (2000). For a presentation of numerous examples using a resampling computer program see Simon (1992).

PAIRED COMPARISONS*

Often an experiment is, or can be, designed so that the observations are taken in pairs. The two units of a pair are chosen in advance so as to be as nearly alike as possible in all respects other than the characteristic to be measured; then one member of each pair is assigned at random to treatment A, and the other to treatment B. For instance, the experimenter may wish to compare the effects of two different treatments on a particular type of device, material, or process. The word "treatments" here is to be understood in a broad sense; the two "treatments" may be different operators, different environmental conditions to which a material may be exposed, merely two different methods of measuring one

*From Mary Gibbons Natrella, *Experimental Statistics, Handbook 91*. United States Department of Commerce, National Bureau of Standards, 1963.

of the properties, or two different laboratories in an interlaboratory test of a particular process of measurement or manufacture. Since the comparison of the two treatments is made within pairs, two advantages result from such pairing. First, the effect of extraneous variation is reduced and there is consequent increase in the precision of the comparison and in its sensitivity to real differences between the treatments with respect to the measured characteristic. Second, the test may be carried out under a wide range of conditions representative of actual use without sacrifice of sensitivity and precision, thereby assuring wider applicability of any conclusions reached.

Example: capacity of batteries

The data below are measurements of the capacity (in ampere hours) of paired batteries, one from each of two different manufacturers.

A	B	DIFFERENCE
146	141	5
141	143	–2
135	139	–4
142	139	3
140	140	0
143	141	2
138	138	0
137	140	–3
142	142	0
136	138	–2
Avg. Diff. –0.1		Sigma Diff. 2.807

The data from such paired tests are commonly evaluated using the t-test. The results are shown below:

t-Test: paired two sample for means

	A	B
Mean	140	140.1
Variance	12	2.766667
Observations	10	10
Pearson Correlation	0.597792	
Hypothesized Mean Difference	0	
df	9	
t Stat	–0.11267	
P(T<=t) one-tail	0.456384	
t Critical one-tail	1.833114	
P(T<=t) two-tail	0.912767	
t Critical two-tail	2.262159	

The t statistic for the mean difference (Δ) is

$$t = \frac{\Delta}{S_{difference} / \sqrt{n-1}}$$

In the table above, as with most statistical software programs, Microsoft Excel automatically determines the probability associated with the t statistic, in this case 0.912767 for a two-tailed test. Since this value exceeds our level of significance of 0.05, we conclude that the observed difference is not statistically significant. The same result could also have been obtained using tables of the t-distribution for $t = -0.11$ at 9 degrees of freedom.

GOODNESS OF FIT TESTS

A common assumption of many statistical tests is that the sample is taken from a population with a known distribution. For example, the Confidence Intervals, t-tests, and F-tests described earlier all require normal (or approximately normal) distributions of the samples. Goodness of Fit tests provide a means to determine whether samples meet the general conditions of any known distribution.

It is not uncommon for students to learn graphical techniques, known as *probability plotting*, to investigate distributional assumptions. In this method, the sample data is plotted on special graph paper that is scaled according to the assumed distribution. If the data meets the conditions of the assumed distribution, then the data will graph as a straight line. While this technique is somewhat simple, it lacks an objective test: how much of a departure from the straight line is permitted? As such, these graphical techniques provide limited practical value by themselves.

Perhaps the best known Goodness of Fit test is the Chi-Square test. It can be used to test the distribution of any known distribution (Duncan, 1986). Montgomery (2002) provides a number of examples testing both discrete and continuous distributions.

The Chi-Square test works as follows, using a set of random sample of n data points from the population:

1. *Determine the number of intervals.* A commonly-used method employs the square root of the number of data values.
2. *Determine the expected frequency within each interval (E_i).* Interval width may vary, and is generally defined to achieve equal expected frequencies across all intervals. Duncan (1986) recommends that if the expected frequency of a class interval is less than five, then the interval should be combined with a neighboring interval (and the total number of intervals reduced by one). Montgomery (2002) notes minimum frequencies of three to five are commonly used.
3. *Determine the width of each interval, and the interval boundaries, based on the assumed probability distribution.*
4. *Count the number of observed data points (O_i) within each interval (i = 1 to k).*

5. *Compute the chi-square test statistic*

$$X_0^2 = \sum_{i=1}^{k} \frac{(O_i - E_i)^2}{E_i}$$

6. *If H_0 is true (i.e. the population follows the assumed distribution), then the calculated chi-square values will tend to follow a chi-square distribution with k–p–1 degrees of freedom, where p is the number of parameters estimated by the sample statistics. Reject the null hypothesis H_0 if the calculated value exceeds the test statistic:*

$$X_0^2 > X_{\alpha,k-p-1}^2$$

Let's look at an example of testing for normality using the data shown in Table VI.4.

Table VI.4. Example data.

9.6	9.9	10.1	9.8	9.9
9.8	9.8	10.1	9.8	9.9
9.8	9.9	10	9.6	10.2
9.7	9.4	9.6	10	9.8
9.9	10.1	10.4	9	10.7
10.2	10.6	9.8	9.5	9.4
10	9.6	10.1	10.3	9.4
9.5	10.6	9.9	10.9	10.3
9.7	10.4	10.6	9.8	10.4
9.7	9.3	9.9	9.9	9.3
9.7	9.8	9.7	9.7	9.8

1. *Determine the number of intervals:*

$$X^2 = \sum^{k} (O_i - E_i)^2$$

2. *Determine the expected frequency within each interval (E_i):*
   - For 8 intervals of equal frequency, the expected percentage in each interval: = 100% ÷ 8 = 12.5% = 0.125
   - For 60 data values, the expected number of data points in each interval : $E_i = 0.125 * 60 = 7.5$, for i=1 to 8

3. *Determine the width of each interval, and the interval boundaries, based on the assumed probability distribution:*
   - For the Normal distribution, we can use the Normal Distribution Table (Appendix 4) to define the z-values corresponding to the minimum and maximum values of each interval, as shown in Table VI.5. For example, the boundaries of cell 5 are defined at the .500 and the .625 percentiles, or z = 0 and z = 0.31864. These are converted to actual units (Table VI.6) using the calculated mean and standard deviation of the data as follows:

$$z = \frac{x - \bar{x}}{\sigma} \longrightarrow x = (z * \sigma) + \bar{x}$$

Table VI.5. Cell boundaries (standard units).

CELL	PERCENTILE AT LOWER BOUNDARY	Z-MIN	Z-MAX
1	0	$-\infty$	–1.15035
2	0.125	–1.15035	–0.67449
3	0.25	–0.67449	–0.31864
4	0.375	–0.31864	0
5	0.5	0	0.31864
6	0.625	0.31864	0.67449
7	0.75	0.67449	1.15035
8	0.875	1.15035	$+\infty$

4. *Count the number of observed data points (O_i) within each interval (i = 1 to k).*

Table VI.6. Cell boundaries (real units).

CELL	LOWER CELL BOUNDARY	UPPER CELL BOUNDARY	OBSERVED COUNT O_i
1		9.4183013	9
2	9.41830125	9.6079113	6
3	9.60791127	9.7497024	6
4	9.74970236	9.8766667	10
5	9.87666667	10.003631	11
6	10.003631	10.145422	5
7	10.1454221	10.335032	5
8	10.3350321		8

5. *Compute the test statistic:*

$$X_0^2 = \sum_{i=1}^{k} \frac{(O_i - E_i)^2}{E_i}$$

$$X_0^2 = \frac{(9-7.5)^2 + (6-7.5)^2 + (6-7.5)^2 + (10-7.5)^2 + (11-7.5)^2 + (5-7.5)^2 + (5-7.5)^2 + (8-7.5)^2}{7.5}$$

$$X_0^2 = 5.067$$

6. *If H_0 is true (i.e. the population follows the assumed distribution), then the calculated chi-square values will tend to follow a chi-square distribution with k–p–1 degrees of freedom, where p is the number of parameters estimated by the sample statistics. Reject the null hypothesis H_0 if the calculated value exceeds the test statistic:*

$$X_0^2 > X_{\alpha, k-p-1}^2$$

- The Normal distribution requires two parameters: the sample mean and the sample standard deviation. Thus, p = 2 and with k = 8 intervals the degrees of freedom used in the chi-square tables is ν = 8–2–1 = 5.
- At α = 0.5, the critical value $\chi^2_{0.5,\,5}$ = 11.070 (from Appendix 6).
- Since the calculated value is less than the critical value, we fail to reject the null hypothesis that the population is normally distributed. In other words, we cannot disprove that the Normal distribution is appropriate.

While the Chi-square test is perhaps the most frequently referenced goodness of fit technique in general application statistical textbooks, its popularity in academic circles may indeed be due to its simplicity: authors can easily describe and demonstrate the test without the need for computer software. In practice, however, data is almost exclusively analyzed using computer software, allowing the use of more sophisticated techniques which are better suited for most applications, particularly in testing continuous distributions (Massey, 1951). Two of these techniques are the Kolmogorov-Smirnov test (often referred to simply as the K-S test) and the Anderson-Darling test. Each of these techniques may be used to compare the fit of assumed distributions to sample data.

In both the K-S and the Anderson-Darling test, the null hypothesis is that the data follows the specified distribution. Most statistical software reports a p-value, where the null hypothesis H_0 is rejected (i.e. the distribution is a poor fit) if the p-value is less than a predetermined alpha value (typically 0.05 or 0.10). Distributions can also be compared: the distribution with the larger p-value is the more likely distributions.

"The K-S criterion is based on the expectation that there is likely to be a difference between a discrete distribution generated from data and the continuous distribution from which it was drawn, caused by step difference and random error. As n increases, the size of the difference is expected to decrease. If the measured maximum difference is smaller than that expected, then the probability that the distributions are the same is high.

Note that the K-S criterion is very demanding as n becomes large, because the K-S criterion is scaled by the square root of n, reflecting an expected decrease in the step size error. The random error and outliers then dominate, with outliers having a strong effect on the reported value for alpha (because K-S is a measure of maximum deviation)" (Quality America, 2002).

For the example data provided above, Minitab reports an Anderson-Darling value of 0.150. Quality America's SPC-PC IV software reports a K-S value of 0.167. Using an alpha value of 0.5, each software fails to reject the null hypothesis that the data is drawn from Normally-distributed population. That is, we can't prove the population is non-normal.

ONE-FACTOR ANALYSIS OF VARIANCE (ANOVA)

Example of a one-way ANOVA

The following example will be used to illustrate the interpretation of a single factor analysis of variance. With the widespread availability of computers, few people actually perform such complex calculations by hand. The analysis below was performed using Microsoft Excel. Commonly used statistical methods such as regression and ANOVA are included in most high-end spreadsheets.

The coded results in Table VI.7 were obtained from a single factor, completely randomized experiment, in which the production outputs of three machines (A, B, and C) were to be compared.

Table VI.7. Experimental raw data (coded).

A	B	C
4	2	–3
8	0	1
5	1	–2
7	2	–1
6	4	0

An ANOVA of these results produced the results shown in Table VI.8:

Table VI.8. Results of the analysis.

ANOVA: SINGLE FACTOR						
SUMMARY						
Groups	*Count*	*Sum*	*Average*	*Variance*		
A	5	30.000	6.000	2.500		
B	5	9.000	1.800	2.200		
C	5	–5.000	–1.000	2.500		
ANOVA						
Source of variation	*SS*	*df*	*MS*	*F*	*P-value*	*F crit*
Between groups	124.133	2	62.067	25.861	0.000	3.885
Within groups	28.800	12	2.400			
Total	152.933	14				

The first part of Table VI.8 shows descriptive statistics for the data; the engineer should always look carefully at these easily understood results to check for obvious errors. The results show that the means vary from a low of –1 for machine *C* to a high of 6 for machine A.

ANOVA procedure

ANOVA proceeds as follows:

1. State the null and alternative hypotheses. The ANOVA table tests the hypothesis H_0: All means are equal versus H_a: At least two of the means are different.
2. Choose the level of significance. For this analysis a significance level α=0.05 was selected.
3. Compute the *F* statistic, the ratio of the mean square between groups to the mean square within groups.
4. Assuming that the observations are random samples from normally distributed populations with equal variances, and that the hypothesis is true, the critical value of F is found in Table 7 or 8 in the Appendix. The numerator will have the degrees of freedom shown in the *df* column for the Between Groups row. The denominator will have the degrees of freedom shown in the *df* column for the Within Groups row.
5. If the computed $F > F_{1-\alpha}$ then reject the null hypothesis and conclude the alternate hypothesis. Otherwise fail to reject the null hypothesis.

The ANOVA table shows that for these data F computed is 62.067/2.4=25.861 and F critical at α=0.05 with numerator df=2 and denominator df=12 is 3.885.* Since 25.861>3.885 the null hypothesis is rejected and the conclusion is made that the machines produce different results. Note that all that is known is that at least the two extreme machines (A and C) are different. The ANOVA does *not* tell if A and B or B and C are significantly different. There are methods which can make this determination, such as *contrasts*. The reader is referred to a text on design of experiments, e.g., Montgomery (2000) for additional information.

Performing ANOVA manually

On rare occasions (such as taking a CQE exam), the engineer may find that computers are not available and the analysis must be performed "by hand." The analysis is illustrated below.

		Total	N	Sum of Squares
Treatment A	4, 8, 5, 7,6	30	5	190
Treatment B	2, 0, 1, 2, 4	9	5	25
Treatment C	–3, 1, –2, –1, 0	–5	5	15
	Totals	34	15	230

$$\text{Total sum of squares} = 230 - \frac{(34)^2}{15} = 152.933$$

$$\text{Treatment sum of squares} = \frac{(30)^2}{5} + \frac{(9)^2}{5} + \frac{(-5)^2}{5} - \frac{(34)^2}{15} = 124.133$$

$$\begin{aligned}&\text{Error sum of squares}\\&= \text{Total sum of squares - Treatment sum of squares}\\&= 152.933 - 124.133 = 28.8\end{aligned}$$

These values are placed in the Sum of Squares (SS) column in the ANOVA table (Table VI.8). The remainder of the ANOVA table is obtained through simple division.

*Referring to the critical value is actually unnecessary; the P-value of 0.000 indicates that the probability of getting an F value as large as that computed is less than 1 in 1,000.

CHI-SQUARE CONTINGENCY TABLES

Making comparisons using chi-square tests

In quality engineering, there are many instances when the engineer wants to compare the percentage of items distributed among several categories. The categories might include operators, methods, materials, or any other grouping of interest. From each of the groups a sample is taken, evaluated, and placed into one of several categories (e.g., high quality, marginal quality, reject quality). The results can be presented as a table with m rows representing the groups of interest and k columns representing the categories. Such tables can be analyzed to answer the question "Do the groups *differ* with regard to the proportion of items in the categories?" The chi-square statistic can be used for this purpose.

Example of chi-square test

The following example is from Natrella (1963). Rejects of metal castings were classified by cause of rejection for three different weeks, as given in the following tabulation. The question to be answered is the following: Does the distribution of rejects differ from week to week?

	CAUSE OF REJECTION							
	Sand	Misrun	Shift	Drop	Core-break	Broken	Other	Total
Week 1	97	8	18	8	23	21	5	180
Week 2	120	15	12	13	21	17	15	213
Week 3	82	4	0	12	38	25	19	180
Total	299	27	30	33	82	63	39	573

Chi-square (χ^2) is computed by first finding the expected frequencies in each cell. This is done using the equation:

$$\text{Frequency expected} = f_e = \frac{\text{Row sum} * \text{column sum}}{\text{overall sum}}$$

For week 1, e.g., the frequency expected of sand rejects was (180*299)/573=93.93. The table below shows the frequency expected for the remainder of the cells.

	Sand	Misrun	Shift	Drop	Corebreak	Broken	Other
Week 1	93.93	8.48	9.42	10.37	25.76	19.79	12.25
Week 2	111.15	10.04	11.15	12.27	30.48	23.42	14.50
Week 3	93.93	8.48	9.42	10.37	25.76	19.79	12.25

The next step was to compute χ^2 as follows:

$$\chi^2 = \sum_{\text{over all cells}} \frac{(\text{Frequency expected - Frequency observed})^2}{\text{Frequency expected}}$$

$$= \frac{(93.93-97)^2}{93.93} + \cdots + \frac{(12.25-19)^2}{12.25} = 45.60$$

Next, a value of α=0.10 was used for this example. The degrees of freedom for the χ^2 test was $(k-1)(m-1) = 12$. Referring to Table 6 in the Appendix the critical value of $\chi^2 = 18.55$ was found for these values. Since the computed value of χ^2 exceeded the critical value, it was concluded that the weeks differed with regard to proportions of various types of defectives.

VI.E MEASURING AND MODELING RELATIONSHIPS BETWEEN VARIABLES

Correlation analysis (the study of the strength of the linear relationships among variables) and regression analysis (modeling the relationship between one or more independent variables and a dependent variable) are activities of considerable importance in quality engineering. A regression problem considers the frequency distributions of one variable when another is held fixed at each of several levels. A correlation problem considers the joint variation of two variables, neither of which is restricted by the experimenter. Correlation and regression analyses are designed to assist the engineer in studying cause and effect. They may be employed in all stages of the problem-solving and planning process. Of course, statistics cannot alone establish cause and effect. Proving cause and effect requires sound scientific understanding of the situation at hand. The statistical methods described in this section assist the engineer in performing this task.

LINEAR MODELS

A linear model is simply an expression of a type of association between two variables, x and y. A *linear relationship* simply means that a change of a given size in x produces a proportionate change in y. Linear models have the form:

$$y = a + bx \qquad \text{(VI.27)}$$

where a and b are constants. The equation simply states that when x changes by one unit, y will change by b units. This relationship can be shown graphically.

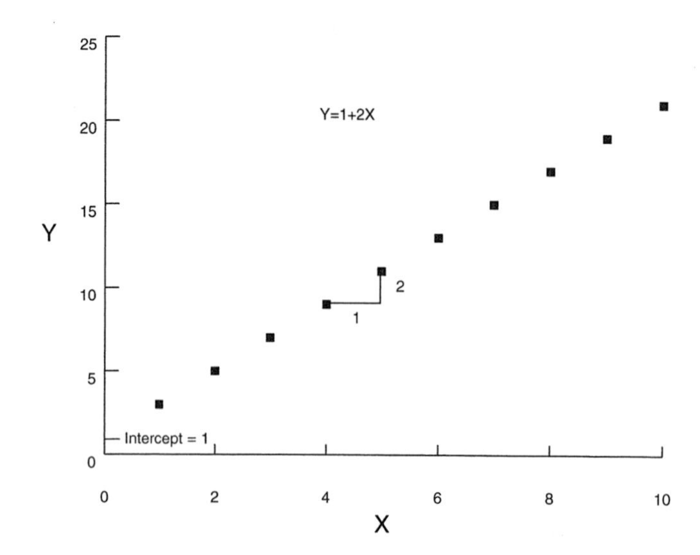

Figure VI.23. Scatter diagram of a linear relationship.

In Figure VI.23, a=1 and b=2. The term a is called the intercept and b is called the slope. When x=0, *y* is equal to the intercept. Figure VI.23 depicts a perfect linear fit; e.g., if *x* is known we can determine y exactly. Of course, perfect fits are virtually unknown when real data are used. In practice we must deal with error in *x* and *y*. These issues are discussed below.

Many types of associations are non-linear. For example, over a given range of *x* values, *y* might increase, and for other *x* values, *y* might decrease. This *curvilinear relationship* is shown in Figure VI.24.

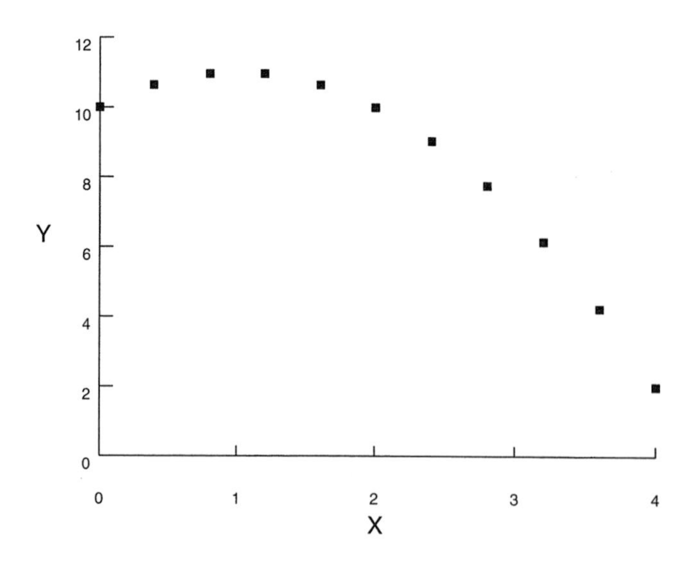

Figure VI.24. Scatter diagram of a curvilinear relationship.

Here we see that y increases when x increases and is less than 1, and decreases as x increases when x is greater than 1. Curvilinear relationships are valuable in the design of robust systems (see VI.F). A wide variety of processes produces such relationships.

It is often helpful to convert these non-linear forms to linear form for analysis using standard computer programs or scientific calculators. Several such transformations are shown in Table VI.9.

Table VI.9. Some linearizing transformations.

(Source: Experimental Statistics, NBS Handbook 51, p. 5–31)

IF THE RELATIONSHIP IS OF THE FORM:	PLOT THE TRANSFORMED VARIABLES		CONVERT STRAIGHT LINE CONSTANTS (B_0 AND B_1) TO ORIGINAL CONSTANTS	
	Y_T	X_T	b_0	b_1
$Y = a + \frac{b}{X}$	Y	$\frac{1}{X}$	a	b
$\frac{1}{Y} = a + bX$	$\frac{1}{Y}$	X	a	b
$Y = \frac{X}{a + bX}$	$\frac{X}{Y}$	X	a	b
$Y = ab^X$	$\log Y$	X	$\log a$	$\log b$
$Y = ae^{bx}$	$\log Y$	X	$\log a$	$b \log e$
$Y = aX^b$	$\log Y$	$\log X$	$\log a$	b
$Y = a + bX^n$ where n is known	Y	X^n	a	b

Fit the straight line $Y_T = b_0 + b_1 X_T$ using the usual linear regression procedures (see below). In all formulas, substitute Y_T for Y and X_T for X. A simple method for selecting a transformation is to simply program the transformation into a spreadsheet and run regressions using every transformation. Then select the transformation which gives the largest value for the statistic R^2.

There are other ways of analyzing non-linear responses. One common method is to break the response into segments that are piecewise linear, and then to analyze each piece separately. For example, in Figure VI.24 y is roughly linear and increasing over the range $0<x<1$ and linear and decreasing over the range $x>1$. Of course, if the engineer has access to powerful statistical software, non-linear forms can be analyzed directly.

When conducting regression and correlation analysis we can distinguish two main types of variables. One type we call *predictor variables* or *independent variables*, the other *response variables* or *dependent variables*. By predictor or independent variable we usually mean variables that can either be set to a desired variable (e.g., oven temperature) or else take values that can be observed but not controlled (e.g., outdoors ambient humidity). As a result of changes that are deliberately made, or simply take place in the predictor variables, an effect is transmitted to the response variables (e.g., the grain size of a composite material). We are usually interested in discovering how changes in the predictor variables effect the values of the response variables. Ideally, we hope that a small number of predictor variables will "explain" nearly all of the variation in the response variables.

In practice, it is sometimes difficult to draw a clear distinction between independent and dependent variables. In many cases it depends on the objective of the investigator. For example, a quality engineer may treat ambient temperature as a predictor variable in the study of paint quality and as the response variable in a study of clean room particulates. However, the above definitions are useful in planning engineering studies.

Another idea important to studying cause and effect is that of the *data space* of the study. The data space of a study refers to the region bounded by the range of the independent variables under study. In general, predictions based on values outside the data space studied, called *extrapolations*, are little more than speculation and not advised. Figure VI.25 illustrates the concept of data space for two independent variables. Defining the data space can be quite tricky when large numbers of independent variables are involved.

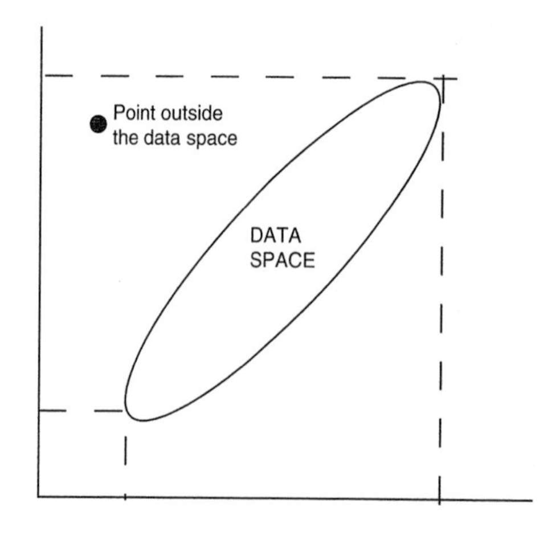

Figure VI.25. Data space.

While the numerical analysis of data provides valuable information, it should always be supplemented with graphical analysis as well. Scatter diagrams are one very useful supplement to regression and correlation analysis. The following illustrates the value of supplementing numerical analysis with scatter diagrams.

I		II		III		IV	
X	Y	X	Y	X	Y	X	Y
10	8.04	10	9.14	10	7.46	8	6.58
8	6.95	8	8.14	8	6.77	8	5.76
13	7.58	13	8.74	13	12.74	8	7.71
9	8.81	9	8.77	9	7.11	8	8.84
11	8.33	11	9.26	11	7.81	8	8.47
14	9.96	14	8.10	14	8.84	8	7.04
6	7.24	6	6.13	6	6.08	8	5.25
4	4.26	4	3.10	4	5.39	19	12.50
12	10.84	12	9.13	12	8.15	8	5.56
7	4.82	7	7.26	7	6.42	8	7.91
5	5.68	5	4.74	5	5.73	8	6.89

Statistics for Processes I–IV

$n = 11$

$\overline{X} = 9.0$

$\overline{Y} = 7.5$

best fit line: $\text{Y} = 3 + 0.5\text{X}$

standard error of slope: 0.118

$t = 4.24$

$\sum X - \overline{X} = 110.0$

regression SS = 27.50

residual SS = 13.75

r = 0.82

$r^2 = 0.67$

Figure VI.26. Illustration of the value of scatter diagrams.

(Source: *The Viusual Display of Quantitative Information*, Edward R. Tufte.)

Continued on next page . . .

Figure VI.26—*Continued . . .*

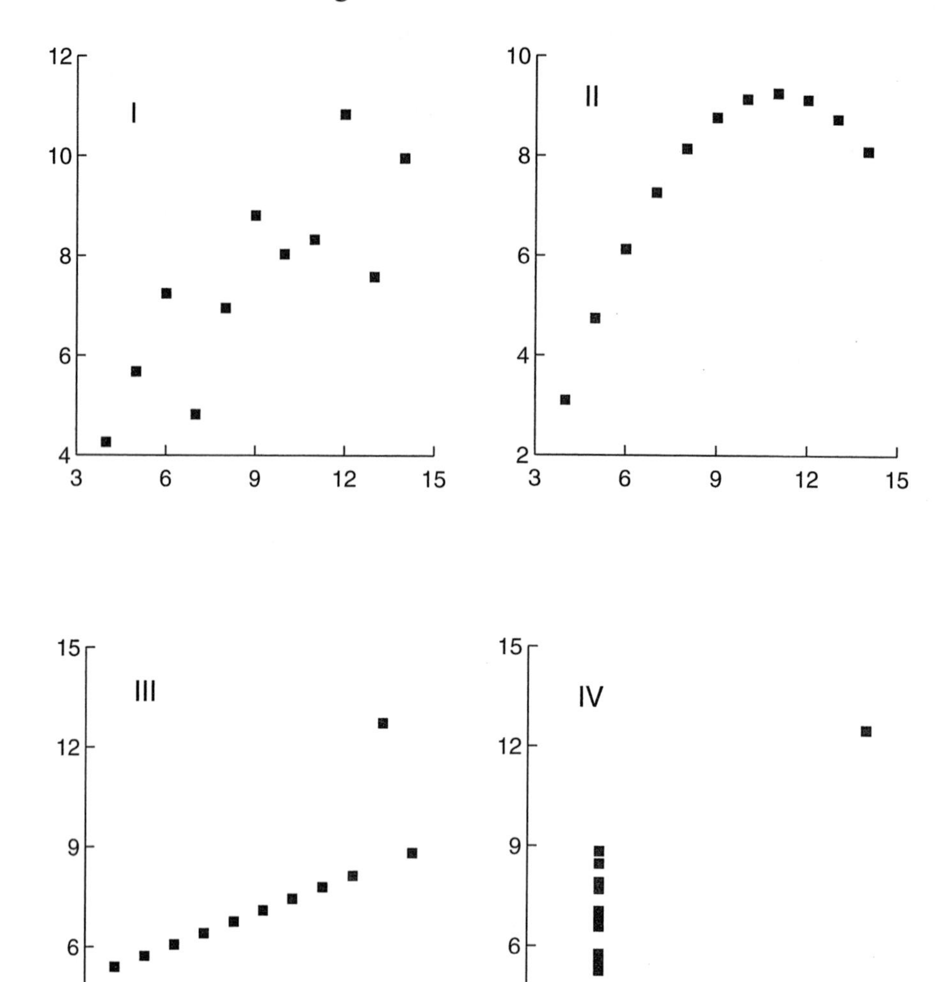

Figure VI.26. Illustration of the value of scatter diagrams.

(Source: *The Viusual Display of Quantitative Information*, Edward R. Tufte.)

In other words, although the scatter diagrams clearly show four distinct processes, the statistical analysis does not. In quality engineering numerical analysis alone is not enough.

LEAST-SQUARES FIT

If all data fell on a perfectly straight line it would be easy to compute the slope and intercept given any two points. However, the situation becomes more complicated when there is "scatter" around the line. That is, for a given value of x, more than one value of y appears. When this occurs, we have error in the model. Figure VI.27 illustrates the concept of error.

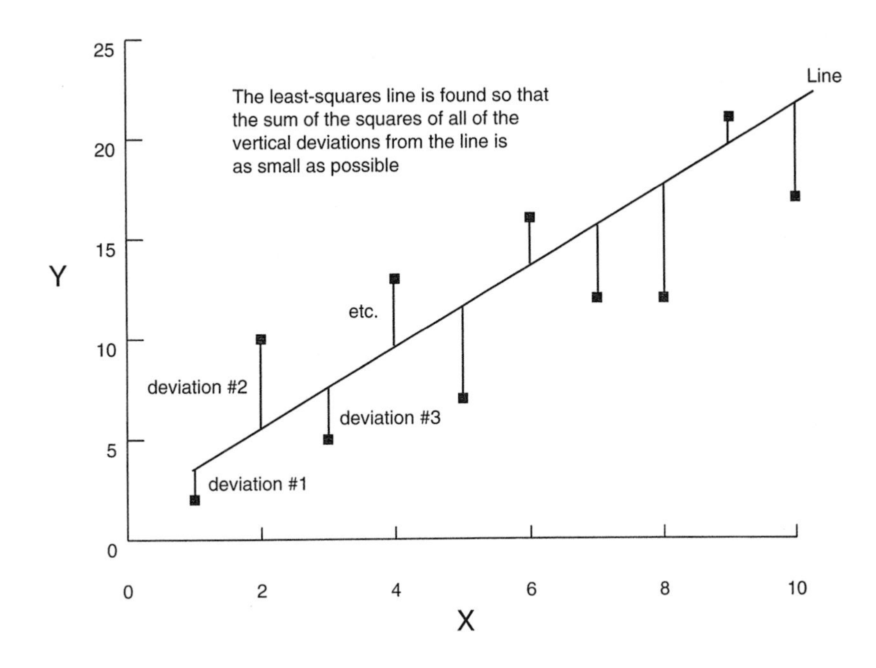

Figure VI.27. Error in the linear model.

The model for a simple linear regression with error is:

$$y = a + bx + \varepsilon \qquad \text{(VI.28)}$$

where ε represents error. Generally, assuming the model adequately fits the data, errors are assumed to follow a normal distribution with a mean of 0 and a constant standard deviation. The standard deviation of the errors is known as the *standard error*. We discuss ways of verifying our assumptions below.

When error occurs, as it does in nearly all "real-world" situations, there are many possible lines which might be used to model the data. Some method must be found which provides, in some sense, a "best fit" equation in these everyday situations. Statisticians have developed a large number of such methods. The method most commonly used in quality engineering finds the straight line that minimizes the sum of the squares of the errors for all of the data points. This method is known as the "least-squares" best-fit line. In other words, the least-squares best-fit line equation is $y_i' = a + bx_i$ where a and b are found so that the sum of the squared deviations from the line is minimized. The best-fit equations for a and b are:

$$b = \frac{\Sigma\left(X_i - \overline{X}\right)\left(Y_i - \overline{Y}\right)}{\Sigma\left(X_i - \overline{X}\right)^2} \qquad \text{(VI.29)}$$

$$a = \overline{Y} - b\overline{X} \qquad \text{(VI.30)}$$

where the sum is taken over all n values. Most spreadsheets and scientific calculators have a built-in capability to compute a and b. As stated above, there are many other ways to compute the slope and intercept (e.g., minimize the sum of the absolute deviations, minimize the maximum deviation, etc.); in certain situations one of the alternatives may be preferred. The reader is advised to consult books devoted to regression analysis for additional information (see, for example, Montgomery 2001).

The reader should note that the fit obtained by regressing x on y will not in general produce the same line as would be obtained by regressing y on x. This is illustrated in Figure VI.28.

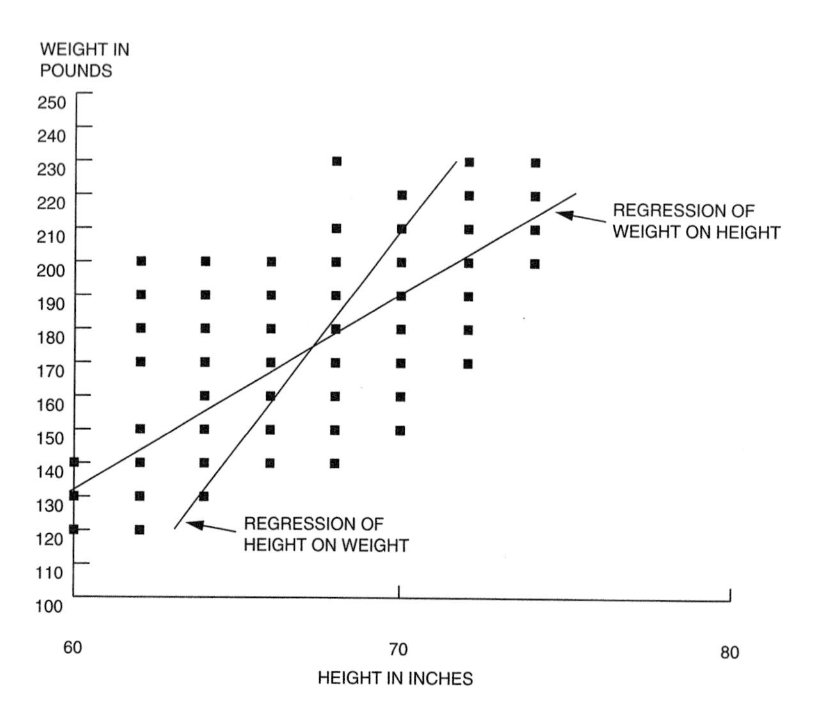

Figure VI.28. Least-squares lines of weight vs. height and height vs. weight.

When weight is regressed on height the equation indicates the average weight (in pounds) for a given height (in inches.) When height is regressed on weight the equation indicates the average height for a given weight. The two lines intersect at the average height and weight.

These examples show how a single independent variable is used to model the response of a dependent variable. This is known as *simple linear regression.* It is also possible to model the dependent variable in terms of two or more independent variables; this is known as *multiple linear regression.* The mathematical model for multiple linear regression has additional terms for the additional independent variables. Equation VI.31 shows a linear model when there are two independent variables.

$$\hat{y} = a + b_1 x_1 + b_2 x_2 + \varepsilon \tag{VI.31}$$

where $x_1 x_2$ are independent variables, b_1 is the coefficient for x_1 and b_2 is the coefficient for x_2.

Example of regression analysis

A restaurant conducted surveys of 42 customers, obtaining customer ratings on staff service, food quality, and overall satisfaction with their visit to the restaurant. Figure VI.29 shows the regression analysis output from a spreadsheet regression function (Microsoft Excel).

SUMMARY OUTPUT						
Regression statistics						
Multiple R	0.847					
R square	0.717					
Adjusted R square	0.703					
Standard error	0.541					
Observations	42					
ANOVA						
	df	*ss*	*ms*	*F*	*Significance F*	
Regression	2	28.97	14.49	49.43	0.00	
Residual	39	11.43	0.29			
Total	41	40.40				
	Coefficients	*Standard error*	*t Stat*	*P-value*	*Lower 95%*	*Upper 95%*
Intercept	–1.188	0.565	–2.102	0.042	–2.331	–0.045
Staff	0.902	0.144	6.283	0.000	0.611	1.192
Food	0.379	0.163	2.325	0.025	0.049	0.710

Figure VI.29. Regression analysis output.

The data consist of two independent variables, staff and food quality, and a single dependent variable, overall satisfaction. The basic idea is that the quality of staff service and the food are *causes* and the overall satisfaction score is an *effect*. The regression output is interpreted as follows:

Multiple R—the multiple correlation coefficient. It is the correlation between y and y'. For the example: multiple R=0.847, which indicates that y and y' are highly correlated, which implies that there is an association between overall satisfaction and the quality of the food and service.

R Square—the square of multiple R. It measures the proportion of total variation about the mean $\bar{Y}$ explained by the regression. For the example: R^2=0.717, which indicates that the fitted equation explains 71.7% of the total variation about the average satisfaction level.

Adjusted R Square—a measure of R^2 "adjusted for degrees of freedom." The equation is

$$\text{Adjusted } R^2 = 1 - (1 - R^2)\left(\frac{n-1}{n-p}\right) \qquad \text{(VI.32)}$$

where p is the number of parameters (coefficients for the xs) estimated in the model. For the example: p=2, since there are two x terms. Some experimenters prefer the adjusted R^2 to the unadjusted R^2, while others see little advantage to it (e.g., Draper and Smith 1981,p. 92).

Standard error—the standard deviation of the residuals. The *residual* is the difference between the observed value of y and the predicted value based on the regression equation.

Observations—refer to the number of cases in the regression analysis, or n.

ANOVA, or ANalysis Of VAriance—a table examining the hypothesis that the variation explained by the regression is zero. If this is so, then the observed association could be explained by chance alone. The rows and columns are those of a standard one-factor ANOVA table (see section F later in this chapter). For this example, the important item is the column labeled "Significance F." The value shown, 0.00, indicates that the probability of getting these results due to chance alone is less than 0.01; i.e., the association is probably not due to chance alone. Note that the ANOVA applies to the entire *model*, not to the individual variables.

The next table in the output examines each of the terms in the linear model separately. The *intercept* is as described above, and corresponds to our term a in the linear equation. Our model uses two independent variables. In our terminology staff=b_1, food=b_2. Thus, reading from the *coefficients* column, the linear model is:

$$\hat{y} = -1.88 + 0.902 \times \text{staff score} + 0.379 \times \text{food score}\ .$$

The remaining columns test the hypotheses that each coefficient in the model is actually zero.

Standard error column—gives the standard deviations of each term, i.e., the standard deviation of the intercept=0.565, etc.

t Stat column—the coefficient divided by the standard error, i.e., it shows how many standard deviations the observed coefficient is from zero.

P-value—shows the area in the tail of a t distribution beyond the computed t value. For most experimental work, a P-value less than 0.05 is accepted as an indication that the coefficient is significantly different than zero. All of the terms in our model have significant P-values.

Lower 95% and Upper 95% columns—a 95% confidence interval on the coefficient. If the confidence interval does not include zero, we will reject the hypothesis that the coefficient is zero. None of the intervals in our example include zero.

CORRELATION ANALYSIS

As mentioned earlier, a correlation problem considers the joint variation of two variables, neither of which is restricted by the experimenter. Unlike regression analysis, which considers the effect of the independent variable(s) on a dependent variable, correlation analysis is concerned with the joint variation of one independent variable with another. In a correlation problem, the engineer has two measurements for each individual item in the sample. Unlike a regression study where the engineer controls the values of the x variables, correlation studies usually involve spontaneous variation in the variables being studied. Correlation methods for determining the strength of the linear relationship between two or more variables are among the most widely applied statistical techniques. More advanced methods exist for studying situations with more than two variables (e.g., canonical analysis, factor analysis, principal components analysis, etc.); however, with the exception of multiple regression, our discussion will focus on the linear association of two variables at a time.

In most cases, the measure of correlation used by engineers is the statistic r, sometimes referred to as *Pearson's product-moment correlation.* Usually x and y are assumed to have a bivariate normal distribution. Under this assumption r is a sample statistic which estimates the population correlation parameter ρ. One interpretation of r is based on the linear regression model described earlier, namely that r^2 is the proportion of the total variability in the y data which can be explained by the linear regression model. The equation for r is:

$$r = \frac{s_{xy}}{s_x s_y} = \frac{n\sum xy - \sum x \sum y}{\sqrt{\left[n\sum x^2 - \left(\sum x\right)^2\right]\left[n\sum y^2 - \left(\sum y\right)^2\right]}} \tag{VI.33}$$

and, of course, r^2 is simply the square of r. r is bounded at -1 and $+1$. When the assumptions hold, the significance of r is tested by the regression ANOVA.

Interpreting r can become quite tricky, so scatter plots should always be used (see V.C). When the relationship between x and y is non-linear, the "explanatory power" of r is difficult to interpret in precise terms and should be discussed with great care. While it is easy to see the value of very high correlations such as $r=0.99$, it is not so easy to draw conclusions from lower values of r, even when they are statistically significant (i.e., they are significantly different than 0.0). For example, $r=0.5$ does *not* mean the data show half as much clustering as a perfect straight-line fit. In fact, $r=0$ does *not* mean that there is no relationship between the x and y data, as Figure VI.30 shows. When $r>0$, y tends to increase when x increases. When $r<0$, y tends to decrease when x increases.

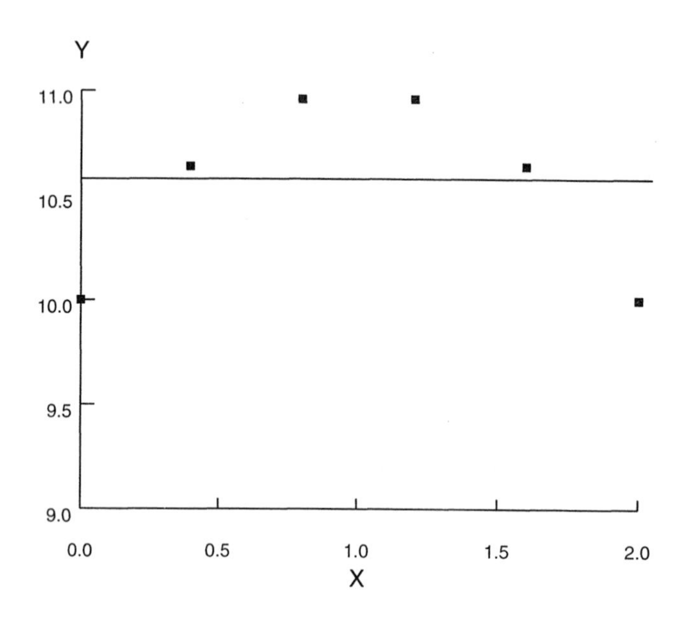

Figure VI.30. Interpreting $r=0$ for curvilinear data.

Although $r=0$, the relationship between x and y is perfect, albeit non-linear.

At the other extreme, $r=1$, a "perfect correlation," does not mean that there is a cause and effect relationship between x and y. For example, both x and y might be determined by a third variable, z. In such situations, z is described as a *lurking variable* which "hides" in the background, unknown to the experimenter. Lurking variables are behind some of the infamous silly associations, such as the association between teacher's pay and liquor sales (the lurking variable is general prosperity).*

*It is possible to evaluate the association of x and y by removing the effect of the lurking variable. This can be done using regression analysis and computing partial correlation coefficients. This advanced procedure is described in most texts on regression analysis.

Establishing causation requires solid scientific understanding. Causation cannot be "proven" by statistics alone. Some statistical techniques, such as path analysis, can help determine if the correlations between a number of variables are consistent with causal assumptions. However, these methods are beyond the scope of this book.

TIME SERIES ANALYSIS

In this section, we will address the subject of time series analysis on a relatively simple level. First, we will look at statistical methods that can be used when we believe the data are from a stable process. This involves analysis of patterns in runs of data in a time-ordered sequence. Next, we discuss the problem of autocorrelation in time series data and provide a method of dealing with this problem.

RUN CHARTS

Run charts are plots of data arranged in time sequence. Analysis of run charts is performed to determine if the patterns can be attributed to common causes of variation, or if special causes of variation were present. Run charts should be used for preliminary analysis of any data measured on a continuous scale that can be organized in time sequence. Run chart candidates include such things as fuel consumption, production throughput, weight, size, etc. Run charts answer the question "was this process in statistical control for the time period observed?" If the answer is "no," then the process was influenced by one or more special causes of variation. If the answer is "yes," then the long-term performance of the process can be estimated using process capability analysis methods. The run chart tests shown are all *non-parametric*, i.e., there are no assumptions made regarding the underlying distribution.

How to prepare and analyze run charts

1. Plot a line chart of the data in time sequence.
2. Find the median of the data. This can be easily done by using the line chart you constructed in the above step. Simply place a straightedge or a piece of paper across the top of the chart, parallel to the bottom axis. Lower the straightedge until half of the data points appear above the straightedge, or on it. Draw a horizontal line across the chart at that point and label the line "Median" or $\tilde{X}$. This procedure is shown in Figure VI.31.

As you might expect, run charts are evaluated by examining the "runs" on the chart. A "run" is a time-ordered sequence of points. There are several different statistical tests that can be applied to the runs.

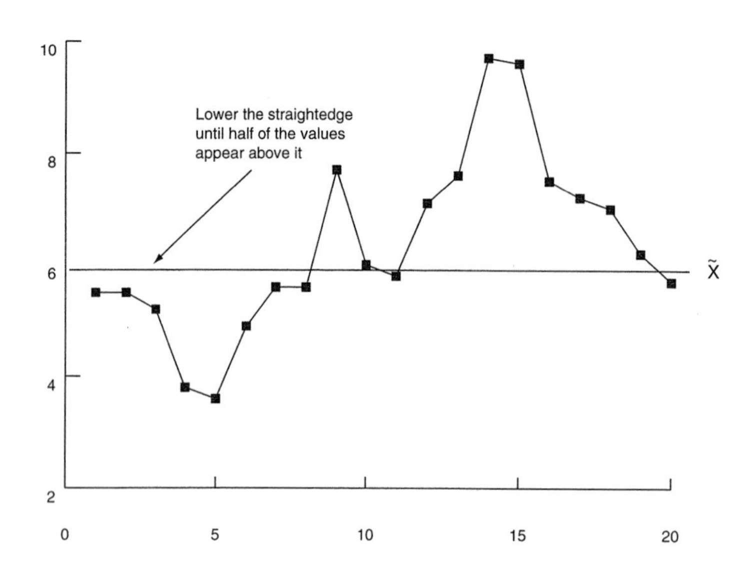

Figure VI.31. Using a straightedge to find the median.

Run length

A *run to the median* is a series of consecutive points on the same side of the median. Unless the process is being influenced by special causes, it is unlikely that a long series of consecutive points will all fall on the same side of the median. Thus, checking run length is one way of checking for special causes of variation. The length of a run is found by simply counting the number of consecutive points on the same side of the median. However, it may be that some values are exactly equal to the median. If only one value is exactly on the median line, ignore it. There will always be at least one value exactly on the median if you have an odd number of data points. If more than one value is on the line, assign them to one side or the other in a way that results in 50% being on one side and 50% on the other. On the run chart, mark those that will be counted as above the median with an *a* and those that will be counted below the median with a *b*. The run length concept is illustrated in Figure VI.32.

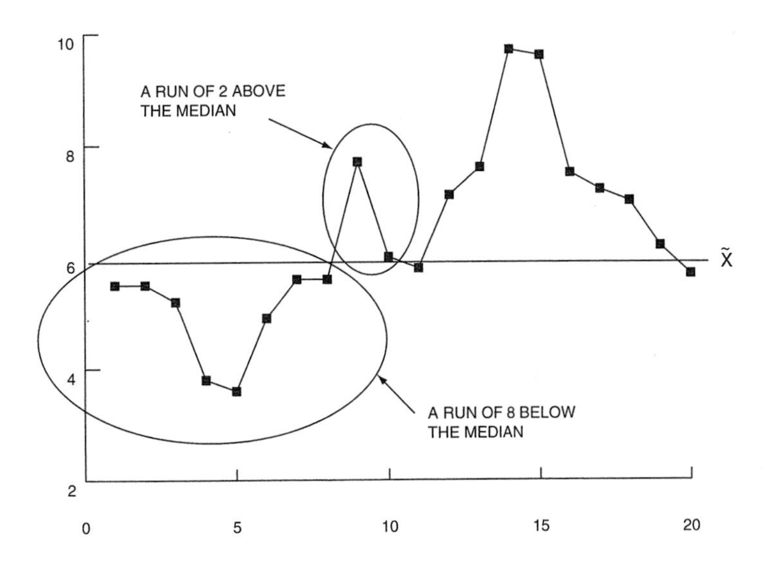

Figure VI.32. Determination of run length.

After finding the longest run, compare the length of the longest run to the values in Table VI.10. If the longest run is longer than the maximum allowed, then the process was probably influenced by a special cause of variation (α=0.05). With the example, there are 20 values plotted and the longest run was 8. Table VI.10 indicates that a run of 7 would not be too unusual for 20 plotted points, but a run of 8 would be. Since our longest run is 8, we conclude that a special cause of variation is indicated and conduct an investigation to identify the special cause.

Table VI.10. Maximum run length.

NUMBER OF PLOTTED VALUES	MAXIMUM RUN LENGTH
10	5
15	6
20	7
30	8
40	9
50	10

Number of runs

The number of runs we expect to find from a controlled process can also be mathematically determined. A process that is not being influenced by special causes will not have either too many runs or too few runs. The number of runs is found by simple counting. Referring to Figure VI.33, we see that there are 5 runs.

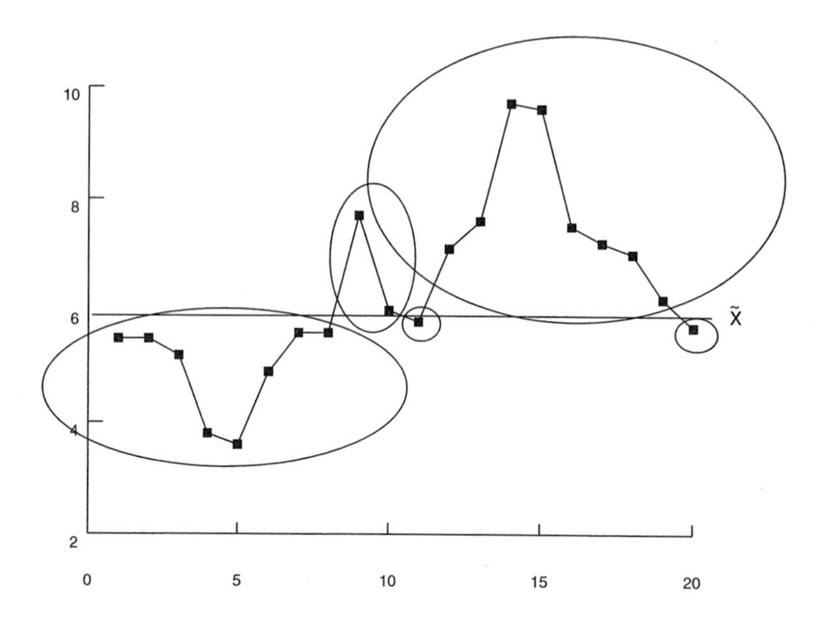

Figure VI.33. Determination of number of runs.

Table VI.11 is used to evaluate the number of runs. If you have fewer runs than the smallest allowed or more runs than the largest allowed then there is a high probability (α=0.05) that a special cause is present. With the example, we have 20 values plotted and 5 runs. Table VI.11 indicates that for 20 plotted points, 6 to 15 runs are expected, so we conclude that a special cause was present.

Table VI.11. Limits on the number of runs.

Number of Plotted Values	Smallest Run Count	Largest Run Count
10	3	8
12	3	10
14	4	11
16	5	12
18	6	13
20	6	15

Continued on next page . . .

Table VI.11—*Continued . . .*

Number of Plotted Values	Smallest Run Count	Largest Run Count
22	7	16
24	8	17
26	9	18
28	10	19
30	11	20
32	11	22
34	12	23
36	13	24
38	14	25
40	15	26
42	16	27
44	17	28
46	17	30
48	18	31
50	19	32

Trends

The run chart should not have any unusually long series of consecutive increases or decreases. If it does, then a trend is indicated and it is probably due to a special cause of variation (α=0.05). Compare the longest count of consecutive increases or decreases to the longest allowed shown in Table VI.12; if your count exceeds the table value then it is likely that a special cause of variation caused the process to drift.

Table VI.12. Maximum consecutive increases/decreases.

NUMBER OF PLOTTED VALUES	MAXIMUM CONSECUTIVE INCREASES/DECREASES
5 to 8	4
9 to 20	5
21 to 100	6
101 or more	7

Figure VI.34 shows the analysis of trends. Note that the trend can extend on both sides of the median; i.e., for this particular run test the median is ignored.

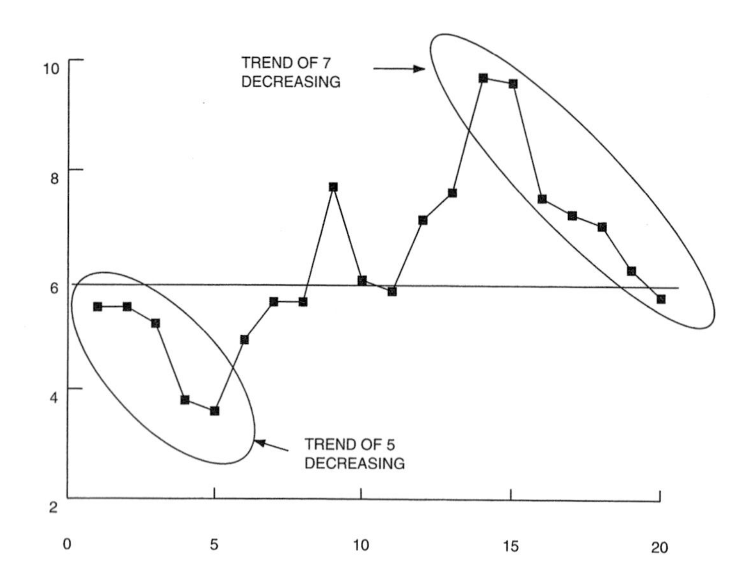

Figure VI.34. Determination of trend length.

When counting increases or decreases, ignore "no change" values. For example, the trend length in the series 2, 3, 3, 5, 6 is four.

Pointers for using run charts

Run charts should not be used if too many of the numbers are the same. As a rule of thumb, don't use run charts if more than 30% of the values are the same. For example, in the data set 1, 2, 3, 3, 6, 7, 7, 11, 17, 19, the number 3 appears twice and the number 7 appears twice. Thus, 4 of the 10, or 40% of the values are the same.

Run charts are preliminary analysis tools, so if you have continuous data in time-order always sketch a quick run chart before doing any more complex analysis. Often the patterns on a run chart will point you in the right direction without any further work.

Run charts are one of the least sensitive SPC techniques. They are unable to detect "freaks," i.e., single points dramatically different from the rest. Thus, run charts may fail to find a special cause even if a special cause were present. In statistical parlance, run charts tend to have large type II errors; i.e., they have a high probability of accepting the hypothesis of no special cause even when the special cause actually exists. Use run charts to aid in troubleshooting. The different run tests indicate different types of special causes. A long run on the same side of the median indicates a special cause that created a process shift. A long series of consecutively increasing or decreasing values indicates a special cause that created a trend. Too many runs often indicates mixture of several sources of variation in the sample. Too few runs often occur in conjunction with a process shift or trend. If you have too few runs and they are not caused by a process shift or trend, then too few runs may indicate mixture that follows a definite pattern (e.g., an operator who is periodically relieved).

TREND ANALYSIS

In the discussion of regression analysis presented above, we discussed the analysis of residuals as a means of testing the adequacy of our linear model. It is also possible to model time-series data using linear regression, where $X_1=1$ represents time period #1, $X_2=2$ represents time period #2, and so on. When this is done with time series data, such as business and economic data, it often happens that the residuals are *autocorrelated*, or *serially correlated.* This simply means that if we let x be the residual for current time period and y be the residual for the previous time period, the correlation coefficient r between x and y would be significantly greater than zero. When autocorrelation exists, it causes a number of problems with regression analysis, e.g.:

- the regression coefficients are often quite inefficient;
- the error variance is seriously underestimated;
- the error variance of the regression coefficients may seriously underestimate the true variance;
- the confidence intervals and significance tests based on the t and F distributions are no longer strictly applicable.

These are obviously serious problems and must be addressed. The best way to address them is to eliminate the autocorrelation at its source. Autocorrelation is often a symptom indicating that one or more key variables are missing from the model. If these variables could be identified the model could be corrected. The new model would represent a better understanding of the dynamics of the time series; the autocorrelation disappearing is a side benefit.

Testing for autocorrelation

Before discussing mathematical means of dealing with autocorrelation, we will present a means for detecting significant autocorrelation. The simple linear regression model introduced earlier can be rewritten for time series data as follows:

$$Y_t = \beta_0 + \beta_1 X_t + \varepsilon_t$$

$$\varepsilon_t = \rho \varepsilon_{t-1} + \mu_t \qquad \text{(VI.34)}$$

where ρ is a parameter such that $|\rho|<1$, μ_t are independent and normally distributed with a mean zero and a variance σ^2.

Note that this model is identical to the simple linear regression model presented earlier except for the structure of the error term. The new model has an error term with two components. The term ρ is called the *autocorrelation parameter* and it represents some fraction of the previous error (if $\rho > 0$), μ_t is the error in the current time-period estimate. If there is no autocorrelation, then $\rho=0$ and the model is the usual simple regression model. Thus, our first order of business is to test the hypothesis that $\rho=0$.

Since most time series data of interest to quality engineers, including business and economic time series data, tend to show positive autocorrelation, we construct the following one-sided hypothesis test:

$$H_0\colon \rho = 0 \qquad\qquad H_a\colon \rho > 0 \tag{VI.35}$$

The test statistic we will use for testing this hypothesis is known as the *Durbin-Watson* statistic, denoted *D*. *D* is obtained by calculating the ordinary residual:

$$e_t = y_t - \hat{y}_t \tag{VI.36}$$

and then:

$$D = \frac{\sum_{t=2}^{n} (e_t - e_{t-1})^2}{\sum_{t=1}^{n} e_t^2} \tag{VI.37}$$

Note that the sum in the numerator of the equation runs from t=2, not from t=1. As usual, n represents the number of cases. (Note: if you wish to test H_a: $\rho<0$, then use 4-D instead of D as the test statistic.)

An exact test procedure is not available, but Durbin and Watson have obtained lower and upper bounds d_L *and* d_U such that a value of D outside these bounds leads to a definite decision. The decision rule is:

$$\text{If } D > d_U, \text{ conclude } H_0$$

$$\text{If } D < d_L, \text{ conclude } H_a \tag{VI.38}$$

$$\text{If } d_L \le D \le d_U, \text{ the test is inconclusive}$$

Table 11 in the Appendix contains the bounds for D.

*Example of serial correlation analysis**

The Blaisdell Company wished to predict its sales by using industry sales as a predictor variable. The market research analyst was concerned whether or not the error terms are positively autocorrelated. He used simple least-squares linear regression to fit a regression line to the data in Table VI.13, letting x be industry sales and y be sales of his company. The results are shown at the bottom of the table.

Table VI.13. Data for Blaisdell Company example, regression results, and Durbin-Watson test calculations.

From *Applied Linear Statistical Models*, copyright 1996 by McGraw Hill.

Year and Quarter	t	(1) Company Sales ($ millions) Y_t	(2) Industry Sales ($ millions) X_t	(3) Residual e_t	(4) e_t-e_{t-1}	(5) $(e_t-e_{t-1})^2$	(6) e_t^2
1983: 1	1	20.96	127.3	−.026052	—	—	.0006787
2	2	21.40	130.0	−.062015	−.035963	.0012933	.0038459
3	3	21.96	132.7	.022021	.084036	.0070620	.0004849
4	4	21.52	129.4	.163754	.141733	.0200882	.0268154
1984: 1	5	22.39	135.0	.046570	−.117184	.0137321	.0021688
2	6	22.76	137.1	.046377	−.000193	.0000000	.0021508
3	7	23.48	141.2	.043617	−.002760	.0000076	.0019024
4	8	23.66	142.8	−.058435	−.102052	.0104146	.0034146
1985: 1	9	24.10	145.5	−.094399	−.035964	.0012934	.0089112
2	10	24.01	145.3	−.149142	−.054743	.0029968	.0222433
3	11	24.54	148.3	−.147991	.001151	.0000013	.0219013
4	12	24.30	146.4	−.053054	.094937	.0090130	.0028147
1986: 1	13	25.00	150.2	−.022928	.030126	.0009076	.0005257
2	14	25.64	153.1	.105852	.128780	.0165843	.0112046
3	15	26.36	157.3	.085464	−.020388	.0004157	.0073041
4	16	26.98	160.7	.106102	.020638	.0004259	.0112576
1987: 1	17	27.52	164.2	.029112	−.076990	.0059275	.0008475
2	18	27.78	165.6	.042316	.013204	.0001743	.0017906
3	19	28.24	168.7	−.044160	−.086476	.0074781	.0019501
4	20	28.78	171.7	−.033009	.011151	.0001243	.0010896
Total						.0979400	.1333018

$$\hat{Y} = -1.4548 + .17628X$$

$$s\{b_0\} = .21415 \quad s\{b_1\} = .00144$$

$$MSE = .00741$$

*Source: Neter, J., Kutner, M.H., Nachtscheim, C.J., Wasserman, W. 1996. *Applied Linear Statistical Models*, 4th ed., New York, NY: McGraw Hill.

When the residuals (column 3) are plotted against time, the plot suggests that there is serial correlation in the data (Figure VI.35).

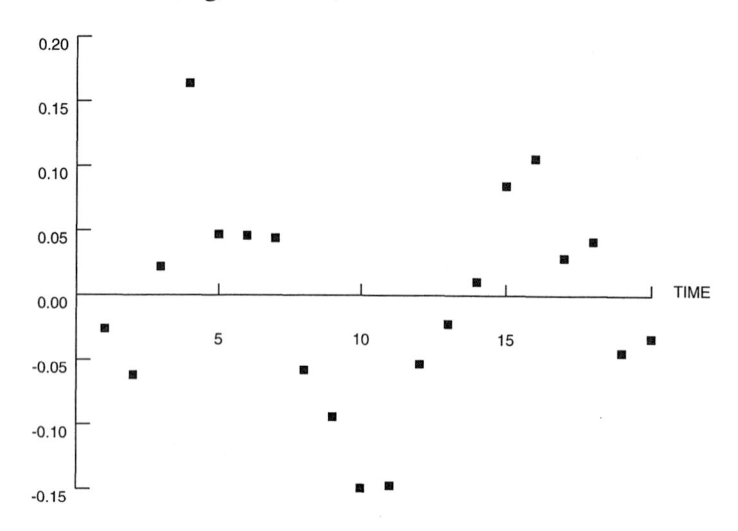

Figure VI.35. Residuals plotted against time.
From *Applied Linear Statistical Models*, figure 13.3. Copyright © 1990 by Richard D. Irwin.

The analyst wished to confirm this suspicion by using the Durbin-Watson test. Columns 4, 5, and 6 of Table VI.13 contain the necessary calculations for computing the test statistic D.

$$D = \frac{\sum_{t=2}^{n} (e_t - e_{t-1})^2}{\sum_{t=1}^{n} e_t^2} = \frac{\text{sum of col. 5}}{\text{sum of col. 6}} = \frac{0.09794}{0.1333018} = 0.735 \qquad \text{(VI.39)}$$

Using a level of significance of α=0.01, he found in Table 11 of the Appendix for n=20 and p-1=1:

$$d_L = 0.95$$
$$d_U = 1.15$$

Since D=0.735 falls below d_L = 0.95, the analyst accepts H_a and concludes that the error terms are positively autocorrelated. The parameter p is the number of parameters in the model: 1–p is the number of independent variables in the regression, so in this case, 1–p=1, since in Simple Linear Regression, there is only one independent variable, x.

Methods of analysis when autocorrelation is present

There are many ways of analyzing autocorrelated data. We will present one of the simpler methods here. The reader should consult texts dedicated to statistics for additional information [Neter, Wasserman and Kutner (1996) and Montgomery, Peck and Vining (2001)].

First differences procedure

The first differences procedure begins with the premise that, since the autocorrelation parameter ρ is frequently large, and since the SSE as a function of ρ tends to be relatively flat when ρ is large, we simply let $\rho=1$. When we do this we can perform a regression analysis using the transformed variables:

$$Y_t' = \beta_t' X_t' + \mu_t$$

where: (VI.40)

$$Y_t' = Y_t - Y_{t-1}$$

$$X_t' = X_t - X_{t-1}$$

The terms Y_t-Y_{t-1} and X_t-X_{t-1} are known as *first differences*. They are obtained by subtracting the first value from the second, the second from the third, and so on. The transformed data are shown in Table VI.14.

Table VI.14. First differences and regression results with first differences procedure.

From *Applied Linear Statistical Models*, copyright 1996 by McGraw Hill.

t	(1) Y_t	(2) X_t	(3) $Y_t'=Y_t-Y_{t-1}$	(4) $X_t'=X_t-X_{t-1}$
1	20.96	127.3	—	—
2	21.40	130.0	.44	2.7
3	21.96	132.7	.56	2.7
4	21.52	129.4	–.44	–3.3
5	22.39	135.0	.87	5.6
6	22.76	137.1	.37	2.1
7	23.48	141.2	.72	4.1
8	23.66	142.8	.18	1.6
9	24.10	145.5	.44	2.7
10	24.01	145.3	–.09	–0.2
11	24.54	148.3	.53	3.0
12	24.30	146.4	–.24	–1.9
13	25.00	150.2	.70	3.8
14	25.64	153.1	.64	2.9
15	26.36	157.3	.72	4.2
16	26.98	160.7	.62	3.4

Continued on next page . . .

Table VI.14—*Continued . . .*

t	(1) Y_t	(2) X_t	(3) $Y_t'=Y_t-Y_{t-1}$	(4) $X_t'=X_t-X_{t-1}$
17	27.52	164.2	.54	3.5
18	27.78	165.6	.26	1.4
19	28.24	168.7	.46	3.1
20	28.78	171.7	.54	3.0

$$\hat{Y}' = .16849X'$$

$$s\{b_1'\} = .005096 \qquad MSE = .00482$$

Note that the model has no intercept term; i.e., unlike the previous regression equations, there is no a term in the equation. Regression equations fitted without the intercept term are known as *regressions through the origin*. The model is:

$$Y_t = \beta_1 x_i + \varepsilon_i \tag{VI.41}$$

and the least-square equation estimating β_1 is:

$$b_1 = \frac{\sum x_i y_i}{\sum x_i^2} \tag{VI.42}$$

where b_1 is the least-squares estimate of β_1. The fitted equation can be used to transform back to the original data as follows:

$$\begin{aligned} &\hat{Y} = b_0 + b_1 x \\ &\text{where} \\ &\text{b}_0 = \overline{Y} - b_1' \overline{X} \\ &b_1 = b_1' \end{aligned} \tag{VI.43}$$

Example

Table VI.14 contains the first differences from the data for the previous example; the least-squares equation for the transformed variables is:

$$\hat{Y}' = 0.16849X' \tag{VI.44}$$

To determine if the procedure has removed the autocorrelation we use the Durbin-Watson test. Since the first differences procedure sometimes produces *negative* autocorrelations, it is appropriate to conduct two-sided hypothesis tests with first differences data.

Also, to use the Durbin-Watson test we must fit the regression using an intercept term (which is required by the Durbin-Watson test, even though we are concerned with the no-intercept model). The test statistic D is computed in the same way as before, and it is $D = 1.75$. This indicates that the error terms are uncorrelated at either α=0.01 or α=0.02. When testing for negative autocorrelation remember to use D=4–D=2.25.

We can now return to the original data by using Equation VI.45:

$$\hat{Y} = -0.30349 + 0.16849X \qquad \text{(VI.45)}$$

where $b_0 = 24.569 - 0.16849(147.62) = -0.30349$ and the slope is the same for both the transformed and original data models. The standard deviation of the slope remains unchanged at 0.005096.

PATTERN ANALYSIS

The experimenter should carefully examine the residuals. The residuals represent the variation "left over" after subtracting the variation explained by the model. The examination of residuals is done to answer the question: "What might explain the rest of the variation?" Potential clues to the answer might arise if a pattern can be detected. For example, the experimenter might notice that the residuals tend to be associated with certain experimental conditions, or they might increase or decrease over time. Other clues might be obtained if certain residuals are *outliers*, i.e., errors much larger than would be expected from chance alone. Residuals that exhibit patterns or which contain outliers are evidence that the linear model is incorrect. There are many reasons why this might be so. The response might be non-linear. The model may leave out important variables. Or, our assumptions may not be valid.

There are four common ways of plotting the residuals:

1. Overall
2. In time sequence (if the order is known)
3. Against the predicted values
4. Against the independent variables

Overall plot of residuals

When the assumptions are correct, we expect to see residuals that follow an approximately normal distribution with zero mean. An overall plot of the residuals, such as a histogram, can be used to evaluate this. It is often useful to plot *standardized residuals* rather than actual residuals. Standardized residuals are obtained by dividing each residual by the standard error; the result is the residual expressed in standard deviations. Knowledge of the normal distribution can be brought to bear in the interpretation of such plots; e.g., approximately 95% of the results should fall between –2 and +2. It may be easier to judge the normality of the residuals by using normal probability plotting paper. Figure VI.36 shows a histogram of the standardized residuals for our example.

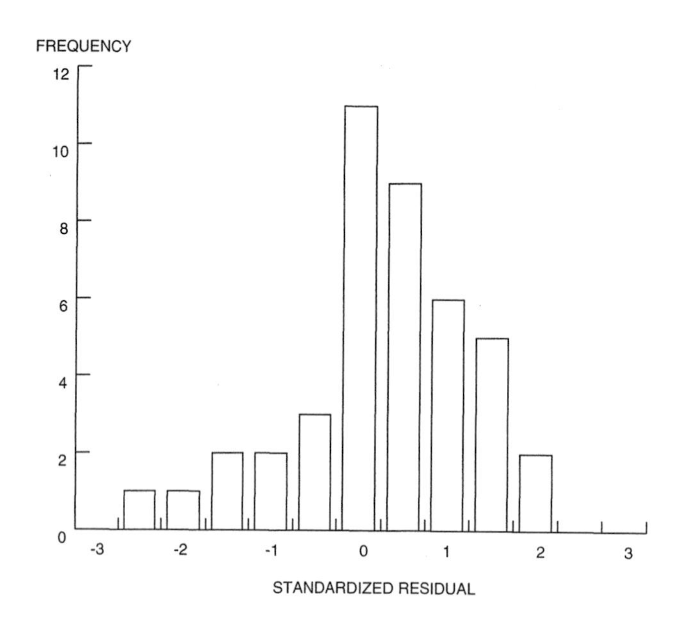

Figure VI.36. Histogram of standardized residuals.

When performing the other three types of analysis on the list, the experimenter should look for any non-randomness in the patterns. Figure VI.37 illustrates some common patterns of residuals behavior.

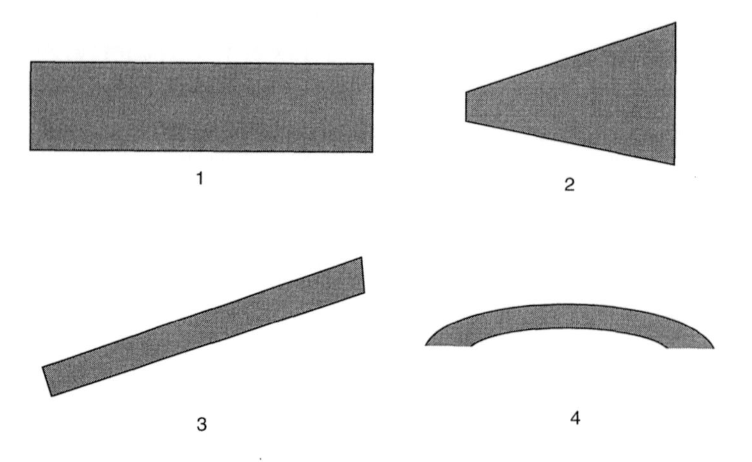

Figure VI.37. Residuals patterns.

Pattern #1 is the overall impression that will be conveyed when the model fits satisfactorily. Pattern #2 indicates that the size of the residual increases with respect to time or the value of the independent variable. It suggests the need for performing a transformation of the y values prior to performing the regression analysis. Pattern #3 indicates that the linear effect of the independent variable was not removed, perhaps due to an error in calculations. Pattern #4 appears when a linear model is fitted to curvilinear data. The solution is to perform a linearizing transformation of the *y*'s, or to fit the appropriate non-linear model.

In addition to the above, the engineer should always bring his scientific knowledge of the process to bear on the problem. Patterns may become apparent when the residuals are related to known laws of science, even though they are not obvious using statistical rules only. This is especially true when analyzing outliers, i.e., standardized residuals greater than 2.5 or so.

VI.F DESIGNING EXPERIMENTS

Designed experiments play an important role in quality improvement. This section will introduce the basic concepts involved and it will contrast the statistically designed experiment with the "one variable at a time" approach that has been used traditionally. Also briefly discussed are the concepts involved in Taguchi methods, statistical methods named after their creator, Dr. Genichi Taguchi.

THE TRADITIONAL APPROACH VS. STATISTICALLY DESIGNED EXPERIMENTS

The traditional approach, which most of us learned in high school science class, is to hold all variables constant except one. When this approach is used we can be sure that the variation is due to a cause and effect relationship. However, this approach suffers from a number of problems:

- It usually isn't possible to hold all other variables constant.
- There is no way to account for the effect of joint variation of independent variables, such as interaction.
- There is no way to account for experimental error, including measurement variation.

The statistically designed experiment usually involves varying two or more variables simultaneously and obtaining multiple measurements under the same experimental conditions. The advantage of the statistical approach is three-fold:

1. Interactions can be detected and measured. Failure to detect interactions is a major flaw in the "one variable at a time" approach.
2. Each value does the work of several values. A properly designed experiment allows you to use the same observation to estimate several different effects. This translates directly to cost savings when using the statistical approach.
3. Experimental error is quantified and used to determine the confidence the experimenter has in his conclusions.

TERMINOLOGY

Much of the early work on the design of experiments involved agricultural studies. The language of experimental design still reflects these origins. The experimental area was literally a piece of ground. A block was a smaller piece of ground with fairly uniform properties. A plot was smaller still and it served as the basic unit of the design. As the plot was planted, fertilized and harvested it could be split simply by drawing a line. A treatment was actually a treatment, such as the application of fertilizer. Unfortunately for the quality engineer, these terms are still part of the language of experiments. The engineer must do his or her best to understand quality improvement experimenting using these terms. Natrella (1963) recommends the following:

> Experimental area can be thought of as the scope of the planned experiment. For us, a block can be a group of results from a particular operator, or from a particular machine, or on a particular day—any planned natural grouping which should serve to make results from one block more alike than results from different blocks. For us, a treatment is the factor being investigated (material, environmental condition, etc.) in a single factor experiment. In factorial experiments (where several variables are being investigated at the same time) we speak of a treatment combination and we mean the prescribed levels of the factors to be applied to an experimental unit. For us, a yield is a measured result and, happily enough, in chemistry it will sometimes be a yield.

Definitions

A designed experiment is an experiment where one or more variables, called independent variables, believed to have an effect on the experimental outcome are identified and manipulated according to a predetermined plan. Data collected from a designed experiment can be analyzed statistically to determine the effect of the independent variables, or combinations of more than one independent variable. An experimental plan must also include provisions for dealing with extraneous variables, that is, variables not explicitly identified as independent variables.

Response variable—The variable being investigated, also called the *dependent variable*.

Primary variables—(aka main factors, main effects, inner array) The controllable variables believed most likely to have an effect. These may be quantitative, such as temperature, pressure, or speed, or they may be qualitative such as vendor, production method, or operator.

Background variables—(aka subsidiary factors, outer array) Variables, identified by the designers of the experiment, which may have an effect but either can not or should not be deliberately manipulated or held constant. The effect of background variables can contaminate primary variable effects unless they are properly handled. The most common method of handling background variables is blocking (blocking is described later in this section).

Experimental error—In any given experimental situation, a great many variables may be potential sources of variation. So many, in fact, that no experiment could be designed that deals with every possible source of variation explicitly. Those variables that are not considered explicitly are analogous to the common causes of variation described in VI.G. They represent the "noise level" of the process and their effects are kept from contaminating the primary variable effects by *randomization*. Randomization is a term meant to describe a procedure that assigns test units to test conditions in such a way that any given unit has an equal probability of being processed under a given set of test conditions.

Interaction—A condition where the effect of one factor depends on the level of another factor. Interaction is illustrated in Figure VI.38.

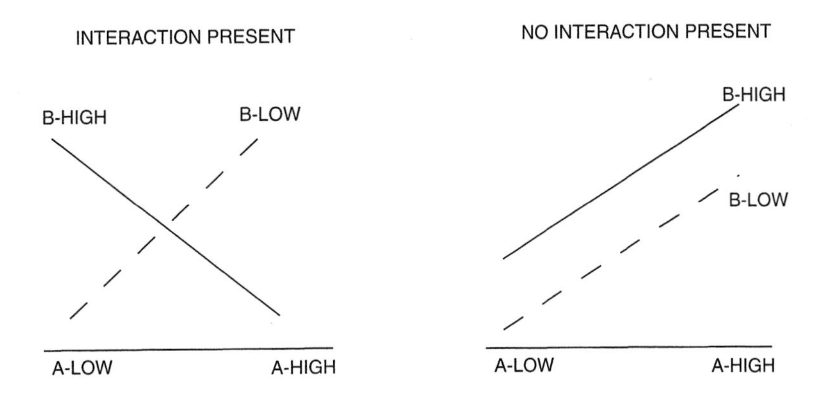

Figure VI.38. Illustration of interaction.

POWER AND SAMPLE SIZE

In designed experiments, the term *power of the test* refers to the probability that the F test will lead to accepting the alternative hypothesis when in fact the alternative hypothesis holds, i.e., 1–ß. To determine power probabilities we use the non-central F distribution. Charts have been prepared to simplify this task (see Appendix Table 17). The tables are indexed by values of ϕ. When all sample sizes are of equal size n, ϕ is computed using Equation VI.46.

$$\phi = \frac{1}{\sigma}\sqrt{\frac{n}{r}\Sigma(\mu_i - \mu_.)^2}$$

where: (VI.46)

$$\mu_. = \frac{\Sigma \mu_{i.}}{r}$$

r is the number of factor levels being studied.

The charts in Appendix Table 17 are used as follows:

1. Each page refers to a different v_1=r–1, the number of degrees of freedom for the numerator of the F statistic.
2. Two levels of significance are shown, α = 0.01 and α = 0.05. The left set of curves are used for α = 0.05 and the right set when α = 0.01.
3. There are separate curves for selected values of $v_2=\Sigma n-r$, the number of degrees of freedom for the denominator of the F statistic.
4. The X scale is in units of ϕ.
5. The Y scale gives the power, 1–β.

Example

Consider the curve on the second page of Table 17 for α=0.05, v_1=3, v_2=12. This ANOVA tests the hypothesis that four (v_1=4–1=3) populations have equal means with sample sizes of n=4 from each population (v_2=16–4=12). Reading above ϕ=2, we see that the chance of recognizing that the four populations do not actually have equal means when ϕ=2 is 0.82. It must be understood that there are many combinations of four unequal means that would produce ϕ=2.

DESIGN CHARACTERISTICS

Good experiments don't just happen; they are a result of careful planning. A good experimental plan depends on the following (Natrella 1963):

- The purpose of the experiment
- Physical restrictions on the process of taking measurements
- Restrictions imposed by limitations of time, money, material, and personnel.

The engineer must explain clearly why the experiment is being done, why the experimental treatments were selected, and how the completed experiment will accomplish the stated objectives. The experimental plan should be in writing and it should be endorsed by all key participants. The plan will include a statement of the objectives of the experiment, the experimental treatments to be applied, the size of the experiment, the time frame, and a brief discussion of the methods to be used to analyze the results. Two concepts are of particular interest to the quality engineer: replication and randomization.

Replication—The collection of more than one observation for the same set of experimental conditions. Replication allows the experimenter to estimate experimental error. If variation exists when all experimental conditions are held constant, the cause must be something other than the variables being controlled by the experimenter. Experimental error can be estimated without replicating the entire experiment. If a process has been in statistical control for a period of time, experimental error can be estimated from the control chart. Replication also serves to decrease bias due to uncontrolled factors.

Randomization—In order to eliminate bias from the experiment, variables not specifically controlled as factors should be randomized. This means that allocations of specimens to treatments should be made using some mechanical method of randomization, such as a random numbers table. Randomization also assures valid estimates of experimental error.

TYPES OF DESIGNS

Experiments can be designed to meet a wide variety of experimental objectives. A few of the more common types of experimental designs are defined here.

Fixed-effects model—An experimental model where all possible factor levels are studied. For example, if there are three different materials, all three are included in the experiment.

Random-effects model—An experimental model where the levels of factors evaluated by the experiment represent a sample of all possible levels. For example, if we have three different materials but only use two materials in the experiment.

Mixed model—An experimental model with both fixed and random effects.

Completely randomized design—An experimental plan where the order in which the experiment is performed is completely random, e.g.,

LEVEL	TEST SEQUENCE NUMBER
A	7, 1, 5
B	2, 3, 6
C	8, 4

Randomized-block design—An experimental design is one where the experimental observations are divided into "blocks" according to some criterion. The blocks are filled sequentially, but the order within the block is filled randomly. For example, assume we are conducting a painting test with different materials, material A and material B. We have four test pieces of each material. Ideally we would like to clean all of the pieces at the same time to assure that the cleaning process doesn't have an effect on our results; but what if our test requires that we use a cleaning tank that cleans two test pieces at a time? The tank load then becomes a "blocking factor." We will have four blocks, which might look like this:

MATERIAL	TANK LOAD	TEST PIECE NUMBER
A	1	7
B		1
B	2	5
A		2
B	3	3
A		6
B	4	4
A		8

Since each material appears exactly once per cleaning tank load we say the design is *balanced.* The material totals or averages can be compared directly. The reader should be aware that statistical designs exist to handle more complicated "unbalanced designs."

Latin-square designs—Designs where each treatment appears once and only once in each row and column. A Latin-square plan is useful when it is necessary or desirable to allow for two specific sources on non-homogeneity in the conditions affecting test results. Such designs were originally applied in agricultural experimentation when the two sources of non-homogeneity were the two directions on the field and the "square" was literally a square piece of ground. Its usage has been extended to many other applications where there are two sources of non-homogeneity that may affect experimental results—for example, machines, positions, operators, runs, days. A third variable is then associated with the other two in a prescribed fashion. The use of Latin squares is restricted by two conditions:

1. the number of rows, columns and treatments must all be the same;
2. there must be no interactions between row and column factors.

Natrella (1963, p. 13–30) provides the following example of a Latin square. Suppose we wish to compare four materials with regard to their wearing qualities. Suppose further that we have a wear-testing machine which can handle four samples simultaneously. Two sources of inhomogeneity might be the variations from run to run, and the variation among the four positions on the wear machine. In this situation, a 4 x 4 Latin square will enable us to allow for both sources of inhomogeneity if we can make four runs. The Latin square plan is as follows: (The four materials are labeled A, B, C, D).

Run	Position Number			
	(1)	(2)	(3)	(4)
1	A	B	C	D
2	B	C	D	A
3	C	D	A	B
4	D	A	B	C

Figure VI.39. A 4 x 4 Latin square.

The procedure to be followed in using a given Latin square is as follows:

1. Permute the columns at random;
2. Permute the rows at random;
3. Assign letters randomly to the treatments.

EXAMPLES OF APPLYING COMMON DOE METHODS USING SOFTWARE

This section includes examples of the most commonly used design of experiment methods using software. Whenever possible the examples employ popular software, such as Microsoft Excel. For detailed mathematical background on these methods, the reader is referred to any of the many fine books on the subject.* The Six Sigma Suite, a full-featured commercial software including tools for SPC, 7MP and DOE, is available from http://www.qualityamerica.com. A statistical analysis shareware package for Windows operating systems can be downloaded from http://www.dagonet.com/scalc.htm.

TWO WAY ANOVA WITH NO REPLICATES

When experiments are conducted which involve two factors, and it is not possible to obtain repeat readings for a given set of experimental conditions, a two-way analysis of variance may be used. The following example assumes that experimental treatments are assigned at random. Note that if the factors involved are each tested at only two levels, the full factorial analysis method described below could also be used.

Example of two way Anova with no replicates

An experiment was conducted to evaluate the effect of different detergents and water temperatures on the cleanliness of ceramic substrates. The experimenter selected three different detergents, based on their pH levels, and conducted a series of experiments at four different water temperatures. Cleanliness was quantified by measuring the contamination of a distilled water beaker after rinsing the parts cleaned using each treatment combination. The coded data are shown in Table VI.15.

---

*Montgomery, Douglas C., 1996; Box, Hunter and Hunter, 1978; Hicks, Charles R., 1993.

Table VI.15. Cleaning experiment raw data.

	DETERGENT A	DETERGENT B	DETERGENT C
Cold	15	18	10
Cool	12	14	9
Warm	10	18	7
Hot	6	12	5

Table VI.16. Cleaning experiment two-way ANOVA output from Microsoft Excel. (Two-factor without replication.)

SUMMARY OUTPUT						
	Count	*Sum*	*Average*	*Variance*		
Cold water	3	43	14.333333	16.33333		
Cool water	3	35	11.666667	6.333333		
Warm water	3	35	11.666667	32.33333		
Hot water	3	23	7.6666667	14.33333		
Detergent A	4	43	10.75	14.25		
Detergent B	4	62	15.5	9		
Detergent C	4	31	7.75	4.916667		
ANOVA						
Source of variation	*SS*	*df*	*MS*	*F*	*P-value*	*F crit*
Rows	68	3	22.666667	8.242424	0.015043179	4.757055
Columns	122.1666667	2	61.083333	22.21212	0.001684751	5.143249
Error	16.5	6	2.75			
Total	206.6666667	11				

Part one of the Excel output provides descriptive statistics on the different treatment levels. The ANOVA table is shown in part two. Note that in the previously presented raw data table the rows represent the different temperatures and the columns the different detergents. Because there are no replicates, Excel is not able to provide an estimate of the interaction of detergent and water temperature. If it is suspected that an interaction may be present, then the experiment should be replicated to estimate this effect. For this experiment, any P-value less than 0.05 would indicate a significant effect. The ANOVA table indicates that there are significant differences between the different detergents and the different water temperatures. To identify *which* differences are significant the experimenter can examine the means of the different detergents and water temperatures using *t*-tests. (Excel's data analysis tools add-in includes these tests). Be aware that the Type I error is affected by conducting multiple *t*-tests. If the Type I error on a single *t*-test is α, then the overall Type I error for k such tests is $1-(1-\alpha)^k$. For example, if $\alpha=0.01$ and three pairs of means are examined, then the combined Type I error for all three t-tests is $1-(1-0.01)^3 = 1-(0.99)^3 = 0.03$. Statistical methods exist that guarantee an overall level of Type I error for simultaneous comparisons (Hicks, 1973; pp. 31–38).

TWO WAY ANOVA WITH REPLICATES

If two factors are investigated which might interact with one another, and more than one result can be obtained for each combination of experimental treatments, then two-way analysis of variance with replicates may be used for the analysis. Spreadsheets such as Microsoft Excel include functions that perform this analysis.

Example of two way ANOVA with replicates

An investigator is interested in improving a process for bonding photoresist to copper clad printed circuit boards. Two factors are to be evaluated: the pressure used to apply the photoresist material and the preheat temperature of the photoresist. Three different pressures and three different temperatures are to be evaluated; the number of levels need not be the same for each factor and there is no restriction on the total number of levels. Each experimental combination of variables is repeated five times. Note that while Excel requires equal numbers of replicates for each combination of treatments, most statistical analysis packages allow different sample sizes to be used. The experimenter recorded the number of photoresist defects per batch of printed wiring boards. The coded data are shown in Table VI.17.

Table VI.17. Photoresist experiment raw data. ANOVA results.

	HIGH PRESSURE	MED PRESSURE	LOW PRESSURE
High temp	39	32	18
	30	31	20
	35	28	21
	43	28	25
	25	29	26
Med temp	38	10	22
	31	15	28
	31	25	29
	30	31	26
	35	36	20
Low temp	30	21	25
	35	22	24
	36	25	20
	37	24	21
	39	27	21

These data were analyzed using Excel's two-way ANOVA with replicates function. The results are shown in Table VI.18.

Table VI.18. Cleaning experiment two-way ANOVA output from Microsoft Excel. (Two-factor with replication.)

SUMMARY OUTPUT				
	High pressure	*Med pressure*	*Low pressure*	*Total*
High temp				
Count	5	5	5	15
Sum	172	148	110	430
Average	34.4	29.6	22	28.66667
Variance	50.8	3.3	11.5	46.66667
Med temp				
Count	5	5	5	15
Sum	165	117	125	407
Average	33	23.4	25	27.13333
Variance	11.5	117.3	15	59.98095
Low temp				
Count	5	5	5	15
Sum	177	119	111	407
Average	35.4	23.8	22.2	27.13333
Variance	11.3	5.7	4.7	43.26667
Total				
Count	15	15	15	
Sum	514	384	346	
Average	34.26666667	25.6	23.06666667	
Variance	22.06666667	44.68571429	10.92380952	

ANOVA						
Source of variation	*SS*	*df*	*MS*	*F*	*P-value*	*F crit*
Sample	23.5111111	2	11.7555556	0.45781	0.6363	3.259444
Columns	1034.84444	2	517.422222	20.1506	1.34E-06	3.259444
Interaction	139.555556	4	34.8888889	1.35872	0.267501	2.633534
Within	924.4	36	25.6777778			
Total	2122.31111	44				

As before, part one of the Excel output provided descriptive statistics on the different treatment levels. The ANOVA table is shown in part two. Because there were now replicates, Excel was able to provide an estimate of the interaction of pressure and temperature. For this experiment, the experimenter decided that any P-value less than 0.05 would indicate a significant effect. The ANOVA table p-value of less than 0.001 indicates that there were significant differences between the different columns (pressure), but the p-value of 0.6363 indicated that there was not a significant difference between the rows (temperature). The interaction of pressure and temperature was also not significant, as indicated by the p-value of 0.267501.

Since the p-value indicated that at least one difference was significant, the largest difference of 34.26666667–23.06666667 = 11.2 is known to be significant. To identify *which other* differences are significant the experimenter can examine the means of the different pressures using *t*-tests. (Excel's data analysis tools add-in includes these tests). Be aware that the Type I error is affected by conducting multiple *t*-tests. If the Type I error on a single *t*-test is α, then the overall Type I error for k such tests is $1-(1-\alpha)^k$. For example, if $\alpha=0.01$ and three pairs of means are examined, then the combined Type I error for all three *t*-tests is $1-(1-0.01)^3 = 1-(0.99)^3 = 0.03$.

FULL AND FRACTIONAL FACTORIAL

Full factorial experiments are those where at least one observation is obtained for every possible combination of experimental variables. For example, if A has 2 levels, B has 3 levels and *C* has 5 levels, a full factorial experiment would have at least 2x3x5=30 observations.

Fractional factorial or *fractional replicate* are experiments where there are some combinations of experimental variables where observations were not obtained. Such experiments may not allow the estimation of every interaction. However, when carefully planned, the experimenter can often obtain all of the information needed at a significant savings.

Analyzing factorial experiments

A simple method exists for analyzing the common 2^n experiment. The method, known as the Yates method, can be easily performed with a pocket calculator or programmed into a spreadsheet. It can be used with any properly designed 2^n experiment, regardless of the number of factors being studied.

To use the Yates algorithm, the data are first arranged in standard order (of course, the actual running order is random). The concept of standard order is easier to understand if demonstrated. Assume that we have conducted an experiment with three factors, A, B, and C. Each of the three factors is evaluated at two levels, which we will call low and high. A factor held at a low level will be identified with a "–" sign, one held at a high level will be identified with a "+" sign. The eight possible combinations of the three factors are identified using the scheme shown in the table below.

I.D.	A	B	C
(1)	–	–	–
a	+	–	–
b	–	+	–
ab	+	+	–
c	–	–	+
ac	+	–	+
bc	–	+	+
abc	+	+	+

Note that the table begins with all factors at their low level. Next, the first factor is high and all others are low. When a factor is high, it is shown in the identification column, otherwise it is not. E.g., whenever "*a*" appears it indicates that factor A is at its high level. To complete the table you simply note that as each factor is added to the table it is "multiplied" by each preceeding row. Thus, when *b* is added it is multiplied by *a*, giving the row *ab*. When *c* is added it is multiplied by, in order, *a*, *b*, and *ab*, giving the remaining rows in the table. (As an exercise, the reader should add a fourth factor D to the above table. Hint: the result will be a table with eight more rows.) Once the data are in standard order, add a column for the data and one additional column for each variable; e.g., for our three variables we will add four columns.

I.D.	A	B	C	DATA	1	2	3
(1)	–	–	–	–2	–7	21	–17
a	+	–	–	–5	28	–38	–15
b	–	+	–	15	–29	–5	–55
ab	+	+	–	13	–9	–10	1
c	–	–	+	–12	–3	35	–59
ac	+	–	+	–17	–2	20	–5
bc	–	+	+	–2	–5	1	–15
abc	+	+	+	–7	–5	–0	–1

Record the data in the data column (if the experiment has been replicated, record the totals). Now record the sum of the data values in the first two rows i.e., *(1)+a* in the first cell of the column labeled column 1. Record the sum of the next two rows in the second cell (i.e., *b+ab*). Continue until the top half of column 1 is completed. The lower half of column 1 is completed by subtracting one row from the next; e.g., the fifth value in column 1 is found by subtracting –5–2=–7. After completing column 1 the same process is completed for column 2, using the values in column 1. Column 3 is created using the value in column 2. The result is shown above. The purpose of the three columns is in providing the input needed to calculate the Effect column, as shown in the example that follows.

Example of Yates method

The table below shows sample data from an actual experiment. The experiment involved a target shooter trying to improve the number of targets hit per box of 25 shots. Three variables were involved: *a*=the gauge of the shotgun (12-gauge and 20-gauge), *b*=the shot size (6 shot and 8 shot), and *c*=the length of the handle on the target thrower (short or long). The shooter ran the experiment twice. The column labeled "1st" is the number of hits the first time the combination was tried. The column labeled "2nd" is the number of hits the second time the combination was tried. The Yates analysis begins with the sums shown in the column labeled Sum.

ID	1st	2nd	Sum	1	2	3	Effect	df	SS	MS	F ratio
1	22	19	41	86	167	288	18	Avg.			
a	21	24	45	81	121	20	2.5	1	25.00	25.00	3.64
b	20	18	38	58	9	0	0	1	0.00	0.00	0.00
ab	21	22	43	63	11	4	0.5	1	1.00	1.00	0.15
c	12	15	27	4	–5	–46	–5.75	1	132.25	132.25	19.24
ac	12	19	31	5	5	2	0.25	1	0.25	0.25	0.04
bc	13	15	28	4	1	10	1.25	1	6.25	6.25	0.91
abc	20	15	35	7	3	2	0.25	1	0.25	0.25	0.04
Error								8	55.00	6.88	
Total	141	147						15	220.00		

The first row in the Effect column is simply the value in the first row of column 3 (288) divided by the number of samples ($r*2^n=2*2^3=16$), where r is the number of replicates (in this case r=2), and n is the number of factors (in this case n=3). The first row in the Effects column is thus the average of the samples. Subsequent rows in the Effect column are found by dividing the numbers in column 3 by $r*2^{n-1}$. The Effect column provides the impact of the given factor on the response; thus, the shooter hit, on average, 2.5 more targets per box when shooting a 12-gauge than he did when shooting a 20-gauge.

The next question is whether or not these differences are statistically significant; i.e., could they be due to chance alone? To answer this question we will use the F-ratio of the effect MS for each factor to the error MS. The degrees of freedom (df) for each effect is simply 1 (the number of factor levels minus 1), the total df is N–1, and the error df is the total df minus the sum of the factor dfs. The sum of squares (SS) for each factor is the column 3 value squared divided by $r*2^n$; e.g., $SS_A=20^2/16=25$. The total SS is the sum of the individual values squared minus the first row in column 3 squared divided by $r*2^n$; e.g.,

$$\left(22^2+21^2+\ldots+15^2\right)-\frac{288^2}{16}=220 .$$

The error SS is the total SS minus the factor SS. The MS and F columns are computed using the same approach as shown above for one-way ANOVA. For the example the F-ratio for factor *c* (thrower) is significant at $\alpha<0.01$ and the F-ratio for factor *a* (gauge) is significant at $\alpha<0.10$; no other F-ratios are significant.

EMPIRICAL MODEL BUILDING AND SEQUENTIAL LEARNING

If you are new to design of experiments and empirical model building, a metaphor may prove helpful. Imagine that you suddenly wake up in a strange wilderness. You don't know where you are, but you'd like to climb to the top of the nearest hill to see if there are any signs of civilization. What would you do?

A first step might be to take a good look around you. Is there anything you should know before starting out? You would probably pay particular attention to things that might be dangerous. If you are in a jungle these might be dangerous animals, quicksand, and other things to avoid. You'd also look for things that could be used for basic survival, such as food, shelter, and clothing. You may wish to establish a "base camp" where you can be assured that all the basic necessities are available; a safe place to return to if things get a bit too exciting. In empirical modeling we also need to begin by becoming oriented with the way things are, before we proceed to changing them. We will call this *knowledge discovery* activity phase 0.

Now that you have a feel for your current situation and you feel confident that you know something about where you are, you may begin planning your trip to the highest hill. Before starting out you will probably try to determine what you will need to make the trip. You are only interested in things that are truly important. However, since you are new to jungle travel, you decide to make a few short trips to be sure that you have what you need. For your first trip you pack up every conceivable item and set out. In all likelihood you will discover that you have more than you need. Those things that are not important you will leave at your camp. As part of your short excursions you also learn something about the local terrain close to your camp, not much, of course, but enough to identify which direction is uphill. This phase is equivalent to a *screening experiment,* which we call Phase I.

You now feel that you are ready to begin your journey. You take only those things you will need and head out into the jungle in the uphill direction. From time to time you stop to get your bearings and to be sure that you are still moving in the right direction. We call this hill-climbing *steepest ascent,* or Phase II.

At some point you notice that you are no longer moving uphill. You realize that this doesn't mean that you are at the highest point in your area of the jungle, only that you are no longer moving in the right direction. You decide to stop and make camp. The next morning you begin to explore the local area more carefully, making a few short excursions from your camp. The jungle is dense and you learn that the terrain in the immediate vicinity is irregular, sometimes steep, sometimes less steep. This is in contrast to the smooth and consistent uphill slope you were on during your ascent. We call this phase of your journey the *factorial experiment,* or Phase III.

Now you decide that a more organized approach will be needed to locate the nearby peak. You break out the heavy artillery, the GPS you've been carrying since the beginning! (one of those cheap ones that doesn't have built-in maps) You take several altitude readings from near your camp, and others at a carefully measured distance on all major compass headings. Each time you carefully record the altitude on a hand-drawn map. You use the map to draw contour lines of equal altitude and eventually a picture emerges that clearly shows the location of the top of the hill. This is the *composite design phase,* which we call Phase IV.

At last you reach the top of the hill. You climb to the top of a tree and are rewarded with a spectacular view, the best for miles around. You decide that you love the view so much, you will build your home on this hill and live there permanently. You make your home sturdy and strong, able to withstand the ravages of wind and weather that are sure to come to your little corner of the jungle. In other words, your home design is *robust,* or impervious to changes in its environment. We call the activity of building products and processes that are insensitive to changes in their operating parameters *robust product and process design,* which is Phase V of the journey.

Now that this little tale has been told, let's go on to the real thing, improving your products, processes, and services!

PHASE 0: GETTING YOUR BEARINGS

"Where are we anyway?"

Before any experimentation can begin the team should get an idea of the major problems, important measures of performance, costs, time and other resources available for experimentation, etc. Methods and techniques for conducting Phase 0 research are described in chapters V and VI. The author recommends that SPC be applied to the process before experimentation. SPC allows the separation of factors into the categories of special and common causes. The process of discovering which variables belong to which class is extremely valuable in development of an experimental plan.

The central premise of the approach described in this section is that learning is, by its very nature, a sequential process. The experimenter, be it an individual or a team, begins with relatively little specific knowledge and proceeds to gain knowledge by conducting experiments on the process. As new knowledge is acquired, the learner is better able to determine which step is most appropriate to take next. In other words, experimentation always involves guesswork; but guesses become more educated as experimental data become available for analysis.

This approach is in contrast to the classical approach where an effort is made to answer all conceivably relevant questions in one large experiment. The classical approach to experimentation was developed primarily for agricultural experiments. Quality Engineering applications are unlike agricultural applications in many ways, especially in that results become available quickly. The approach described here takes advantage of this to accelerate and direct learning.

An example from electronic manufacturing will be used. At the outset, a team of personnel involved in a soldering process received a mission from another team that had been evaluating problems for the factory as a whole. The factory team had learned that a leading reason for customer returns was solder problems. Another team discovered that the solder area spent more resources in terms of floor space than other areas; a major usage of floor space was for the storage of defective circuit boards and the repair of solder defects. Thus, the solder process improvement team was formed and asked to find ways to eliminate solder defects if possible, or to at least reduce them by a factor of 10. Team members included a Quality Engineering technical leader, a process engineer, an inspector, a production operator, and a product engineer.

The team spent several meetings reviewing Pareto charts and problem reports. It also performed a process audit which uncovered several obvious problems. When the problems were repaired the team conducted a process capability study, which revealed a number of special causes of variation, which were investigated and corrected. Over a four-month period, this preliminary work resulted in a 50% reduction in the number of solder defects, from about 160 defects per standard unit to the 70–80 defect range. The productivity of the solder area nearly doubled as a result of these efforts. While impressive, the results were still well short of the 10X minimum improvement the team was asked to deliver.

PHASE I: THE SCREENING EXPERIMENT

"What's important here?"

At this point the process was stable and the team was ready to move from the process control stage to the process improvement stage. This involved conducting designed experiments to measure important effects. The solder team decided to list as many items as possible that might be causing solder problems. Since many variables had already been studied as part of the Phase 0 work, the list was not unreasonably long. The team looked at ways to control the variables listed and was able to develop methods for eliminating the effects of many variables on their list. The remaining list included the following factors:

VARIABLE	LOW LEVEL (–)	HIGH LEVEL (+)
A: Pre-baking of boards in an oven	No	Yes
B: Pre-heat time	10 seconds	20 seconds
C: Pre-heat temperature	150 °F	200 °F
D: Distance from pre-heat element to board surface	25cm	50 cm
E: Line speed	3 fpm	5 fpm
F: Solder temperature	495 °F	505 °F
G: Circuit density	Low	High
H: Was the board in a fixture?	No	Yes

This information was used to create an experimental design using a statistical software package. There are many packages on the market that perform similar analyses to the one shown here.

Since this was only to be a screening experiment, the team was not interested in obtaining estimates of factor interactions. The focus was to identify important main effects. The software allowed selection from among several designs. The Quality Engineer decided upon the design which would estimate the main effects with the smallest number of test units. This design involved testing 16 units. The data matrix produced by the computer is shown in Table VI.19. The run order had been randomized by the computer. If the experiment could not be conducted in that particular order, the computer software would allow the data to be run in blocks and it would adjust the analysis accordingly. The program also tells us that the design is of resolution IV, which means that main effects are not confounded with each other or any two factor interactions. (See Appendix 3 for more details on *confounding*.)

Table VI.19. Screening experiment layout. Data matrix (randomized).

RUN	A	B	C	D	E	F	G	H	RESPONSE
1	+	–	–	–	–	+	+	+	65
2	+	–	+	+	–	+	–	–	85
3	+	+	–	–	+	+	–	–	58
4	–	+	–	–	+	–	+	+	57
5	–	–	–	–	–	–	–	–	63
6	+	+	+	+	+	+	+	+	75
7	–	+	–	+	–	+	+	–	77
8	–	+	+	–	–	+	–	+	60
9	+	–	+	–	+	–	–	+	67
10	+	+	+	–	–	–	+	–	56
11	–	–	+	–	+	+	+	–	63
12	–	–	–	+	+	+	–	+	81
13	+	+	–	+	–	–	–	+	73
14	+	–	–	+	+	–	+	–	87
15	–	+	+	+	+	–	–	–	75
16	–	–	+	+	–	–	+	+	84

In Table VI.19 the "–" indicates that the variable was run at its low level, while a "+" sign indicate that it was run at its high level. For example, the unit for run #16 was processed as follows:

- ✓ Pre-baking = No
- ✓ Pre-heat time = 10 seconds
- ✓ Pre-heat temperature = 200 degrees F
- ✓ Distance from preheat element to board surface = 50 cm
- ✓ Line speed = 3 fpm
- ✓ Solder temperature = 495 degrees F
- ✓ Circuit density = High
- ✓ Fixture used = Yes
- ✓ Defects per standard unit = 82

Experimental data were collected using the randomized run order recommended by the software. The "response" column are data that were recorded in terms of defective solder joints per "standard unit", where a standard unit represented a circuit board with a median number of solder joints.* The results are shown in Table VI.20.

---

*Technically, a Poisson model would be the correct choice here. However, use of a normal model, which the analysis assumes, is reasonably accurate for defect counts of this magnitude. The team also evaluated the variance, more specifically, the log of the variance. The variances at each factor combination did not differ significantly and are not shown here.

Table VI.20. Results of experimental data analysis. Fractional factorial fit.

Estimated effects and coefficients for response (coded units)

Term	*Effect*	*Coef*	*StDev coef*	*T*	*P*
Constant		70.375	0.6597	106.67	0.000
A	−0.750	−0.375	0.6597	−0.57	0.588
B	8.000	4.000	0.6597	6.06	0.001
C	−0.500	−0.250	0.6597	−0.38	0.716
D	−18.500	−9.250	0.6597	−14.02	0.000
E	0.000	0.000	0.6597	0.00	1.000
F	−0.250	−0.125	0.6597	−0.19	0.855
G	−0.250	−0.125	0.6597	−0.19	0.855
H	0.250	0.125	0.6597	0.19	0.855

ANOVA for defects (coded units)

Source of variation	*df*	*Seq. SS*	*Adj. SS*	*Adj. MS*	*F*	*P-value*
Main effects	8	1629.00	1629.00	203.625	29.24	0.000
Residual error	7	48.75	48.75	6.964		
Total	15	1677.75				

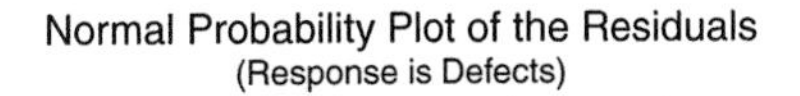

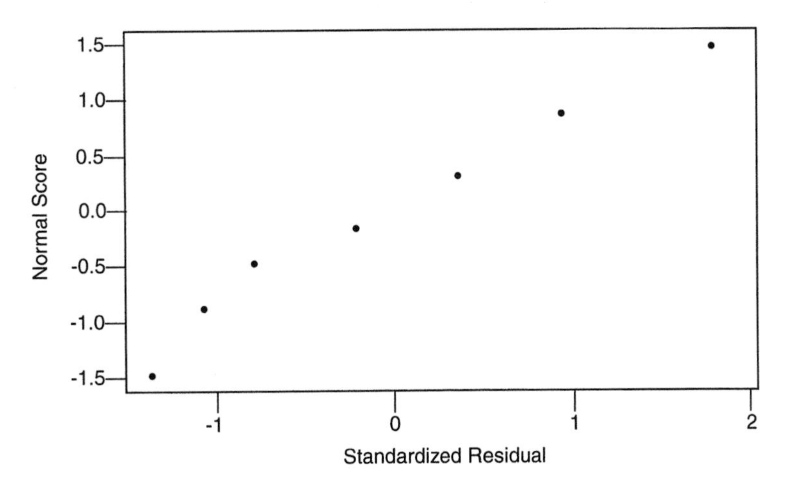

Figure VI.40. Residuals from experimental model.

A model that fit the data well would produce residuals that fall along a straight line. The Quality Engineer concluded that the fit of the model was adequate.

The analysis indicated that factors B (pre-heat time) and D (distance from pre-heat element to board surface) produced significant effects. Figure VI.40 shows a normal probability plot of the experimental effects. This figure shows the coefficients column from

Table VI.20 plotted on a normal probability scale. If the factor's effect was due to chance variation it would plot close to the line representing normal variation. In Figure VI.41 the effects of B and D are shown to be further from the line than can be accounted for by random variation.

The effects of the significant factors are graphed in response units in Figure VI.41.

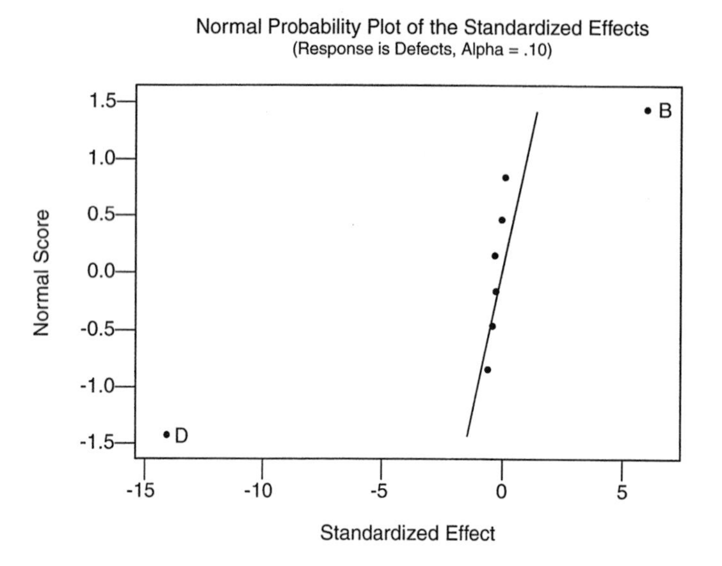

Figure VI.41. Significant factor effects.

Since the response is a defect count, the graph indicates that the low level of factor D gave better results, while the high level of factor B gave the better results. This can also be seen by examination of the coefficients for the variables. When D was low the average defect rate was 18.5 defects per unit better than when D was high; when B was high the average defect rate was 8 defects per unit better than when B was low.

The team met to discuss these results. They decided to set all factors that were not found to be statistically significant to the levels that cost the least to operate, and to set factors B and D at their midpoints. The process would be monitored at these settings for a while to determine that the results were similar to what the team expected based on the experimental analysis. While this was done, another series of experiments would be planned to further explore the significant effects uncovered by the screening experiment.

PHASE II: STEEPEST ASCENT (DESCENT)

"Which way is up?"

Based on the screening experiment, the linear model for estimating the defect rate was found from the coefficients in Table VI.20 to be

$$\text{Defect Rate} = 70.375 + 4B - 9.25D \qquad \text{(VI.47)}$$

The team wanted to conduct a series of experiments to evaluate factors B and D. The Phase I experiment revealed the direction and ratio at which B and D should be changed to get the most rapid improvement in the defect rate, i.e., the direction of steepest ascent (where "ascent" means improvement in the measurement of interest). To calculate a series of points along the direction of steepest ascent, start at the center of the design and change the factors in proportion to the coefficients of the fitted equation; i.e., for every 4 unit increase in factor B, factor D is decreased 9.25 units. For the data at hand, the center of the experiment and unit sizes are shown in Table VI.21.

Table VI.21. Unit sizes and center of experiment.

FACTOR	UNIT SIZE	CENTER
B	5	15 seconds
D	12.5	37.5 cm

A test unit was produced at the center value of B and D. The team decided that they would reduce the pre-heat time (B) in increments of 5 seconds (1 unit), while lowering the distance from the heating element (D) by increments of (9.25/4) x 12.5cm = 28.75 cm. This resulted in a single experiment where B = 20 seconds, D = 8.75 cm. The result was 52 defects per unit. However, despite the improved solder defect performance, the team noted that at the short distance the board was beginning to scorch. This necessitated that the team abandon the steepest ascent path. They conducted a series of experiments where board scorching was examined at different distances from the pre-heating element (factor D); they determined that a distance of at least 15cm was required to be confident that scorching would be avoided. To allow a margin of safety, the team set the distance D at 20cm. They then proceeded to increase pre-heat time in 5 second intervals, producing one board at each pre-heat setting. The results are shown in Table VI.22.

Table VI.22. Data for experiments on path of steepest descent.

RUN	B (sec.)	D (cm)	AVERAGE DEFECTS
1	15	37.5	70
2	20	8.75	52
3	25	20	51
4	30	20	31
5	35	20	18
6	40	20	12
7	45	20	10
8	50	20	13

These data are presented graphically in Figure VI.42.

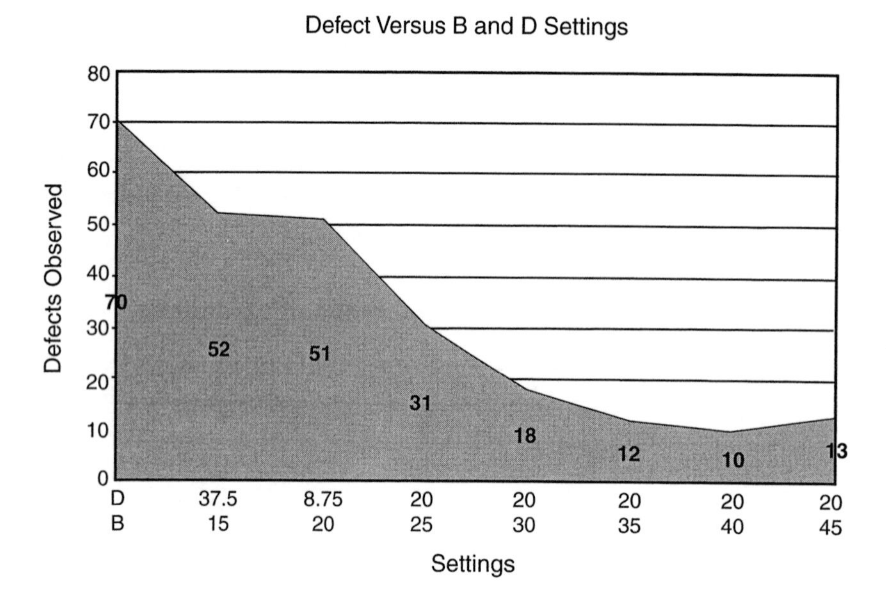

Figure VI.42. Steepest descent results.

With the distance fixed at 20 cm from the preheat element to the board surface, the best results were obtained with a pre-heat time of 40 seconds. Beyond that the defect rate was greater.

PHASE III: THE FACTORIAL EXPERIMENT

The team decided to conduct a factorial experiment near the best settings to explore that experimental region more thoroughly. To do so, they decided to run a factorial experiment which would allow them to estimate the two-factor BD interaction as well as the main effects. They also wished to determine any "curvature" in the area. This required that more than two levels be explored (only linear estimates are possible with two-level designs). Finally, the team wanted to obtain an estimate of the experimental error in the region; this required replicating the experiment. The design selected is shown in Table VI.23.

Code numbers used for the computer are shown in parentheses. The runs marked 0, 0 were center points. Note that each combination (i.e., set of plus and minus signs or zeros) was repeated three times. The team decided to center the design at the B value found in steepest B = 45 seconds. The interval for D was reduced to 2.5 cm and the experiment was centered one interval above D=20 (i.e., at D = 22.5).

Table VI.23. Replicated full-factorial design with center points.

RUN	B	D
1	40 (–1)	20.0 (–1)
2	45 (0)	22.5 (0)
3	50 (1)	25.0 (1)
4	40 (–1)	25.0 (1)
5	40 (–1)	20.0 (–1)
6	45 (0)	22.5 (0)
7	40 (–1)	25.0 (1)
8	40 (–1)	25.0 (1)
9	50 (1)	20.0 (–1)
10	50 (1)	25.0 (1)
11	40 (–1)	20.0 (–1)
12	40 (–1)	20.0 (–1)
13	50 (1)	25.0 (1)
14	50 (1)	20.0 (–1)
15	45 (0)	22.5 (0)

Table VI.24. Results of full factorial experiment with center points and replicates.

RUN	B	D	RESULT
1	40 (-1)	20.0 (-1)	11
2	45 (0)	22.5 (0)	9
3	50 (1)	25.0 (1)	11
4	40 (-1)	25.0 (1)	15
5	50 (1)	20.0 (-1)	12
6	45 (0)	22.5 (0)	10
7	40 (-1)	25.0 (1)	17
8	40 (-1)	25.0 (1)	15
9	50 (1)	20.0 (-1)	11
10	50 (1)	25.0 (1)	11
11	40 (-1)	20.0 (-1)	13
12	40 (-1)	20.0 (-1)	13
13	50 (1)	25.0 (1)	11
14	50 (1)	20.0 (-1)	11
15	45 (0)	22.5 (0)	10

Fractional factorial fit

Estimated effects and coefficients for defects (coded units)					
Term	*Effect*	*Coef*	*StDev coef*	*T*	*P*
Constant		12.583	0.2357	53.39	0.000
A	–2.833	–1.417	0.2357	–6.01	0.000
B	1.500	0.750	0.2357	3.18	0.010
A*B	–1.833	–0.917	0.2357	–3.89	0.003
Ct Pt		–2.917	0.5270	–5.53	0.000

Looking at the P column all terms in the model are significant (any P value below 0.05 indicates a significant effect.) This analysis is confirmed by the ANOVA table (Table VI.25).

Table VI.25. ANOVA for factorial experiment with center points.

ANOVA for defects (coded units)						
Source of variation	*df*	*Seq. SS*	*Adj. SS*	*Adj. MS*	*F*	*P-value*
Main effects	2	30.833	30.833	15.4167	23.13	0.000
2-way interactions	1	10.083	10.083	10.0833	15.13	0.003
Curvature	1	20.417	20.417	20.4167	30.63	0.000
Residual error	10	6.667	6.667	0.6667		
Pure error	10	6.667	6.667	0.6667		
Total	14	68.000				

Looking at the P column of the ANOVA table, it can be seen that main effects, the 2-way interaction, and "curvature" are all significant ($P < 0.05$). Curvature is measured by comparing the average response at the center points with the responses at the corner points of the design. The fact that curvature is significant means that we are no longer experimenting in a linear region of the responses. This means that the original coefficients, which were based on the linear model, are no longer adequate. Upon seeing these results, the Quality Engineer decided that it was necessary to proceed to Phase IV to better investigate the response region and to try to locate a stationary optimum.

PHASE IV: THE COMPOSITE DESIGN

The Quality Engineer decided to try using a design known as a *composite design* or *central composite design* to obtain additional information on the region where the process was operating. This design involved augmenting the corner points and center point of the previous factorial experiment with additional points, as shown in Figure VI.43. The points extended the design beyond the levels previously designed by the high and low values for each factor. The team decided that they could allow the distance to be decreased somewhat below the 20cm "minimum" distance because they had added a 5cm margin of safety. They also noted that they were now taking relatively small experimental steps compared to the large jumps they took during steepest ascent.

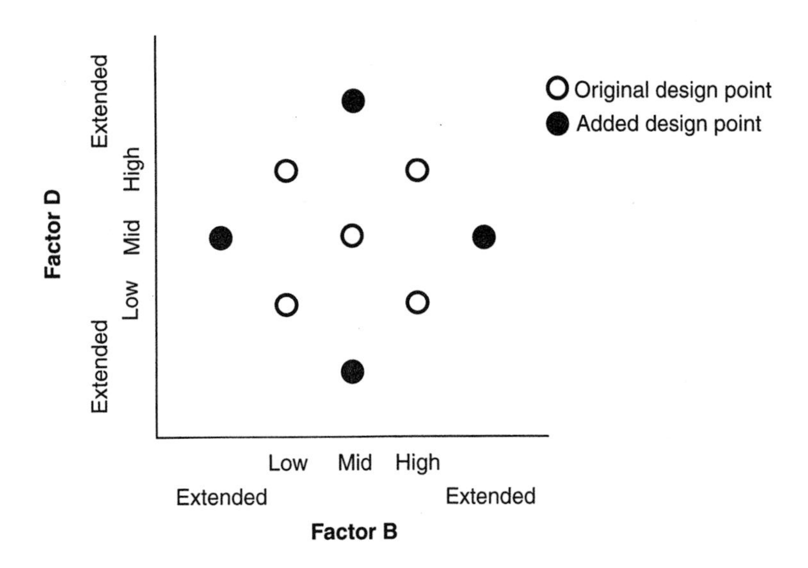

Figure VI.43. Central composite design for solder process.

DOE software finds the coefficients of the equation that describes a complex region for the responses. The equation being fitted is the following:

$$y = \beta_0 + \beta_1 x_1 + \beta_2 x_2 + \beta_{11} x_1^2 + \beta_{22} x_2^2 + \beta_{12} x_1 x_2 + \varepsilon \qquad \text{(VI.48)}$$

The region described by this equation may contain a maximum, a minimum, or a "saddle point." At a maximum any movement away from the stationary point will cause the response to decrease. At the minimum any movement away from the stationary point will cause the response to increase. At a saddle point moving away from the stationary value of one variable will cause a decrease, while moving away from the stationary value of the other variable will cause an increase. Some DOE software will give the values of X and Y at the stationary point, and the nature of the stationary point (max, min, or saddle).

Other DOE software display two and three dimensional drawings that graphically describe the region of experimentation. It is usually not difficult to interpret the response surface drawings.

The data collected by the team are shown in Table VI.26. Note that the data are shown in standard order, but the run order was random.

Table VI.26. Central composite design experiment and data.

B	D	DEFECTS
–1.41421	0.00000	16
1.00000	1.00000	11
0.00000	0.00000	9
0.00000	–1.41421	11
1.00000	–1.00000	9
1.41421	0.00000	4
0.00000	0.00000	10
0.00000	0.00000	10
0.00000	1.41421	15
0.00000	0.00000	9
0.00000	0.00000	10
–1.00000	1.00000	15
–1.00000	–1.00000	13

The computer analysis of these data is shown in Table VI.27.

Table VI.27. Analysis of central composite design.
Estimated regression coefficients for Y.

Estimated regression coefficients for defects				
Term	*Coef*	*StDev coef*	*T*	*P*
Constant	9.600	0.5880	16.326	0.000
B	–3.121	0.4649	–6.714	0.000
D	1.207	0.4649	2.597	0.036
B*B	0.325	0.4985	0.652	0.535
D*D	1.825	0.4985	3.661	0.008
B*D	0.000	0.6574	0.000	1.000
S = 1.315		R–Sq = 90.3%		R–Sq(adj) = 83.4%

The P-values indicate that all terms except the B^2 term and the interaction term are significant.

ANOVA for defects						
Source of variation	*df*	*Seq. SS*	*Adj. SS*	*Adj. MS*	*F*	*P*
Regression	5	112.821	112.8211	22.5642	13.05	0.002
Linear	2	89.598	89.5980	44.7990	25.91	0.001
Square	2	23.223	23.2231	11.6115	6.72	0.024
Interaction	1	0.000	0.0000	0.0000	0.00	1.000
Residual Error	7	12.102	12.1020	1.7289		
Lack-of-Fit	3	10.902	10.9020	3.6340	12.11	0.018
Pure Error	4	1.200	1.2000	0.3000		
Total	12	124.923				

Unusual observations for defects					
Observation	*Defects*	*Fit*	*StDev Fit*	*Residual*	*St Resid*
6	4.000	5.836	1.039	−1.836	−2.28R

R denotes an observation with a large standardized residual. The team confirmed the defect count for observation #6.

The ANOVA indicates that the lack-of-fit is not significant. It also indicates that the interaction term is not significant and could be removed from the model, gaining a degree of freedom for estimating the error.

The response surface 3D and contour plots are shown in Figure VI.44a and VI.44b.

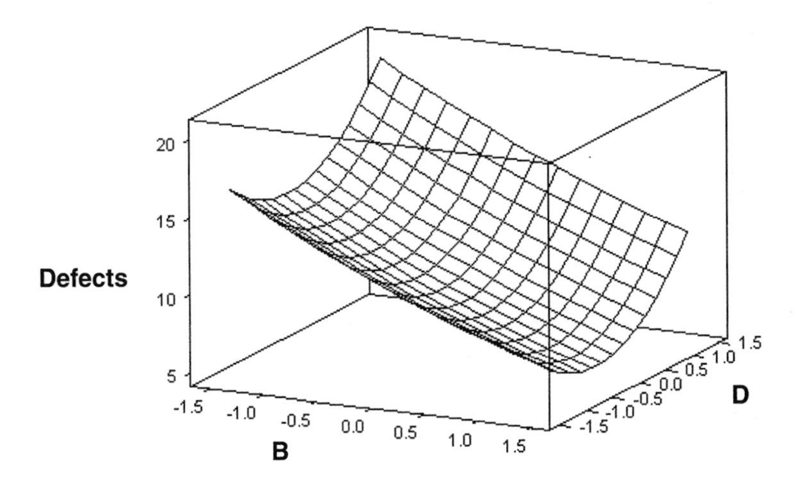

Figure VI.44a. Response surface plot for defect data.

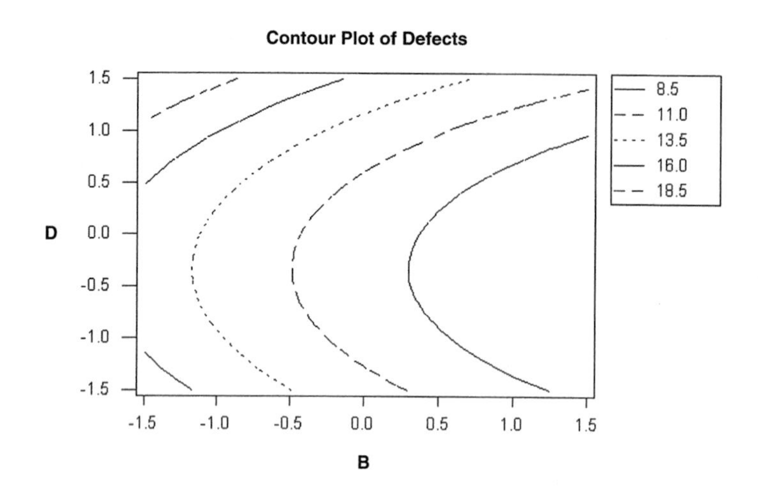

Figure VI.44b. Contour plot for defect data.

The analysis could become somewhat more advanced if the Quality Engineer chose to perform a *Canonical analysis* to investigate the nature of the response surface in greater detail. Canonical analysis involves finding a *stationary point S* and performing a coordinate system transformation to eliminate the cross-product and first order terms. The techniques for performing this analysis are described in a number of advanced texts (Box, G.E.P., and Draper 1987; Meyers and Montgomery, 1995). However, it is obvious from the contour plot and the 3D response surface plot that there may still be some room to improve by holding D constant and gradually increasing B.

At this point the team decided that they had reached a point of diminishing returns as far as the current process was concerned. The data indicated that the existing process for wave solder can, if properly controlled, manage to produce 10 or fewer defects per standard unit at the center of the last design. This was about 16 times better than the process was producing at the outset of the project and about 7 times better than the average result of the first experiment.

The team, guided by the Quality Engineer, decided to set the process at the center point of the last experiment (B=0, D=0) and to implement Evolutionary Operation (EVOP) to pursue further optimization. EVOP involves running a series of designed experiments on production units, with operating personnel making small changes (Box, G.E.P., and Draper, 1969). By restricting EVOP to small process changes the risk of producing scrap is reduced. Although the movements in process settings are small, the cumulative improvement in performance can be substantial. The apparent gradual slope of the defect rate in the +B direction also made it unlikely that the process would "fall off of a cliff" during EVOP.

The Quality Engineer helped set up EVOP on the process and train supervisory and hourly personnel in its use. She also agreed to provide ongoing support in the form of periodic visits and availability should questions arise. The team decided that after turning over process improvement to operating personnel, they would look at ways of maintaining their gains, while simultaneously investigating basic process and product design changes to obtain further improvement.

PHASE V: ROBUST PRODUCT AND PROCESS DESIGN

Maintaining gains involves, among other things, creating processes and products that operate close to their optimum conditions even when changes occur. Robust design can begin with careful study of the contour plot. Note that if you start at B = D = 0 and move from along a line from left to right the response changes relatively slowly. However, if you move from the center along a line from lower to upper, the defect rate increases rapidly. Robust process control planning should take such non-linearity into account. If there is a need to change factor B or D, they should be changed in a way that avoids increasing the defect rate. This does *not* mean that all changes should be forbidden; after all, without change there can be no learning or improvement. However, changes should be monitored (as with EVOP) to provide a filter between the customer and the production of non-conforming product that may occur during the learning process.

More formally, robust design can be integrated into experimental design. The methods described by Genichi Taguchi are a well-known approach to integrating DOE and product and process design. While there has been much criticism of Taguchi's statistical approach, there is a broad consensus that his principles of robust parameter design are both valid and valuable contributions to engineering science.

TAGUCHI ROBUSTNESS CONCEPTS

This section will introduce some of the special concepts introduced by Dr. Genichi Taguchi of Japan. A complete discussion of Taguchi's approach to designed experiments is beyond the scope of this course. However, many of Taguchi's ideas are useful in that they present an alternative way of looking at quality in general.

Introduction

Quality is defined as the loss imparted to the society from the time a product is shipped (Taguchi 1986). Taguchi divides quality control efforts into two categories: on-line quality control and off-line quality control.

On-line quality control—involves diagnosis and adjusting of the process, forecasting and correction of problems, inspection and disposition of product, and follow-up on defectives shipped to the customer.

Off-line quality control—quality and cost control activities conducted at the product and the process design stages in the product development cycle. There are three major aspects to off-line quality control:

1. **System design**—is the process of applying scientific and engineering knowledge to produce a basic functional prototype design. The prototype model defines the initial settings of product or process design characteristics.
2. **Parameter design**—is an investigation conducted to identify settings that minimize (or at least reduce) the performance variation. A product or a process can perform its intended function at many settings of its design characteristics. However, variation in the performance characteristics may change with different settings. This variation increases both product manufacturing and lifetime costs. The term *parameter design* comes from an engineering tradition of referring to product characteristics as product parameters. An exercise to identify optimal parameter settings is therefore called *parameter design*.
3. **Tolerance design**—is a method for determining tolerances that minimize the sum of product manufacturing and lifetime costs. The final step in specifying product and process designs is to determine tolerances around the nominal settings identified by parameter design. It is still a common practice in industry to assign tolerances by convention rather than scientifically. Tolerances that are too narrow increase manufacturing costs and tolerances that are too wide increase performance variation and the lifetime cost of the product.

Expected loss—the monetary losses an arbitrary user of the product is likely to suffer at an arbitrary time during the product's life span due to performance variation. Taguchi advocates modeling the loss function so the issue of parameter design can be made more concrete. The most often-used model of loss is the quadratic loss function illustrated in Figure VI.45. Note that the loss from operating the process is found by integrating the process pdf over the dollar-loss function. Under this model there is always a benefit from moving the process mean closer to the target value and reducing variation in the process.

Of course, there is often a cost associated with these two activities. Weighing the cost/benefit ratio is possible when viewed from this perspective.

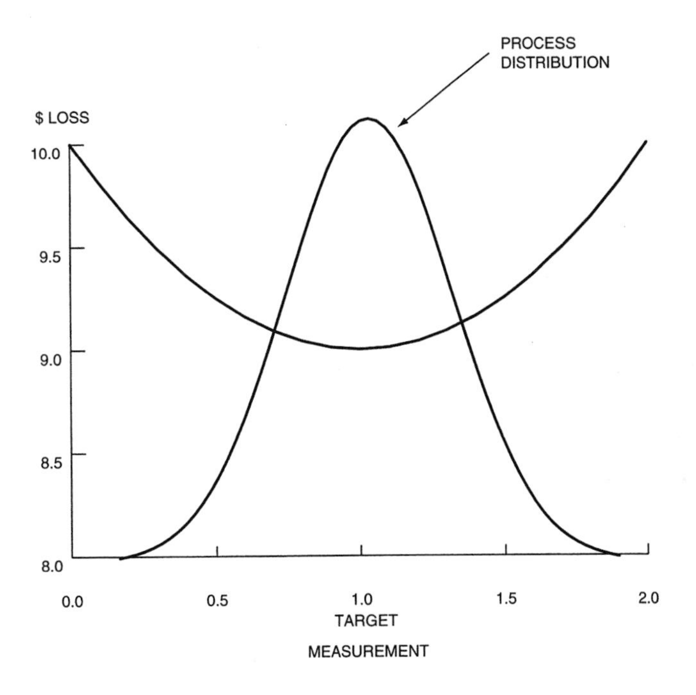

Figure VI.45. Taguchi's quadratic loss function.

Note the contrast between the quadratic loss function and the conceptual loss function implicit in the traditional management view. The traditional management approach to loss is illustrated in Figure VI.46.

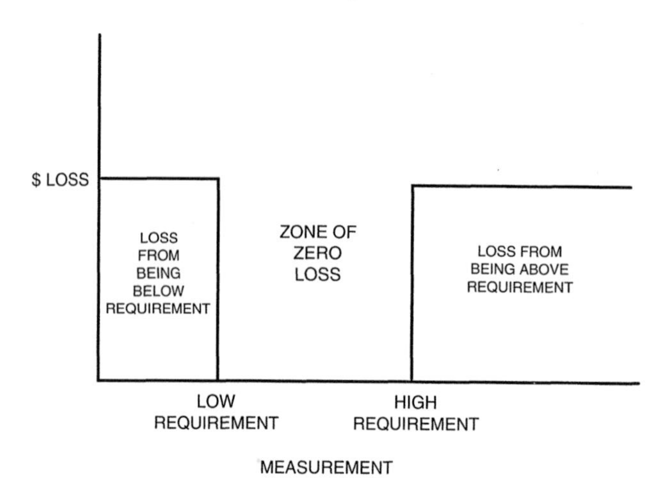

Figure VI.46. Traditional approach to loss.

Interpretation—there is no loss as long as a product or servicemeets requirements. There is no "target" or "optimum"; just barely meeting requirements is as good as operating anywhere else within the zone of zero loss. Deviating a great deal from requirements incurs the same loss as being just barely outside the prescribed range. The process distribution is irrelevant as long as it meets the requirements.

Note that under this model of loss there is no incentive for improving a process that meets the requirements since there is no benefit, i.e., the loss is zero. Thus, cost>benefit for any process that meets requirements. This effectively destroys the idea of continuous improvement and leads to the acceptance of an "acceptable quality level" as an operating standard.

Noise—the term used to describe all those variables, except design parameters, that cause performance variation during a product's life span and across different units of the product. Sources of noise are classified as either external sources or internal sources.

External sources of noise—variables external to a product that affect the product's performance.

Internal sources of noise—the deviations of the actual characteristics of a manufactured product from the corresponding nominal settings.

Performance statistics—estimate the effect of noise factors on the performance characteristics. Performance statistics are chosen so that maximizing the performance measure will minimize expected loss. Many performance statistics used by Taguchi use "signal to noise ratios" which account jointly for the levels of the parameters and the variation of the parameters.

Summary of the Taguchi method

The Taguchi method for identifying settings of design parameters that maximize a performance statistic is summarized by Kackar (1985):

- Identify initial and competing settings of the design parameters, and identify important noise factors and their ranges.
- Construct the design and noise matrices, and plan the parameter design experiment.
- Conduct the parameter design experiment and evaluate the performance statistic for each test run of the design matrix.
- Use the values of the performance statistic to predict new settings of the design parameters.
- Confirm that the new settings do indeed improve the performance statistic.

VI.G STATISTICAL PROCESS CONTROL (SPC)

TERMS AND CONCEPTS

Distributions

A central concept in statistical process control is that every measurable phenomenon is a statistical distribution. In other words, an observed set of data constitutes a sample of the effects of unknown common causes. It follows that, after we have done everything to eliminate special causes of variations, there will still remain a certain amount of variability exhibiting the state of control. Figure VI.47 illustrates the relationships between common causes, special causes, and distributions.

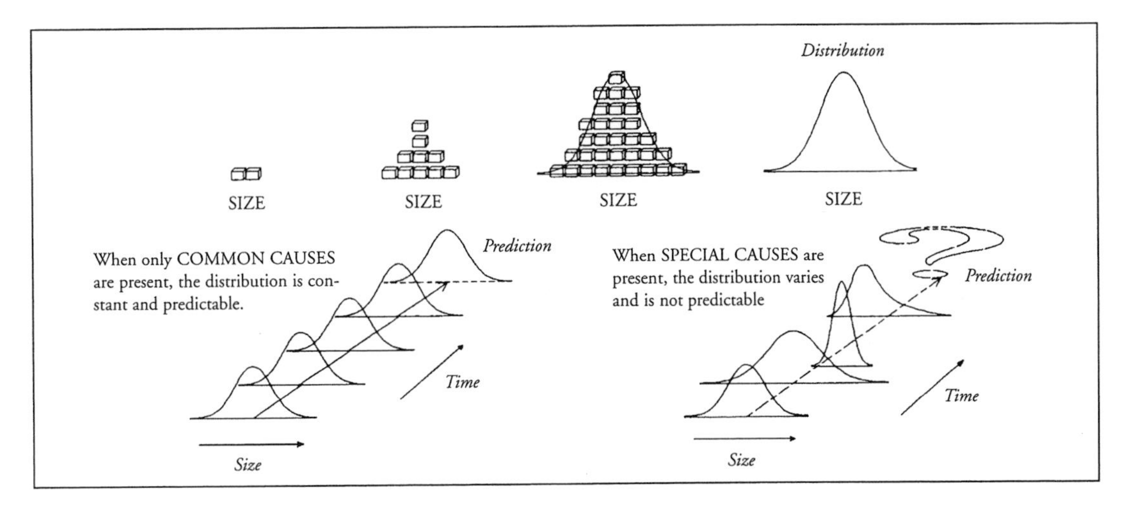

Figure VI.47a. Common causes, special causes, and distributions. *(Figure V.24 repeated.)*

From *Pyzdek's Guide to SPC-Volume One: Fundamentals*, p. 43. Copyright © 1998 by Thomas Pyzdek.

There are three basic properties of a distribution: location, spread, and shape. The *location* refers to the typical value of the distribution, such as the mean. The *spread* of the distribution is the amount by which smaller values differ from larger ones. The standard deviation and variance are measures of distribution spread. The *shape* of a distribution is its pattern—peakedness, symmetry, etc. A given phenomenon may have any one of a number of distribution shapes; e.g., the distribution may be bell-shaped, rectangular-shaped, etc.

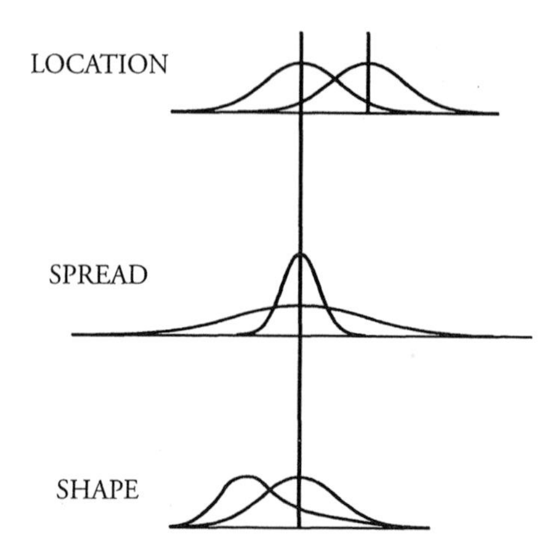

Figure VI.47b. How distributions vary from one another. *(Figure V.25 repeated.)*
From *Pyzdek's Guide to SPC-Volume One: Fundamentals*, p. 44. Copyright © 1998 by Thomas Pyzdek.

*Central limit theorem**

The central limit theorem can be stated as follows:

> Irrespective of the shape of the distribution of the population or universe, the distribution of average values of samples drawn from that universe will tend toward a normal distribution as the sample size grows without bound.

It can also be shown that the average of sample averages will equal the average of the universe and that the standard deviation of the averages equals the standard deviation of the universe divided by the square root of the sample size. Shewhart performed experiments that showed that small sample sizes were needed to get approximately normal distributions from even wildly non-normal universes. Figure VI.48 was created by Shewhart using samples of four measurements.

The practical implications of the central limit theorem are immense. Consider that without the central limit theorem effects, we would have to develop a separate statistical model for every non-normal distribution encountered in practice. This would be the only

*Material from section VI.A repeated for the convenience of the reader.

way to determine if the system were exhibiting chance variation. Because of the central limit theorem we can use *averages* of small samples to evaluate *any* process using the normal distribution. The central limit theorem is the basis for the most powerful of statistical process control tools, Shewhart control charts.

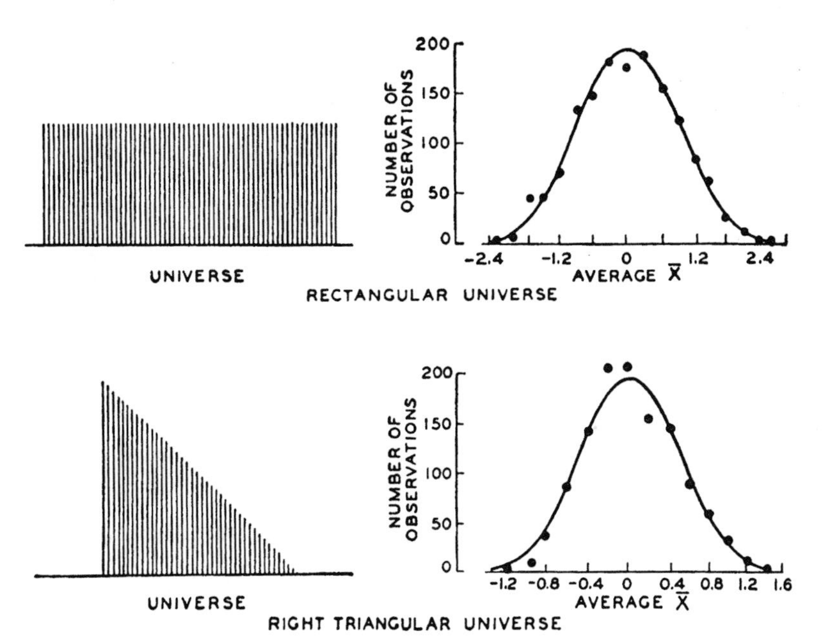

Figure VI.48. Illustration of the central limit theorem. *(Figure VI.2 repeated.)*

From Shewhart, Figure 59. Used by permission of the publisher.

OBJECTIVES AND BENEFITS

Without SPC, the basis for decisions regarding quality improvement are based on intuition, after-the-fact product inspection, or seat-of-the-pants "data analysis." SPC provides a scientific basis for decisions regarding process improvement.

Prevention versus detection

A process control system is essentially a feedback system that links process outcomes with process inputs. There are four main elements involved: the process itself, information about the process, action taken on the process, and action taken on the output from the process. The way these elements fit together is shown in Figure VI.49.

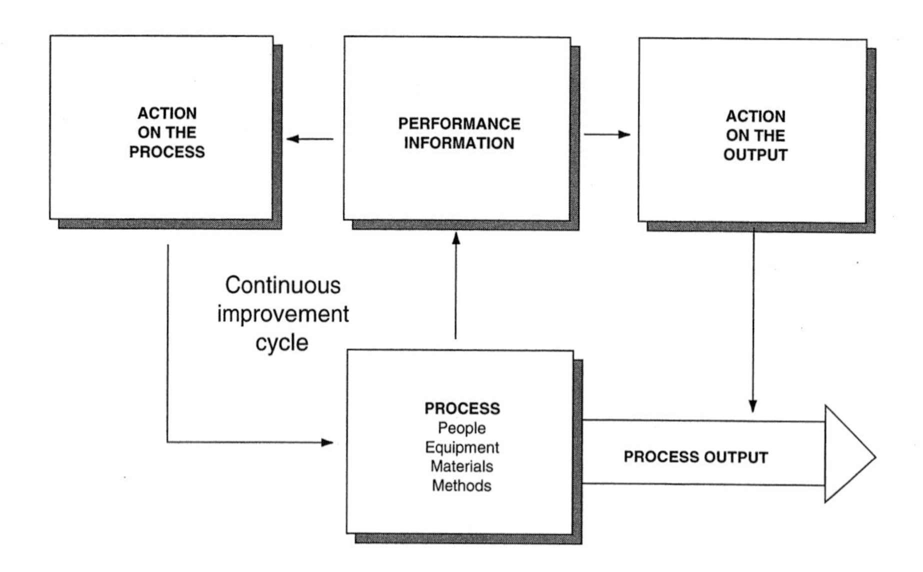

Figure VI.49. A process control system.

By the process, we mean the whole combination of people, equipment, input materials, methods, and environment that work together to produce output. The performance information is obtained, in part, from evaluation of the process output. The output of a process includes more than product; it also includes information about the operating state of the process such as temperature, cycle times, etc. Action taken on a *process* is future-oriented in the sense that it will affect output yet to come. Action on the *output* is past-oriented because it involves detecting out-of-specification output that has already been produced.

There has been a tendency in the past to concentrate attention on the detection-oriented strategy of product inspection. With this approach, we wait until an output has been produced, then the output is inspected and either accepted or rejected. SPC takes you in a completely different direction: improvement in the future. A key concept is *the smaller the variation around the target, the better.* Thus, under this school of thought, it is not enough to merely meet the requirements; continuous improvement is called for even if the requirements are already being met. The concept of never-ending, continuous improvement is at the heart of SPC.

COMMON AND SPECIAL CAUSES OF VARIATION*

Shewhart (1931, 1980) defined *control* as follows:

> A phenomenon will be said to be controlled when, through the use of past experience, we can predict, at least within limits, how the phenomenon may be expected to vary in the future. Here it is understood that prediction within limits means that we can state, at least approximately, the probability that the observed phenomenon will fall within the given limits.

The critical point in this definition is that control is not defined as the complete absence of variation. Control is simply a state where all variation is *predictable* variation. A controlled process isn't necessarily a sign of good management, nor is an out-of-control process necessarily producing non-conforming product.

In all forms of prediction there is an element of chance. For our purposes, we will call any unknown random cause of variation a *chance cause* or a *common cause*; the terms are synonymous and will be used as such. If the influence of any particular chance cause is very small, and if the number of chance causes of variation are very large and relatively constant, we have a situation where the variation is predictable within limits. You can see from the definition above that a system such as this qualifies as a controlled system. Where Dr. Shewhart used the term chance cause, Dr. W. Edwards Deming coined the term *common cause* to describe the same phenomenon. Both terms are encountered in practice.

Needless to say, not all phenomena arise from constant systems of common causes. At times, the variation is caused by a source of variation that is not part of the constant system. These sources of variation were called *assignable causes* by Shewhart, *special causes* of variation by Dr. Deming. Experience indicates that special causes of variation can usually be found without undue difficulty, leading to a process that is less variable.

Statistical tools are needed to help us effectively identify the effects of special causes of variation. This leads us to another definition:

Statistical process control—the use of valid analytical statistical methods to identify the existence of special causes of variation in a process.

The basic rule of statistical process control is as follows:

> Variation from common-cause systems should be left to chance, but special causes of variation should be identified and eliminated.

*See also V.C

This is Shewhart's original rule. However, the rule should not be misinterpreted as meaning that variation from common causes should be ignored. Rather, common-cause variation is explored "off-line." That is, we look for long-term process improvements to address common-cause variation.

Figure VI.50 illustrates the need for statistical methods to determine the category of variation.

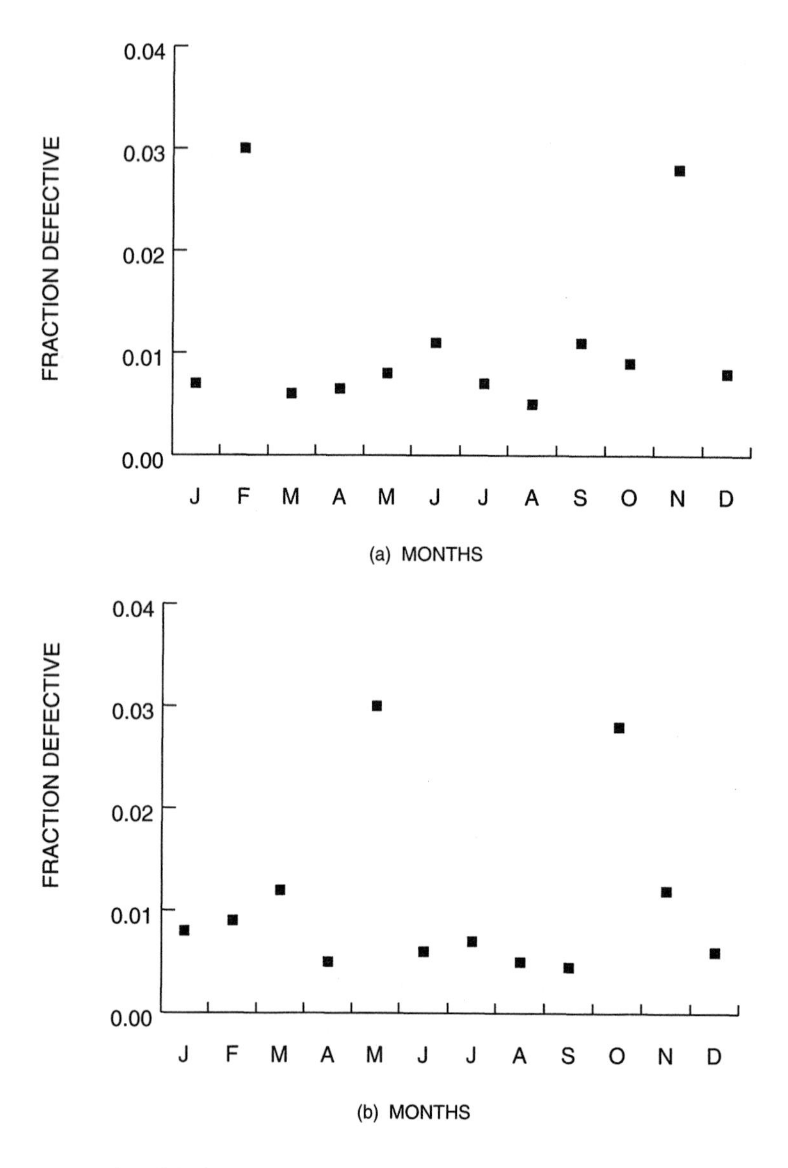

Figure VI.50. Should these variations be left to chance? *(Figure V.22 repeated.)*

From Shewhart, p. 13. Used by permission of the publisher.

The answer to the question "should these variations be left to chance?" can only be obtained through the use of statistical methods. Figure VI.51 illustrates the basic concept.

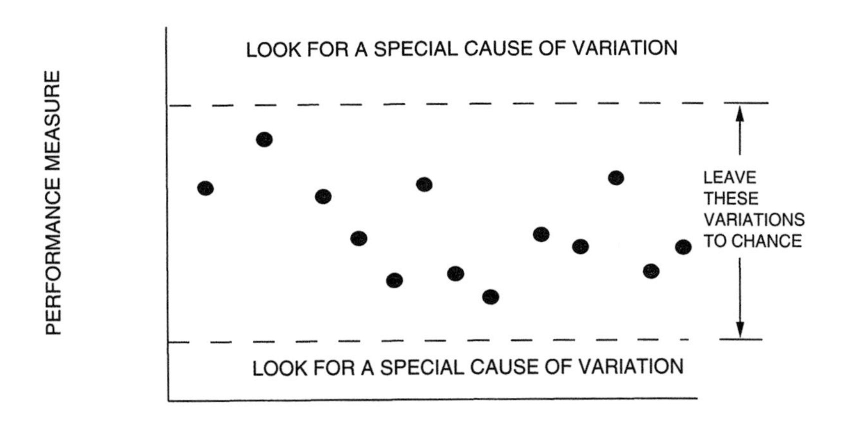

Figure VI.51. Types of variation.

In short, variation between the two "control limits" designated by the dashed lines will be deemed as variation from the common-cause system. Any variability beyond these fixed limits will be assumed to have come from special causes of variation. We will call any system exhibiting only common-cause variation "statistically controlled." It must be noted that the control limits are not simply pulled out of the air; they are calculated from actual process data using valid statistical methods. Figure VI.50 is shown below as Figure VI.52, only with the control limits drawn on it; notice that process (a) is exhibiting variations from special causes, while process (b) is not. This implies that the type of action needed to reduce the variability in each case are of a different nature. Without statistical guidance there could be endless debate over whether special or common causes were to blame for variability.

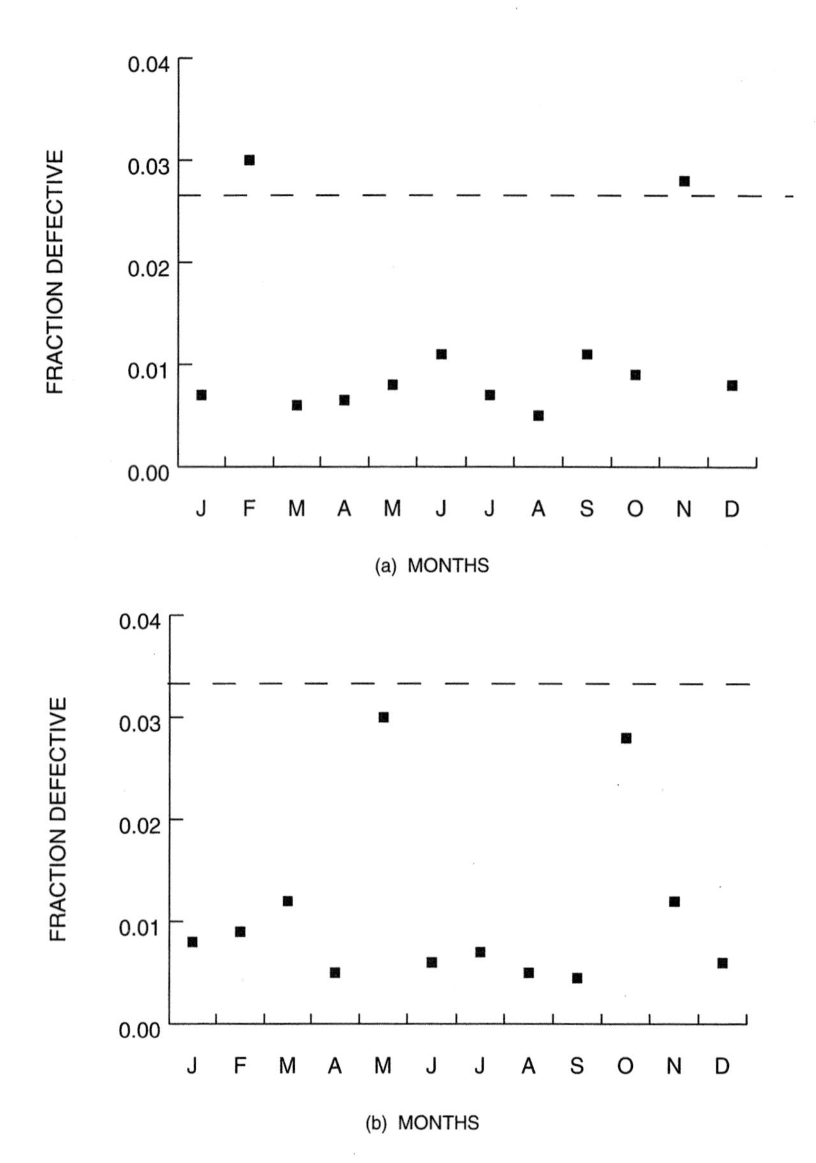

Figure VI.52. Charts from Figure VI.50 with control limits shown.

From Shewhart, p. 13. Used by permission of the publisher..

TYPES OF CONTROL CHARTS

There are two broad categories of control charts: those for use with continuous data (e.g., measurements) and those for use with attributes data (e.g., counts). This section describes the various charts used for these different data.

VARIABLE CHARTS

In SPC the mean, range, and standard deviation are the statistics most often used for analyzing measurement data. Control charts are used to monitor these statistics. An out-of-control point for any of these statistics is an indication that a special cause of variation is present and that an immediate investigation should be made to identify the special cause.

Average and range control charts

Average charts are statistical tools used to evaluate the central tendency of a process over time. Range charts are statistical tools used to evaluate the dispersion or spread of a process over time.

Average charts answer the question "Has a special cause of variation caused the central tendency of this process to change over the time period observed?" *Range charts* answer the question "Has a special cause of variation caused the process distribution to become more or less consistent?" Average and range charts can be applied to any continuous variable like weight, size, etc.

The basis of the control chart is the *rational subgroup*. Rational subgroups are composed of items which were produced under essentially the same conditions (see "practical subgroup sampling, later this section, for additional discussion). The average and range are computed for each subgroup separately, then plotted on the control chart. Each subgroup's statistics are compared to the control limits, and patterns of variation between subgroups are analyzed.

Subgroup equations for average and range charts

$$\overline{X} = \frac{\text{sum of subgroup measurements}}{\text{subgroup size}} \qquad \text{(VI.49)}$$

$$R = \text{Largest in subgroup - Smallest in subgroup} \qquad \text{(VI.50)}$$

Control limit equations for averages and ranges charts

Control limits for both the averages and the ranges charts are computed such that it is highly unlikely that a subgroup average or range from a stable process would fall outside of the limits. All control limits are set at plus and minus three standard deviations from the center line of the chart. Thus, the control limits for subgroup averages are plus and minus three standard deviations of the mean from the grand average; the control limits for the subgroup ranges are plus and minus three standard deviations of the range from the average range. These control limits are quite robust with respect to non-normality in the process distribution.

To facilitate calculations, constants are used in the control limit equations. Table 13 in the Appendix provides control chart constants for subgroups of 25 or less. The derivation of the various control chart constants is shown in Burr (1976, pp. 97–105).

Control limit equations for range charts

$$\overline{R} = \frac{\text{sum of subgroup ranges}}{\text{number of subgroups}} \tag{VI.51}$$

$$LCL = D_3\overline{R} \tag{VI.52}$$

$$UCL = D_4\overline{R} \tag{VI.53}$$

Control limit equations for averages charts using R-bar

$$\overline{\overline{X}} = \frac{\text{sum of subgroup averages}}{\text{number of subgroups}} \tag{VI.54}$$

$$LCL = \overline{\overline{X}} - A_2\overline{R} \tag{VI.55}$$

$$UCL = \overline{\overline{X}} + A_2\overline{R} \tag{VI.56}$$

Example of averages and ranges control charts

Table VI.28 contains 25 subgroups of five observations each.

Table VI.28. Data for averages and ranges control charts.

Sample 1	Sample 2	Sample 3	Sample 4	Sample 5	AVERAGE	RANGE
110	93	99	98	109	101.8	17
103	95	109	95	98	100.0	14
97	110	90	97	100	98.8	20
96	102	105	90	96	97.8	15
105	110	109	93	98	103.0	17
110	91	104	91	101	99.4	19
100	96	104	93	96	97.8	11
93	90	110	109	105	101.4	20
90	105	109	90	108	100.4	19
103	93	93	99	96	96.8	10
97	97	104	103	92	98.6	12
103	100	91	103	105	100.4	14
90	101	96	104	108	99.8	18
97	106	97	105	96	100.2	10
99	94	96	98	90	95.4	9
106	93	104	93	99	99.0	13
90	95	98	109	110	100.4	20
96	96	108	97	103	100.0	12
109	96	91	98	109	100.6	18
90	95	94	107	99	97.0	17
91	101	96	96	109	98.6	18
108	97	101	103	94	100.6	14
96	97	106	96	98	98.6	10
101	107	104	109	104	105.0	8
96	91	96	91	105	95.8	14

The control limits are calculated from this data as follows:

Ranges control chart example

$$\overline{R} = \frac{\text{sum of subgroup ranges}}{\text{number of subgroups}} = \frac{369}{25} = 14.76$$

$$LCL_R = D_3\overline{R} = 0 \times 14.76 = 0$$

$$UCL_R = D_4\overline{R} = 2.115 \times 14.76 = 31.22$$

Since it is not possible to have a subgroup range less than zero, the LCL is not shown on the control chart for ranges.

Averages control chart example

$$\overline{\overline{X}} = \frac{\text{sum of subgroup averages}}{\text{number of subgroups}} = \frac{2,487.5}{25} = 99.5$$

$$LCL_{\overline{X}} = \overline{\overline{X}} - A_2\overline{R} = 99.5 - 0.577 \times 14.76 = 90.97$$

$$UCL_{\overline{X}} = \overline{\overline{X}} + A_2\overline{R} = 99.5 + 0.577 \times 14.76 = 108.00$$

The completed averages and ranges control charts are shown in Figure VI.53.

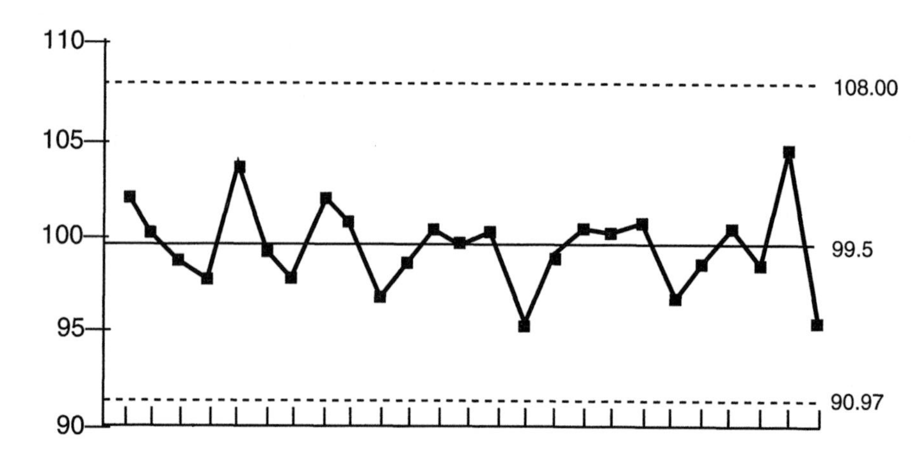

Figure VI.53a. Completed averages control chart—*Continued on next page . . .*

Figure VI.53—*Continued . . .*

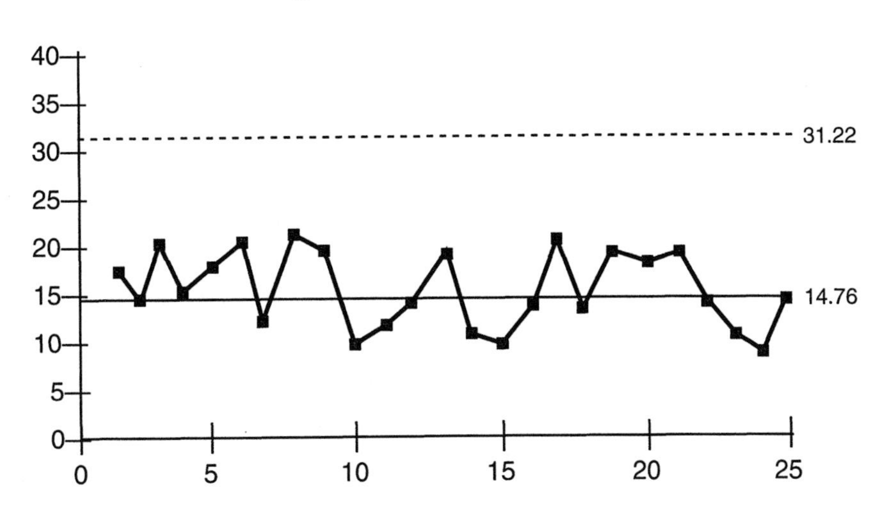

Figure VI.53b. Completed ranges control chart.

The above charts show a process in statistical control. This merely means that we can predict the limits of variability for this process. To determine the capability of the process with respect to requirements one must use the methods described in sectionVI.H.

Averages and standard deviation (sigma) control charts

Averages and standard deviation control charts are conceptually identical to averages and ranges control charts. The difference is that the subgroup standard deviation is used to measure dispersion rather than the subgroup range. The subgroup standard deviation is statistically more efficient than the subgroup range for subgroup sizes greater than 2. This efficiency advantage increases as the subgroup size increases. However, the range is easier to compute and easier for most people to understand. In general, this author recommends using subgroup ranges unless the subgroup size is 10 or larger. However, if the analyses are to be interpreted by statistically knowledgeable personnel and calculations are not a problem, the standard deviation chart may be preferred for all subgroup sizes.

Subgroup equations for average and sigma charts

$$\overline{X} = \frac{\text{sum of subgroup measurements}}{\text{subgroup size}} \tag{VI.57}$$

$$s = \sqrt{\frac{\sum_{i=1}^{n}\left(x_i - \overline{X}\right)^2}{n-1}} \tag{VI.58}$$

The standard deviation, *s*, is computed separately for each subgroup, using the subgroup average rather than the grand average. This is an important point; using the grand average would introduce special cause variation if the process were out of control, thereby underestimating the process capability, perhaps significantly.

Control limit equations for averages and sigma charts

Control limits for both the averages and the sigma charts are computed such that it is highly unlikely that a subgroup average or sigma from a stable process would fall outside of the limits. All control limits are set at plus and minus three standard deviations from the center line of the chart. Thus, the control limits for subgroup averages are plus and minus three standard deviations of the mean from the grand average. The control limits for the subgroup sigmas are plus and minus three standard deviations of sigma from the average sigma. These control limits are quite robust with respect to non-normality in the process distribution.

To facilitate calculations, constants are used in the control limit equations. Table 13 in the Appendix provides control chart constants for subgroups of 25 or less.

Control limit equations for sigma charts based on s-bar

$$\bar{s} = \frac{\text{sum of subgroup sigmas}}{\text{number of subgroups}} \quad \text{(VI.59)}$$

$$LCL = B_3\bar{s} \quad \text{(VI.60)}$$

$$UCL = B_4\bar{s} \quad \text{(VI.61)}$$

Control limit equations for averages charts based on s-bar

$$\bar{\bar{X}} = \frac{\text{sum of subgroup averages}}{\text{number of subgroups}} \quad \text{(VI.62)}$$

$$LCL = \bar{\bar{X}} - A_3\bar{s} \quad \text{(VI.63)}$$

$$UCL = \bar{\bar{X}} + A_3\bar{s} \quad \text{(VI.64)}$$

Example of averages and standard deviation control charts

To illustrate the calculations and to compare the range method to the standard deviation results, the data used in the previous example will be reanalyzed using the subgroup standard deviation rather than the subgroup range.

Table VI.29. Data for averages and sigma control charts.

Sample 1	Sample 2	Sample 3	Sample 4	Sample 5	AVERAGE	SIGMA
110	93	99	98	109	101.8	7.396
103	95	109	95	98	100.0	6.000
97	110	90	97	100	98.8	7.259
96	102	105	90	96	97.8	5.848
105	110	109	93	98	103.0	7.314
110	91	104	91	101	99.4	8.325
100	96	104	93	96	97.8	4.266
93	90	110	109	105	101.4	9.290
90	105	109	90	108	100.4	9.607
103	93	93	99	96	96.8	4.266
97	97	104	103	92	98.6	4.930
103	100	91	103	105	100.4	5.550
90	101	96	104	108	99.8	7.014
97	106	97	105	96	100.2	4.868
99	94	96	98	90	95.4	3.578
106	93	104	93	99	99.0	6.042
90	95	98	109	110	100.4	8.792
96	96	108	97	103	100.0	5.339
109	96	91	98	109	100.6	8.081
90	95	94	107	99	97.0	6.442
91	101	96	96	109	98.6	6.804
108	97	101	103	94	100.6	5.413
96	97	106	96	98	98.6	4.219
101	107	104	109	104	105.0	3.082
96	91	96	91	105	95.8	5.718

The control limits are calculated from this data as follows:

Sigma control chart

$$\bar{s} = \frac{\text{sum of subgroup sigmas}}{\text{number of subgroups}} = \frac{155.45}{25} = 6.218$$

$$LCL_s = B_3\bar{s} = 0 \times 6.218 = 0$$

$$UCL_s = B_4\bar{s} = 2.089 \times 6.218 = 12.989$$

Since it is not possible to have a subgroup sigma less than zero, the LCL is not shown on the control chart for sigma.

Averages control chart

$$\bar{\bar{X}} = \frac{\text{sum of subgroup averages}}{\text{number of subgroups}} = \frac{2,487.5}{25} = 99.5$$

$$LCL_{\bar{X}} = \bar{\bar{X}} - A_3\bar{s} = 99.5 - 1.427 \times 6.218 = 90.63$$

$$UCL_{\bar{X}} = \bar{\bar{X}} + A_3\bar{s} = 99.5 + 1.427 \times 6.218 = 108.37$$

The completed averages and sigma control charts are shown in Figure VI.54. Note that the control limits for the averages chart are only slightly different than the limits calculated using ranges.

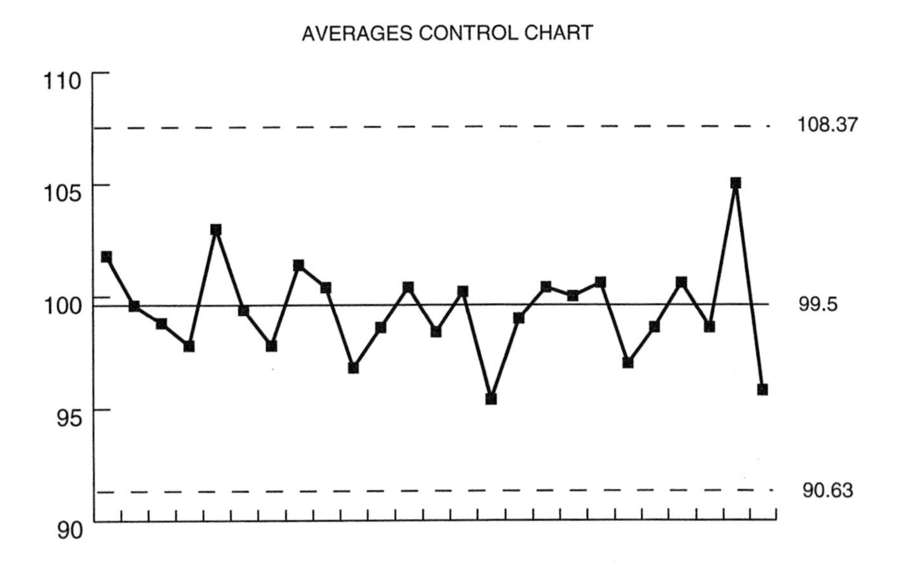

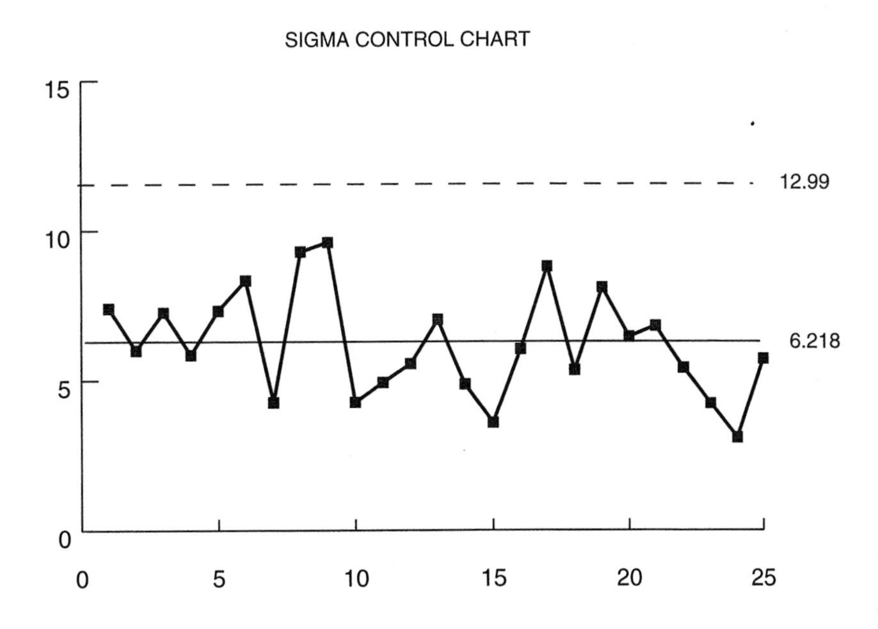

Figure VI.54. Completed average and sigma control charts.

Note that the conclusions reached are the same as when ranges were used.

Control charts for individuals measurements (X charts)

Individuals control charts are statistical tools used to evaluate the central tendency of a process over time. They are also called *X charts* or *moving range charts*. Individuals control charts are used when it is not feasible to use averages for process control. There are many possible reasons why averages control charts may not be desirable: observations may be expensive to get (e.g., destructive testing), output may be too homogeneous over short time intervals (e.g., pH of a solution), the production rate may be slow and the interval between successive observations long, et cetera. Control charts for individuals are often used to monitor batch processes, such as chemical processes, where the within-batch variation is so small relative to between-batch variation that the control limits on a standard $\overline{X}$ chart would be too close together. Range charts are used in conjunction with individuals charts to help monitor dispersion.*

Calculations for moving ranges charts

As with averages and ranges charts, the range is computed as shown above,

$$R = \text{Largest in subgroup} - \text{Smallest in subgroup}$$

Where the subgroup is a consecutive pair of process measurements (see "rational subgroup sampling", later this section, for additional information on the meaning of the term "subgroup" as applied to individuals charts). The range control limit is computed as was described for averages and ranges charts, using the D_4 constant for subgroups of 2, which is 3.267. I.e.,

$$\text{LCL} = 0 \text{ (for } n = 2)$$
$$\text{UCL} = 3.267 \text{ x R-Bar}$$

Control limit equations for individuals charts

$$\overline{X} = \frac{\text{sum of measurements}}{\text{number of measurements}} \quad \text{(VI.65)}$$

$$LCL = \overline{X} - E_2\overline{R} = \overline{X} - 2.66 \times \overline{R} \quad \text{(VI.66)}$$

$$UCL = \overline{X} + E_2\overline{R} = \overline{X} + 2.66 \times \overline{R} \quad \text{(VI.67)}$$

Where E_2=2.66 is the constant used when individual measurements are plotted, and $\overline{R}$ is based on subgroups of n = 2.

*There is considerable debate over the value of moving R charts. Academic researchers have failed to show statistical value in them. However, many practitioners (including the author) believe that moving R charts provide valuable additional information that can be used in troubleshooting.

Example of individuals and moving ranges control charts

Table VI.30 contains 25 measurements. To facilitate comparison, the measurements are the first observations in each subgroup used in the previous average/ranges and average/standard deviation control chart examples.

Table VI.30. Data for individuals and moving ranges control charts.

SAMPLE 1	RANGE
110	None
103	7
97	6
96	1
105	9
110	5
100	10
93	7
90	3
103	13
97	6
103	6
90	13
97	7
99	2
106	7
90	16
96	6
109	13
90	19
91	1
108	17
96	12
101	5
96	5

Continued at right . . .

The control limits are calculated from this data as follows:

Moving ranges control chart control limits

$$\overline{R} = \frac{\text{sum of ranges}}{\text{number of ranges}} = \frac{196}{24} = 8.17$$

$$LCL_R = D_3\overline{R} = 0 \times 8.17 = 0$$

$$UCL_R = D_4\overline{R} = 3.267 \times 8.17 = 26.69$$

Since it is not possible to have a subgroup range less than zero, the LCL is not shown on the control chart for ranges.

Individuals control chart control limits

$$\overline{X} = \frac{\text{sum of measurements}}{\text{number of measurements}} = \frac{2,475}{25} = 99.0$$

$$LCL_X = \overline{X} - E_2\overline{R} = 99.0 - 2.66 \times 8.17 = 77.27$$

$$UCL_X = \overline{X} + E_2\overline{R} = 99.0 + 2.66 \times 8.17 = 120.73$$

The completed individuals and moving ranges control charts are shown in Figure VI.55.

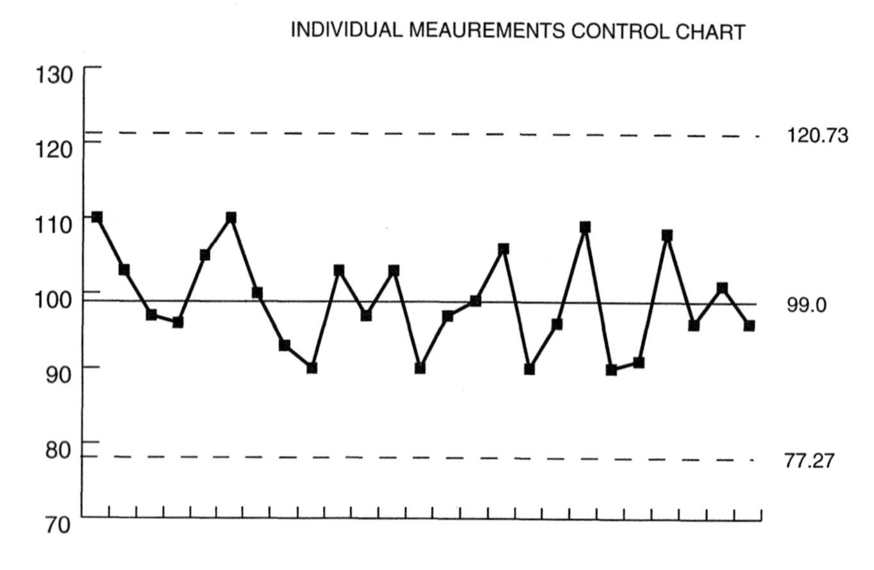

Figure VI.55a. Completed individuals range control chart. *Continued on next page . . .*

Figure VI.55—*Continued . . .*

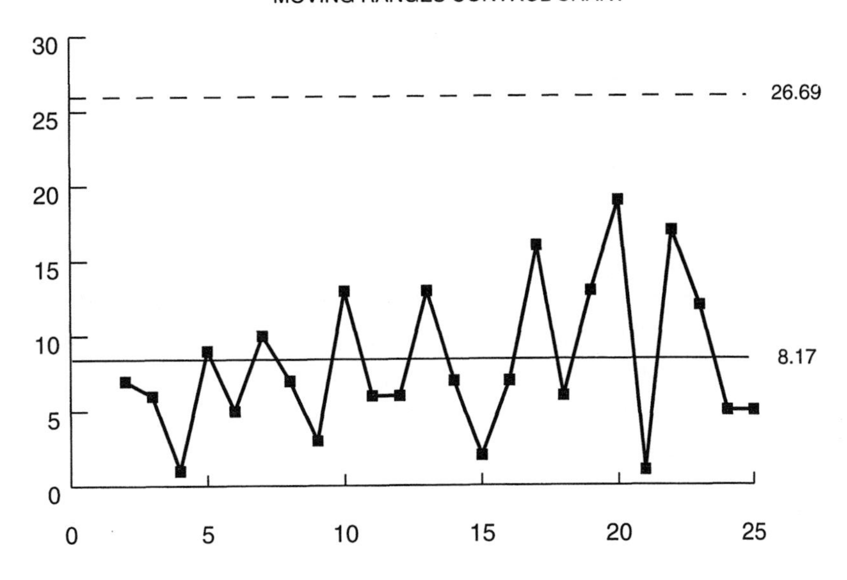

Figure VI.55b. Completed moving range control chart.

In this case, the conclusions are the same as with averages charts. However, averages charts always provide tighter control than *X* charts. In some cases, the additional sensitivity provided by averages charts may not be justified on either an economic or an engineering basis. When this happens, the use of averages charts will merely lead to wasting money by investigating special causes that are of minor importance.

ATTRIBUTE CHARTS

Control charts for proportion defective (p charts)

P charts are statistical tools used to evaluate the proportion defective, or proportion non-conforming, produced by a process.

P charts can be applied to any variable where the appropriate performance measure is a unit count. *P* charts answer the question "Has a special cause of variation caused the central tendency of this process to produce an abnormally large or small number of defective units over the time period observed?"

P chart control limit equations

Like all control charts, *p* charts consist of three guidelines: center line, a lower control limit, and an upper control limit. The center line is the average proportion defective and the two control limits are set at plus and minus three standard deviations. If the process is in statistical control, then virtually all proportions should be between the control limits and they should fluctuate randomly about the center line.

$$p = \frac{\text{subgroup defective count}}{\text{subgroup size}} \tag{VI.68}$$

$$\bar{p} = \frac{\text{sum of subgroup defective counts}}{\text{sum of subgroup sizes}} \tag{VI.69}$$

$$LCL = \bar{p} - 3\sqrt{\frac{\bar{p}(1-\bar{p})}{n}} \tag{VI.70}$$

$$UCL = \bar{p} + 3\sqrt{\frac{\bar{p}(1-\bar{p})}{n}} \tag{VI.71}$$

In the above equations, n is the subgroup size. If the subgroup sizes varies, the control limits will also vary, becoming closer together as n increases.

Analysis of p charts

As with all control charts, a special cause is probably present if there are any points beyond either the upper or the lower control limit. Analysis of p chart patterns between the control limits is extremely complicated if the sample size varies because the distribution of p varies with the sample size.

Example of p chart calculations

The data in Table VI.31 were obtained by opening randomly selected crates from each shipment and counting the number of bruised peaches. There are 250 peaches per crate. Normally, samples consist of one crate per shipment. However, when part-time help is available, samples of two crates are taken.

Table VI.31. Raw data for *p* chart.

SHIPMENT #	CRATES	PEACHES	BRUISED	p
1	1	250	47	0.188
2	1	250	42	0.168
3	1	250	55	0.220
4	1	250	51	0.204
5	1	250	46	0.184
6	1	250	61	0.244
7	1	250	39	0.156
8	1	250	44	0.176
9	1	250	41	0.164
10	1	250	51	0.204
11	2	500	88	0.176
12	2	500	101	0.202
13	2	500	101	0.202
14	1	250	40	0.160
15	1	250	48	0.192
16	1	250	47	0.188
17	1	250	50	0.200
18	1	250	48	0.192
19	1	250	57	0.228
20	1	250	45	0.180
21	1	250	43	0.172
22	2	500	105	0.210
23	2	500	98	0.196
24	2	500	100	0.200
25	2	500	96	0.192
	TOTALS	8,000	1,544	

Using the above data the center line and control limits are found as follows:

$$p = \frac{\text{subgroup defective count}}{\text{subgroup size}}$$

these values are shown in the last column of Table VI.31.

$$\bar{p} = \frac{\text{sum of subgroup defective counts}}{\text{sum of subgroup sizes}} = \frac{1{,}544}{8{,}000} = 0.193,$$

which is constant for all subgroups.

n=250 (*1 crate*):

$$LCL = \bar{p} - 3\sqrt{\frac{\bar{p}(1-\bar{p})}{n}} = 0.193 - 3\sqrt{\frac{0.193 \times (1-0.193)}{250}} = 0.118$$

$$UCL = \bar{p} + 3\sqrt{\frac{\bar{p}(1-\bar{p})}{n}} = 0.193 + 3\sqrt{\frac{0.193 \times (1-0.193)}{250}} = 0.268$$

n=500 (*2 crates*):

$$LCL = 0.193 - 3\sqrt{\frac{0.193 \times (1-0.193)}{500}} = 0.140$$

$$UCL = 0.193 + 3\sqrt{\frac{0.193 \times (1-0.193)}{500}} = 0.246$$

The control limits and the subgroup proportions are shown in Figure VI.56.

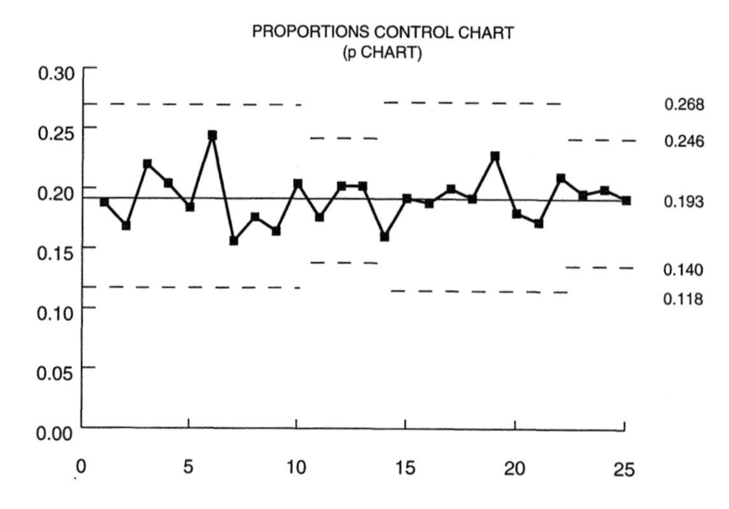

Figure VI.56. Completed *p* control chart.

Pointers for using p charts

Determine if "moving control limits" are really necessary. It may be possible to use the average sample size (total number inspected divided by the number of subgroups) to calculate control limits. For instance, with our example the sample size doubled from 250 peaches to 500 but the control limits hardly changed at all. Table VI.32 illustrates the different control limits based on 250 peaches, 500 peaches, and the average sample size which is 8,000÷25 = 320 peaches.

Table VI.32. Effect of using average sample size.

SAMPLE SIZE	LOWER CONTROL LIMIT	UPPER CONTROL LIMIT
250	0.1181	0.2679
500	0.1400	0.2460
320	0.1268	0.2592

Notice that the conclusions regarding process performance are the same when using the average sample size as they are using the exact sample sizes. This is usually the case if the variation in sample size isn't too great. There are many rules of thumb, but most of them are extremely conservative. The best way to evaluate limits based on the average sample size is to check it out as shown above. SPC is all about improved decision-making. In general, use the most simple method that leads to correct decisions.

Control charts for count of defectives (np charts)

Np charts are statistical tools used to evaluate the count of defectives, or count of items non-conforming, produced by a process. *Np* charts can be applied to any variable where the appropriate performance measure is a unit count and the subgroup size is held constant. Note that wherever an *np* chart can be used, a *p* chart can be used too.

Control limit equations for np charts

Like all control charts, *np* charts consist of three guidelines: center line, a lower control limit, and an upper control limit. The center line is the average count of defectives-per-subgroup and the two control limits are set at plus and minus three standard deviations. If the process is in statistical control, then virtually all subgroup counts will be between the control limits, and they will fluctuate randomly about the center line.

$$np = \text{subgroup defective count} \qquad \text{(VI.72)}$$

$$n\bar{p} = \frac{\text{sum of subgroup defective counts}}{\text{number of subgroups}} \qquad \text{(VI.73)}$$

$$LCL = n\bar{p} - 3\sqrt{n\bar{p}(1-\bar{p})} \quad \text{(VI.74)}$$

$$UCL = n\bar{p} + 3\sqrt{n\bar{p}(1-\bar{p})} \quad \text{(VI.75)}$$

Note that

$$\bar{p} = \frac{n\bar{p}}{n} \quad \text{(VI.76)}$$

Example of np chart calculation

The data in Table VI.33 were obtained by opening randomly selected crates from each shipment and counting the number of bruised peaches. There are 250 peaches per crate (constant n is required for np charts).

Table VI.33. Raw data for np chart.

SHIPMENT NUMBER	BRUISED PEACHES
1	20
2	28
3	24
4	21
5	32
6	33
7	31
8	29
9	30
10	34
11	32
12	24
13	29
14	27
15	37

Continued at right . . .

SHIPMENT NUMBER	BRUISED PEACHES
16	23
17	27
18	28
19	31
20	27
21	30
22	23
23	23
24	27
25	35
26	29
27	23
28	23
29	30
30	28
TOTAL	838

Using the above data the center line and control limits are found as follows:

$$n\bar{p} = \frac{\text{sum of subgroup defective counts}}{\text{number of subgroups}} = \frac{838}{30} = 27.93$$

$$LCL = n\bar{p} - 3\sqrt{n\bar{p}(1-\bar{p})} = 27.93 - 3\sqrt{27.93 \times \left(1 - \frac{27.93}{250}\right)} = 12.99$$

$$UCL = n\bar{p} + 3\sqrt{n\bar{p}(1-\bar{p})} = 27.93 + 3\sqrt{27.93 \times \left(1 - \frac{27.93}{250}\right)} = 42.88$$

The control limits and the subgroup defective counts are shown in Figure VI.57.

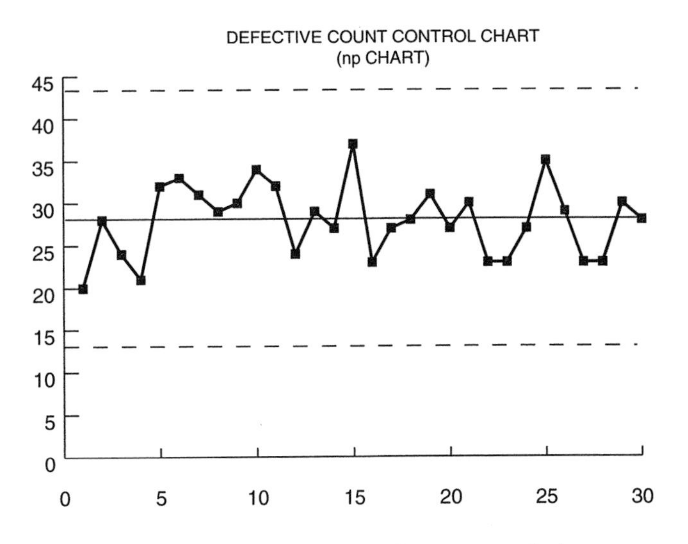

Figure VI.57. Completed *np* control chart.

Control charts for average occurrences-per-unit (u charts)

U charts are statistical tools used to evaluate the average number of occurrences-per-unit produced by a process. *U* charts can be applied to any variable where the appropriate performance measure is a count of how often a particular event occurs. *U* charts answer the question "Has a special cause of variation caused the central tendency of this process to produce an abnormally large or small number of occurrences over the time period observed?" Note that, unlike *p* or *np* charts, *u* charts do not necessarily involve counting physical items. Rather, they involve counting of *events*. For example, when using a *p* chart one would count bruised peaches. When using a *u* chart they would count the *bruises*.

Control limit equations for u charts

Like all control charts, u charts consist of three guidelines: center line, a lower control limit, and an upper control limit. The center line is the average number of occurrences-per-unit and the two control limits are set at plus and minus three standard deviations. If the process is in statistical control then virtually all subgroup occurrences-per-unit should be between the control limits and they should fluctuate randomly about the center line.

$$u = \frac{\text{subgroup count of occurrences}}{\text{subgroup size in units}} \tag{VI.77}$$

$$\bar{u} = \frac{\text{sum of subgroup occurrences}}{\text{sum of subgroup sizes in units}} \tag{VI.78}$$

$$LCL = \bar{u} - 3\sqrt{\frac{\bar{u}}{n}} \tag{VI.79}$$

$$UCL = \bar{u} + 3\sqrt{\frac{\bar{u}}{n}} \tag{VI.80}$$

In the above equations, n is the subgroup size in units. If the subgroup size varies, the control limits will also vary.

One way of helping to determine whether or not a particular set of data is suitable for a u chart or a p chart is to examine the equation used to compute the center line for the control chart. If the unit of measure is the same in both the numerator and the denominator, then a p chart is indicated; otherwise a u chart is indicated. For example, if

$$\text{Center Line} = \frac{\text{bruises per crate}}{\text{number of crates}},$$

then the numerator is in terms of bruises while the denominator is in terms of crates, indicating a u chart.

The unit size is arbitrary, but once determined it cannot be changed without recomputing all subgroup occurrences-per-unit and control limits. For example, if the occurrences were accidents and a unit was 100,000 hours worked, then a month with 250,000 hours worked would be 2.5 units and a month with 50,000 hours worked would be 0.5 units. If the unit size were 200,000 hours then the two months would have 1.25 and 0.2 units respectively. The equations for the center line and control limits would "automatically" take into account the unit size, so the control charts would give identical results regardless of which unit size is used.

Analysis of u charts

As with all control charts, a special cause is probably present if there are any points beyond either the upper or the lower control limit. Analysis of u chart patterns between the control limits is extremely complicated when the sample size varies and is usually not done.

Example of u chart

The data in Table VI.34 were obtained by opening randomly selected crates from each shipment and counting the number of bruised peaches. There are 250 peaches per crate. Our unit size will be taken as one full crate; i.e., we will be counting crates rather than the peaches themselves. Normally, samples consist of one crate per shipment. However, when part-time help is available, samples of two crates are taken.

Table VI.34. Raw data for u chart.

SHIPMENT NO.	UNITS (CRATES)	FLAWS	FLAWS-PER-UNIT
1	1	47	47
2	1	42	42
3	1	55	55
4	1	51	51
5	1	46	46
6	1	61	61
7	1	39	39
8	1	44	44
9	1	41	41
10	1	51	51
11	2	88	44
12	2	101	50.5
13	2	101	50.5
14	1	40	40
15	1	48	48
16	1	47	47
17	1	50	50
18	1	48	48
19	1	57	57
20	1	45	45
21	1	43	43
22	2	105	52.5
23	2	98	49
24	2	100	50
25	2	96	48
TOTALS	32	1,544	

Using the above data the center line and control limits are found as follows:

$$u = \frac{\text{subgroup count of occurrences}}{\text{subgroup size in units}},$$

these values are shown in the last column of Table VI.34.

$$\bar{u} = \frac{\text{sum of subgroup count of occurrences}}{\text{sum of subgroup unit sizes}} = \frac{1{,}544}{32} = 48.25,$$

which is constant for all subgroups.

n=1 unit:

$$LCL = \bar{u} - 3\sqrt{\frac{\bar{u}}{n}} = 48.25 - 3\sqrt{\frac{48.25}{1}} = 27.411$$

$$UCL = \bar{u} + 3\sqrt{\frac{\bar{u}}{n}} = 48.25 + 3\sqrt{\frac{48.25}{1}} = 69.089$$

n=2 units:

$$LCL = 48.25 - 3\sqrt{\frac{48.25}{2}} = 33.514$$

$$UCL = 48.25 + 3\sqrt{\frac{48.25}{2}} = 62.986$$

The control limits and the subgroup occurrences-per-unit are shown in Figure VI.58.

The reader may note that the data used to construct the *u* chart were the same as that used for the *p* chart, except that we considered the counts as being counts of occurrences (flaws) instead of counts of physical items (bruised peaches). The practical implications of using a *u* chart when a *p* chart should have been used, or vice versa, are usually not serious. The decisions based on the control charts will be quite similar in most cases encountered in quality engineering regardless of whether a *u* or a *p* chart is used.

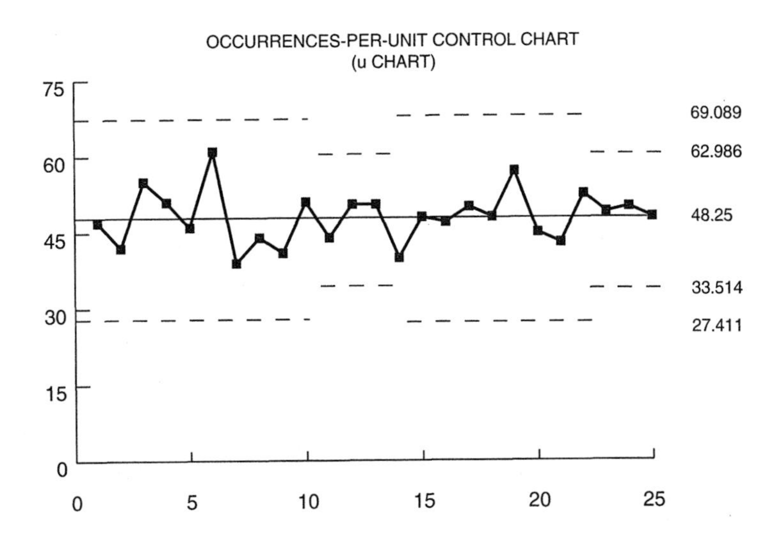

Figure VI.58. Completed *u* control chart.

Control charts for counts of occurrences-per-unit (c charts)

C charts are statistical tools used to evaluate the number of occurrences-per-unit produced by a process. *C* charts can be applied to any variable where the appropriate performance measure is a count of how often a particular event occurs and samples of constant size are used. *C* charts answer the question "Has a special cause of variation caused the central tendency of this process to produce an abnormally large or small number of occurrences over the time period observed?" Note that, unlike *p* or *np* charts, *c* charts do not involve counting physical items. Rather, they involve counting of *events*. For example, when using an *np* chart one would count bruised peaches. When using a *c* chart they would count the *bruises*.

Control limit equations for c charts

Like all control charts, *c* charts consist of three guidelines: center line, a lower control limit, and an upper control limit. The center line is the average number of occurrences-per-unit and the two control limits are set at plus and minus three standard deviations. If the process is in statistical control then virtually all subgroup occurrences-per-unit should be between the control limits and they should fluctuate randomly about the center line.

$$\bar{c} = \frac{\text{sum of subgroup occurrences}}{\text{number of subgroups}} \quad \text{(VI.81)}$$

$$LCL = \bar{c} - 3\sqrt{\bar{c}} \quad \text{(VI.82)}$$

$$UCL = \bar{c} + 3\sqrt{\bar{c}} \quad \text{(VI.83)}$$

One way of helping to determine whether or not a particular set of data is suitable for a *c* chart or an *np* chart is to examine the equation used to compute the center line for the control chart. If the unit of measure is the same in both the numerator and the denominator, then a *p* chart is indicated, otherwise a *c* chart is indicated. For example, if

$$\text{Center Line} = \frac{\text{bruises}}{\text{number of crates}},$$

then the numerator is in terms of bruises while the denominator is in terms of crates, indicating a *c* chart.

The unit size is arbitrary but, once determined, it cannot be changed without recomputing all subgroup occurrences-per-unit and control limits.

Analysis of c charts

As with all control charts, a special cause is probably present if there are any points beyond either the upper or the lower control limit. Analysis of *c* chart patterns between the control limits is shown later in this section.

Example of c chart

The data in Table VI.35 were obtained by opening randomly selected crates from each shipment and counting the number of flaws. There are 250 peaches per crate. Our unit size will be taken as one full crate, i.e., we will be counting crates rather than the peaches themselves. Every subgroup consists of one crate. If the subgroup size varied, a *u* chart would be used.

Table VI.35. Raw data for *c* chart.

SHIPMENT #	FLAWS
1	27
2	32
3	24
4	31
5	42
6	38
7	33
8	35
9	35
10	39
11	41
12	29
13	34
14	34
15	43

Continued at right . . .

SHIPMENT #	FLAWS
16	29
17	33
18	33
19	38
20	32
21	37
22	30
23	31
24	32
25	42
26	40
27	21
28	23
29	39
30	29
TOTAL	1,006

Using the above data the center line and control limits are found as follows:

$$\bar{c} = \frac{\text{sum of subgroup occurrences}}{\text{number of subgroups}} = \frac{1{,}006}{30} = 33.53$$

$$LCL = \bar{c} - 3\sqrt{\bar{c}} = 33.53 - 3\sqrt{33.53} = 16.158$$

$$UCL = \bar{c} + 3\sqrt{\bar{c}} = 33.53 + 3\sqrt{33.53} = 50.902$$

The control limits and the occurrence counts are shown in Figure VI.59.

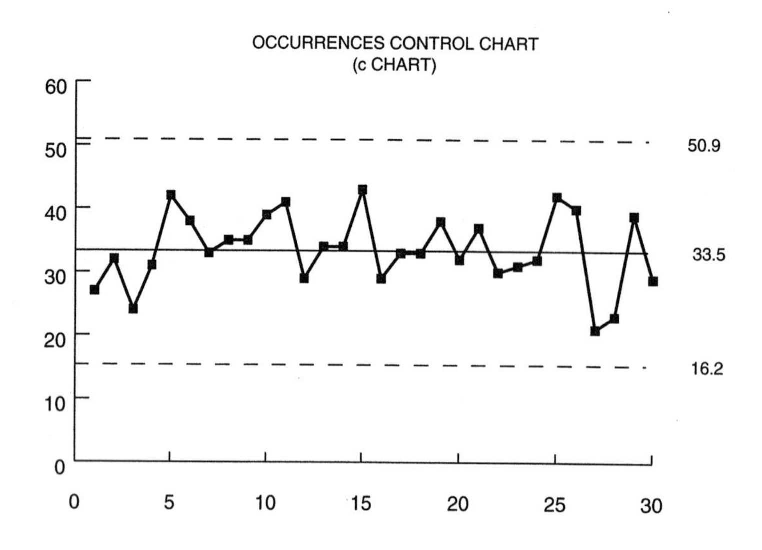

Figure VI.59. Completed *c* control chart.

IMPLEMENTATION

Assuming that the organization's leadership has created an environment where open and honest communication can flourish, SPC implementation becomes a matter of selecting processes for applying the SPC approach and then selecting variables within each process. This section describes an approach to this activity.

Variable selection: preparing the process control plan

Process control plans should be prepared for each key process. The plans should be prepared by teams of people who understand the process. The team should begin by creating a flowchart of the process using the process elements determined in creating the house of quality (see the QFD discussion in I.H). The flowchart will show how the process elements relate to one another and it will help in the selection of control points. It will also show the point of delivery to the customer, which is usually an important control point. Note that the customer may be an internal customer.

For any given process there can be a number of different types of process elements. Some process elements are *internal* to the process, others *external.* The rotation speed of a drill is an internal process element, while the humidity in the building is external. Some process elements, while important, are easy to hold constant at a given value so that they do not change unless deliberate action is taken. We will call these *fixed* elements. Other process elements vary of their own accord and must be watched; we call these *variable* elements. The drill rotation speed can be set in advance, but the line voltage for the drill press may vary, which causes the drill speed to change in spite of its initial setting (a good example of how a correlation matrix might be useful). Figure VI.60 provides a planning guide based on the internal/external and fixed/variable classification scheme. Of course, other classification schemes may be more suitable on a given project and the engineer is encouraged to develop the approach that best serves his or her needs. For convenience, each class is identified with a Roman numeral; I = fixed–internal, II = fixed–external, III = variable–internal and IV = variable–external.

	INTERNAL	EXTERNAL
FIXED	**I**	**II**
	• Setup approval • Periodic audits • Preventive maintenance	• Audit • Certification
VARIABLE	**III**	**IV**
	• Control charts • Fool-proofing product • Fool-proofing process • Sort the output	• Supplier SPC • Receiving inspection • Supplier sorting • Fool-proof product

Figure VI.60. Guide to selecting and controlling process variables.

In selecting the appropriate method of control for each process element, pay particular attention to those process elements which received high importance rankings in the house of quality analysis. In some cases an important process element is very expensive to control. When this happens, look at the QFD correlation matrix (I.H) or the statistical correlation matrix (VI.E) for possible assistance. The process element may be correlated with other process elements that are less costly to control. Either correlation matrix will also help you to minimize the number of control charts. It is usually unnecessary to keep control charts on several variables that are correlated with one another. In these cases, it may be possible to select the process element that is least expensive (or most sensitive) to monitor as the control variable.

As Figure VI.60 indicates, control charts are not always the best method of controlling a given process element. In fact, control charts are seldom the method of choice. When process elements are important we would prefer that they *not vary at all!* Only when this can not be accomplished economically should the engineer resort to the use of control charts to monitor the element's variation. Control charts may be thought of as a control mechanism of last resort. Control charts are useful only when the element being monitored can be expected to exhibit measurable and "random-looking" variation when the process is properly controlled. A process element that always checks "10" if everything is okay is not a good candidate for control charting. Nor is one that checks "10" or "12," but never anything else. Ideally, the measurements being monitored with variables control charts will be capable of taking on any value; i.e., the data will be continuous. Discrete measurement data can be used if it's not too discrete; indeed, all real-world data are somewhat discrete. As a rule of thumb, at least ten different values should appear in the data set and no one value should comprise more than 20% of the data set. When the measurement data becomes too discrete for SPC, monitor it with check sheets or simple time-ordered plots.

Of course, the above discussion applies to measurement data. Attribute control charts can be used to monitor process elements that are discrete counts.

Any process control plan must include instructions on the action to be taken if problems appear. This is particularly important where control charts are being used for process control. Unlike process control procedures such as audits or setup approvals, it is not always apparent just what is wrong when a control chart indicates a problem. The investigation of special causes of variation usually consists of a number of predetermined actions (such as checking the fixture or checking a cutting tool) followed by notifying someone if the items checked don't reveal the source of the problem. Also verify that the arithmetic was done correctly and that the point was plotted in the correct position on the control chart.

The reader may have noticed that Figure VI.60 includes "sort output" as part of the process control plan. Sorting the output implies that the process is not capable of meeting the customer's requirements, as determined by a process capability study and the application of Deming's all-or-none rules. However, even if sorting is taking place, SPC is still advisable. SPC will help assure that things don't get any worse. SPC will also reveal improvements that may otherwise be overlooked. The improvements may result in a process that is good enough to eliminate the need for sorting.

Control chart selection

Selecting the proper control chart for a particular data set is a simple matter if approached properly. The proper approach is illustrated in Figure VI.61.

To use the decision tree, begin at the leftmost node and determine if the data are measurements or counts. If measurements, then select the control chart based on the subgroup size. If the data are counts, then determine if the counts are of occurrences or pieces. An aid in making this determination is to examine the equation for the process average. If the numerator and denominator involve the same units, then a p or np chart is indicated. If different units of measure are involved, then a u or c chart is indicated. For example, if the average is in accidents-per-month, then a c or u chart is indicated because the numerator is in terms of accidents but the denominator is in terms of time.

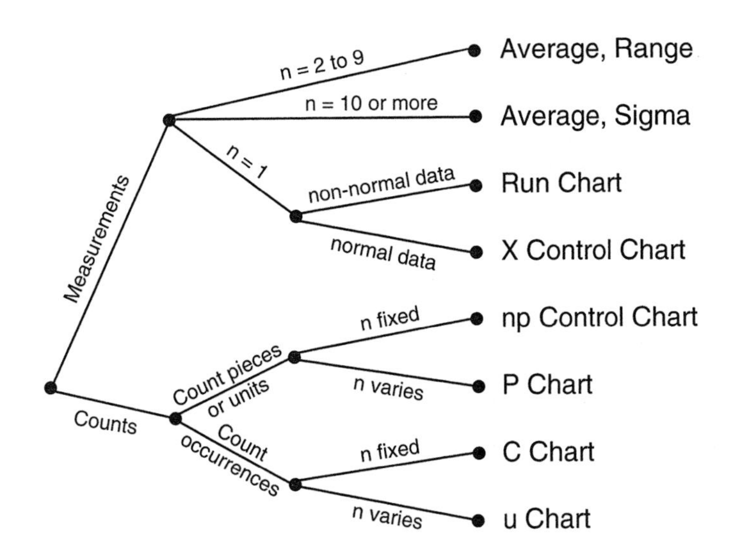

Figure VI.61. Control chart selection decision tree.

Rational subgroup sampling

The basis of all control charts is the *rational subgroup*. Rational subgroups are composed of items which were produced under essentially the same conditions. The statistics, for example, the average and range, are computed for each subgroup separately, then plotted on the control chart. When possible, rational subgroups are formed by using consecutive units. Each subgroup's statistics are compared to the control limits, and patterns of variation between subgroups are analyzed. Note the sharp contrast between this approach and the *random sampling* approach used for enumerative statistical methods.

The idea of rational subgrouping becomes a bit fuzzy when dealing with x charts, or individuals control charts. The reader may well wonder about the meaning of the term subgrouping when the "subgroup" is a single measurement. The basic idea underlying control charts of all types is to identify the capability of the process. The mechanism by which this is accomplished is careful formation of rational subgroups as defined above. When possible, rational subgroups are formed by using consecutive units. The measure of process variability, either the subgroup standard deviation or the subgroup range, is the basis of the control limits for averages. Conceptually, this is akin to basing the control limits on short-term variation. These control limits are used to monitor variation over time.

As far as possible, this approach also forms the basis of establishing control limits for individual measurements. This is done by forming quasi-subgroups using pairs of consecutive measurements. These "subgroups of 2" are used to compute ranges. The ranges are used to compute the control limits for the individual measurements.

Control charts interpretation

Control charts provide the operational definition of the term *special cause.* A special cause is simply anything which leads to an observation beyond a control limit. However, this simplistic use of control charts does not do justice to their power. Control charts are running records of the performance of the process and, as such, they contain a vast store of information on potential improvements. While some guidelines are presented here, control chart interpretation is an art that can only be developed by looking at many control charts and probing the patterns to identify the underlying system of causes at work.

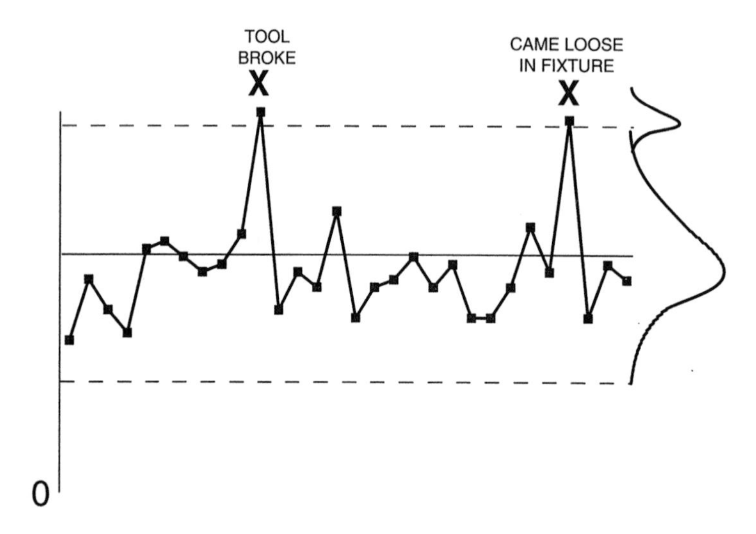

Figure VI.62. Control chart patterns: *freaks.*

Freak patterns are the classical special cause situation. Freaks result from causes that have a large effect but that occur infrequently. When investigating freak values look at the cause-and-effect diagram for items that meet these criteria. The key to identifying freak causes is timeliness in collecting and recording the data. If you have difficulty, try sampling more frequently.

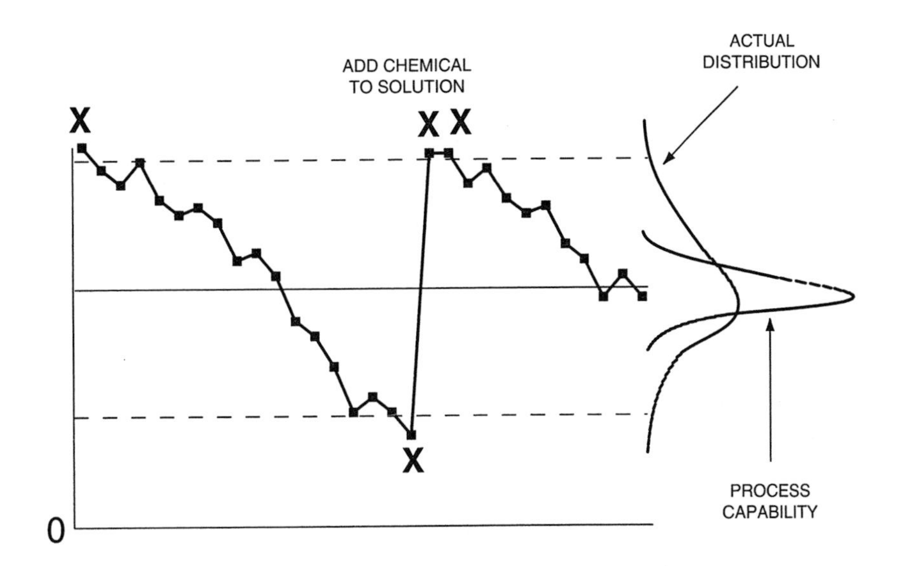

Figure VI.63. Control chart patterns: *drift.*

Drift is generally seen in processes where the current process value is partly determined by the previous process state. For example, if the process is a plating bath, the content of the tank cannot change instantaneously, instead it will change gradually. Another common example is tool wear: the size of the tool is related to its previous size. Once the cause of the drift has been determined, the appropriate action can be taken. Whenever economically feasible, the drift should be eliminated, e.g., install an automatic chemical dispenser for the plating bath, or make automatic compensating adjustments to correct for tool wear. Note that the total process variability increases when drift is allowed, which adds cost. When this is not possible, the control chart can be modified in one of two ways:

1. Make the slope of the center line and control limits match the natural process drift. The control chart will then detect departures from the natural drift.
2. Plot *deviations* from the natural or expected drift.

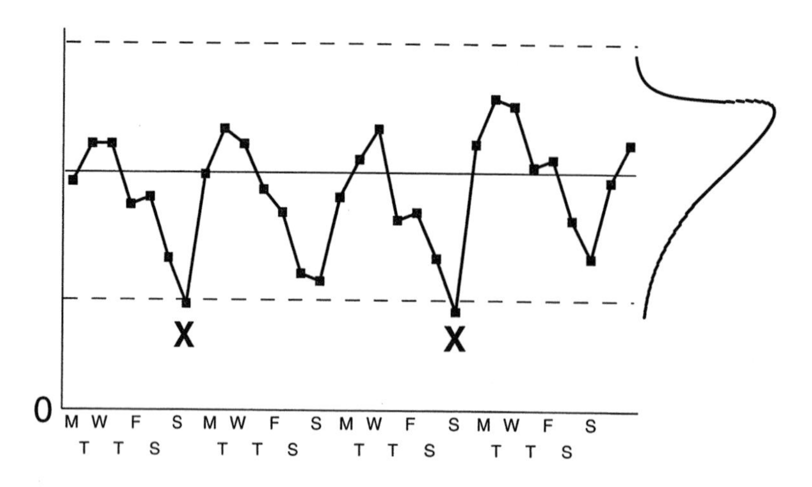

Figure VI.64. Control chart patterns: *cycles.*

Cycles often occur due to the nature of the process. Common cycles include hour of the day, day of the week, month of the year, quarter of the year, week of the accounting cycle, etc. Cycles are caused by modifying the process inputs or methods according to a regular schedule. The existence of this schedule and its effect on the process may or may not be known in advance. Once the cycle has been discovered, action can be taken. The action might be to adjust the control chart by plotting the control measure against a variable base. For example, if a day-of-the-week cycle exists for shipping errors because of the workload, you might plot shipping errors per 100 orders shipped instead of shipping errors per day. Alternatively, it may be worthwhile to change the system to smooth out the cycle. Most processes operate more efficiently when the inputs are relatively stable and when methods are changed as little as possible.

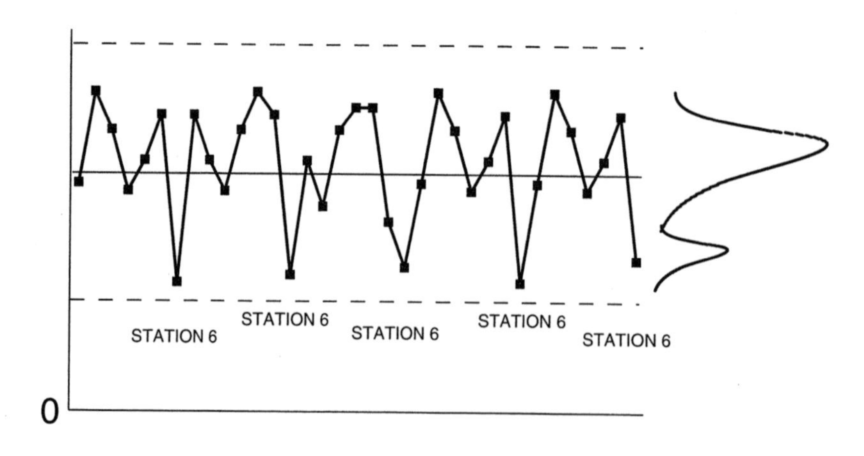

Figure VI.65. Control chart patterns: *repeating patterns.*

A controlled process will exhibit only "random looking" variation. A pattern where every n^{th} item is different is, obviously, non-random. These patterns are sometimes quite subtle and difficult to identify. It is sometimes helpful to see if the average fraction defective is close to some multiple of a known number of process streams. For example, if the machine is a filler with 40 stations, look for problems that occur 1/40, 2/40, 3/40, etc., of the time.

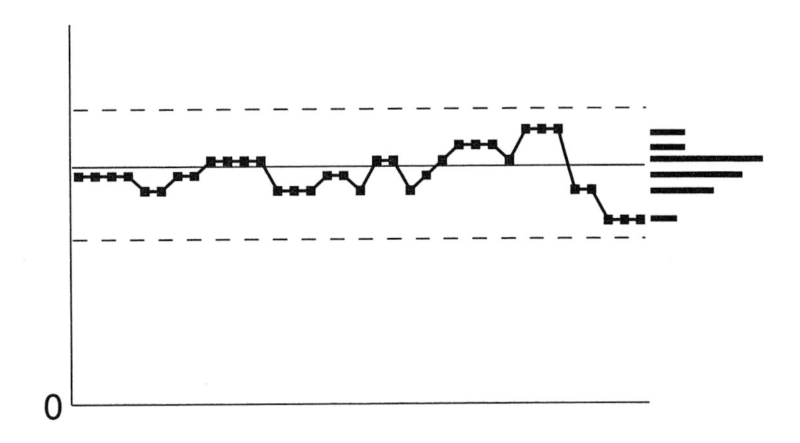

Figure VI.66. Control chart patterns: *discrete data.*

When plotting measurement data the assumption is that the numbers exist on a continuum; i.e., there will be many different values in the data set. In the real world, the data are never completely continuous. It usually doesn't matter much if there are, say, 10 or more different numbers. However, when there are only a few numbers that appear over-and-over it can cause problems with the analysis. A common problem is that the R chart will underestimate the average range, causing the control limits on both the average and range charts to be too close together. The result will be too many "false alarms" and a general loss of confidence in SPC.

The usual cause of this situation is inadequate gage resolution. The ideal solution is to obtain a gage with greater resolution. Sometimes the problem occurs because operators, inspectors, or computers are rounding the numbers. The solution here is to record additional digits.

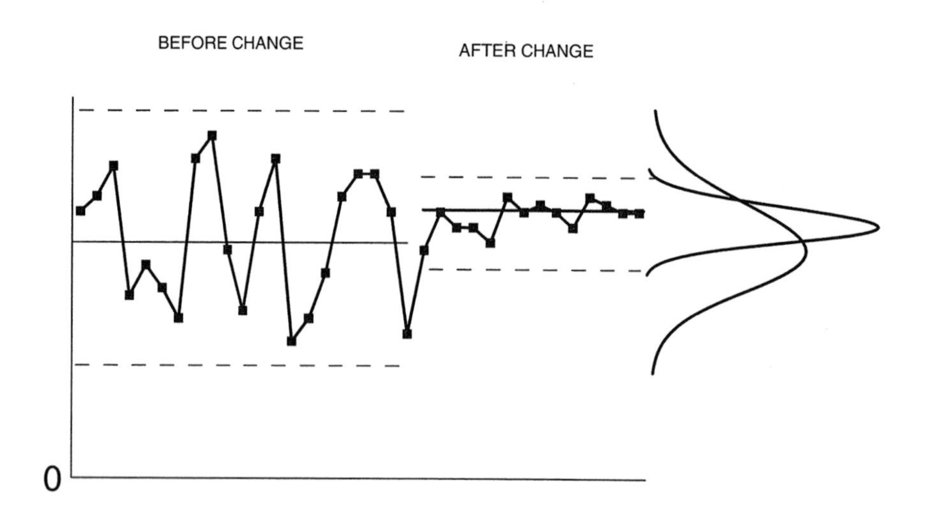

Figure VI.67. Control chart patterns: *planned changes.*

The reason SPC is used is to accelerate the learning process and to eventually produce an improvement. Control charts serve as historical records of the learning process and they can be used by others to improve other processes. When an improvement is realized the change should be written on the old control chart; its effect will show up as a less variable process. These charts are also useful in communicating the results to leaders, suppliers, customers, and others interested in quality improvement.

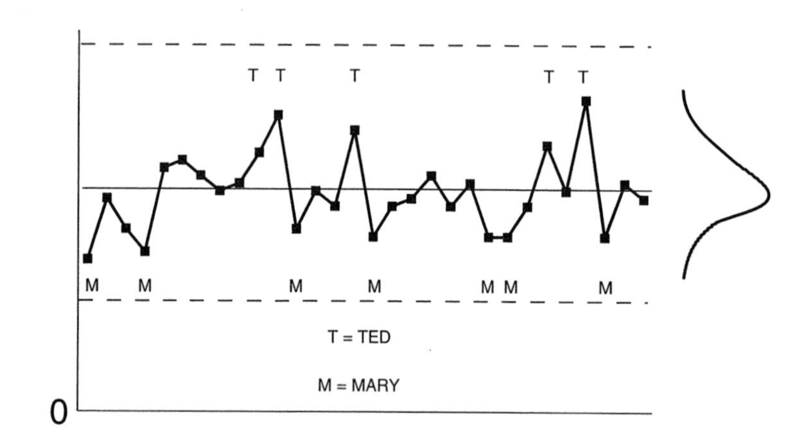

Figure VI.68. Control chart patterns: *suspected differences.*

Seemingly random patterns on a control chart are evidence of unknown causes of variation, which is not the same as *uncaused* variation. There should be an ongoing effort to reduce the variation from these so-called common causes. Doing so requires that the unknown causes of variation be identified. One way of doing this is a retrospective evaluation of control charts. This involves brainstorming and preparing cause and effect diagrams, then relating the control chart patterns to the causes listed on the diagram. For example, if "operator" is a suspected cause of variation, place a label on the control chart points produced by each operator. If the labels exhibit a pattern, there is evidence to suggest a problem. Conduct an investigation into the reasons and set up controlled experiments (prospective studies) to test any theories proposed. If the experiments indicate a true cause and effect relationship, make the appropriate process improvements. Keep in mind that a statistical *association* is not the same thing as a causal *correlation*. The observed association must be backed up with solid subject-matter expertise and experimental data.

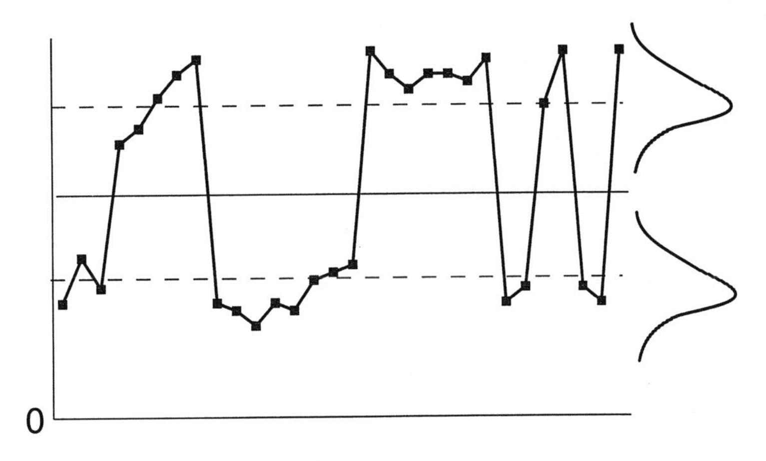

Figure VI.69. Control chart patterns: *mixture.*

Mixture exists when data from two different cause-systems are plotted on a single control chart. It indicates a failure in creating rational subgroups. The underlying differences should be identified and corrective action taken. The nature of the corrective action will determine how the control chart should be modified.

Mixture example #1

The mixture represents two different operators who can be made more consistent. A single control chart can be used to monitor the new, consistent process.

Mixture example #2

The mixture is in the number of emergency room cases received on Saturday evening, versus the number received during a normal week. Separate control charts should be used to monitor patient-load during the two different time periods.

Rules for determining statistical control: run tests

If the process is stable, then the distribution of subgroup averages will be approximately normal. With this in mind, we can also analyze the *patterns* on the control charts to see if they might be attributed to a special cause of variation. To do this, we divide a normal distribution into zones, with each zone one standard deviation wide. Figure VI.70 shows the approximate percentage we expect to find in each zone from a stable process.

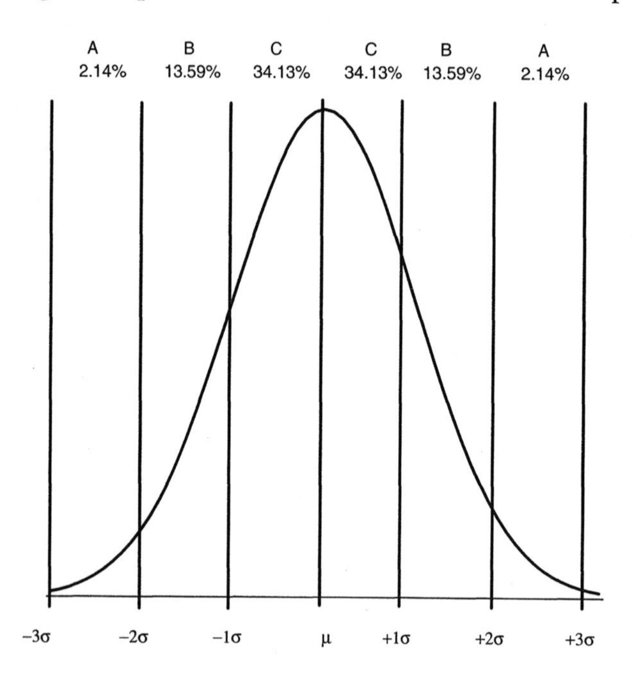

Figure VI.70. Percentiles for a normal distribution.

Zone *C* is the area from the mean to the mean plus or minus one sigma, zone B is from plus or minus one to plus or minus two sigma, and zone A is from plus or minus two to plus or minus three sigma. Of course, any point beyond three sigma (i.e., outside of the control limit) is an indication of an out-of-control process.

Since the control limits are at plus and minus three standard deviations, finding the one and two sigma lines on a control chart is as simple as dividing the distance between the grand average and either control limit into thirds, which can be done using a ruler. This divides each half of the control chart into three zones. The three zones are labeled A, B, and C as shown on Figure VI.71.

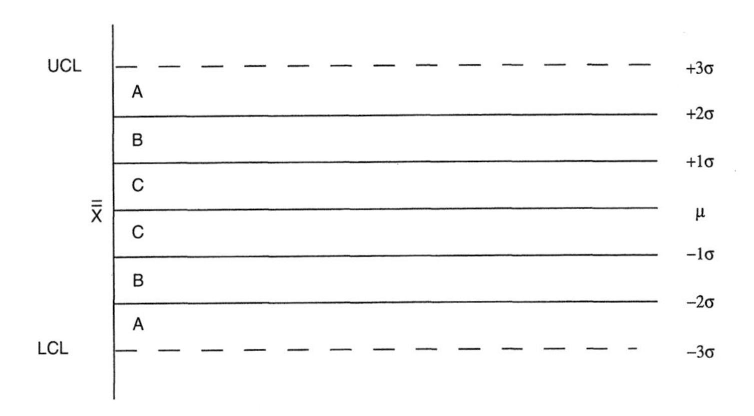

Figure VI.71. Zones on a control chart.

Based on the expected percentages in each zone, sensitive run tests can be developed for analyzing the patterns of variation in the various zones. Remember, the existence of a non-random pattern means that a special cause of variation was (or is) probably present. The averages, *np* and *c* control chart run tests are shown in Figure VI.72.

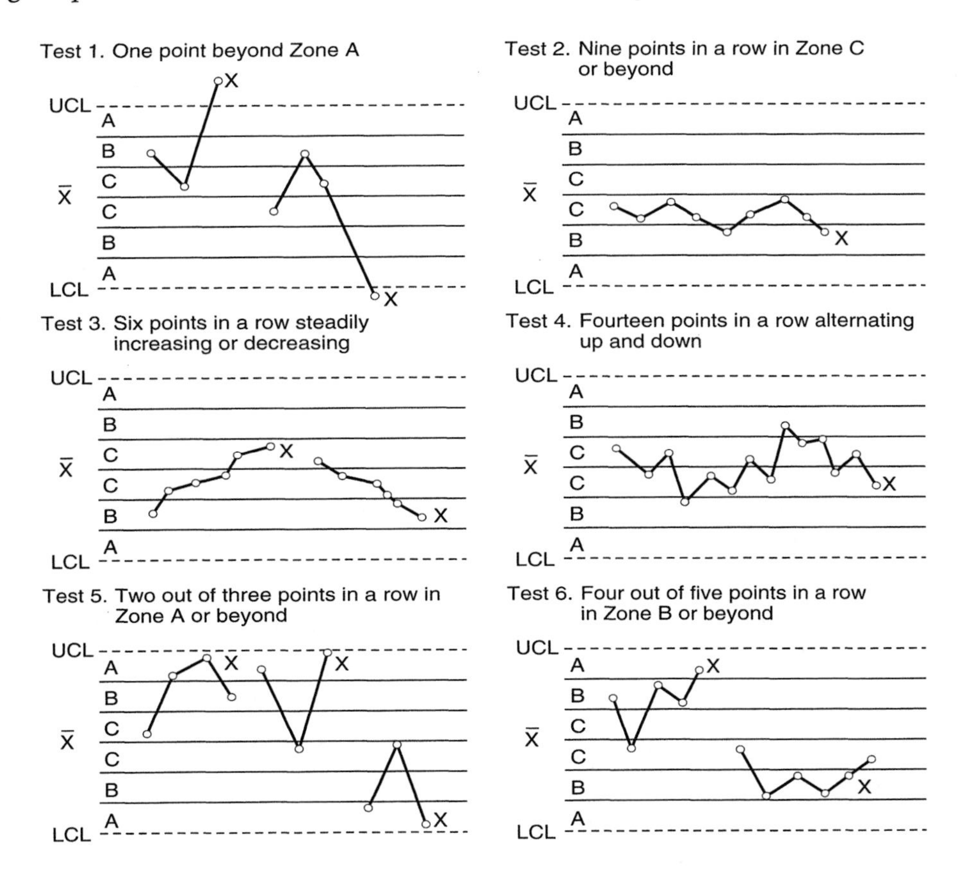

Figure VI.72—*Continued on next page . . .*

Figure VI.72—*Continued . . .*

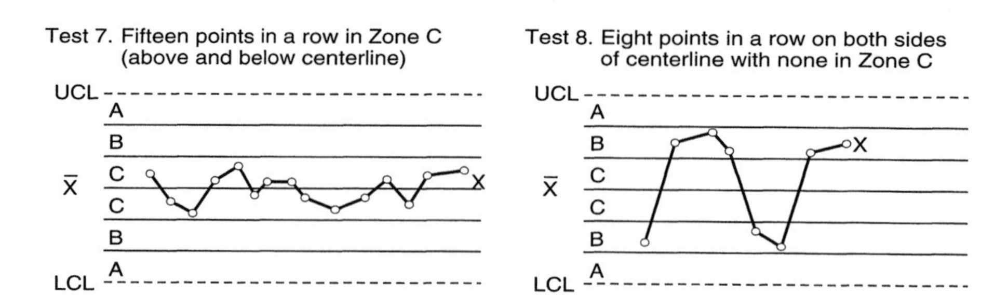

Figure VI.72. Tests for out of control patterns on control charts.

From "The Shewhart Control Chart—Tests for Special Causes," *Journal of Quality Technology*, 16(4), p 238.

Note that, when a point responds to an out-of-control test it is marked with an "X" to make the interpretation of the chart easier. Using this convention, the patterns on the control charts can be used as an aid in troubleshooting.

Tampering effects and diagnosis

Tampering occurs when adjustments are made to a process that is in statistical control. Adjusting a controlled process will always increase process variability, an obviously undesirable result. The best means of diagnosing tampering is to conduct a process capability study (see VI.H) and to use a control chart to provide guidelines for adjusting the process.

Perhaps the best analysis of the effects of tampering is from Deming (1986). Deming describes four common types of tampering by drawing the analogy of aiming a funnel to hit a desired target. These "funnel rules" are described by Deming (1986, p. 328):

1. "Leave the funnel fixed, aimed at the target, no adjustment.
2. "At drop k (k = 1, 2, 3, ...) the marble will come to rest at point z_k, measured from the target. (In other words, z_k is the error at drop k.) Move the funnel the distance $-z_k$ from the last position. Memory 1.
3. "Set the funnel at each drop right over the spot z_k, measured from the target. No memory.
4. "Set the funnel at each drop right over the spot (z_k) where it last came to rest. No memory."

Rule #1 is the best rule for stable processes. By following this rule, the process average will remain stable and the variance will be minimized. Rule #2 produces a stable output, but one with twice the variance of rule #1. Rule #3 results in a system that "explodes"; i.e., a symmetrical pattern will appear with a variance that increases without bound. Rule #4 creates a pattern that steadily moves away from the target, without limit.

At first glance, one might wonder about the relevance of such apparently abstract rules. However, upon more careful consideration, one finds many practical situations where these rules apply.

Rule #1 is the ideal situation and it can be approximated by using control charts to guide decision-making. If process adjustments are made only when special causes are indicated and identified, a pattern similar to that produced by rule #1 will result.

Rule #2 has intuitive appeal for many people. It is commonly encountered in such activities as gage calibration (check the standard once and adjust the gage accordingly) or in some automated equipment (using an automatic gage, check the size of the last feature produced and make a compensating adjustment). Since the system produces a stable result, this situation can go unnoticed indefinitely. However, as shown by Taguchi, increased variance translates to poorer quality and higher cost.

The rationale that leads to rule #3 goes something like this: "A measurement was taken and it was found to be 10 units above the desired target. This happened because the process was set 10 units too high. I want the average to equal the target. To accomplish this I must try to get the next unit to be 10 units too low." This might be used, for example, in preparing a chemical solution. While reasonable on its face, the result of this approach is a wildly oscillating system.

A common example of rule #4 is the "train-the-trainer" method. A master spends a short time training a group of "experts," who then train others, who train others, et cetera. An example is on-the-job training. Another is creating a setup by using a piece from the last job. Yet another is a gage calibration system where standards are used to create other standards, which are used to create still others, and so on. Just how far the final result will be from the ideal depends on how many levels deep the scheme has progressed.

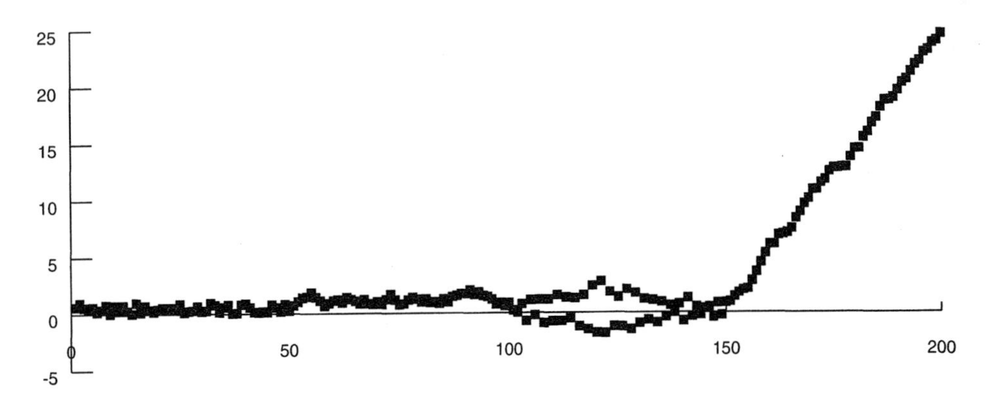

Figure VI.73. Funnel rule simulation results.

SHORT-RUN SPC

Introduction

A starting place for understanding SPC for short and small runs is to define our terms. The question "what *is* a short run?" will be answered for our purposes as an environment that has a large number of jobs per operator in a production cycle, each job involving different product. A production cycle is typically a week or a month. A *small run* is a situation where only a very few products of the same type are to be produced. An extreme case of a small run is the one-of-a-kind product, such as the Hubble Space Telescope. Short runs need not be small runs; a can manufacturing line can produce over 100,000 cans in an hour or two. Likewise small runs are not neessarily short runs; the Hubble Space Telescope took over 15 years to get into orbit (and even longer to get into orbit and working properly!) However, it is possible to have runs that are both short *and* small. Programs such as Just-In-Time inventory control (JIT) are making this situation more common all of the time.

Process control for either small or short runs involve similar strategies. Both situations involve markedly different approaches than those used in the classical mass-production environment. Thus, this chapter will treat both the small run and the short run situations simultaneously. You should, however, select the SPC tool that best fits your particular situation.

Strategies for short and small runs

Juran's famous trilogy separates quality activities into three distinct phases (Juran, 1988):

- Planning
- Control
- Improvement

Figure VI.74 provides a graphic portrayal of the Juran trilogy.

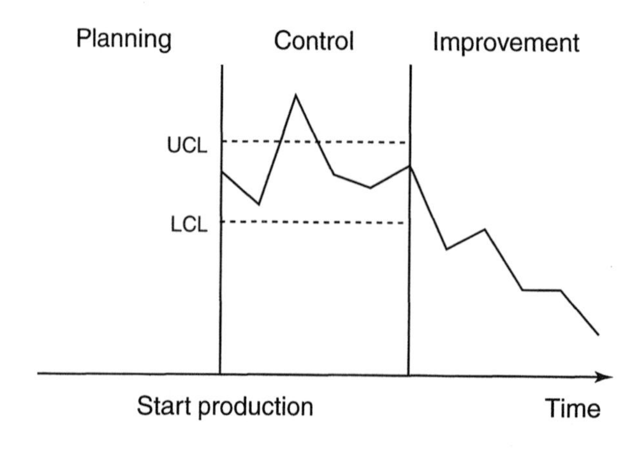

Figure VI.74. Juran's Trilogy

When faced with small or short runs, the emphasis should be placed in the planning phase. As much as possible needs to be done *before* any product is made, because it simply isn't possible to waste time or materials "learning from mistakes" made during production. It is also helpful to realize that the Juran trilogy is usually applied to *products*, while SPC applies to *processes*. It is quite possible that the element being monitored with SPC is a process element and not a product feature at all. In this case there really is no "short run."

A common problem with application of SPC to short/small runs is that people fail to realize the limitations of SPC in this application. Even the use of SPC to *long production runs* will benefit from a greater emphasis on pre-production planning. In the best of all worlds, SPC will merely confirm that the correct process has been selected and controlled in such a way that it consistently produces well-designed parts at very close to the desired target values for every dimension.

Preparing the short-run Process Control Plan (PCP)

Plans for short runs require a great deal of up-front attention. The objective is to create a list of as many potential sources of variation as possible and to take action to deal with them *before* going into production. One of the first steps to be taken is to identify which processes may be used to produce a given part; this is called the "Approved Process List." Analogously, parts that can be produced by a given process should also be identified; this is called the "Approved Parts List." These determinations are made based on process capability studies. Because short runs usually involve less than the recommended number of pieces, the acceptability criteria is usually modified. When less than 50 observations are used to determine the capability, I recommend that the capability indices described in chapter VI.H be modified by using a $\pm 4\sigma$ acceptable process width (instead of $\pm 3\sigma$) and a minimum acceptable C_{pk} of 1.5 (instead of 1.33). Don't bother making formal capability estimates until you have at least 30 observations. (You will see below that these observations need not always be from 30 separate parts.)

When preparing for short runs, it often happens that actual production parts are not available in sufficient quantity for process capability studies. One way of dealing with this situation is to study process elements separately and to then sum the variances from all of the known elements to obtain an estimate of the best overall variance a given process will be able to produce.

For example, in an aerospace firm that produced conventional guided missiles, each missile contained thousands of different parts. In any given month, the number of missiles produced was small. Thus, the CNC machine shop, and the rest of the plant, were faced with a small/short run situation. However, it was not possible to do separate pre-production capability studies of each part. The approach used instead was to design a special test part that would provide estimates of the machine's ability to produce every basic type of characteristic (flatness, straightness, angularity, location, etc.). Each CNC machine produced a number of these test parts under controlled conditions and the results were plotted on a Short-Run X and R chart using the exact method (described later in this chapter). The studies were repeated periodically for each machine.

These studies provided pre-production estimates of the machine's ability to produce different characteristics. However, these estimates were always *better* than the process would be able to do with actual production parts. Actual pro duction would involve different operators, tooling, fixtures, materials, and other common and special causes not evaluated by the *machine capability study.* Preliminary Approved Parts Lists and Preliminary Approved Process Lists were created from the capability analysis using the more stringent acceptability criteria described above (C_{pk} at least 1.5 based on a $\pm 4\sigma$ process spread). When production commenced, the actual results of the production runs were used instead of the estimates based on special runs. Once sufficient data were available, the parts were removed from the preliminary lists and placed on the appropriate permanent lists.

When creating Approved Parts and Approved Process lists always use the most stringent product requirements to determine the process requirement. For example, if a process will be used to drill holes in 100 different parts with hole location tolerances ranging from 0.001 inches to 0.030 inches, the *process requirement* is 0.001 inches. The process capability estimate is based on its ability to hold the 0.001 inch tolerance.

The approach used is summarized as follows:

1. Get the process into statistical control.
2. Set the control limits *without regard to the requirement.*
3. Based on the calculated process capability, determine if the most stringent product requirement can be met.

Process audit

The requirements for all processes should be documented. A process audit checklist should be prepared and used to determine the condition of the process prior to production. The audit can be performed by the operator himself, but the results should be documented. The audit should cover known or suspected sources of variation. These include such things as the production plan, condition of fixtures, gage calibration, the resolution of the gaging being used, obvious problems with materials or equipment, operator changes, and so on. SPC can be used to monitor the results of the process audits over time. For example, an audit score can be computed and tracked using an individuals control chart or a demerit control chart.

Selecting process control elements

Many short-run SPC programs bog down because the number of control charts being used grows like Topsy. Before anyone knows what is happening the walls are plastered with charts that few understand and nobody uses. The operators and inspectors wind up spending more time filling out paperwork than they spend on true value-added work. Eventually the entire SPC program collapses under its own weight.

One reason for this is that people tend to focus their attention on the *product* rather than on the *process*. Control elements are erroneously selected because they are functionally important. A great fear is that an important product feature will be produced out of specification and that it will slip by unnoticed. This is a misunderstanding of the purpose of SPC, which is to provide a means of *process* control; SPC is not intended to be a substitute for inspection or testing. This is the guiding rule of selecting control items for SPC:

> SPC control items should be selected to provide a maximum amount of information on the state of the process at a minimum cost.

Fortunately, most process elements are correlated with one another. Because of this, one process element may provide information not only about itself, but about several other process elements as well. This means that a small number of process control elements will often explain a large portion of the process variance.

Although sophisticated statistical methods exist to help determine which process components explain the most variance, common sense and knowledge of the process can often do as well, if not better. The key is to think about the process carefully.

- What are the "generic process elements" that affect all parts?
- How do the process elements combine to affect the product?
- Do several process elements affect a single product feature?
- Do changes in one process element automatically cause changes in some other process elements?
- What process elements or product features are most sensitive to unplanned changes?

Example #1

The CNC machines mentioned earlier were extremely complex. A typical machine had dozens of different tools and produced hundreds of different parts with thousands of characteristics. However, the SPC team reasoned that the machines themselves involved only a small number of "generic operations": select a tool, position the tool, rotate the tool, move the part, remove metal. Further study revealed that nearly all of the problems encountered after the initial setup involved only the ability of the machine to position the tool precisely. A control plan was created that called for monitoring no more than one variable for each axis of movement. The features selected were those farthest from the machine's "home position" and involving the most difficult to control operations, not necessaily functionally important features. Often a single feature provided control of more than one axis of movement; for example, the location of a single hole provides information on the location of the tool in both the X and Y directions. As a result of this system no part had more than four features monitored with control charts, even though many parts had thousands of features. Subsequent evaluation of the accumulated data by a statistician revealed that the choices made by the team explained over 90% of the process variance.

Example #2

A wave solder machine was used to solder printed circuit boards for a manufacturer of electronic test equipment. After several months of applying SPC, the SPC team evaluated the data and decided that they needed only a single measure of product quality for SPC purposes: defects per 1,000 solder joints. A single control chart was used for dozens of different circuit boards. They also determined that most of the process variables being checked could be eliminated.

The only process variables monitored in the future would be flux density, solder chemistry (provided by the vendor), solder temperature, and final rinse contamination. Historic data showed that one of these variables was nearly always out-of-control when process problems were encountered. Other variables were monitored with periodic audits but not charted.

Notice that in both of these examples all of the variables being monitored were related to the *process*, even though some of them were product features. The terms "short run" and "small run" refer to the product variables only; the process is in continuous operation so its run size and duration is neither short nor small. It makes soldered joints continuously.

The single part process

The ultimate small run is the single part. A great deal can be learned by studying single pieces, even if your situation involves more than one part. The application of SPC to single pieces may seem incongruous. Yet when we consider that the "P" in SPC stands for process and not product, perhaps it is possible after all. Even the company producing one-of-a-kind products usually does so with the same equipment, employees, facilities, suppliers, etc. In other words, they use the same process to produce different products. Also, they usually produce products that are similar, even though not identical. This is also to be expected. It would be odd indeed to find a company making microchips one day and baking bread the next. The processes are too dissimilar.

This discussion implies that the key to controlling the quality of single parts is to concentrate on the process elements rather than on the product features. This is the same rule we applied above to larger runs. In fact, it's a good rule to apply to all SPC applications, regardless of the number of parts being produced! Consider a company manufacturing communications satellites. The company produces a satellite every year or two. The design and complexity of each satellite is quite different than any other. How can SPC be applied at this company? A close look at a satellite will reveal immense complexity. The satellite will have thousands of terminals, silicon solar cells, solder joints, fasteners, and so on. Hundreds, even thousands, of people are involved in the design, fabrication, testing, and assembly. In other words, there are processes that involve massive amounts of repetition. The processes include engineering (errors per engineering drawing); terminal manufacture (size, defect rates); solar cell manufacture (yields, electrical properties); soldering (defects per 1,000 joints; strength); fastener installation quality (torque) and so on.

Another example of a single-piece run is software development. The "part" in this case is the working copy of the software delivered to the customer. Only a single unit of product is involved. How can we use SPC here?

Again, the answer comes when we direct our attention to the underlying process. Any marketable software product will consist of thousands, perhaps millions, of bytes of finished machine code. This code will be compiled from thousands of lines of source code. The source code will be arranged in modules; the modules will contain procedures; the procedures will contain functions; and so on. Computer science has developed a number of ways of measuring the quality of computer code. The resulting numbers, called computer software metrics, can be analyzed using SPC tools just like any other numbers. The processes that produced the code can thus be measured, controlled and improved. If the process is in statistical control, the process elements, such as programmer selection and training, coding style, planning, procedures, etc. must be examined. If the process is not in statistical control, the special cause of the problem must be identified.

As discussed earlier, although the single part process is a small run, it isn't necessarily a short run. By examining the process rather than the part, improvement possibilities will begin to suggest themselves. The key is to find the process and to define its elements so they may be measured, controlled, and improved.

Other elements of the PCP

In addition to the selection of process control elements, the PCP should also provide information on the following:

- the method of inspection
- dates and results of measurement error studies
- dates and results of process capability studies
- subgroup sizes and methods of selecting subgroups
- sampling frequency
- required operator certifications
- pre-production checklists
- setup approval procedures
- notes and suggestions regarding previous problems

In short, the PCP provides a complete, detailed roadmap that describes how process integrity will be measured and maintained. By preparing a PCP the inputs to the process are controlled ahead of time, thus assuring that the *outputs* from the process will be consistently acceptable.

Short-run SPC techniques

Short production runs are a way of life with many manufacturing companies. In the future, this will be the case even more often. The trend in manufacturing has been toward smaller production runs with product tailored to the specific needs of individual customers. The days of "the customer can have any color, as long as it's black" have long since passed.

Classical SPC methods, such as X and R charts, were developed in the era of mass production of identical parts. Production runs often lasted for weeks, months, or even years. Many of the "SPC rules of thumb" currently in use were created for this situation. For example, there is the rule that control limits not be calculated until data is available from at least 25 subgroups of 5. This may not have been a problem in 1930, but it certainly is today. In fact, many *entire production runs* involve fewer parts than required to start a standard control chart! Many times the usual SPC methods can be modified slightly to work with short and small runs. For example, X and R control charts can be created using moving averages and moving ranges. However, there are special SPC methods that are particularly well-suited to application on short or small runs.

Variables data

Variables data involve measurements on a continuous scale such as size, weight, pH, temperature, etc.. In theory, data are variables data if no two values are exactly the same. In practice, this is seldom the case. As a rough rule-of-thumb you can consider data to be variables data if at least ten different values occur and repeated values make up no more than 20% of the data set. If this is not the case, your data may be too discrete to use standard control charts. Consider trying an attribute procedure such as the demerit charts described later in this section. We will discuss the following approaches to SPC for short or small runs:

1. **Exact method**—Tables of special control chart constants are used to create X, X and R charts that compensate for the fact that a limited number of subgroups are available for computing control limits. The exact method is also used to compute control limits when using a code value chart or stabilized X or X and R charts (see below). The exact method allows the calculation of control limits that are correct when only a small amount of data is available. As more data becomes available the exact method updates control limits until, finally, no further updates are required and standard control chart factors can be used.

2. **Code value charts**—These control charts are created by subtracting nominal or other target values from actual measurements. These charts are often standardized so that measurement units are converted to whole numbers. For example, if measurements are in thousandths of an inch, a reading of 0.011 inches above nominal would be recorded simply as "11." Code value charts enable the user to plot several parts from a given process on a single chart, or to plot several features from a single part on the same control chart. The Exact Method can be used to adjust the control limits when code value charts are created with limited data.
3. **Stabilized control charts for variables**—Statisticians have known about normalizing transformations for many years. This approach can be used to create control charts that are independent of the unit of measure and scaled in such a way that several different characteristics can be plotted on the same control chart. Since stabilized control charts are independent of the unit of measure, they can be thought of as true process control charts. The Exact Method adjusts the control limits for stabilized charts created with limited data.

Exact method of computing control limits for short and small runs

This procedure applies to short runs or any situation where a small number of subgroups will be used to set up a control chart. It consists of three stages: (1) finding the process (establishing statistical control); (2) setting limits for the remainder of the initial run; and (3) setting limits for future runs. The procedure correctly compensates for the uncertainties involved when computing control limits with small amounts of data.

Stage one: find the process

1. Collect an initial sample of subgroups (g). The factors for the recommended minimum number of subgroups are shown in Table 18 in the Appendix in a larger font size. If it is not possible to get the minimum number of subgroups, use the appropriate control chart constant for the number of subgroups you actually have.
2. Using Table 18 in the Appendix compute the Range chart control limits using the equation Upper Control Limit for Ranges (UCL_R) = $D_{4F} \times \overline{R}$. Compare the subgroup ranges to the UCL_R and drop any out-of-control groups after identifying and eliminating the special cause. Repeat the process until all remaining subgroup ranges are smaller than UCL_R.
3. Using the $\overline{R}$ value found in step #2, compute the control limits for the averages or individuals chart. The control limits are found by adding and subtracting $A_{2F} \times \overline{R}$ from the overall average. Drop any subgroups that have out-of-control averages and recompute. Continue until all remaining values are within the control limits. Go to Stage Two.

Stage two: set limits for remainder of the initial run

1. Using Table 18 in the Appendix compute the control limits for the remainder of the run. Use the A_{2S} factors for the $\overline{X}$ chart and the D_{4S} factors for the R chart; g = the number of groups used to compute Stage One control limits.

Stages One and Two can be repeated periodically during the run, e.g., after the 5th, 10th, 15th, etc. subgroup has been produced.

Stage three: set limits for a future run

1. After the run is complete, combine the raw data from the entire run and perform the analysis as described in Stage One above. Use the results of this analysis to set limits for the next run, following the Stage Two procedure. If more than 25 groups are available, use a standard table of control chart constants, such as Table 13 in the Appendix.

Notes on the exact method

1. Stage Three assumes that there are no special causes of variation between runs. If there are, the process may go out of control when using the Stage Three control limits. In these cases, remove the special causes. If this isn't possible, apply this procedure to each run separately (i.e., start over each time).
2. This approach will lead to the use of standard control chart tables when enough data is accumulated.
3. The constants for subgroups of five are from Hillier (1969). The minimum number of subgroups recommended for groups of size two to five is from Proschan and Savage (1960). The control chart constants for the first stage are A_{2F} and D_{4F} (the "F" subscript stands for first stage); for the second stage use A_{2S} and D_{4S} (the "S" stands for second stage.) These factors correspond to the A_2 and D_4 factors usually used, except that they are adjusted for the small number of subgroups actually available.

Setup approval procedure

The following procedure can be used to determine if a setup is acceptable using a relatively small number of sample units.

1. After the first-article approval, run n =3 to 10 pieces *without adjusting the process.*
2. Compute the average and the range of the sample.
3. Compute $T = \left| \dfrac{\text{average - target}}{range} \right|$, use absolute values (i.e., ignore any minus signs.) The target value is usually the specification midpoint or nominal.

4. If T is less than the critical T in the table below, accept the setup. Otherwise, adjust the setup to bring it closer to the target. NOTE: There is approximately 1 chance in 20 that an on-target process will fail this test.

n	3	4	5	6	7	8	9	10
Critical T	0.885	0.529	0.388	0.312	0.263	0.230	0.205	0.186

Example

Assume we wish to use SPC for a process that involves producing a part in lots of 30 parts each. The parts are produced approximately once each month. The control feature on the part is the depth of a groove and we will be measuring every piece. We decide to use subgroups of size three and to compute the Stage One control limits after the first five groups. The measurements obtained are shown in Table VI.36.

Table VI.36. Raw data for example of exact method.

SUBGROUP	SAMPLE NUMBER				
NUMBER	1	2	3	$\overline{X}$	R
1	0.0989	0.0986	01031	0.1002	0.0045
2	0.0986	0.0985	0.1059	0.1010	0.0074
3	0.1012	0.1004	0.1000	0.1005	0.0012
4	0.1023	0.1027	0.1000	0.1017	0.0027
5	0.0992	0.0997	0.0988	0.0992	0.0009

Using the data in Table VI.36, we can compute the grand average and average range as

$$\text{Grand average} = 0.10053$$

$$\text{Average range } (\overline{R}) = 0.00334$$

From Table 18 in the Appendix, we obtain the first stage constant for the range chart of $D_{4F} = 2.4$ in the row for g =5 groups and a subgroup size of 3. Thus,

$$UCL_R = D_{4F} \text{ x } \overline{R} = 2.4 \text{ x } 0.00334 = 0.0080$$

All of the ranges are below this control limit, so we can proceed to the analysis of the averages chart. For the averages chart we get:

$$LCL_{\overline{X}} = \text{grand average} - A_{2F} \text{ x } \overline{R}$$
$$= 0.10053 - 1.20 \text{ x } 0.00334 = 0.09652$$

$$UCL_{\overline{X}} = \text{grand average} + A_{2F} \text{ x } \overline{R}$$
$$= 0.10053 + 1.20 \text{ x } 0.00334 = 0.10454$$

All of the subgroup averages are between these limits. Now setting limits for the remainder of the run we use $D_{4S} = 3.4$ and $A_{2S} = 1.47$. This gives

$$UCL_R = 0.01136$$
$$LCL_{\bar{X}} = 0.09562$$
$$UCL_{\bar{X}} = 0.10544$$

If desired, this procedure can be repeated when a larger number of subgroups becomes available, say 10 groups. This would provide somewhat better estimates of the control limits, but it involves additional administrative overhead. When the entire run is finished you will have 10 subgroups of 3 per subgroup. The data from all of these subgroups should be used to compute Stage One and Stage Two control limits. The resulting Stage Two control limits would then be applied to the *next run* of this part number.

By applying this method in conjunction with the code value charts or stabilized charts described below, the control limits can be applied to the next part(s) produced by this process (assuming the part-to-part difference can be made negligible). Note that if the standard control chart factors were used the limits for *both* stages would be

$$UCL_R = 0.00860$$
$$LCL_{\bar{X}} = 0.09711$$
$$UCL_{\bar{X}} = 0.10395$$

As the number of subgroups available for computing the control limits increases, the "short run" control limits approach the standard control limits. However, if the standard control limits are used when only small amounts of data are available, there is a greater chance of erroneously rejecting a process that is actually in controL

Code value charts

This procedure allows the control of multiple features with a single control chart. It consists of making a simple transformation to the data, namely

$$\hat{x} = \frac{X - \text{Target}}{\text{unit of measure}} \quad \text{(VI.84)}$$

The resulting x values are used to compute the control limits and as plotted points on the X and R charts. This makes the target dimension irrelevant for the purposes of SPC and makes it possible to use a single control chart for several different features or part numbers.

Example

A lathe is used to produce several different sizes of gear blanks, as is indicated in Figure VI.75.

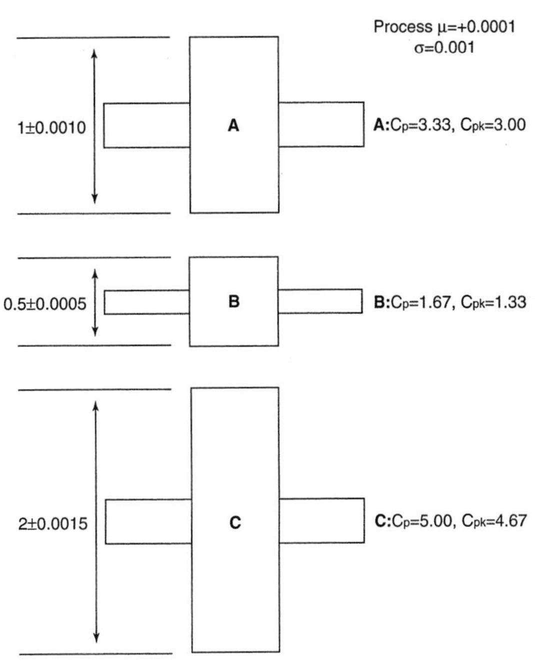

Figure VI.75. Some of the gear blanks to be machined.

Product engineering wants all of the gear blanks to be produced as near as possible to their nominal size. Process engineering believes that the process will have as little deviation for larger sizes as it does for smaller sizes. Quality engineering believes that the inspection system will produce approximately the same amount of measurement error for larger sizes as for smaller sizes. Process capability studies and measurement error studies support these conclusions. I hope you are starting to get the idea that a number of assumptions are being made and that they must be valid before using code value charts.

Based on these conclusions, the code value chart is recommended. By using the code value chart, the amount of paperwork will be reduced and more data will be available for setting control limits. Also, the process history will be easier to follow since the information won't be fragmented among several different charts. The data in Table VI.37 show some of the early results.

Table VI.37. Deviation from target in hundred-thousandths.

PART	NOMINAL	NO.	SAMPLE NUMBER 1	2	3	$\overline{X}$	R
A	1.0000	1	4	3	25	10.7	22
		2	3	3	39	15.0	36
		3	16	12	10	12.7	6
B	0.5000	4	21	24	10	18.3	14
		5	6	8	4	6.0	4
		6	19	7	21	15.7	14
C	2.0000	7	1	11	4	5.3	10
		8	1	25	8	11.3	24
		9	6	8	7	7.0	2

Note that the process must be able to produce the *tightest tolerance* of ±0.0005 inches (gear blank B.) The capability analysis should indicate its ability to do this; i.e., C_{pk} should be at least 1.33 based on the tightest tolerance. It will *not* be allowed to drift or deteriorate when the less stringently toleranced parts are produced. Process control is independent of the product requirements. Permitting the process to degrade to its worst acceptable level (from the product perspective) creates engineering nightmares when the more tightly toleranced parts come along again. It also confuses and demoralizes operators and others trying to maintain high levels of quality. Operators should be trained to control their own processes, which requires that they understand the relationship between product specifications and process control limits. They should also appreciate that any variation from the target value incurs some loss, even if the part is still within the specification limits.

The control chart of the data in Table VI.37 is shown in Figure VI.76. Since only nine groups were available, the exact method was used to compute the control limits. Note that the control chart shows the deviations on the $\overline{X}$ and R chart axes, not the actual measured dimensions; e.g., the value of Part A, subgroup #1, sample #1 was +0.00004" from the target value of 1.0000" and it is shown as a deviation of +4 hundred-thousandths; i.e., the part checked 1.00004". The Stage One control chart shows that the process is in statistical control, but it is producing parts that are consistently too large—regardless of the nominal dimension. If the process were on target, the grand average would be very close to 0. The setup problem would have been detected by the second subgroup if the setup approval procedure described earlier in this chapter had been followed. This ability to see process performance across different part numbers is one of the advantages of Code Value charts. It is good practice to actually identify the changes in part numbers on the charts, as is done in Figure VI.76.

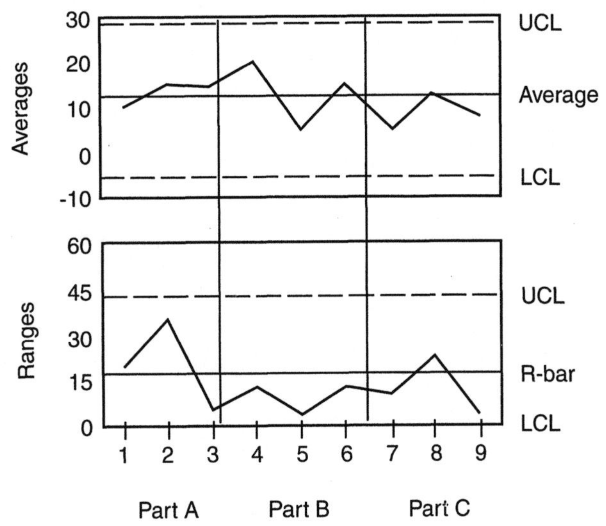

Figure VI.76. Code value chart of Table VI.37.

Stabilized control charts for variables

All control limits, for standard sized runs or short and small runs, are based on methods that determine if a process statistic falls within limits that might be expected from chance variation alone. In most cases, the statistic is based on actual measurements from the process and it is in the same unit of measure as the process measurements. As we saw with code value charts, it is sometimes useful to transform the data in some way. With code value charts we used a simple transformation that removed the effect of changing nominal and target dimensions. While useful, this approach still requires that all measurements be in the same units of measurement, e.g., all inches, all grams, etc.. For example, all of the variables on the control chart for the different gear blanks had to be in units of hundred-thousandths of an inch. If we had wanted to plot, for example, the perpendicularity of two surfaces on the gear blank we would have needed a separate control chart because the units would be in degrees instead of inches. Stabilized control charts for variables overcome the units of measure problem by converting all measurements into standard, non-dimensional units. Such "standardizing transformations" are not new; they have been around for many years and they are commonly used in all types of statistical analyses. The two transformations we will be using here are shown in Equations VI.83 and VI.84.

$$(\overline{X} - \text{grand average})/\overline{R} \qquad \text{(VI.85)}$$

$$R/\overline{R} \qquad \text{(VI.86)}$$

As you can see, Equation VI.85 involves subtracting the grand average from each subgroup average (or from each individual measurement if the subgroup size is one) and dividing the result by $\overline{R}$ Equation VI.86 divides each subgroup range by the average range. Since the numerator and denominator of both equations are in the same unit of measurement, the unit of measurement cancels and we are left with a number that is in terms of the number of average ranges, $\overline{R}$'s. It turns out that control limits are also in the same units; i.e., to compute standard control limits we simply multiply $\overline{R}$ by the appropriate table constant to determine the width between the control limits.

Hillier (1969) noted that this is equivalent to using the transformations shown in Equations VI.87 and VI.88 with control limits set at

$$-A_2 \leq \left(\overline{X} - \text{grand average}\right)/\overline{\text{R}} \leq \text{A}_2 \tag{VI.87}$$

for the individuals or averages chart. Control limits are

$$D_3 \leq R/\overline{R} \leq D_4 \tag{VI.88}$$

for the range chart. Duncan (1986) described a similar transformation for attribute charts, p charts in particular, and called the resulting chart a "stabilized p chart." We will call charts of the transformed variables data stabilized charts as well.

Stabilized charts allow you to plot multiple units of measurement on the same control chart. The procedure described in this book for stabilized variables charts requires that all subgroups be of the same size*. When using stabilized charts, the control limits are always fixed. The raw data are "transformed" to match the scale determined by the control limits. When only limited amounts of data are available, the constants in Table 18 of the Appendix should be used for computing control limits for stabilized variables charts. As more data become available, the Appendix Table 18 constants approach the constants in standard tables of control chart factors. Table VI.38 summarizes the control limits for stabilized averages, stabilized ranges, and stabilized individuals control charts.

*The procedure for stabilized attribute charts, described later in this section, allows varying subgroup sizes.

Table VI.38. Control limits for stabilized charts.

STAGE	AVAILABLE GROUPS		CHART		
			$\overline{X}$	R	x
One	25 or less	LCL	$-A_{2F}$	None	$-A_{2F}$
		Average	0	1	0
		UCL	$+A_{2F}$	D_{4F}	$+A_{2F}$
Two	25 or less	LCL	$-A_{2S}$	None	$-A_{2S}$
		Average	0	1	0
		UCL	$+A_{2S}$	D_{4S}	$+A_{2S}$
One or Two	More than 25	LCL	$-A_2$	D_3	-2.66
		Average	0	1	0
		UCL	$+A_2$	D_4	+2.66

The values for A_2, D_3 and D_4 can be found in Table 13 of the Appendix.

Example

A circuit board is produced on an electroplating line. Three parameters are considered important for SPC purposes: lead concentration of the solder plating bath, plating thickness, and resistance. Process capability studies have been done using more than 25 groups; thus, from Table VI.38, the control limits are

$$-A_2 \leq \overline{X} \leq A_2$$

for the averages control chart, and

$$D_3 \leq R \leq D_4$$

for the ranges control chart. The actual values of the constants A_2, D_3, and D_4 depend on the subgroup size, for subgroups of three A_2=1.023, D_3=O and D_4=2.574.

The capabilities are given in Table VI.39.

Table VI.39. Process capabilities for example.

FEATURE CODE	FEATURE	GRAND AVG.	AVG. RANGE
A	Lead %	10%	1%
B	Plating thickness	0.005"	0.0005"
C	Resistance	0.1Ω	0.0005Ω

A sample of three will be taken for each feature. The three lead concentration samples are taken at three different locations in the tank. The results of one such set of sample measurements are shown in Table VI.40, along with their stabilized values.

Table VI.40. Sample data for example.

NUMBER	LEAD % (A)	THICKNESS (B)	RESISTANCE (C)
1	11%	0.0050"	0.1000Ω
2	11%	0.0055"	0.1010Ω
3	8%	0.0060"	0.1020Ω
$\overline{X}$	10%	0.0055"	0.1010Ω
R	3%	0.0010"	0.0020Ω
$\frac{(x-\overline{x})}{\overline{R}}$	0	1	2
$\frac{R}{\overline{\overline{R}}}$	3	2	4

On the control chart *only the extreme values are plotted.* Figure VI.76 shows a stabilized control chart for several subgroups. Observe that the feature responsible for the plotted point is written on the control chart; features are identified by the feature code in Table VI.39. If a long series of largest or smallest values come from the same feature, it is an indication that the feature has changed. If the process is in statistical control for all features, the feature responsible for the extreme values will vary randomly. Table 19 in the Appendix can be used to determine if the same feature is showing up too often. See earlier this section for a description of the run test procedure.

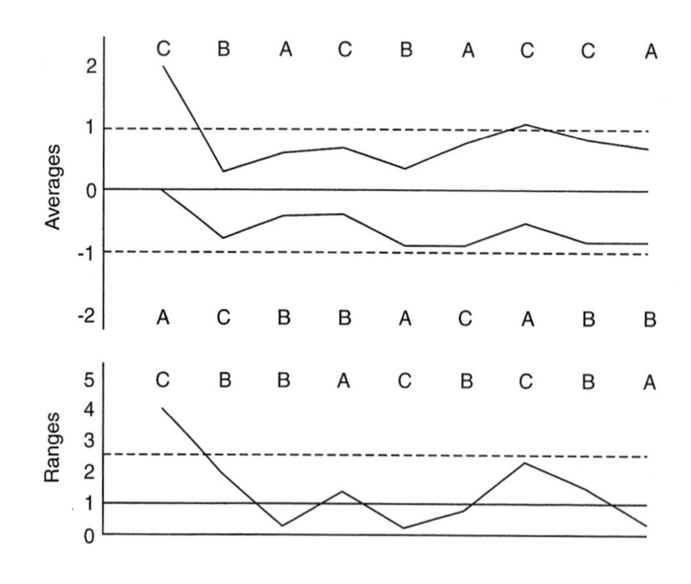

Figure VI.77 Stabilized control chart for variables.

When using stabilized charts it is possible to have a single control chart accompany a particular part or lot of parts through the entire production sequence. For example, the circuit boards described above could have a control chart that shows the results of process and product measurement for characteristics at all stages of production. The chart would then show the "processing history" for the part or lot. The advantage would be a coherent log of the production of a given part. Table VI.41 illustrates a process control plan that could possibly use this approach.

Table VI.41. PWB Fab process capabilities and SPC plan.

OPERATION	FEATURE	$\overline{X}$	$\overline{R}$	n
Clean	Bath Ph	7.5	0.1	3/hr
	Rinse contamination	100 ppm	5 ppm	3/hr
	Cleanliness quality rating	78	4	3 pcs/hr
Laminate	Riston thickness	1.5 min.	0.1mm	3 pcs/hr
	Adhesion	7 in.–lbs.	0.2 in.–lbs.	3 pcs/hr
Plating	Bath lead %	10%	1%	3/hr
	Thickness	0.005"	0.0005"	3 pcs/hr
	Resistance	0.1Ω	0.0005Ω	3 pcs/hr

A caution is in order if the processing history approach is used. When small and short runs are common, the history of a given process can be lost among the charts of many different parts. This can be avoided by keeping a separate chart for each distinct process; this involves additional paperwork, but it might be worth the effort. If the additional paperwork burden becomes large, computerized solutions may be worth investigating.

Attribute SPC for small and short runs

When data is difficult to obtain, as it usually is when small or short runs are involved, you should use variables SPC if at all possible. A variables measurement on a continuous scale contains more information than a discrete attributes classification provides. For example, say a machine is cutting a piece of metal tubing to length. The specifications call for the length to be between 0.990" and 1.010" with the preferred length being 1.000" exactly. There are two methods available for checking the process. Method #1 involves measuring the length of the tube with a micrometer and recording the result to the nearest 0.001". Method #2 involves placing the finished part into a "go/no-go gage." With method #2, a part that is shorter than 0.990" will go into the "no-go" portion of the gage, while a part that is longer than 1.010" will fail to go into the "go" portion of the gage. With method #1, we can determine the size of the part to within 0.001" (assuming the measurement system was properly studied). With method #2, we can only determine the size of the part to within 0.020"; i.e. either it is within the size tolerance, it's too short, or it's too long. If the process could hold a tolerance of *less than 0.020"*, method #1 would provide the necessary information to hold the process to the variability it is capable of holding. Method #2 would not detect a process drift until out of tolerance parts were actually produced.

Another way of looking at the two different methods is to consider each part as belonging to a distinct category, determined by the part's length. We see that method #1 allows us to place any part that is within tolerance into one of twenty categories. When out of tolerance parts are considered, method #1 is able to place parts into even more than twenty different categories. Method #1 also tells us if the part is in the best category, namely within ±0.001" of 1.000"; if not, we know how far the part is from the best category. With method #2, we can place a given part into only three categories: too short, within tolerance, or too long. A part that is far too short will be placed in the same category as a part that is only slightly short. A part that is barely within tolerance will be placed in the same category as a part that is exactly on target.

SPC of attributes data from short runs

Special methods must be used for attributes data used to control short run processes. We will describe two such methods:

- Stabilized attribute control charts
- Demerit control charts

Stabilized attribute control charts

When plotting attribute data statistics from short run processes we encounter two difficulties:

1. Varying subgroup sizes
2. A small number of subgroups per production runs

Item #1 results in messy charts with different control limits for each subgroup, distorted chart scales that mask significant variations, and chart patterns that are difficult to interpret because they are affected by both sample size changes and true process changes. Item #2 makes it difficult to track long-term process trends because the trends are broken up among many different control charts for individual parts. For these reasons, many people believe that SPC is not practical unless large and long runs are involved. This is not the case. In most cases, stabilized attribute charts can be used to eliminate these problems. Although somewhat more complicated than classical control charts, stabilized attribute control charts offer a way of realizing the benefits of SPC with processes that are difficult to control any other way.

Stabilized attribute charts may be used if a process is producing part features that are essentially the same from one part number to the next. Production lot sizes and sample sizes can vary without visibly affecting the chart.

Example #1

A lathe is being used to machine terminals of different sizes. Samples (of different sizes) are taken periodically and inspected for burrs, nicks, tool marks and other visual defects.

Example #2

A printed circuit board hand assembly operation involves placing electrical components into a large number of different circuit boards. Although the boards differ markedly from one another, the hand assembly operation is similar for all of the different boards.

Example #3

A job-shop welding operation produces small quantities of "one order only" items. However, the operation always involves joining parts of similar material and similar size. The process control statistic is weld imperfections per 100 inches of weld.

The techniques used to create stabilized attribute control charts are all based on corresponding classical attribute control chart methods. There are four basic types of control charts involved:

1. Stabilized p charts for proportion of defective units per sample.
2. Stabilized np charts for the number of defective units per sample.
3. Stabilized c charts for the number of defects per unit.
4. Stabilized u charts for the average number of defects per unit.

All of these charts are based on the transformation

$$Z = \frac{\text{sample statistic} - \text{process average}}{\text{process standard deviation}} \tag{VI.89}$$

In other words, stabilized charts are plots of the number of standard deviations (plus or minus) between the sample statistic and the long-term process average. Since control limits are conventionally set at ± 3 standard deviations, stabilized control charts always have the lower control limit at -3 and the upper control limit at $+3$. Table VI.42 summarizes the control limit equations for stabilized control charts for attributes.

Table VI.42. Stabilized attribute chart statistics.

ATTRIBUTE	CHART	SAMPLE STATISTIC	PROCESS AVERAGE	PROCESS STD. DEV.	Z
Proportion defective units	p chart	p	$\bar{p}$	$\sqrt{\bar{p}(1-\bar{p})}$	$(p-\bar{p})/\sigma$
Number of defective units	np chart	np	$\overline{np}$	$\sqrt{\overline{np}(1-\bar{p})}$	$(np-\overline{np})/\sigma$
Defects per unit	c chart	c	$\bar{c}$	$\sqrt{\bar{c}}$	$(c-\bar{c})/\sigma$
Average defects per unit	u chart	u	$\bar{u}$	$\sqrt{\bar{u}/n}$	$(u-\bar{u})/\sigma$

When applied to long runs, stabilized attribute charts are used to compensate for varying sample sizes; process averages are assumed to be constant. However, stabilized attribute charts can be created even if the process average varies. This is often done when applying this technique to short runs of parts that vary a great deal in average quality. For example, a wave soldering process used for several missiles had boards that varied in complexity from less than 100 solder joints to over 1,500 solder joints. Tables VI.27 and VI.28 show how the situation is handled to create a stabilized u chart. The unit size is 1,000 leads, set arbitrarily. It doesn't matter what the unit size is set to, the calculations will still produce the correct result since the actual number of leads is divided by the unit size selected. However, the unit size must be the same for all boards; e.g., if the unit is 1,000 leads, a board with 100 leads is 1/10 unit, one with 2,000 leads is 2 units. $\bar{u}$ is the average number of defects per 1,000 leads.

Table VI.43. Data from a wave solder process.

MISSILE	BOARD	LEADS	UNITS/BOARD	$\bar{u}$
Phoenix	A	1,650	1.65	16
	B	800	0.80	9
	C	1,200	1.20	9
TOW	D	80	0.08	4
	E	50	0.05	2
	F	100	0.10	1

Example #4

From the process described in Table VI.43, a sample of 10 TOW missile boards of type E are sampled. Three defects were observed in the sample. Using Tables VI.30 and VI.31, we compute Z for the subgroup as follows:

$\sigma = \sqrt{\bar{u}/n}$ We get $\bar{u}$=2 from Table VI.43.

$$n = \frac{50 \times 10}{1000} = 0.5 \text{ units}$$

$$\sigma = \sqrt{2/0.5} = \sqrt{4} = 2$$

$$u = \frac{\text{number of defects}}{\text{number of units}} = \frac{3}{0.5} = 6 \text{ defects per unit}$$

$$Z = \frac{u - \bar{u}}{\sigma} = \frac{6-2}{2} = \frac{4}{2} = 2$$

Since Z is between –3 and +3, we conclude that the process has not gone out of control; i.e., it is not being influenced by a special cause of variation.

Table VI.44 shows the data for several samples from this process. The resulting control chart is shown in Figure VI.78. Note that the control chart indicates that the process was better than average when it produced subgroups 2 and 3 and perhaps 4. Negative Z values mean that the defect rate is below (better than) the long-term process average. Groups 7–8 show an apparent deterioration in the process with group 7 being out of control. Positive Z values indicate a defect rate above (worse than) the long term process average.

Table VI.44. Stabilized u chart data for wave solder.

NO.	BOARD	$\bar{u}$	UNITS	# SAMPLED	n	σ	DEFECTS	u	Z
1	E	2	0.05	10	0.50	2.00	3	6.00	2.00
2	A	16	1.65	1	1.65	3.11	8	4.85	–3.58
3	A	16	1.65	1	1.65	3.11	11	6.67	–3.00
4	B	9	0.80	1	0.80	3.35	0	0.00	–2.68
5	F	1	0.10	2	0.20	2.24	1	5.00	1.79
6	E	2	0.05	5	0.25	2.83	2	8.00	2.12
7	C	9	1.20	1	1.20	2.74	25	20.83	4.32
8	D	4	0.08	5	0.40	3.16	5	12.50	2.69
9	B	9	0.80	1	0.80	3.35	7	8.75	–0.07
10	B	9	0.80	1	0.80	3.35	7	8.75	–0.07

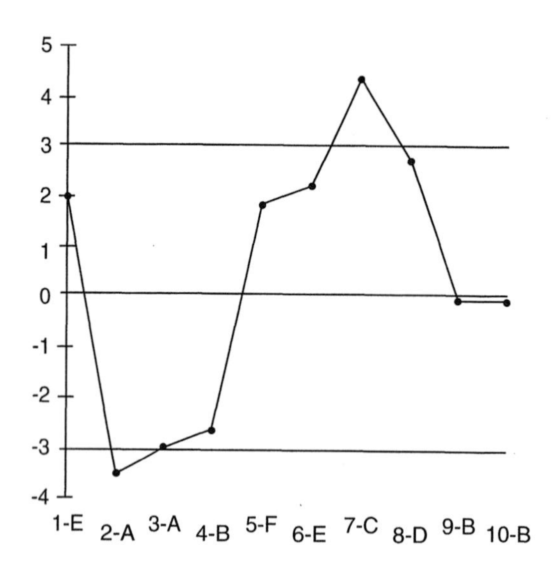

Figure VI.78. Control chart of Z values from Table VI.44.

The ability to easily see process trends and changes like these, in spite of changing part numbers and sample sizes, is the big advantage of stabilized control charts. The disadvantages of stabilized control charts include the following:

1. They convert a number that is easy to understand, the number of defects or defectives, into a confusing statistic with no intuitive meaning to operating personnel.
2. They involve tedious calculation.

Item #1 can only be corrected by training and experience applying the technique. Item #2 can be handled with computers; the calculations are simple to perform with a spreadsheet. Table VI.44 can be used as a guide to setting up the spreadsheet. Inexpensive programmable calculators can be used to perform the calculations right at the process, thus making the results available immediately.

Demerit control charts

As described above, there are two kinds of data commonly used to perform SPC: variables data and attributes data. When short runs are involved, we can seldom afford the information loss that results from using attribute data. However, there are ways of extracting additional information from attribute data:

1. Making the attribute data "less discrete" by adding more classification categories.
2. Assigning weights to the categories to accentuate different levels of quality.

Consider a process that involves fabricating a substrate for a hybrid microcircuit. The surface characteristics of the substrate are extremely important. The "ideal part" will have a smooth surface, completely free of any visible flaws or blemishes. However, parts are sometimes produced with stains, pits, voids, cracks and other surface defects. Although imperfect, most of the less than ideal parts are still acceptable to the customer.

If we were to apply conventional attribute SPC methods to this process the results would probably be disappointing. Since very few parts are actually rejected as unacceptable, a standard p chart or stabilized p chart would probably show a flat line at "zero defects" most of the time, even though the quality level might be less than the ideal. Variables SPC methods can't be used because attributes data such as "stains" are not easily measured on a variables scale. Demerit control charts offer an effective method of applying SPC in this situation. To use demerit control charts we must determine how many imperfections of each type are found in the parts. Weights are assigned to the different categories. The quality score for a given sample is the sum of the weights times the frequencies of each category. Table VI.45 illustrates this approach for the substrate example.

Table VI.45. Demerit scores for substrates.

SUBGROUP NUMBER →		1		2		3	
Attribute	**Weight**	**Freq.**	**Score**	**Freq.**	**Score**	**Freq.**	**Score**
Light stain	1	3	3				
Dark stain	5			1	5	1	5
Small blister	1			2	2	1	1
Medium blister	5	1	5				
Pit: 0.01–0.05 mm	1					3	3
Pit: 0.06–0.10	5			2	10		
Pit: larger than 0.10 mm	10	1	10				
TOTAL DEMERITS →		18		17		9	

If the subgroup size is kept constant, the average for the demerit control chart is computed as follows (Burr, 1976):

$$\text{Average} = \overline{D} = \frac{\text{sum of subgroup demerits}}{\text{number of subgroups}}$$

Control limits are computed in two steps. First, compute the average defect rate for each category. For example, we might have the following categories and weights:

CATEGORY	WEIGHT
Major	10
Minor	5
Incidental	1

We could compute three average defect rates, one each for major, minor and incidental. Let's designate these as

$$\bar{c}_1 = \text{Average number of major defects per subgroup}$$
$$\bar{c}_2 = \text{Average number of minor defects per subgroup}$$
$$\bar{c}_3 = \text{Average number of incidental defects per subgroup}$$

The corresponding weights are $W_1 = 10$, $W_2 = 5$, $W_3 = 1$. Using this notation, we compute the demerit standard deviation for this three category example as

$$\sigma_D = \sqrt{W_1^2\bar{c}_1 + W_2^2\bar{c}_2 + W_3^2\bar{c}_3} \qquad \text{(VI.90)}$$

For the general case, the standard deviation is

$$\sigma_D = \sqrt{\sum_{i=1}^{k} W_i^2\bar{c}_i} \qquad \text{(VI.91)}$$

The control limits are

$$LCL = \bar{D} - 3\sigma_D \qquad \text{(VI.92)}$$

$$UCL = \bar{D} + 3\sigma_D \qquad \text{(VI.93)}$$

Simplified quality score charts

The above procedure, while correct, may sometimes be too burdensome to implement effectively. When this is the case, a simplified approach may be used. The simplified approach is summarized as follows:

1. Classify each part in the subgroup into one of the following classes (points are arbitrary)

CLASS	DESCRIPTION	POINTS
A	Preferred quality. All product features at or very near targets.	10
B	Acceptable quality. Some product features have departed significantly from target quality levels, but they are a safe distance from the reject limits.	5
C	Marginal quality. One or more product features are in imminent danger of exceeding reject limits.	1
D	Reject quality. One or more product features fail to meet minimum acceptability requirements.	0

2. Plot the total scores for each subgroup, keeping the subgroup sizes constant.
3. Treat the total scores as if they were variables data and prepare an individuals and moving range control chart or an X and R chart. These charts are described in most texts on SPC.

Conclusion

Small runs and short runs are common in modem business environments. Different strategies are needed to deal with these situations. Advance planning is essential. Special variables techniques were introduced which compensate for small sample sizes and short runs by using special tables or mathematically transforming the statistics and charts. Attribute short run SPC methods were introduced that make process patterns more evident when small runs are produced. Demerit and scoring systems were introduced that extract more information from attribute data.

PRE-CONTROL

The PRE-Control method was originally developed by Dorian Shainin in the 1950s. According to Shainin, PRE-Control is a simple algorithm for controlling a process based on the tolerances. It assumes the process is producing product with a measurable and adjustable quality characteristic which varies according to some distribution. It makes no assumptions concerning the actual shape and stability of the distribution. Cautionary zones are designated just inside each tolerance extreme. A new process is qualified by taking consecutive samples of individual measurements until five in a row fall within the central zone before two in a row fall into the cautionary zones. To simplify the application, PRE-Control charts are often color-coded. On such charts the central zone is colored green, the cautionary zones yellow, and the zone outside of the tolerance red. PRE-Control is not equivalent to SPC. SPC is designed to identify special causes of variation; PRE-Control starts with a process that is known to be capable of meeting the tolerance and assures that it does so. SPC and process capability analysis should always be used before PRE-Control is applied.*

Once the process is qualified, it is monitored by taking periodic samples consisting of two individuals each (called the A,B pair). Action is taken only if both A and B are in the cautionary zone. Processes must be requalified after any action is taken.

*Considerable controversy surrounds the use of PRE-Control. However, the discussions tend towards considerable statistical sophistication and will not be reviewed here. The reader should keep in mind that PRE-Control should not be considered a *replacement* for SPC.

Setting up PRE-Control

Figure VI.79 illustrates the PRE-Control zones for a two-sided tolerance (i.e., a tolerance with both a lower specification limit and an upper specification limit).

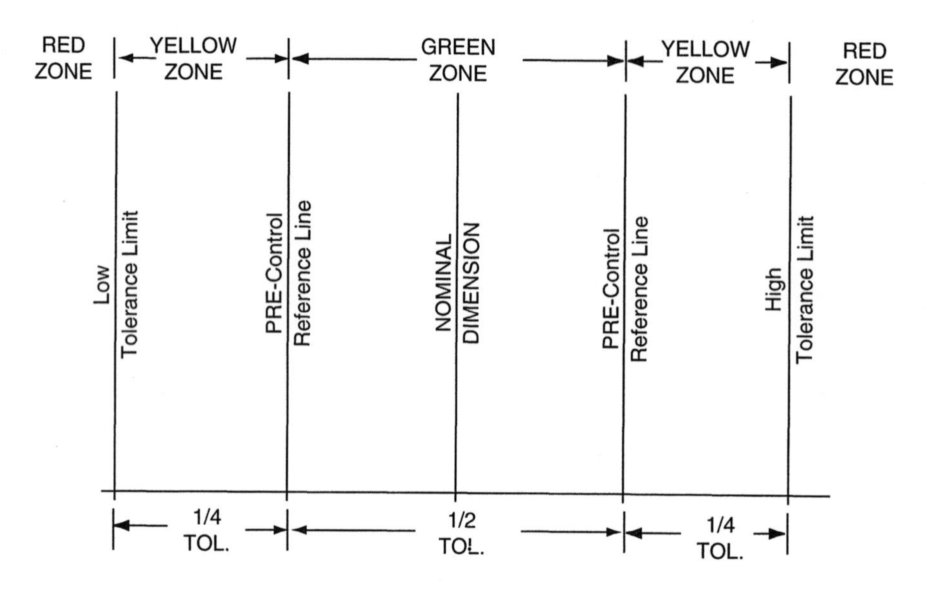

Figure VI.79. PRE-Control zones (2-sided tolerance).

Figure VI.80 illustrates the PRE-Control zones for a one-sided tolerance (i.e., a tolerance with only a lower specification limit or only an upper specification limit). Examples of this situation are flatness, concentricity, runout and other total indicator reading type features.

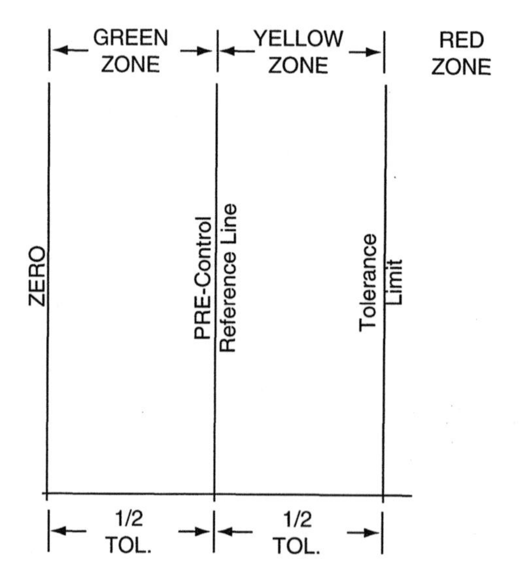

Figure VI.80. PRE-Control zones (1-sided tolerance).

Figure VI.81 illustrates the PRE-Control zones for characteristics with minimum or maximum specification limits. Examples of this situation are tensile strength, contamination levels, etc. In this situation place one reference line a quarter of the way from the tolerance limit toward the best sample produced during past operations.

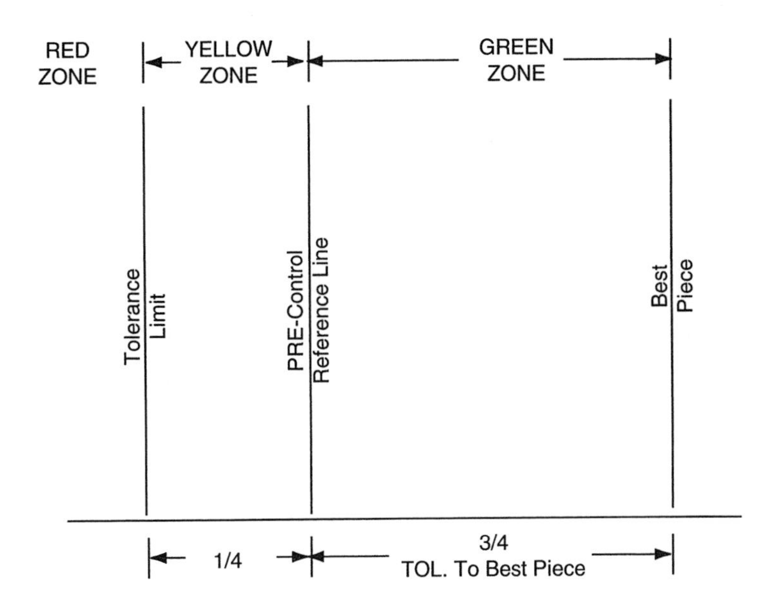

Figure VI.81. PRE-Control zones (minimum/maximum specifications).

Using PRE-Control

The first step is *setup qualification.* To begin, measure every piece produced until you obtain five greens in a row. If one yellow is encountered, restart the count. If two yellows in a row or any reds are encountered, adjust the process and restart the count. This step replaces first-piece inspection.

After setup qualification you will enter the *run* phase. Measure two consecutive pieces periodically (the A,B pair). If both are yellow on the same side, adjust. If yellow on opposite sides, call for help to reduce the variability of the process. If either are red, adjust. In the case of two yellows, the adjustment must be made immediately to prevent nonconforming work. In the case of red, stop; nonconforming work is already being produced. Segregate all nonconforming product according to established procedures.

Shainin and Shainin(1988) recommend adjusting the inspection frequency such that six A,B pairs are measured on average between each process adjustment. A simple formula for this is shown in Equation VI.94.

$$\text{minutes between measurements} = \text{hours between adjustments} \times 10 \quad \text{(VI.94)}$$

VI.H ANALYZING PROCESS CAPABILITY

Process capability analysis is a two-stage process that involves:

1. Bringing a process into a state of statistical control for a reasonable period of time.
2. Comparing the long-term process performance to management or engineering requirements.

Process capability analysis can be done with either attribute data or continuous data *if and only if the process is in statistical control,* and has been for a reasonable period of time.*

Application of process capability methods to processes that are not in statistical control results in unreliable estimates of process capability and should never be done.

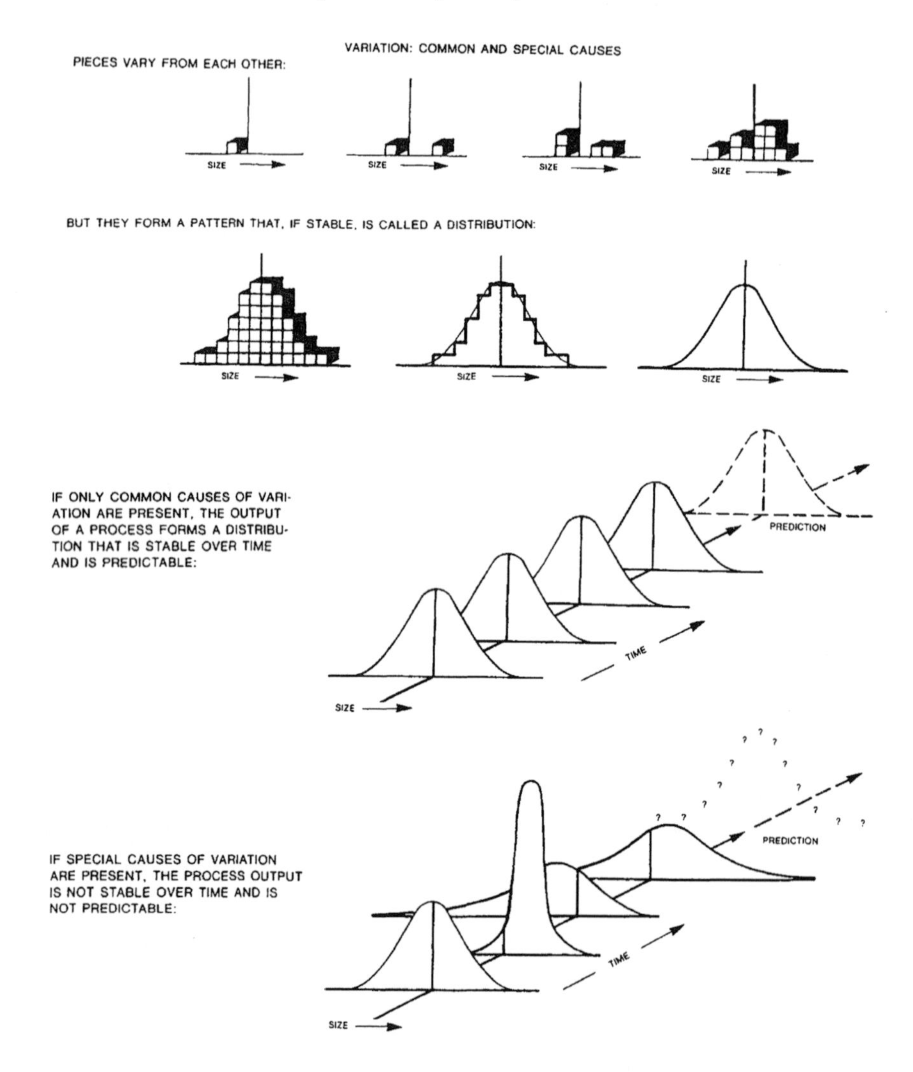

Figure VI.82. Process control concepts illustrated.

From *Continuing Process Control and Process Capability Improvement*, p. 4a. Copyright 1983. Used by permission of the publisher, Ford Motor Company, Dearborn, Michigan.

*Occasional freak values from known causes can usually be ignored.

HOW TO PERFORM A PROCESS CAPABILITY STUDY

This section presents a step-by-step approach to process capability analysis (Pyzdek 1985.) The approach makes frequent reference to materials presented elsewhere in this book.

1. **Select a candidate for the study.**
   This step should be institutionalized. A goal of any organization should be ongoing process improvement. However, because a company has only a limited resource base and can't solve all problems simultaneously, it must set priorities for its efforts. The tools for this include Pareto analysis and fishbone diagrams.
2. **Define the process.**
   It is all too easy to slip into the trap of solving the wrong problem. Once the candidate area has been selected in step 1, define the scope of the study. A process is a unique combination of machines, tools, methods, and personnel engaged in adding value by providing a product or service. Each element of the process should be identified at this stage. This is not a trivial exercise. The input of many people may be required. There are likely to be a number of conflicting opinions about what the process actually involves.
3. **Procure resources for the study.**
   Process capability studies disrupt normal operations and require significant expenditures of both material and human resources. Since it is a project of major importance, it should be managed as such. All of the usual project management techniques should be brought to bear. This includes planning, scheduling, and management status reporting.
4. **Evaluate the measurement system.**
   Using the techniques described in section III.D, evaluate the measurement system's ability to do the job. Again, be prepared to spend the time necessary to get a valid means of measuring the process before going ahead.
5. **Prepare a control plan.**
   The purpose of the control plan is twofold: 1) isolate and control as many important variables as possible and, 2) provide a mechanism for tracking variables that can not be completely controlled. The object of the capability analysis is to determine what the process can do if it is operated the way it is designed to be operated. This means that such obvious sources of *potential* variation as operators and vendors will be controlled while the study is conducted. In other words, a single well-trained operator will be used and the material will be from a single vendor.

   There are usually some variables that are important, but that are not controllable. One example is the ambient environment, such as temperature, barometric pressure, or humidity. Certain process variables may degrade as part of the normal operation; for example, tools wear and chemicals are depleted. These variables should still be tracked using logsheets and similar tools.

6. **Select a method for the analysis.**
   The SPC method will depend on the decisions made up to this point. If the performance measure is an attribute, one of the attribute charts will be used. Variables charts will be used for process performance measures assessed on a continuous scale. Also considered will be the skill level of the personnel involved, need for sensitivity, and other resources required to collect, record, and analyze the data.
7. **Gather and analyze the data.**
   Use one of the control charts described in this chapter, plus common sense. It is usually advisable to have at least two people go over the data analysis to catch inadvertent errors in transcribing data or performing the analysis.
8. **Track down and remove special causes.**
   A special cause of variation may be obvious, or it may take months of investigation to find it. The effect of the special cause may be good or bad. Removing a special cause that has a bad effect usually involves eliminating the cause itself. For example, if poorly trained operators are causing variability the special cause is the training system (not the operator) and it is eliminated by developing an improved training system or a process that requires less training. However, the removal of a beneficial special cause may actually involve incorporating the special cause into the normal operating procedure. For example, if it is discovered that materials with a particular chemistry produce better product the special cause is the newly discovered material and it can be made a common cause simply by changing the specification to assure that the new chemistry is always used.
9. **Estimate the process capability.**
   One point can not be overemphasized: *the process capability can not be estimated until a state of statistical control has been achieved!* After this stage has been reached, the methods described later in this chapter may be used. After the numerical estimate of process capability has been arrived at, it must be compared to management's goals for the process, or it can be used as an input into economic models. Deming's all-or-none rules (see III.C) provide a simple model that can be used to determine if the output from a process should be sorted 100% or shipped as-is.
10. **Establish a plan for continuous process improvement.**
    Once a stable process state has been attained, steps should be taken to maintain it and improve upon it. SPC is just one means of doing this. Far more important than the particular approach taken is a company environment that makes continuous improvement a normal part of the daily routine of everyone.

STATISTICAL ANALYSIS OF PROCESS CAPABILITY DATA

This section presents several methods of analyzing the data obtained from a process capability study.

Control chart method: attributes data

1. Collect samples from 25 or more subgroups of consecutively produced units. Follow the guidelines presented in steps 1–10 above.
2. Plot the results on the appropriate control chart (e.g., *c* chart). If all groups are in statistical control, go to step #3. Otherwise identify the special cause of variation and take action to eliminate it. Note that a special cause might be beneficial. Beneficial activities can be "eliminated" as special causes by doing them all of the time. A special cause is "special" only because it comes and goes, not because its impact is either good or bad.
3. Using the control limits from the previous step (called operation control limits), put the control chart to use for a period of time. Once you are satisfied that sufficient time has passed for most special causes to have been identified and eliminated, as verified by the control charts, go to step #4.
4. The process capability is estimated as the control chart *centerline*. The centerline on attribute charts is the long-term expected quality level of the process, e.g., the average proportion defective. This is the level created by the common causes of variation.

If the process capability doesn't meet management requirements, take immediate action to modify the process for the better. "Problem solving" (e.g., studying each defective) won't help, and it may result in tampering. Whether it meets requirements or not, always be on the lookout for possible process improvements. The control charts will provide verification of improvement.

Control chart method: variables data

1. Collect samples from 25 or more subgroups of consecutively produced units, following the 10-step plan described above.
2. Plot the results on the appropriate control chart (e.g., $\overline{X}$ and R chart). If all groups are in statistical control, go to step #3. Otherwise identify the special cause of variation and take action to eliminate it.
3. Using the control limits from the previous step (called operation control limits), put the control chart to use for a period of time. Once you are satisfied that sufficient time has passed for most special causes to have been identified and eliminated, as verified by the control charts, estimate process capability as described below.

The process capability is estimated from the process average and standard deviation, where the standard deviation is computed based on the average range or average standard deviation. When statistical control has been achieved, the capability is the level created by the common causes of process variation. The formulas for estimating the process standard deviation are:

R chart method:

$$\hat{\sigma} = \overline{R}/d_2 \qquad \text{(VI.95)}$$

s chart method:

$$\hat{\sigma} = \overline{s}/c_4 \qquad \text{(VI.96)}$$

The values d_2 and c_4 are constants from Table 13 in the Appendix.

PROCESS CAPABILITY INDEXES

Only now can the process be compared to engineering requirements. One way of doing this is by calculating "Capability Indexes." Several popular capability indexes are given in Table VI.46.

Table VI.46. Process capability analysis.

$C_P = \dfrac{\text{engineering tolerance}}{6\hat{\sigma}}$	(VI.97)
$C_R = 100 \times \dfrac{6\hat{\sigma}}{\text{engineering tolerance}}$	(VI.98)
$C_M = \dfrac{\text{engineering tolerance}}{8\hat{\sigma}}$	(VI.99)
$Z_U = \dfrac{\text{upper specification} - \overline{\overline{X}}}{\hat{\sigma}}$	(VI.100)

Continued on next page . . .

Table VI.46—*Continued . . .*

$$Z_L = \frac{\overline{\overline{X}} \text{ - lower specification}}{\hat{\sigma}}$$	(VI.101)
$$Z_{MIN.} = Minimum\{Z_L, Z_U\}$$	(VI.102)
$$C_{PK} = \frac{Z_{MIN}}{3}$$	(VI.103)
$$C_{pm} = \frac{C_p}{\sqrt{1 + \frac{(\mu - T)^2}{\hat{\sigma}^2}}}$$	(VI.104)

INTERPRETING CAPABILITY INDEXES

Perhaps the biggest drawback of using process capability indexes is that they take the analysis a step away from the data. The danger is that the analyst will lose sight of the purpose of the capability analysis, which is to improve quality. To the extent that capability indexes help accomplish this goal, they are worthwhile. To the extent that they distract from the goal, they are harmful. The quality engineer should continually refer to this principle when interpreting capability indexes.

C_P—This is one of the first capability indexes used. The "natural tolerance" of the process is computed as 6σ. The index simply makes a direct comparison of the process natural tolerance to the engineering requirements. Assuming the process distribution is normal and the process average is exactly centered between the engineering requirements, a C_P index of 1 would give a "capable process." However, to allow a bit of room for process drift, the generally accepted minimum value for C_P is 1.33. In general, the larger C_P is, the better.

The C_P index has two major shortcomings. First, it can't be used unless there are both upper and lower specifications. Second, it does not account for process centering. If the process average is not exactly centered relative to the engineering requirements, the C_P index will give misleading results. In recent years, the C_P index has largely been replaced by C_{PK} (see below).

C_R—The C_R index is equivalent to the C_P index. The index simply makes a direct comparison of the process to the engineering requirements. Assuming the process distribution is normal and the process average is exactly centered between the engineering requirements, a C_R index of 100% would give a "capable process." However, to allow a bit of room for process drift, the generally accepted maximum value for C_R is 75%. In general, the smaller C_R is, the better. The C_R index suffers from the same shortcomings as the C_P index.

C_M—The C_M index is generally used to evaluate machine capability studies, rather than full-blown process capability studies. Since variation will increase when other sources of process variation are added (e.g., tooling, fixtures, materials, etc.), C_M uses an 8 sigma spread rather than a 6 sigma spread to represent the natural tolerance of the process.

Z_U—The Z_U index measures the process location (central tendency) relative to its standard deviation and the upper requirement. If the distribution is normal, the value of Z_U can be used to determine the percentage above the upper requirement by using Table 4 in the Appendix. The method is the same as described in VI.C using the Z statistic, simply use Z_U instead of using Z.

In general, the bigger Z_U is, the better. A value of at least +3 is required to assure that 0.1% or less defective will be produced. A value of +4 is generally desired to allow some room for process drift.

Z_L—The Z_L index measures the process location relative to its standard deviation and the lower requirement. If the distribution is normal, the value of Z_L can be used to determine the percentage above the upper requirement by using Table 4 in the Appendix. The method is the same as described in VI.C using the Z transformation, except that you use $-Z_L$ instead of using Z.

In general, the bigger Z_L is, the better. A value of at least +3 is required to assure that 0.1% or less defective will be produced. A value of +4 is generally desired to allow some room for process drift.

Z_{MIN}—The value of Z_{MIN} is simply the smaller of the Z_L or the Z_U values. It is used in computing C_{PK}.

C_{PK}—The value of C_{PK} is simply Z_{MIN} divided by 3. Since the smallest value represents the nearest specification, the value of C_{PK} tells you if the process is truly capable of meeting requirements. A C_{PK} of at least +1 is required, and +1.33 is preferred. Note that C_{PK} is closely related to C_P; the difference between C_{PK} and C_P represents the potential gain to be had from centering the process.

C_{PM}—A C_{PM} of at least 1 is required, and 1.33 is preferred. C_{PM} is closely related to C_P. The difference represents the potential gain to be obtained by moving the process mean closer to the target. Unlike C_{PK}, the target need not be the center of the specification range.

EXAMPLE OF CAPABILITY ANALYSIS USING NORMALLY DISTRIBUTED VARIABLES DATA

Assume we have conducted a capability analysis using X bar and R charts with subgroups of 5. Also assume that we found the process to be in statistical control with a grand average of 0.99832 and an average range of 0.2205. From the table of d_2 values (Appendix Table 13), we find d_2 is 2.326 for subgroups of 5. Thus, using Equation VI.95,

$$\hat{\sigma} = \frac{0.2205}{2.326} = 0.00948$$

Before we can analyze process capability, we must know the requirements. For this process the requirements are a lower specification of 0.980 and an upper specification of 1.020 (1.000±0.020). With this information, plus the knowledge that the process performance has been in statistical control, we can compute the capability indexes for this process.

$$C_P = \frac{\text{engineering tolerance}}{6\hat{\sigma}} = \frac{1.020 - 0.9800}{6 \times 0.00948} = 0.703$$

$$C_R = 100 \times \frac{6\hat{\sigma}}{\text{engineering tolerance}} = 100 \times \frac{6 \times 0.00948}{0.04} = 142.2\%$$

$$C_M = \frac{\text{engineering tolerance}}{8\hat{\sigma}} = \frac{0.04}{8 \times 0.00948} = 0.527$$

$$Z_U = \frac{\text{upper specification - }\overline{\overline{X}}}{\hat{\sigma}} = \frac{1.020 - 0.99832}{0.00948} = 2.3$$

$$Z_L = \frac{\overline{\overline{X}}\text{ - lower specification}}{\hat{\sigma}} = \frac{0.99832 - 0.980}{0.00948} = 1.9$$

$$Z_{MIN} = \text{Minimum}\{1.9, 2.3\} = 1.9$$

$$C_{PK} = \frac{Z_{MIN}}{3} = \frac{1.9}{3} = 0.63$$

Assuming that the target is precisely 1.000, we compute

$$C_{pm} = \frac{C_p}{\sqrt{1+\frac{\left(\overline{\overline{X}}-T\right)^2}{\hat{\sigma}^2}}} = \frac{0.703}{\sqrt{1+\frac{(0.99832-1.000)^2}{0.00948^2}}} = 0.692$$

DISCUSSION

C_P—(0.703) Since the minimum acceptable value for this index is 1, the 0.703 result indicates that this process can not meet the requirements. Furthermore, since the C_P index doesn't consider the centering process, we know that the process can't be made acceptable by merely adjusting the process closer to the center of the requirements. Thus, we can expect the Z_L, Z_U, and Z_{MIN} values to be unacceptable too.

C_R—(142.2%) This value always gives the same conclusions as the C_P index. The number itself means that the "natural tolerance" of the process uses 142.2% of the engineering requirement, which is, of course, unacceptable.

C_M—(0.527) The C_M index should be 1.33 or greater. Obviously it is not. If this were a machine capability study the value of the C_M index would indicate that the machine was incapable of meeting the requirement.

Z_U—(+2.3) We desire a Z_U of at least +3, so this value is unacceptable. We can use Z_U to estimate the percentage of production that will exceed the upper specification. Referring to Table 4 in the Appendix we find that approximately 1.1% will be oversized.

Z_L—(+1.9) We desire a Z_L of at least +3, so this value is unacceptable. We can use Z_L to estimate the percentage of production that will be below the lower specification. Referring to Table 4 in the Appendix we find that approximately 2.9% will be undersized. Adding this to the 1.1% oversized and we estimate a total reject rate of 4.0%. By subtracting this from 100% we get the projected yield of 96.0%.

Z_{MIN}—(+1.9) The smaller of Z_L and Z_U. Since neither of these two results were acceptable, Z_{MIN} cannot be acceptable.

C_{PK}—(0.63) The value of C_{PK} is only slightly smaller than that of C_P. This indicates that we will not gain much by centering the process. The actual amount we would gain can be calculated by assuming the process is exactly centered at 1.000 and recalculating Z_{MIN}. This gives a predicted total reject rate of 3.6% instead of 4.0%.

MOTOROLA'S SIX SIGMA PROGRAM

In 1988, Motorola, Inc. was honored with the Malcolm Baldrige National Quality Award. One of the objectives of the Baldrige award is to identify those excellent firms that are worthy role models to other businesses. One of Motorola's innovations that attracted a great deal of attention was their six sigma program. Six sigma is, basically, a process quality goal. As such it falls into the category of a process capability technique.

As shown above, the traditional quality paradigm defined a process as capable if the process natural spread, plus and minus three sigma, was less than the engineering tolerance. Under the assumption of normality, this translates to a process yield of 99.73%. A later refinement considered the process location as well as its spread and tightened the minimum acceptable so that the process was at least four sigma from the nearest engineering requirement. Motorola's six sigma asks that processes operate such that the nearest engineering requirement is at least six sigma from the process mean.

Motorola's six sigma program also applies to attribute data. This is accomplished by converting the six sigma requirement to equivalent conformance levels. This is illustrated in Figure VI.83.

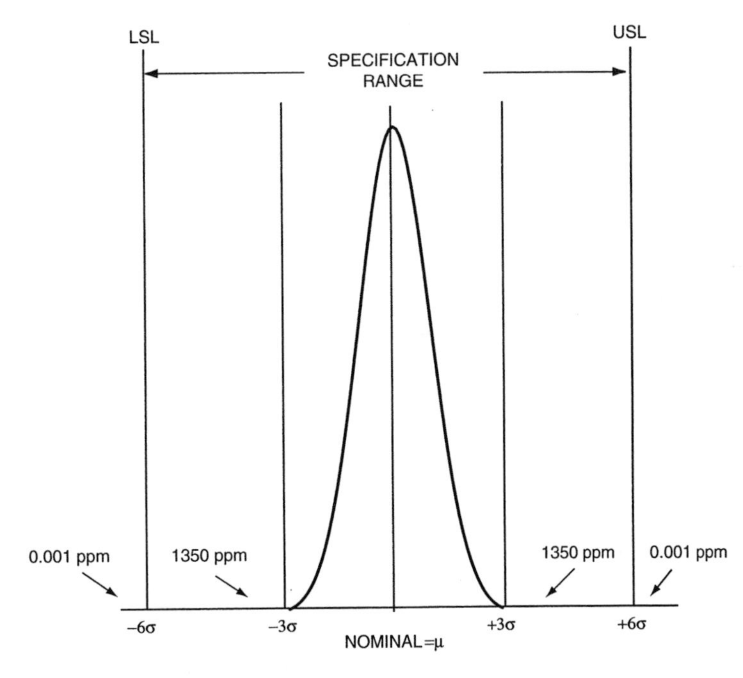

Figure VI.83. Sigma levels and equivalent conformance rates.

One of Motorola's most significant contributions was to change the discussion of quality from one where quality levels were measured in percents (parts-per-hundred), to a discussion of parts-per-million or even parts-per-billion. Motorola correctly pointed out that modern technology was so complex that old ideas about "acceptable quality levels" were no longer acceptable.*

One puzzling aspect of the "official" six sigma literature is that it states that a process operating at six sigma will produce 3.4 ppm non-conformances. However, if a normal distribution table is consulted (very few go out to six sigma), one finds that the expected non-conformances are 0.002 ppm (2 ppb). The difference occurs because Motorola presumes that the process mean can drift 1.5 sigma in either direction. The area of a normal distribution beyond 4.5 sigma from the mean is indeed 3.4 ppm. Since control charts will easily detect any process shift of this magnitude in a single sample, the 3.4 ppm represents a very conservative upper bound on the non-conformance rate.

PROCESS PERFORMANCE INDICES**

Some have advocated computing the process capability indices even when the process is not in a state of statistical control. Although the formulas used are the same, the standard deviation used is not based on the control chart. Instead the formula for the standard deviation used is the one shown in Equation VI.105.

$$s=\sqrt{\sum_{i=1}^{n}\frac{\left(x_i-\bar{x}\right)^2}{n-1}} \qquad \text{(VI.105)}$$

When this analysis is performed on processes that are not in statistical control the indices are called process performance indices (PPIs). They cannot properly be called process capability indices (PCIs) since when σ is computed from Equation VI.105, it includes variability from special causes of variation as well as variation from common causes. When the special causes are identified and eliminated the process will do better, which implies that the process capability is better than the process performance index indicates.

A PPI is intended to show how the process actually performed, rather than how well it can perform under properly controlled conditions. Thus, in a sense, PPIs are designed to describe the past while PCIs are designed to predict the future. The difference between a PPI and its corresponding PCI is a measure of the potential improvement possible from eliminating special causes.

---

*Other quality experts, including this author, believe that Motorola didn't go far enough. In fact, even "zero defects" falls short. Defining quality as only the lack of non-conforming product reflects a limited view of quality.

**Thomas Pyzdek, "Process Capability Analysis using Personal Computers," *Quality Engineering*, 4, No. 3, (1992), 432–433.

I have serious reservations about using PPIs. First of all, if the process is not in a state of statistical control the future can't be predicted from the past; that's what statistical control means! An out-of-control process has no underlying distribution, it is a mixture of distributions in unknown proportions. And since the special causes are unknown, there is no basis for predicting which distribution will appear next or how long it will last. PPIs assume the process distribution is normal, even though we don't even have a process distribution!

That being the case, what's the point of a PPI? So what if a PPI indicated that for a particular run the process met the requirements, or didn't? Since the future can't be predicted, the only possible reason for caring is the quality of the current production lot. Using a PPI to estimate lot quality is futile because of the vagaries of estimating parameters with statistics, fitting curves, finding the relationship between the specification zone and the performance statistic's distribution, etc. Even the much maligned acceptance sampling approach provides a much better way to estimate the quality of a production lot from an out of control process than a PPI. And a simple, but expensive, sorting operation will tell you *exactly* what it did.

Ascribing meaning to an index, either a PPI or a PCI, always depends implicitly on knowing the underlying distribution. This knowledge can be based on engineering knowledge (preferably) or on empirical evidence. However, if the process is not in control then, by definition, there is no "distribution" per se. There are two or more distributions mixed together. The causes of the distributions are not generally known for out-of-control processes. Given the difficulties associated with capability analysis of homogeneous output, analysis with unknown proportions of output from unknown distributions is virtually impossible.

If PCIs suffer because they only measure process capability indirectly, how much worse are PPIs which can't even be indirectly related to yields. This is because they are computed from data that represent a mixture of several unknown process distributions. To compute the yield, you would need to know the characteristics of each separate process and their representation in the sample. If you knew *that,* you could probably identify what the special causes were and you could remove them, achieve a state of statistical control, and compute a PCI instead of a PPI.

It is best to spend your time finding and correcting the special causes; then you won't need to use PPIs.

♦♦♦

APPENDIX

1

Quality Engineer Certification (CQE) Body of Knowledge

Please Note: The Body of Knowledge for certification is affected by new technologies, policies, and the changing dynamics of manufacturing and service industries. Changed versions of the examination based on the current Body of Knowledge are used at each offering.

The following is an outline of the topics that constitute the Body of Knowledge for Quality Engineering.

I. MANAGEMENT AND LEADERSHIP IN QUALITY ENGINEERING (19 Questions)

A. **Professional Conduct and ASQ Code of Ethics**
Identify appropriate behaviors for situations requiring ethical decisions. (Evaluation)

B. **Management Systems for Improving Quality**
Define, select and apply elements such as policy deployment, benchmarking, goal setting, planning and scheduling, project management, quality information systems, etc. (Analysis)

C. **Leadership Principles and Techniques**
Describe and apply principles and techniques for developing, building, and organizing teams, and leading quality initiatives. (Application)

D. **Facilitation Principles and Techniques**
Describe facilitator roles and responsibilities in the team environment. Define and apply brainstorming, nominal group technique (NGT), conflict resolution, etc. (Application)

E. **Training**
Identify training needs, describe elements of training programs and material development, and apply methods for determining effectiveness. (Application)

F. **Cost of Quality**
Describe and apply cost of quality concepts, including quality cost categories, data collection, reporting, etc. Evaluate cost of quality and interpret results. (Evaluation)

G. **Quality Philosophies and Approaches (e.g., Juran, Deming, Taguchi, Ishikawa)**

1. **Benefits of quality**
Describe the advantages of managing for quality and using quality techniques, both in theory and in practice. (Comprehension)
2. **History of quality**
Describe how and why modern quality has evolved, with emphasis on the forces, significant events, and leading contributors that have shaped quality practices. NOTE: Specific dates will **not** be addressed. (Comprehension)
3. **Definitions of quality**
Differentiate between various definitions of quality such as fitness-for-use, the Taguchi loss function etc. (Comprehension)

H. **Customer Relations, Expectations, Needs, and Satisfaction**
Define, apply, and analyze the results of customer relation measures such as quality function deployment (QFD), customer satisfaction surveys, etc. (Analysis)

I. **Supplier Relations and Management Methodologies**
Define, select, and apply methodologies such as qualification, certification, evaluation, ratings, performance improvement, etc. (Analysis)

II. QUALITY SYSTEMS DEVELOPMENT, IMPLEMENTATION, AND VERIFICATION (19 Questions)

A. **Elements of a Quality System**
Identify and integrate the common elements such as design control, process control, quality costs, auditing, test procedures, etc. (Synthesis)

B. **Documentation Systems**
Identify, define, and apply the elements of a documentation system such as configuration management, document control, etc. (Application)

C. **Domestic and International Standards and Specifications**
Identify the content and applicability of ASQ and ISO quality standards. NOTE: Industry-specific standards will not be included. (Comprehension)

D. **Quality Audits**

1. **Types and purpose of quality audits**
Identify and apply the concepts and techniques of various types of quality audits such as product, process, system, registration, certification, management, compliance, 1st, 2nd, 3rd-party, etc. (Application)

2. **Roles and responsibilities of individuals involved in the audit process**
   Identify and define roles and responsibilities for audit participants such as audit team, client, auditee, etc. (Comprehension)
3. **Quality audit planning, preparation, and execution**
   Identify, describe and apply the steps of planning, preparation, and execution of a quality audit. (Application)
4. **Audit reporting and followup**
   Identify, describe, and apply the steps of audit reporting and followup, including the need for and verification of audit corrective action, etc. (Application)

III. PLANNING, CONTROLLING, AND ASSURING PRODUCT AND PROCESS QUALITY (33 Questions)

A. **Processes for Planning Product and Service Development**

1. **Classification of quality characteristics**
   Identify, define and classify quality characteristics, including seriousness classification of defects. (Application)
2. **Design inputs and design review**
   Identify sources of design input such as customer needs, regulatory requirements, etc. Identify and apply common elements of the design review process, including roles and responsibilities of participants. (Application)
3. **Validation and qualification methods**
   Identify and describe how validation and qualification methods are used for process, product, and service designs. (Application)
4. **Interpretation of technical drawings and specifications**
   Interpret basic technical drawings including characteristics such as views, title blocks, dimensioning, tolerancing, GDandT symbols, etc. Interpret specification requirements in relation to product and process characteristics. (Application)
5. **Determining product and process control methods**
   Identify and apply methods such as job instructions, process control points, etc. (Application)

B. **Material Control**

1. **Material identification, status, and traceability**
   Describe and apply these methods. NOTE: Product recall procedures will **not** be included. (Application)
2. **Sample integrity**
   Describe the importance of establishing and maintaining sample integrity and select the appropriate techniques for avoiding contamination, misidentification, etc. (Application)

3. **Material segregation**
   Describe the importance of and apply methods for material segregation. (Application)
4. **Material Review Board (MRB)**
   Describe the purpose and function of an MRB, including appropriate disposition decisions. (Analysis)

C. **Acceptance Sampling**
1. **General concepts**
   Use, interpret, and apply lot-by-lot protection, average quality protection, producer's and consumer's risk, operating characteristic (OC) curves, attributes and variables sampling plans, etc. (Application)
2. **Definitions of AQL, LTPD, AOQ, AOQL**
   Interpret and describe these terms. (Comprehension)
3. **ANSI/ASQC Z1.4, ANSI/ASQC Z1.9 Standards**
   Use, interpret, and apply these standards. (Analysis)
4. **Acceptance sampling plans**
   Use, interpret, and apply single, double, multiple, sequential, and continuous sampling, including Dodge-Romig. (Analysis)

D. **Measurement Systems**
1. **Terms and definitions**
   Interpret and describe precision, accuracy, metrology, etc. (Comprehension)
2. **Destructive and nondestructive measurement and test methods**
   Distinguish between these methods and apply them appropriately. (Analysis)
3. **Selection of measurement tools, gages, and instruments**
   Select and describe appropriate uses of inspection tools such as gage blocks, calipers, micrometers, optical comparators, etc. (Application)
4. **Measurement system analysis**
   Calculate, analyze, and interpret repeatability and reproducibility, measurement correlation, capability, bias, linearity, etc., including both conventional and control chart methods. (Analysis)
5. **Metrology**
   Interpret, describe, and evaluate traceability to calibration standards, measurement error, calibration systems, control and integrity of standards and measurement devices. (Evaluation)

IV. **RELIABILITY AND RISK MANAGEMENT (11 Questions)**

A. **Terms and Definitions**
   Identify and define basic reliability measures and terms such as MTTF, MTBF, MTTR, availability, failure rate, etc. (Comprehension)

B. **Reliability Life Characteristic Concepts**
Identify and interpret elements of reliability life characteristics of the bathtub curve. (Comprehension)

C. **Design of Systems for Reliability**
Compute and evaluate reliability for redundant, series, and parallel systems. (Evaluation)

D. **Reliability and Maintainability**

1. **Prediction**
Compute, classify, and apply reliability and maintainability characteristics such as MTTF, MTBF, MTTR, availability, failure rate, etc. (Application)
2. **Prevention**
Identify and apply methods to maintain and improve process and product reliability. (Application)
3. **Maintenance scheduling**
Identify, classify, and describe methods of predictive and preventive maintenance. (Application)

E. **Reliability Failure Analysis and Reporting**
Analyze reliability failure information and evaluate possible actions to improve or correct performance. (Evaluation)

F. **Reliability / Safety / Hazard Assessment Tools**

1. **Failure mode and effects analysis (FMEA)**
Define, construct, and interpret FMEAs. (Application)
2. **Failure mode and effects criticality analysis (FMECA)**
Define, construct, and interpret FMECAs. (Application)
3. **Fault-tree analysis (FTA)**
Define, construct, and interpret FTAs. (Application)

V. PROBLEM SOLVING AND QUALITY IMPROVEMENT **(25 Questions)**

A. **Approaches**
Describe and classify the implementation steps of quality improvement models such as Kaizen, PDSA, continuous improvement, etc. (Application)

B. **Management and Planning Tools**
Select, construct, apply, and interpret affinity diagrams, tree diagrams, process decision program charts, matrix diagrams, interrelationship digraphs, prioritization matrices, and activity network diagrams. (Analysis)

C. **Quality Tools**
Select, construct, apply, and interpret flowcharts, Pareto charts, and cause and effect diagrams. Select, apply, and interpret, control charts, check sheets, scatter diagrams, and histograms. [NOTE: The **mechanics** of these tools are covered in Section VI as follows: control charts (VI.G.4 and 5), check sheets (VI.B.3), scatter diagrams (VI.B.6.a.), histograms (VI.B.6.b.)] (Analysis)

D. **Corrective Action**
Identify elements of the corrective action process including problem identification, root cause analysis, correction, recurrence control, and verification of effectiveness, and determine root causes and appropriate corrective actions. (Analysis)

E. **Preventive Action**
Describe and apply preventive action concepts and techniques such as error-proofing, poka-yoke, robust design, etc., and analyze the effectiveness of their implementation. (Analysis)

F. **Overcoming Barriers to Quality Improvement**
Identify barriers and their causes, evaluate their impact, and describe methods for overcoming them. (Evaluation)

VI. QUANTITATIVE METHODS (53 Questions)

A. **Concepts of Probability and Statistics**

1. **Terms**
Describe population, parameter, statistic, random sample, expected value, etc., and compute expected value. (Application)

2. **Drawing valid statistical conclusions**
Distinguish between enumerative and analytical studies and evaluate the validity of conclusions based on statistical assumptions and the robustness of the technique used. (Evaluation)

3. **Central limit theorem and sampling distribution of the mean**
Define and apply these concepts. (Application)

4. **Basic probability concepts**
Describe and apply concepts such as independence, mutually exclusive, multiplication rules, complementary probability, joint occurrence of events, etc. NOTE: Bayes' Theorem will **not** be included. (Application)

B. **Collecting and Summarizing Data**

1. **Types of data**
Identify, define, classify, and compare continuous (variables) and discrete (attributes) data. (Application)

2. **Measurement scales**
Define and apply nominal, ordinal, interval, and ratio measurement scales. (Application)

3. **Methods for collecting data**
Define and apply methods for collecting data such as check sheets, coding data, automatic gaging, etc. (Application)

4. **Techniques for assuring data accuracy and integrity**
   Define and apply techniques for assuring data accuracy and integrity such as random sampling, stratified sampling, sample homogeneity, etc. NOTE: Sample integrity is covered in III.B.2, not here. (Application)
5. **Descriptive statistics**
   Define, compute, and interpret measures of dispersion and central tendency, and construct and interpret frequency distributions and cumulative frequency distributions. NOTE: Geometric and harmonic means will not be included. (Synthesis)
6. **Graphical methods**
   a. **Depicting relationships**
      Construct, apply, and interpret diagrams and charts such as stem-and-leaf plots, box-and-whisker plots, run charts, scatter diagrams, etc. (Analysis)
   b. **Depicting distributions**
      Construct, apply, and interpret diagrams such as histograms, normal probability plots, Weibull plots, etc. (Analysis)

C. **Properties and Applications of Probability Distributions**
1. **Discrete distributions**
   Describe and apply binomial, Poisson, hypergeometric, and multinomial distributions. (Analysis)
2. **Continuous distributions**
   Describe and apply uniform, normal, bivariate normal, exponential, log normal, Weibull, Chi-square, Student's t, and F distributions. (Analysis)

D. **Statistical Decision-Making**
1. **Point and interval estimation**
   Define and interpret the efficiency and bias of estimators. Compute, draw conclusions from, and interpret statistics such as standard error, tolerance intervals, and confidence intervals. (Analysis)
2. **Hypothesis testing** (NOTE: Nonparametric tests will **not** be included.)
   a. **Tests for means, variances, and proportions**
      Apply parametric hypothesis tests for means, variances, and proportions, and interpret the results. (Analysis)
   b. **Significance level, power, type I, and type II errors**
      Apply and interpret these concepts as they apply to statistical tests. (Analysis)
   c. **Statistical versus practical significance**
      Define and distinguish between statistical and practical significance. (Evaluation)

3. **Paired comparison tests**
   Define, determine applicability, and apply paired comparison parametric hypothesis tests, and interpret the results. (Analysis)
4. **Goodness-of-fit tests**
   Define, determine applicability, and apply Chi-square tests, and interpret the results. (Analysis)
5. **Analysis of variance (ANOVA)**
   Define, determine applicability, and apply analysis of variance, and interpret the results. (Analysis)
6. **Contingency tables**
   Define, determine applicability, and construct a contingency table, and use it to determine statistical significance. (Analysis)

E. **Measuring and Modeling Relationships Between Variables**
1. **Simple and multiple least-squares linear regression**
   Calculate the regression equation. Apply and interpret hypothesis tests for regression statistics. Use the regression model for estimation and prediction, and analyze the uncertainty in the estimate. NOTE: Models that are nonlinear in their parameters will **not** be included. (Evaluation)
2. **Simple linear correlation**
   Calculate and interpret the correlation coefficient and its confidence interval. Apply and interpret a hypothesis test for the correlation coefficient. NOTE: Serial correlation will **not** be included. (Analysis)
3. **Basic time-series analysis**
   Apply basic time-series analyses such as moving average. Interpret time-series graphs to identify trends, seasonal, and cyclical variation, etc. (Analysis)

F. **Designing Experiments**
NOTE: Mixture designs, data transformations, nested designs, and response surface methods will **not** be included.
1. **Terminology**
   Define terms such as independent and dependent variables, factors and levels, response, treatment, error, and replication. (Knowledge)
2. **Planning and organizing experiments**
   Describe and apply the basic elements of experiment planning and organizing, including determining the experiment objective, selecting factors, responses, and measurement methods, choosing the appropriate design, etc. (Evaluation)

3. **Design principles**
   Define and apply the principles of power and sample size, balance, replication, order, efficiency, randomization and blocking, interaction, and confounding. (Application)
4. **Design and analysis of one-factor experiments**
   Construct one-factor experiments such as completely randomized, randomized block, and Latin square designs, and apply computational and graphical methods to analyze and evaluate the significance of results. (Evaluation)
5. **Design and analysis of full-factorial experiments**
   Construct full-factorial designs and apply computational and graphical methods to analyze and evaluate the significance of results. (Evaluation)
6. **Design and analysis of two-level fractional factorial experiments**
   Construct two-level fractional factorial designs (including Taguchi designs) and apply computational and graphical methods to analyze and evaluate the significance of results. NOTE: Higher-order and mixed-level designs will **not** be included. (Evaluation)
7. **Taguchi robustness concepts**
   Identify and describe Taguchi robustness concepts and techniques such as signal-to-noise ratio, controllable and uncontrollable factors, and robustness to external sources of variability. (Comprehension)

G. **Statistical Process Control (SPC)**

1. **Objectives and benefits**
   Identify and describe objectives and benefits of SPC such as assessing process performance, distinguishing special from common causes, etc. (Comprehension)
2. **Selection of variable**
   Identify and select characteristics for monitoring by control chart. (Application)
3. **Rational subgrouping**
   Define and apply the principle of rational subgrouping. (Application)
4. **Selection and application of control charts**
   Identify, select, construct and apply the following control charts: $\overline{X}$ and R, $\overline{X}$ and s, individual and moving range (ImR), moving average and moving range (MamR), median, p, np, c, and u charts. (Synthesis)
5. **Analysis of control charts**
   Interpret control charts and distinguish between common and special causes using rules for determining statistical control. (Evaluation)

6. **PRE-control**
   Define and describe PRE-control and perform PRE-control calculations and analysis. (Analysis)
7. **Short-run SPC**
   Identify, define, and apply short-run SPC methods and techniques. (Analysis)

H. **Analyzing Process Capability**

1. **Designing and conducting process capability studies**
   Identify, describe, and apply the elements of designing and conducting process capability studies, including identifying characteristics, specifications, and/or tolerances, developing sampling plans, establishing statistical control, etc. (Evaluation)
2. **Calculating process performance versus specification**
   Distinguish between natural process limits and specification limits and calculate process performance metrics such as percent defective. (Analysis)
3. **Process capability indices**
   Define, select, and calculate Cp, Cpk, Cpm, and CR and assess process capability. (Evaluation)
4. **Process performance indices**
   Define, select, and calculate Pp and Ppk and assess process performance. (Evaluation)

SIX LEVELS OF COGNITION BASED ON BLOOM'S TAXONOMY (1956)

In addition to **content** specifics, the subtext detail also indicates the intended **complexity level** of the test questions for that topic. These levels are based on "Levels of Cognition" (from Bloom's Taxonomy, 1956) and are presented below in rank order, from least complex to most complex.

Knowledge Level

(Also commonly referred to as recognition, recall, or rote knowledge.) Being able to remember or recognize terminology, definitions, facts, ideas, materials, patterns, sequences, methodologies, principles, etc.

Comprehension Level

Being able to read and understand descriptions, communications, reports, tables, diagrams, directions, regulations, etc.

Application Level
Being able to apply ideas, procedures, methods, formulas, principles, theories, etc., in job-related situations

Analysis
Being able to break down information into its constituent parts and recognize the parts' relationship to one another and how they are organized; identify sublevel factors or salient data from a complex scenario

Synthesis
Being able to put parts or elements together in such a way as to show a pattern or structure not clearly there before; identify which data or information from a complex set is appropriate to examine further or from which supported conclusions can be drawn

Evaluation
Being able to make judgments regarding the value of proposed ideas, solutions, methodologies, etc., by using appropriate criteria or standards to estimate accuracy, effectiveness, economic benefits, etc.

♦ ♦ ♦

APPENDIX

2

Quality Auditing Terminology

Source: *The Complete Guide to the CQA*, S.M. Baysinger, Tucson, AZ: Quality Publishing, Inc., reprinted by permission.

The following were chosen as sources of definitions and terms. They represent a combination of national and international standards used throughout the world and are the bedrock upon which quality standards have been built in the past and which will shape the face of quality models in the future. They include:

ANSI/ASQC A3 (1987)—Quality Systems Terminology
ISO (Draft International Standard) 8402 (1992)—Quality Management and Quality Assurance Vocabulary
ISO 10011-1 (1990)—Guidelines for Auditing Quality Systems
ANSI/ASQC NQA-1 (1986)—Quality Assurance Program Requirements for Nuclear Facilities (Supplement S–1, Terms and Definitions)
ANSI/ASQC Q1 (1986)—Generic Guidelines for Auditing of Quality Systems

Terminology and definitions are presented in the following format: Terminology followed by source (in parenthesis). Where more than one source of definition or terminology exists all sources will be shown and ranked in the order shown above.

Acceptance criteria—Specified limits placed on characteristics of an item, process or service defined in codes, standards, or other requirement doc-uments. (ANSI/ASQC NQA-1-1986)

Accreditation—Certification by a duly recognized body of the facilities, capability, objectivity, competence and integrity of an agency, service, or operational group or individual to provide the specific service(s) or operations(s) needed. (ANSI/ASQC A3-1987)

Assessment—An estimate or determination of the significance, importance, or value of something. (ASQC Quality Auditing Technical Committee)

Audit: (Quality) audit—A systematic and independent examination to determine whether quality activities and related results comply with planned arrangements and whether these arrangements are implemented effectively and are suitable to achieve objectives. (ISO/DIS 8402 (1992))

A planned and documented activity performed to determine by investigation, examination or evaluation of objective evidence the adequacy of and compliance with established procedures, instructions, drawings, and other applicable documents, and the effectiveness of implementation. An audit should not be confused with surveillance or inspection activities performed for the sole purpose of process control or product appearance. (ANSI/ASQC NQA-1 [1986])

A systematic examination of the acts and decisions by people with respect to quality in order to independently verify or evaluate and report degree of compliance to the operational requirements of the quality program, or the specifications or contract requirements of the product or service. (ANSI/ASQC Q1 [1986])

A planned, independent, and documented assessment to determine whether agreed-upon requirements are being met. (ASQC Quality Auditing Technical Committee)

Auditee—An organization being audited. (ISO/DIS 8402 [1992])An organization to be audited. (ISO 10011-1-1990)

Auditing organization—A unit or function that carries out audits through its employees. This organization may be a department of the Auditee, Client or an independent third party. (ANSI/ASQC Q1 [1986])

Auditor—A person who has the qualification(s) to perform quality audits. (ISO 10011-1-1990)

The individual who carries out the audit. The auditor is appointed by the auditing organization selected by the client or by the client directly. A "lead auditor" supervises auditors during an audit as a team leader. (ANSI/ASQC Q1 [1986])

Audit standard—The authentic description of essential characteristics of audits which reflects current thought and practice. (ANSI/ASQC Q1 (1986))

Audit program—The organizational structure, commitment, and documented methods used to plan and perform audits. (ASQC Quality Auditing Technical Committee)

Audit program management—Organization, or function within an organization, given the responsibility to plan and carry out a programmed series of quality system audits. (ISO 10011-3 [1991])

Certificate of conformance—A document signed by an authorized party affirming that a product or service has met the requirements of the relevant specifications, contract or regulation. (ANSI/ASQC A3 [1987])

A document signed or otherwise authenticated by an authorized individual certifying the degree to which items or services meet specified requirements. (ANSI/ASQC NQA-1 [1986])

Certificate of compliance—A document signed by an authorized party that the supplier of a product or service has met the requirements of the rel-evant specifications, contract, or regulation. (ANSI/ASQC A3 [1987])

Certification—The procedure and action by a duly authorized body of determining, verifying, and attesting in writing to the qualifications of personnel, processes procedures, or items in accordance with applicable requirements. (ANSI/ASQC A3 [1987]); (ANSI/ASQC NQA-1 [1986])

Characteristic—Any property or attribute of an item, process or service that is distinct, describable, and measurable. (ANSI/ASQC NQA-1 [1986])

Client—The person or organization requesting the audit. Depending upon the circumstances, the client may be the auditing organization, the auditees or a third party. More specifically, the client can be:

An independent agency wishing to determine whether the quality program provides adequate control of the products being provided, e.g., food and drug, nuclear, and other regulatory bodies.

A potential customer wishing to evaluate the quality program of a potential supplier using their own auditors, a third party, or an independent auditor.

The auditee wishing to have their own quality program evaluated against some quality program standard. (ANSI/ASQC Q1 [1986])

Condition adverse to quality—An all-inclusive term used in reference to any of the following: 1) failures; 2) malfunctions; 3) deficiencies; 4) defectives, and 5) non-conformances. A significant condition adverse to quality is one which, if uncorrect-ed, could have a serious effect on safety or operability. (ANSI.ASQC NQA-1 [1986])

Conformance—An affirmative indication or judgment that a product or service has met the requirements of the relevant specifications, contract, or regulation. The state of meeting the requirements. (ANSI/ASQC A3 [1987])

Convention—A customary practice, rule, or method. (ASQC Quality Auditing Technical Committee)

Compliance—An affirmative indication or judgment that the supplier of a product or service has met the requirements of the relevant specifications, contract, or regulation. The state of meeting the requirements. (ANSI/ASQC A3 [1987])

Corrective action—An action taken to eliminate the causes of an existing non-conform-ity, defect or other undesirable situation(s) in order to prevent recurrence. (ISO/DIS 8402 [1992])

Measures taken to rectify conditions adverse to quality and, when nec-essary, to preclude repetition. (ANSI/ASQC NQA-1 [1986])

Design change—Any revision or alteration of the technical requirements defined by approved and issued design-output documents, and approved and issued changes thereto. (ANSI/ASQC NQA-1 [1986])

Design process—Technical and management processes that commence with identification of design input and that lead to and include the issuance of design-output documents. (ANSI/ASQC NQA-1 [1986])

Design review—A formal, documented, comprehensive and systematic examination of a design to evaluate the design requirements and the capability of the design to meet these requirements and to identify problems and propose solutions. (ANSI/ASQC A3 [1987])

Deviation—A departure from specified requirements. (ANSI/ASQC NQA-1 [1986])

Deviation permit—Written authorization, prior to production or provision of a service, to depart from specified requirements for a specified quantity or a specified time. (ANSI/ASQC A3 [1987])

Disposition of non-conformity—The action to be taken to deal with an existing, non-conforming entity in order to resolve the non-conformity. (ISO/DIS 8402 [1992])

Document—Any written or pictorial information describing, defining, specifying, reporting or certifying activities, requirements, procedures or results. A document is not considered a Quality Assurance Record until it satisfies the definition of a Quality Assurance Record as defined by ANSI/ASQC NQA-1, Supplement S-1. (ANSI/ASQC NQA-1 [1986])

External audit—An audit of those portions of another organization's quality assurance program not under the direct control or within the organizational structure of the auditing organization. (ANSI/ASQC NQA-1 [1986])

Final design—Approved design-output documents and approved changes thereto. (ANSI/ASQC NQA-1 [1986])

Finding—A conclusion of importance based on observation(s). (ASQ's Quality Auditing Technical Committee)

A condition adverse to quality which, if not corrected, the quality of the product or service may continue to suffer. (Author's definition)

Follow-up audit—An audit whose purpose and scope are limited to verifying that corrective action has been accomplished as scheduled and to determining that the action prevented recurrence effectively. (ASQ's Quality Auditing Technical Committee)

Guideline—A suggested practice that is not mandatory in programs intended to comply with a standard. The word "should" denotes a guideline; the word "shall" denotes a requirement. (ANSI/ASQC NQA-1 [1986])

Inspection—Activities, such as measuring, examining, testing, gaging one or more characteristics of a product or service and comparing these with specified requirements to determine conformity. (ANSI/ASQC A3 [1987])

The process of comparing an item, product, or service with the applicable specification. (ANSI/ASQC Q1 [1986])

Examination or measurement to verify whether an item or activity conforms to specified requirements. (ANSI/ASQC NQA-1 [1986])

Inspector—A person who performs inspection activities to verify conformance to specific requirements. (ANSI/ASQC NQA-1 [1986])

Internal audit—An audit of those portions of an organization's quality assurance program retained under its direct control and within its organizational structure. (ANSI/ASQC NQA-1 [1986])

Management and test equipment—Devices or systems used to calibrate, measure, gage, test, or inspect in order to control or acquire data to verify conformance to specified requirements. (ANSI/ASQC NQA-1 [1986])

Nonconformance—A deficiency in characteristic, documentation or procedure that renders the quality of an item or activity unacceptable or indeterminate. (ANSI/ASQC NQA-1 [1986])

Observation—(A quality audit observation is) a statement of fact made during (a) quality audit and substantiated by objective evidence. (ISO/DIS 8402 [1992])

A statement of fact made during an audit and substantiated by objective evidence. (ISO 10011-1 [1990])

A finding (made) during an audit substantiated by evidence. (ANSI/ASQC Q1 [1986])

A detected procedure or program weakness. Often a precursor to a *finding*. (Author's definition)

Objective evidence—Information which can be proven true, based on facts obtained through observation, measurement, test or other means. (ISO/DIS 8402 [1992])

Qualitative or quantitative information, records or statements of fact pertaining to the quality of an item or service or to the existence and implementation of a quality system element, which is based on observation, measurement or test and can be verified. (ISO 10011-1 [1990])

Process Review and Action Team (PRAT)—A Process Review and Action Team (PRAT) has its foundation in the work center, i.e., shop, department, etc. A PRAT usually meets weekly for one to two hours, is small in number (generally 4–6 members per team) and is multi-disciplined in nature. PRAT members generally have an equal vote in the decision making process. (Author's definition)

Procedure—A specified way to perform an activity. (ISO/DIS 8402 [1992])

A document that specifies or describes how an activity is to be performed. (ANSI/ASQC NQA-1 [1986])

Process quality audit—An analysis of elements of a process and appraisal of completeness, correctness of conditions and probable effectiveness. (ANSI/ASQC A3 [1987])

Product quality audit—A quantitative assessment of conformance to required product characteristics. Product quality audits are performed for the following reasons:

To estimate the outgoing quality level of the product or group of products.

To ascertain if the outgoing product meets a predetermined standard level of quality for a product or a group of products.

To estimate the level of quality originally submitted to inspection.

To measure the ability of inspection to make valid quality decisions.

To determine the suitability of the controls. (ANSI/ASQC A3 [1987])

Qualification (personnel)—The characteristics or abilities gained through education, training or experience as measured against established requirements such as standards or tests, that qualify an individual to per-form a required function. (ANSI/ASQC NQA-1 [1986])

Qualification (process)—The process of demonstrating whether an entity is capable of fulfilling specified requirements. (ISO/DIS 8402 [1992])

Qualified procedure—An approved procedure that has been demonstrated to meet the specified requirements for its intended purpose. (ANSI/ASQC NQA-1 [1986])

Quality assurance—All those planned or systematic actions necessary to provide adequate confidence that a product or service will satisfy given requirements for quality. Note: Quality assurance involves making sure that quality is what it should be. This includes a continuing evaluation of adequacy and effectiveness with a view to having timely corrective measures and feedback initiated where necessary. For a specific product or service, quality assurance involves the necessary plans and actions to provide confidence through verifications, audits, and the evaluation of the quality factors that affect the adequacy of the design for intended applications, specification, production, installation, inspection and use of the product or service. Providing assurance may involve producing evidence. (ANSI/ASQC A3 [1987])

All the planned and systematic activities implemented within the quality system, and demonstrated as needed, to provide adequate confidence that an entity will fulfill requirements for quality. (ISO/DIS 8402 [1992])

All those planned and systematic actions necessary to provide adequate confidence that a structure, system or component will perform satisfactorily in service. (ANSI/ASQC NQA-1 [1986])

Quality assurance record—A completed document that furnishes evidence of the quality of items and/or activities affecting quality. (ANSI/ASQC NQA-1 [1986])

Quality control—The operational techniques and the activities used to fulfill requirements of quality. Note: The aim of quality control is to provide quality that is satisfactory (e.g., safe, adequate, dependable and economical). The overall system involves integrating the quality aspects of several related steps including the proper specification of what is wanted; design of the product or service to meet the requirements; production or installation to meet the full intent of the specification; inspection to determine whether the resulting product or service conforms to the applicable specification; and, review of usage to provide the revision of specification.

Effective utilization of these technologies and activities is an essential element in the economic control of quality. (ANSI/ASQCA3 [1987])

The operational techniques and activities that are used to fulfill requirements for quality. (ISO/DIS 8402 [1992])

Quality management—The aspect of the overall management function that determines and implements the quality policy. (ANSI/ASQC A3 [1987])

All activities of the overall management function that determines the quality policy, objectives and responsibilities and implements them by means such as planning, quality control, quality assurance and quality improvement, within the quality system. (ISO/DIS 8402 [1992])

Quality plan—A document setting out the specific quality practices, resources, and sequence of activities relevant to a particular product, project or contract. (ISO/DIS 8402 [1992])

Quality policy—The overall intentions and direction of an organization as regards quality as formally expressed by top management. (ANSI/ASQC A3 [1987])

Quality system—The organizational structure, responsibilities, procedures, processes, and resources for implementing quality management. (ANSI/ASQC A3 [1987])

The organizational structure, responsibilities, procedures, processes, and resources needed to implement quality management. (ISO/DIS 8402 [1992])

The collective plans, activities, and events that are provided to ensure that a product, process or service will satisfy given needs. (ANSI/ASQC Q1 [1986])

Quality system audit—A documented activity performed to verify, by examination and evaluation of objective evidence, that applicable elements of the quality system are suitable and have been developed, documented and effectively implemented in accordance with specified requirements. (ANSI/ASQC A3 [1987])

A systematic and independent examination to determine whether quality activities and related results comply with planned arrangements and whether these arrangements are implemented effectively and are suitable to achieve objectives. (ISO 10011-1 [1990])

A documented activity performed to verify, by examination and evaluations of objective evidence, that applicable elements of the quality system are appropriate and have been developed, documented and effectively implemented in accordance and in conjunction with specified requirements. (ANSI/ASQC Q1 [1986])

Quality system review—A formal evaluation by management of the status and adequacy of the quality system in relation to quality policy and/or new objectives resulting from changing circumstances. (ANSI/ASQC A3 [1987])

Root cause—A fundamental deficiency that results in a non-conformance and must be corrected to prevent recurrence of the same or similar non-conformance. (ASQ's Quality Auditing Technical Committee)

Repair—The action taken on a non-conforming product so that it will fulfill the intended usage requirements, although it may not conform to the originally specified requirements. (ISO/DIS 8402 [1992])

The process of restoring a non-conforming characteristic to a condition such that the capability of an item to function reliably and safely is unimpaired, even though that item still does not conform to the original requirement. (ANSI/ASQC NQA-1 [1986])

Rework—The action taken on a non-conforming product so that it will fulfill the specified requirements. (ISO/DIS 8402 [1992])

The process by which an item is made to conform to original requirements by completion or correction. (ANSI/ASQC NQA-1 [1986])

Right of access—The right of a purchaser (customer) or designated representative to enter the premises of a supplier for the purpose of inspection, surveillance or quality assurance audit. (ANSI/ASQC NQA-1 [1986])

Service—The results generated by activities at the interface between the supplier and the customer and by supplier internal activities, to meet customer needs. (ISO/DIS 8402 [1992])

The performance of activities such as design, fabrication, inspection, nondestructive examination, repair or installation. (ANSI/ASQC NQA-1 [1986])

Special process—A process, the results of which are highly dependent on the control of the process or the skill of the operators, or both, and in which the specified quality cannot be readily determined by inspection or test of the product. (ANSI/ASQC NQA-1 [1986])

Specifications—The document that prescribes the requirements with which the product or service has come to conform. (ANSI/ASQC 3 [1987])

A document stating requirements. (ISO/DIS 8402 [1992])

Statistical process control—The application of statistical techniques to the control of processes. (Note: Statistical process control and statistical quality control are sometimes used synonymously; sometimes statistical process control is considered a subset of statistical quality control concentrating on tools associated with process aspects but not product acceptance measures. These techniques (tools) include the use of frequency distributions, measures of central tendency and dispersion, control charts, acceptance sampling, regression analysis, tests of significance, etc.) (ANSI/ASQC A3 [1987])

Supplier—The organization that provides a product to the customer. (ISO/DIS 8402 [1992])

Any individual or organization who furnishes items or services in accordance with a procurement document. An all-inclusive term used in place of any of the following: vendor, seller, contractor, subcontractor, fabricator, consultant and their subtier levels. (ANSI/ASQC NQA-1 [1986])

Surveillance—The continuous monitoring and verification of the status of an entity and analysis of records to ensure that specified requirements are being fulfilled. (ISO/DIS 8402 [1992])

The act of monitoring or observing to verify whether an item or activity conforms to specified requirements. (ANSI/ASQC NQA-1 [1986])

Survey—An examination of some specific purpose; to inspect or consider carefully; to review in detail. (Note: some authorities use the words "audit" and "survey" inter changeably. Audit implies the existence of some agreed upon criteria against which the plans and execution can be checked. Survey implies the inclusion of matters not covered by agreed-upon criteria.) (ASQ's Quality Auditing Technical Committee)

Testing—A means of determining the capability of an item to meet specified requirements by subjecting the item to a set of physical, chemical, environmental or operating actions or conditions. (ANSI/ASQC A3 [1987])

Total quality management—A management approach of an organization centered on quality, based on the participation of all its members and aiming at long-term success through customer satisfaction, and benefits to the members of the organization and to society. (ISO/DIS 8402 [1992])

Traceability—The ability to trace the history, application or location of an item or activity and like items or activities by means of recorded identification. (ANSI/ASQC A3 [1987])

Use-as-is—A disposition permitted for a non-conforming item when it can be established that the item is satisfactory for its intended use. (ANSI/ASQC NQA-1 [1986])

Verification—The act of reviewing, inspecting, testing, checking, auditing or otherwise establishing and documenting whether items, processes, services or documents conform to specified requirements. (ANSI/ASQC A3 [1987])

Waiver—Written authorization to use or release a quantity of material, component or stores already manufactured but not conforming to the specified requirements. (ANSI/ASQC A3 [1987])

Work instruction—The specific, detailed steps used to perform a particular function. They may be in the form of check lists, method sheets, etc. (Baysinger's definition)

APPENDIX

3

Glossary of Basic Statistical Terms*

Acceptable quality level—The maximum percentage or proportion of variant units in a lot or batch that, for the purposes of acceptance sampling, can be considered satisfactory as a process average.

Analysis of Variance (ANOVA)—A technique which subdivides the total variation of a set of data into meaningful component parts associated with specific sources of variation for the purpose of testing some hypothesis on the parameters of the model or estimating variance components.

Assignable cause—A factor which contributes to variation and which is feasible to detect and identify.

Average Outgoing Quality (AOQ)—The expected quality of outgoing product following the use of an acceptance sampling plan for a given value of incoming product quality.

Average Outgoing Quality Limit (AOQL)—For a given acceptance sampling plan, the maximum AOQ over all possible levels of incoming quality.

Chance causes—Factors, generally numerous and individually of relatively small importance, which contribute to variation, but which are not feasible to detect or identify.

Coefficient of determination—A measure of the part of the variance for one variable that can be explained by its linear relationship with a second variable. Designated by ρ^2 or r^2.

Coefficient of multiple correlation—A number between 0 and 1 that indicates the degree of the combined linear relationship of several predictor variables $X_1, X_2, \ldots, X_p$ to the response variable Y. It is the simple correlation coefficient between predicted and observed values of the response variable.

---

*From *Glossary & Tables for Statistical Quality Control*, prepared by the ASQC Statistics Division.

Coefficient of variation—A measure of relative dispersion that is the standard deviation divided by the mean and multiplied by 100 to give a percentage value. This measure cannot be used when the data take both negative and positive values or when it has been coded in such a way that the value X = 0 does not coincide with the origin.

Confidence limits—The end points of the interval about the sample statistic that is believed, with a specified confidence coefficient, to include the population parameter.

Confounding—When two or more factors or interactions are confounded in an experiment, the confounded terms cannot be estimated independently. Fractional Factorial designs are constructed so that main factors are only confounded with higher order interactions. For example, an eight run experiment with 4 factors at 2 levels each cannot independently estimate each of the 4 main factors (A, B, C, D), the 6 two-factor interactions (AB, AC, AD, BC, BD, CD), the 4 three-factor interactions (ABC, ABD, ACD, BCD), and the four-factor interaction (ABCD). Instead, the design is constructed so that one of the main factors, for example D, is confounded with its complementary three-factor interaction (ABC). Since D is confounded with ABC, if either is identified as statistically significant in the analysis, then we cannot determine (with the given experimental data) if D is significant, or ABC is significant, or if it is a combination of the two that is significant. Furthermore, when D is confounded with ABC (i.e. ABC = D), then A is confounded with BCD, B is confounded with ACD, C is confounded with ABD, AB is confounded with CD, AC is confounded with BD, and AD is confounded with BC.

Consumer's risk (β)—For a given sampling plan, the probability of acceptance of a lot, the quality of which has a designated numerical value representing a level which it is seldom desired to accept. Usually the designated value will be the **Limiting Quality Level (LQL)**.

Correlation coefficient—A number between –1 and 1 that indicates the degree of linear relationship between two sets of numbers:

$$r_{xy} = \frac{s_{xy}}{s_x s_y} = \frac{n\sum XY - \sum X \sum Y}{\sqrt{\left[n\sum X^2 - (\Sigma X)^2\right]\left[n\sum Y^2 - (\Sigma Y)^2\right]}}$$

Defect—A departure of a quality characteristic from its intended level or state that occurs with a severity sufficient to cause an associated product or service not to satisfy intended normal, or reasonably foreseeable, usage requirements.

Defective—A unit of product or service containing at least one defect, or having several imperfections that in combination cause the unit not to satisfy intended normal, or reasonably foreseeable, usage requirements. The word *defective* is appropriate for use when a unit of product or service is evaluated in terms of usage (as contrasted to conformance to specifications).

Double sampling—Sampling inspection in which the inspection of the first sample of size n_1, leads to a decision to accept a lot; not to accept it; or to take a second sample of size n_2, and the inspection of the second sample then leads to a decision to accept or not to accept the lot.

Experiment design—The arrangement in which an experimental program is to be conducted, and the selection of the versions (levels) of one or more factors or factor combinations to be included in the experiment.

Factor—An assignable cause which may affect the responses (test results) and of which different versions (levels) are included in the experiment.

Factorial experiments—Experiments in which all possible treatment combinations formed from two or more factors, each being studied at two or more versions (levels), are examined so that interactions (differential effects) as well as main effects can be estimated.

Frequency distribution—A set of all the various values that individual observations may have and the frequency of their occurrence in the sample or population.

Histogram—A plot of the frequency distribution in the form of rectangles whose bases are equal to the cell interval and whose areas are proportional to the frequencies.

Hypothesis, alternative—The hypothesis that is accepted if the null hypothesis is disproved. The choice of alternative hypothesis will determine whether "one-tail" or "two-tail" tests are appropriate.

Hypothesis, null—The hypothesis tested in tests of significance is that there is no difference (null) between the population of the sample and specified population (or between the populations associated with each sample). The null hypothesis can never be proved true. It can, however, be shown, with specified risks of error, to be untrue; that is, a difference can be shown to exist between the populations. If it is not disproved, one usually acts on the assumption that there is no adequate reason to doubt that it is true. (It may be that there is insufficient power to prove the existence of a difference rather than that there is no difference; that is, the sample size may be too small. By specifying the minimum difference that one wants to detect and β, the risk of failing to detect a difference of this size, the actual sample size required, however, can be determined.)

In-control process—A process in which the statistical measure(s) being evaluated are in a "state of statistical control."

Kurtosis—A measure of the shape of a distribution. A positive value indicates that the distribution has longer tails than the normal distribution (platykurtosis); while a negative value indicates that the distribution has shorter tails (leptokurtosis). For the normal distribution, the kurtosis is 0.

Mean, standard error of—The standard deviation of the average of a sample of size *n*.

$$s_{\bar{X}} = \frac{s_X}{\sqrt{n}}$$

Mean—A measure of the location of a distribution. The centroid.

Median—The middle measurement when an odd number of units are arranged in order of size; for an ordered set $X_1, X_2, \ldots, X_{2k-1}$

$$\text{Med} = X_k$$

When an even number are so arranged, the median is the average of the two middle units; for an ordered set $X_1, X_2, \ldots, X_{2k}$

$$Med = \frac{X_k + X_{k+1}}{2}$$

Mode—The most frequent value of the variable.

Multiple sampling—Sampling inspection in which, after each sample is inspected, the decision is made to accept a lot; not to accepts it; or to take another sample to reach the decision. There may be a prescribed maximum number of samples, after which a decision to accept or not to accept must be reached.

Operating Characteristics Curve (OC Curve)—

1. For isolated or unique lots or a lot from an isolated sequence: a curve showing, for a given sampling plan, the probability of accepting a lot as a function of the lot quality. (Type A)
2. For a continuous stream of lots: a curve showing, for a given sampling plan, the probability of accepting a lot as a function of the process average. (Type B)
3. For continuous sampling plans: a curve showing the proportion of submitted product over the long run accepted during the sampling phases of the plan as a function of the product quality.
4. For special plans: a curve showing, for a given sampling plan, the probability of continuing to permit the process to continue without adjustment as a function of the process quality.

Parameter—A constant or coefficient that describes some characteristic of a population (e.g., standard deviation, average, regression coefficient).

Population—The totality of items or units of material under consideration.

NOTE: The items may be units or measurements, and the population may be real or conceptual. Thus *population* may refer to all the items actually produced in a given day or all that might be produced if the process were to continue *in-control.*

Power curve—The curve showing the relation between the probability $(1-\beta)$ of rejecting the hypothesis that a sample belongs to a given population with a given characteristic(s) and the actual population value of that characteristic(s). NOTE: if β is used instead of $(1-\beta)$, the curve is called an operating characteristic curve (OC curve) (used mainly in sampling plans for quality control).

Process capability—The limits within which a tool or process operate based upon minimum variability as governed by the prevailing circumstances.

NOTE: The phrase "by the prevailing circumstances" indicates that the definition of inherent variability of a process involving only one operator, one source of raw material, etc., differs from one involving multiple operators, and many sources of raw material, etc. If the measure of inherent variability is made within very restricted circumstances, it is necessary to add components for frequently occurring assignable sources of variation that cannot economically be eliminated.

Producer's risk (α)—For a given sampling plan, the probability of not accepting a lot the quality of which has a designated numerical value representing a level which it is generally desired to accept. Usually the designated value will be the **Acceptable Quality Level (AQL)**.

Quality—The totality of features and characteristics of a product or service that bear on its ability to satisfy given needs.

Quality assurance—All those planned or systematic actions necessary to provide adequate confidence that a product or service will satisfy given needs.

Quality control—The operational techniques and the activities which sustain a quality of product or service that will satisfy given needs; also the use of such techniques and activities.

Random sampling—The process of selecting units for a sample of size *n* in such a manner that all combinations of n units under consideration have an equal or ascertainable chance of being selected as the sample.

R (range)—A measure of dispersion which is the difference between the largest observed value and the smallest observed value in a given sample. While the range is a measure of dispersion in its own right, it is sometimes used to estimate the population standard deviation, but is a biased estimator unless multiplied by the factor ($1/d_2$) appropriate to the sample size.

Replication—The repetition of the set of all the treatment combinations to be compared in an experiment. Each of the repetitions is called a *replicate.*

Sample—A group of units, portion of material, or observations taken from a larger collection of units, quantity of material, or observations that serves to provide information that may be used as a basis for making a decision concerning the larger quantity.

Single sampling—Sampling inspection in which the decision to accept or not to accept a lot is based on the inspection of a single sample of size *n.*

Skewness—A measure of the symmetry of a distribution. A positive value indicates that the distribution has a greater tendency to tail to the right (positively skewed or skewed to the right), and a negative value indicates a greater tendency of the distribution to tail to the left (negatively skewed or skewed to the left). Skewness is 0 for a normal distribution.

Standard deviation—

1. σ—population standard deviation. A measure of variability (dispersion) of observations that is the positive square root of the population variance.
2. s—sample standard deviation. A Measure of variability (dispersion) that is the positive square root of the sample variance.

$$\sqrt{\frac{1}{n}\Sigma\left(X_i-\overline{X}\right)^2}$$

Statistic—A quantity calculated from a sample of observations, most often to form an estimate of some population parameter.

Type I error (acceptance control sense)—The incorrect decision that a process is unacceptable when, in fact, perfect information would reveal that it is located within the "zone of acceptable processes."

Type II error (acceptance control sense)—The incorrect decision that a process is acceptable when, in fact, perfect information would reveal that it is located within the "zone of rejectable processes."

Variance—

1. σ^2—population variance. A measure of variability (dispersion) of observations based upon the mean of the squared deviation from the arithmetic mean.
2. s^2—sample variance. A measure of variability (dispersion) of observations in a sample based upon the squared deviations from the arithmetic average divided by the degrees of freedom.

APPENDIX

4

Area Under the Standard Normal Curve

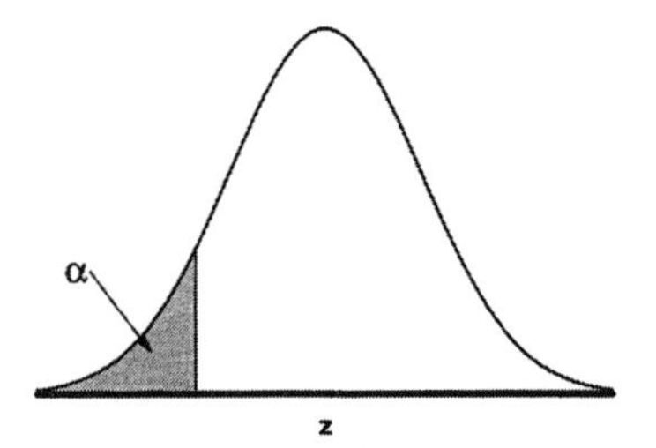

z	0.00	0.01	0.02	0.03	0.04	0.05	0.06	0.07	0.08	0.09
–3.4	0.0003	0.0003	0.0003	0.0003	0.0003	0.0003	0.0003	0.0003	0.0003	0.0002
–3.3	0.0005	0.0005	0.0005	0.0004	0.0004	0.0004	0.0004	0.0004	0.0004	0.0003
–3.2	0.0007	0.0007	0.0006	0.0006	0.0006	0.0006	0.0006	0.0005	0.0005	0.0005
–3.1	0.0010	0.0009	0.0009	0.0009	0.0008	0.0008	0.0008	0.0008	0.0007	0.0007
–3.0	0.0013	0.0013	0.0013	0.0012	0.0012	0.0011	0.0011	0.0011	0.0010	0.0010
–2.9	0.0019	0.0018	0.0018	0.0017	0.0016	0.0016	0.0015	0.0015	0.0014	0.0014
–2.8	0.0026	0.0025	0.0024	0.0023	0.0023	0.0022	0.0021	0.0021	0.0020	0.0019
–2.7	0.0035	0.0034	0.0033	0.0032	0.0031	0.0030	0.0029	0.0028	0.0027	0.0026
–2.6	0.0047	0.0045	0.0044	0.0043	0.0041	0.0040	0.0039	0.0038	0.0037	0.0036
–2.5	0.0062	0.0060	0.0059	0.0057	0.0055	0.0054	0.0052	0.0051	0.0049	0.0048
–2.4	0.0082	0.0080	0.0078	0.0075	0.0073	0.0071	0.0069	0.0068	0.0066	0.0064
–2.3	0.0107	0.0104	0.0102	0.0099	0.0096	0.0094	0.0091	0.0089	0.0087	0.0084
–2.2	0.0139	0.0136	0.0132	0.0129	0.0125	0.0122	0.0119	0.0116	0.0113	0.0110
–2.1	0.0179	0.0174	0.0170	0.0166	0.0162	0.0158	0.0154	0.0150	0.0146	0.0143
–2.0	0.0228	0.0222	0.0217	0.0212	0.0207	0.0202	0.0197	0.0192	0.0188	0.0183
–1.9	0.0287	0.0281	0.0274	0.0268	0.0262	0.0256	0.0250	0.0244	0.0239	0.0233
–1.8	0.0359	0.0351	0.0344	0.0336	0.0329	0.0322	0.0314	0.0307	0.0301	0.0294
–1.7	0.0446	0.0436	0.0427	0.0418	0.0409	0.0401	0.0392	0.0384	0.0375	0.0367
–1.6	0.0548	0.0537	0.0526	0.0516	0.0505	0.0495	0.0485	0.0475	0.0465	0.0455
–1.5	0.0668	0.0655	0.0643	0.0630	0.0618	0.0606	0.0594	0.0582	0.0571	0.0559

Continued on next page . . .

Continued . . .

z	0.00	0.01	0.02	0.03	0.04	0.05	0.06	0.07	0.08	0.09
–1.4	0.0808	0.0793	0.0778	0.0764	0.0749	0.0735	0.0721	0.0708	0.0694	0.0681
–1.3	0.0968	0.0951	0.0934	0.0918	0.0901	0.0885	0.0869	0.0853	0.0838	0.0823
–1.2	0.1151	0.1131	0.1112	0.1093	0.1075	0.1056	0.1038	0.1020	0.1003	0.0985
–1.1	0.1357	0.1335	0.1314	0.1292	0.1271	0.1251	0.1230	0.1210	0.1190	0.1170
–1.0	0.1587	0.1562	0.1539	0.1515	0.1492	0.1469	0.1446	0.1423	0.1401	0.1379
–0.9	0.1841	0.1814	0.1788	0.1762	0.1736	0.1711	0.1685	0.1660	0.1635	0.1611
–0.8	0.2119	0.2090	0.2061	0.2033	0.2005	0.1977	0.1949	0.1922	0.1894	0.1867
–0.7	0.2420	0.2389	0.2358	0.2327	0.2296	0.2266	0.2236	0.2206	0.2177	0.2148
–0.6	0.2743	0.2709	0.2676	0.2643	0.2611	0.2578	0.2546	0.2514	0.2483	0.2451
–0.5	0.3085	0.3050	0.3015	0.2981	0.2946	0.2912	0.2877	0.2843	0.2810	0.2776
–0.4	0.3446	0.3409	0.3372	0.3336	0.3300	0.3264	0.3228	0.3192	0.3156	0.3121
–0.3	0.3821	0.3783	0.3745	0.3707	0.3669	0.3632	0.3594	0.3557	0.3520	0.3483
–0.2	0.4207	0.4168	0.4129	0.4090	0.4052	0.4013	0.3974	0.3936	0.3897	0.3859
–0.1	0.4602	0.4562	0.4522	0.4483	0.4443	0.4404	0.4364	0.4325	0.4286	0.4247
–0.0	0.5000	0.4960	0.4920	0.4880	0.4840	0.4801	0.4761	0.4721	0.4681	0.4641
0.0	0.5000	0.5040	0.5080	0.5120	0.5160	0.5199	0.5239	0.5279	0.5319	0.5359
0.1	0.5398	0.5438	0.5478	0.5517	0.5557	0.5596	0.5636	0.5675	0.5714	0.5753
0.2	0.5793	0.5832	0.5871	0.5910	0.5948	0.5987	0.6026	0.6064	0.6103	0.6141
0.3	0.6179	0.6217	0.6255	0.6293	0.6331	0.6368	0.6406	0.6443	0.6480	0.6517
0.4	0.6554	0.6591	0.6628	0.6664	0.6700	0.6736	0.6772	0.6808	0.6844	0.6879
0.5	0.6915	0.6950	0.6985	0.7019	0.7054	0.7088	0.7123	0.7157	0.7190	0.7224
0.6	0.7257	0.7291	0.7324	0.7357	0.7389	0.7422	0.7454	0.7486	0.7517	0.7549
0.7	0.7580	0.7611	0.7642	0.7673	0.7704	0.7734	0.7764	0.7794	0.7823	0.7852
0.8	0.7881	0.7910	0.7939	0.7967	0.7995	0.8023	0.8051	0.8078	0.8106	0.8133
0.9	0.8159	0.8186	0.8212	0.8238	0.8264	0.8289	0.8315	0.8340	0.8365	0.8389
1.0	0.8413	0.8438	0.8461	0.8485	0.8508	0.8531	0.8554	0.8577	0.8599	0.8621
1.1	0.8643	0.8665	0.8686	0.8708	0.8729	0.8749	0.8770	0.8790	0.8810	0.8830
1.2	0.8849	0.8869	0.8888	0.8907	0.8925	0.8944	0.8962	0.8980	0.8997	0.9015
1.3	0.9032	0.9049	0.9066	0.9082	0.9099	0.9115	0.9131	0.9147	0.9162	0.9177
1.4	0.9192	0.9207	0.9222	0.9236	0.9251	0.9265	0.9279	0.9292	0.9306	0.9319
1.5	0.9332	0.9345	0.9357	0.9370	0.9382	0.9394	0.9406	0.9418	0.9429	0.9441
1.6	0.9452	0.9463	0.9474	0.9484	0.9495	0.9505	0.9515	0.9525	0.9535	0.9545
1.7	0.9554	0.9564	0.9573	0.9582	0.9591	0.9599	0.9608	0.9616	0.9625	0.9633
1.8	0.9641	0.9649	0.9656	0.9664	0.9671	0.9678	0.9686	0.9693	0.9699	0.9706
1.9	0.9713	0.9719	0.9726	0.9732	0.9738	0.9744	0.9750	0.9756	0.9761	0.9767
2.0	0.9772	0.9778	0.9783	0.9788	0.9793	0.9798	0.9803	0.9808	0.9812	0.9817
2.1	0.9821	0.9826	0.9830	0.9834	0.9838	0.9842	0.9846	0.9850	0.9854	0.9857
2.2	0.9861	0.9864	0.9868	0.9871	0.9875	0.9878	0.9881	0.9884	0.9887	0.9890
2.3	0.9893	0.9896	0.9898	0.9901	0.9904	0.9906	0.9909	0.9911	0.9913	0.9916
2.4	0.9918	0.9920	0.9922	0.9925	0.9927	0.9929	0.9931	0.9932	0.9934	0.9936

Continued on next page . . .

Continued . . .

z	0.00	0.01	0.02	0.03	0.04	0.05	0.06	0.07	0.08	0.09
2.5	0.9938	0.9940	0.9941	0.9943	0.9945	0.9946	0.9948	0.9949	0.9951	0.9952
2.6	0.9953	0.9955	0.9956	0.9957	0.9959	0.9960	0.9961	0.9962	0.9963	0.9964
2.7	0.9965	0.9966	0.9967	0.9968	0.9969	0.9970	0.9971	0.9972	0.9973	0.9974
2.8	0.9974	0.9975	0.9976	0.9977	0.9977	0.9978	0.9979	0.9979	0.9980	0.9981
2.9	0.9981	0.9982	0.9982	0.9983	0.9984	0.9984	0.9985	0.9985	0.9986	0.9986
3.0	0.9987	0.9987	0.9987	0.9988	0.9988	0.9989	0.9989	0.9989	0.9990	0.9990
3.1	0.9990	0.9991	0.9991	0.9991	0.9992	0.9992	0.9992	0.9992	0.9993	0.9993
3.2	0.9993	0.9993	0.9994	0.9994	0.9994	0.9994	0.9994	0.9995	0.9995	0.9995
3.3	0.9995	0.9995	0.9995	0.9996	0.9996	0.9996	0.9996	0.9996	0.9996	0.9997
3.4	0.9997	0.9997	0.9997	0.9997	0.9997	0.9997	0.9997	0.9997	0.9997	0.9998

Critical values of the *t*-Distribution

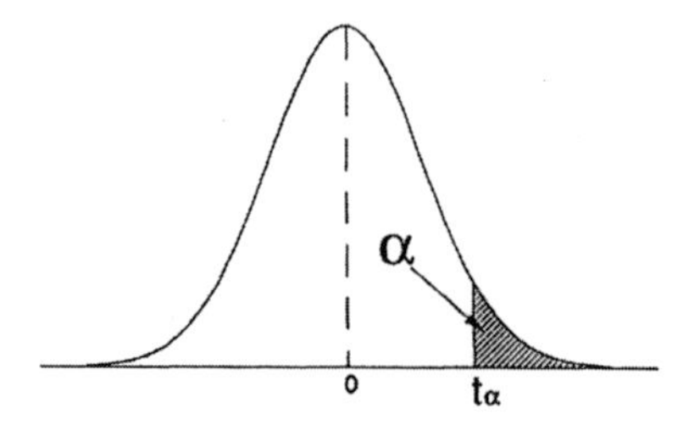

df	α				
	0.1	0.05	0.025	0.01	0.005
1	3.078	6.314	12.706	31.821	63.657
2	1.886	2.920	4.303	6.965	9.925
3	1.638	2.353	3.182	4.541	5.841
4	1.533	2.132	2.776	3.747	4.604
5	1.476	2.015	2.571	3.365	4.032
6	1.440	1.943	2.447	3.143	3.707
7	1.415	1.895	2.365	2.998	3.499
8	1.397	1.860	2.306	2.896	3.355
9	1.383	1.833	2.262	2.821	3.250
10	1.372	1.812	2.228	2.764	3.169
11	1.363	1.796	2.201	2.718	3.106
12	1.356	1.782	2.179	2.681	3.055
13	1.350	1.771	2.160	2.650	3.012
14	1.345	1.761	2.145	2.624	2.977
15	1.341	1.753	2.131	2.602	2.947

Continued on next page . . .

Continued . . .

df	α				
	0.1	**0.05**	**0.025**	**0.01**	**0.005**
16	1.337	1.746	2.120	2.583	2.921
17	1.333	1.740	2.110	2.567	2.898
18	1.330	1.734	2.101	2.552	2.878
19	1.328	1.729	2.093	2.539	2.861
20	1.325	1.725	2.086	2.528	2.845
21	1.323	1.721	2.080	2.518	2.831
22	1.321	1.717	2.074	2.508	2.819
23	1.319	1.714	2.069	2.500	2.807
24	1.318	1.711	2.064	2.492	2.797
25	1.316	1.708	2.060	2.485	2.787
26	1.315	1.706	2.056	2.479	2.779
27	1.314	1.703	2.052	2.473	2.771
28	1.313	1.701	2.048	2.467	2.763
29	1.311	1.699	2.045	2.462	2.756
∞	1.282	1.645	1.960	2.326	2.576

APPENDIX

6

Chi-Square Distribution

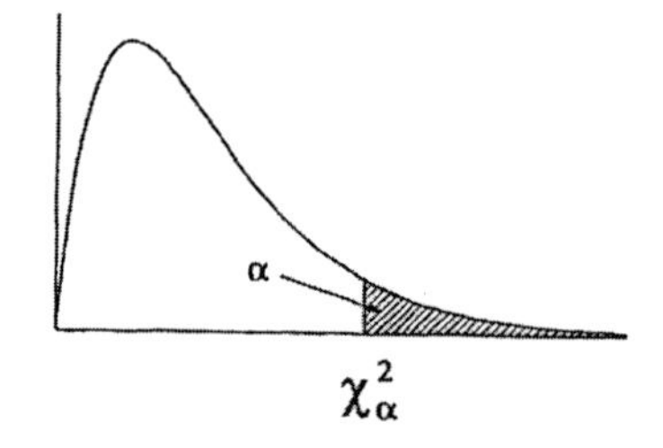

γ	α 0.995	0.99	0.98	0.975	0.95	0.90	0.80	0.75	0.70	0.50
1	0.00004	0.000	0.001	0.001	0.004	0.016	0.064	0.102	0.148	0.455
2	0.0100	0.020	0.040	0.051	0.103	0.211	0.446	0.575	0.713	1.386
3	0.0717	0.115	0.185	0.216	0.352	0.584	1.005	1.213	1.424	2.366
4	0.207	0.297	0.429	0.484	0.711	1.064	1.649	1.923	2.195	3.357
5	0.412	0.554	0.752	0.831	1.145	1.610	2.343	2.675	3.000	4.351
6	0.676	0.872	1.134	1.237	1.635	2.204	3.070	3.455	3.828	5.348
7	0.989	1.239	1.564	1.690	2.167	2.833	3.822	4.255	4.671	6.346
8	1.344	1.646	2.032	2.180	2.733	3.490	4.594	5.071	5.527	7.344
9	1.735	2.088	2.532	2.700	3.325	4.168	5.380	5.899	6.393	8.343
10	2.156	2.558	3.059	3.247	3.940	4.865	6.179	6.737	7.267	9.342
11	2.603	3.053	3.609	3.816	4.575	5.578	6.989	7.584	8.148	10.341
12	3.074	3.571	4.178	4.404	5.226	6.304	7.807	8.438	9.034	11.340
13	3.565	4.107	4.765	5.009	5.892	7.042	8.634	9.299	9.926	12.340
14	4.075	4.660	5.368	5.629	6.571	7.790	9.467	10.165	10.821	13.339
15	4.601	5.229	5.985	6.262	7.261	8.547	10.307	11.037	11.721	14.339

Continued on next page . . .

Continued . . .

γ	α									
	0.995	**0.99**	**0.98**	**0.975**	**0.95**	**0.90**	**0.80**	**0.75**	**0.70**	**0.50**
16	5.142	5.812	6.614	6.908	7.962	9.312	11.152	11.912	12.624	15.338
17	5.697	6.408	7.255	7.564	8.672	10.085	12.002	12.792	13.531	16.338
18	6.265	7.015	7.906	8.231	9.390	10.865	12.857	13.675	14.440	17.338
19	6.844	7.633	8.567	8.907	10.117	11.651	13.716	14.562	15.352	18.338
20	7.434	8.260	9.237	9.591	10.851	12.443	14.578	15.452	16.266	19.337
21	8.034	8.897	9.915	10.283	11.591	13.240	15.445	16.344	17.182	20.337
22	8.643	9.542	10.600	10.982	12.338	14.041	16.314	17.240	18.101	21.337
23	9.260	10.196	11.293	11.689	13.091	14.848	17.187	18.137	19.021	22.337
24	9.886	10.856	11.992	12.401	13.848	15.659	18.062	19.037	19.943	23.337
25	10.520	11.524	12.697	13.120	14.611	16.473	18.940	19.939	20.867	24.337
26	11.160	12.198	13.409	13.844	15.379	17.292	19.820	20.843	21.792	25.336
27	11.808	12.879	14.125	14.573	16.151	18.114	20.703	21.749	22.719	26.336
28	12.461	13.565	14.847	15.308	16.928	18.939	21.588	22.657	23.647	27.336
29	13.121	14.256	15.574	16.047	17.708	19.768	22.475	23.567	24.577	28.336
30	13.787	14.953	16.306	16.791	18.493	20.599	23.364	24.478	25.508	29.336

Continued on next page . . .

Continued . . .

γ	α									
	0.30	0.25	0.20	0.10	0.05	0.025	0.02	0.01	0.005	0.001
1	1.074	1.323	1.642	2.706	3.841	5.024	5.412	6.635	7.879	10.828
2	2.408	2.773	3.219	4.605	5.991	7.378	7.824	9.210	10.597	13.816
3	3.665	4.108	4.642	6.251	7.815	9.348	9.837	11.345	12.838	16.266
4	4.878	5.385	5.989	7.779	9.488	11.143	11.668	13.277	14.860	18.467
5	6.064	6.626	7.289	9.236	11.070	12.833	13.388	15.086	16.750	20.515
6	7.231	7.841	8.558	10.645	12.592	14.449	15.033	16.812	18.548	22.458
7	8.383	9.037	9.803	12.017	14.067	16.013	16.622	18.475	20.278	24.322
8	9.524	10.219	11.030	13.362	15.507	17.535	18.168	20.090	21.955	26.124
9	10.656	11.389	12.242	14.684	16.919	19.023	19.679	21.666	23.589	27.877
10	11.781	12.549	13.442	15.987	18.307	20.483	21.161	23.209	25.188	29.588
11	12.899	13.701	14.631	17.275	19.675	21.920	22.618	24.725	26.757	31.264
12	14.011	14.845	15.812	18.549	21.026	23.337	24.054	26.217	28.300	32.909
13	15.119	15.984	16.985	19.812	22.362	24.736	25.472	27.688	29.819	34.528
14	16.222	17.117	18.151	21.064	23.685	26.119	26.873	29.141	31.319	36.123
15	17.322	18.245	19.311	22.307	24.996	27.488	28.259	30.578	32.801	37.697
16	18.418	19.369	20.465	23.542	26.296	28.845	29.633	32.000	34.267	39.252
17	19.511	20.489	21.615	24.769	27.587	30.191	30.995	33.409	35.718	40.790
18	20.601	21.605	22.760	25.989	28.869	31.526	32.346	34.805	37.156	42.312
19	21.689	22.718	23.900	27.204	30.144	32.852	33.687	36.191	38.582	43.820
20	22.775	23.828	25.038	28.412	31.410	34.170	35.020	37.566	39.997	45.315
21	23.858	24.935	26.171	29.615	32.671	35.479	36.343	38.932	41.401	46.797
22	24.939	26.039	27.301	30.813	33.924	36.781	37.659	40.289	42.796	48.268
23	26.018	27.141	28.429	32.007	35.172	38.076	38.968	41.638	44.181	49.728
24	27.096	28.241	29.553	33.196	36.415	39.364	40.270	42.980	45.559	51.179
25	28.172	29.339	30.675	34.382	37.652	40.646	41.566	44.314	46.928	52.620
26	29.246	30.435	31.795	35.563	38.885	41.923	42.856	45.642	48.290	54.052
27	30.319	31.528	32.912	36.741	40.113	43.195	44.140	46.963	49.645	55.476
28	31.391	32.620	34.027	37.916	41.337	44.461	45.419	48.278	50.993	56.892
29	32.461	33.711	35.139	39.087	42.557	45.722	46.693	49.588	52.336	58.301
30	33.530	34.800	36.250	40.256	43.773	46.979	47.962	50.892	53.672	59.703

APPENDIX

7

F Distribution (α=1%)

$F_{.99}(n_1, n_2)$

n_1 = degrees of freedom for numerator; n_2 = degrees of freedom for denominator

n_2 \ n_1	1	2	3	4	5	6	7	8	9	10
1	4052	4999.5	5403	5625	5764	5859	5928	5982	6022	6056
2	98.50	99.00	99.17	99.25	99.30	99.33	99.36	99.37	99.39	99.40
3	34.12	30.82	29.46	28.71	28.24	27.91	27.67	27.49	27.35	27.23
4	21.20	18.00	16.69	15.98	15.52	15.21	14.98	14.80	14.66	14.55
5	16.26	13.27	12.06	11.39	10.97	10.67	10.46	10.29	10.16	10.05
6	13.75	10.92	9.78	9.15	8.75	8.47	8.26	8.10	7.98	7.87
7	12.25	9.55	8.45	7.85	7.46	7.19	6.99	6.84	6.72	6.62
8	11.26	8.65	7.59	7.01	6.63	6.37	6.18	6.03	5.91	5.81
9	10.56	8.02	6.99	6.42	6.06	5.80	5.61	5.47	5.35	5.26
10	10.04	7.56	6.55	5.99	5.64	5.39	5.20	5.06	4.94	4.85
11	9.65	7.21	6.22	5.67	5.32	5.07	4.89	4.74	4.63	4.54
12	9.33	6.93	5.95	5.41	5.06	4.82	4.64	4.50	4.39	4.30
13	9.07	6.70	5.74	5.21	4.86	4.62	4.44	4.30	4.19	4.10
14	8.86	6.51	5.56	5.04	4.69	4.46	4.28	4.14	4.03	3.94
15	8.68	6.36	5.42	4.89	4.56	4.32	4.14	4.00	3.89	3.80
16	8.53	6.23	5.29	4.77	4.44	4.20	4.03	3.89	3.78	3.69
17	8.40	6.11	5.18	4.67	4.34	4.10	3.93	3.79	3.68	3.59
18	8.29	6.01	5.09	4.58	4.25	4.01	3.84	3.71	3.60	3.51
19	8.18	5.93	5.01	4.50	4.17	3.94	3.77	3.63	3.52	3.43
20	8.10	5.85	4.94	4.43	4.10	3.87	3.70	3.56	3.46	3.37
21	8.02	5.78	4.87	4.37	4.04	3.81	3.64	3.51	3.40	3.31
22	7.95	5.72	4.82	4.31	3.99	3.76	3.59	3.45	3.35	3.26
23	7.88	5.66	4.76	4.26	3.94	3.71	3.54	3.41	3.30	3.21
24	7.82	5.61	4.72	4.22	3.90	3.67	3.50	3.36	3.26	3.17
25	7.77	5.57	4.68	4.18	3.85	3.63	3.46	3.32	3.22	3.13
26	7.72	5.53	4.64	4.14	3.82	3.59	3.42	3.29	3.18	3.09
27	7.68	5.49	4.60	4.11	3.78	3.56	3.39	3.26	3.15	3.06
28	7.64	5.45	4.57	4.07	3.75	3.53	3.36	3.23	3.12	3.03
29	7.60	5.42	4.54	4.04	3.73	3.50	3.33	3.20	3.09	3.00
30	7.56	5.39	4.51	4.02	3.70	3.47	3.30	3.17	3.07	2.98
40	7.31	5.18	4.31	3.83	3.51	3.29	3.12	2.99	2.89	2.80
60	7.08	4.98	4.13	3.65	3.34	3.12	2.95	2.82	2.72	2.63
120	6.85	4.79	3.95	3.48	3.17	2.96	2.79	2.66	2.56	2.47
∞	6.63	4.61	3.78	3.32	3.02	2.80	2.64	2.51	2.41	2.32

Continued on next page . . .

Continued . . .

n_1 = degrees of freedom for numerator

n_2 = degrees of freedom for denominator

n_2 \ n_1	12	15	20	24	30	40	60	120	∞
1	6106	6157	6209	6235	6261	6287	6313	6339	6366
2	99.42	99.43	99.45	99.46	99.47	99.47	99.48	99.49	99.50
3	27.05	26.87	26.69	26.60	26.50	26.41	26.32	26.22	26.13
4	14.37	14.20	14.02	13.93	13.84	13.75	13.65	13.56	13.46
5	9.89	9.72	9.55	9.47	9.38	9.29	9.20	9.11	9.02
6	7.72	7.56	7.40	7.31	7.23	7.14	7.06	6.97	6.88
7	6.47	6.31	6.16	6.07	5.99	5.91	5.82	5.74	5.65
8	5.67	5.52	5.36	5.28	5.20	5.12	5.03	4.95	4.86
9	5.11	4.96	4.81	4.73	4.65	4.57	4.48	4.40	4.31
10	4.71	4.56	4.41	4.33	4.25	4.17	4.08	4.00	3.91
11	4.40	4.25	4.10	4.02	3.94	3.86	3.78	3.69	3.60
12	4.16	4.01	3.86	3.78	3.70	3.62	3.54	3.45	3.36
13	3.96	3.82	3.66	3.59	3.51	3.43	3.34	3.25	3.17
14	3.80	3.66	3.51	3.43	3.35	3.27	3.18	3.09	3.00
15	3.67	3.52	3.37	3.29	3.21	3.13	3.05	2.96	2.87
16	3.55	3.41	3.26	3.18	3.10	3.02	2.93	2.84	2.75
17	3.46	3.31	3.16	3.08	3.00	2.92	2.83	2.75	2.65
18	3.37	3.23	3.08	3.00	2.92	2.84	2.75	2.66	2.57
19	3.30	3.15	3.00	2.92	2.84	2.76	2.67	2.58	2.49
20	3.23	3.09	2.94	2.86	2.78	2.69	2.61	2.52	2.42
21	3.17	3.03	2.88	2.80	2.72	2.64	2.55	2.46	2.36
22	3.12	2.98	2.83	2.75	2.67	2.58	2.50	2.40	2.31
23	3.07	2.93	2.78	2.70	2.62	2.54	2.45	2.35	2.26
24	3.03	2.89	2.74	2.66	2.58	2.49	2.40	2.31	2.21
25	2.99	2.85	2.70	2.62	2.54	2.45	2.36	2.27	2.17
26	2.96	2.81	2.66	2.58	2.50	2.42	2.33	2.23	2.13
27	2.93	2.78	2.63	2.55	2.47	2.38	2.29	2.20	2.10
28	2.90	2.75	2.60	2.52	2.44	2.35	2.26	2.17	2.06
29	2.87	2.73	2.57	2.49	2.41	2.33	2.23	2.14	2.03
30	2.84	2.70	2.55	2.47	2.39	2.30	2.21	2.11	2.01
40	2.66	2.52	2.37	2.29	2.20	2.11	2.02	1.92	1.80
60	2.50	2.35	2.20	2.12	2.03	1.94	1.84	1.73	1.60
120	2.34	2.19	2.03	1.95	1.86	1.76	1.66	1.53	1.38
∞	2.18	2.04	1.88	1.79	1.70	1.59	1.47	1.32	1.00

APPENDIX

8

F Distribution (α=5%)

$F_{.95}(n_1, n_2)$

n_1 = degrees of freedom for numerator

n_2 = degrees of freedom for denominator

n_2 \ n_1	1	2	3	4	5	6	7	8	9	10
1	161.4	199.5	215.7	224.6	230.2	234.0	236.8	238.9	240.5	241.9
2	18.51	19.00	19.16	19.25	19.30	19.33	19.35	19.37	19.38	19.40
3	10.13	9.55	9.28	9.12	9.01	8.94	8.89	8.85	8.81	8.79
4	7.71	6.94	6.59	6.39	6.26	6.16	6.09	6.04	6.00	5.96
5	6.61	5.79	5.41	5.19	5.05	4.95	4.88	4.82	4.77	4.74
6	5.99	5.14	4.76	4.53	4.39	4.28	4.21	4.15	4.10	4.06
7	5.59	4.47	4.35	4.12	3.97	3.87	3.79	3.73	3.68	3.64
8	5.32	4.46	4.07	3.84	3.69	3.58	3.50	3.44	3.39	3.35
9	5.12	4.26	3.86	3.63	3.48	3.37	3.29	3.23	3.18	3.14
10	4.96	4.10	3.71	3.48	3.33	3.22	3.14	3.07	3.02	2.98
11	4.84	3.98	3.59	3.36	3.20	3.09	3.01	2.95	2.90	2.85
12	4.75	3.89	3.49	3.26	3.11	3.00	2.91	2.85	2.80	2.75
13	4.67	3.81	3.41	3.18	3.03	2.92	2.83	2.77	2.71	2.67
14	4.60	3.74	3.34	3.11	2.96	2.85	2.76	2.70	2.65	2.60
15	4.54	3.68	3.29	3.06	2.90	2.79	2.71	2.64	2.59	2.54
16	4.49	3.63	3.24	3.01	2.85	2.74	2.66	2.59	2.54	2.49
17	4.45	3.59	3.20	2.96	2.81	2.70	2.61	2.55	2.49	2.45
18	4.41	3.55	3.16	2.93	2.77	2.66	2.58	2.51	2.46	2.41
19	4.38	3.52	3.13	2.90	2.74	2.63	2.54	2.48	2.42	2.38
20	4.35	3.49	3.10	2.87	2.71	2.60	2.51	2.45	2.39	2.35
21	4.32	3.47	3.07	2.84	2.68	2.57	2.49	2.42	2.37	2.32
22	4.30	3.44	3.05	2.82	2.66	2.55	2.46	2.40	2.34	2.30
23	4.28	3.42	3.03	2.80	2.64	2.53	2.44	2.37	2.32	2.27
24	4.26	3.40	3.01	2.78	2.62	2.51	2.42	2.36	2.30	2.25
25	4.24	3.39	2.99	2.76	2.60	2.49	2.40	2.34	2.28	2.24
26	4.23	3.37	2.98	2.74	2.59	2.47	2.39	2.32	2.27	2.22
27	4.21	3.35	2.96	2.73	2.57	2.46	2.37	2.31	2.25	2.20
28	4.20	3.34	2.95	2.71	2.56	2.45	2.36	2.29	2.24	2.19
29	4.18	3.33	2.93	2.70	2.55	2.43	2.35	2.28	2.22	2.18
30	4.17	3.32	2.92	2.69	2.53	2.42	2.33	2.27	2.21	2.16
40	4.08	3.23	2.84	2.61	2.45	2.34	2.25	2.18	2.12	2.08
60	4.00	3.15	2.76	2.53	2.37	2.25	2.17	2.10	2.04	1.99
120	3.92	3.07	2.68	2.45	2.29	2.17	2.09	2.02	1.96	1.91
∞	3.84	3.00	2.60	2.37	2.21	2.10	2.01	1.94	1.88	1.83

Continued on next page . . .

Continued . . .

n_1 = degrees of freedom for numerator

n_2 = degrees of freedom for denominator

$n_2 \backslash n_1$	12	15	20	24	30	40	60	120	∞
1	243.9	245.9	248.0	249.1	250.1	251.1	252.2	253.3	254.3
2	19.41	19.43	19.45	19.45	19.46	19.47	19.48	19.49	19.50
3	8.74	8.70	8.66	8.64	8.62	8.59	8.57	8.55	8.53
4	5.91	5.86	5.80	5.77	5.75	5.72	5.69	5.66	5.63
5	4.68	4.62	4.56	4.53	4.50	4.46	4.43	4.40	4.36
6	4.00	3.94	3.87	3.84	3.81	3.77	3.74	3.70	3.67
7	3.57	3.51	3.44	3.41	3.38	3.34	3.30	3.27	3.23
8	3.28	3.22	3.15	3.12	3.08	3.04	3.01	2.97	2.93
9	3.07	3.01	2.94	2.90	2.86	2.83	2.79	2.75	2.71
10	2.91	2.85	2.77	2.74	2.70	2.66	2.62	2.58	2.54
11	2.79	2.72	2.65	2.61	2.57	2.53	2.49	2.45	2.40
12	2.69	2.62	2.54	2.51	2.47	2.43	2.38	2.34	2.30
13	2.60	2.53	2.46	2.42	2.38	2.34	2.30	2.25	2.21
14	2.53	2.46	2.39	2.35	2.31	2.27	2.22	2.18	2.13
15	2.48	2.40	2.33	2.29	2.25	2.20	2.16	2.11	2.07
16	2.42	2.35	2.28	2.24	2.19	2.15	2.11	2.06	2.01
17	2.38	2.31	2.23	2.19	2.15	2.10	2.06	2.01	1.96
18	2.34	2.27	2.19	2.15	2.11	2.06	2.02	1.97	1.92
19	2.31	2.23	2.16	2.11	2.07	2.03	1.98	1.93	1.88
20	2.28	2.20	2.12	2.08	2.04	1.99	1.95	1.90	1.84
21	2.25	2.18	2.10	2.05	2.01	1.96	1.92	1.87	1.81
22	2.23	2.15	2.07	2.03	1.98	1.94	1.89	1.84	1.78
23	2.20	2.13	2.05	2.01	1.96	1.91	1.86	1.81	1.76
24	2.18	2.11	2.03	1.98	1.94	1.89	1.84	1.79	1.73
25	2.16	2.09	2.01	1.96	1.92	1.87	1.82	1.77	1.71
26	2.15	2.07	1.99	1.95	1.90	1.85	1.80	1.75	1.69
27	2.13	2.06	1.97	1.93	1.88	1.84	1.79	1.73	1.67
28	2.12	2.04	1.96	1.91	1.87	1.82	1.77	1.71	1.65
29	2.10	2.03	1.94	1.90	1.85	1.81	1.75	1.70	1.64
30	2.09	2.01	1.93	1.89	1.84	1.79	1.74	1.68	1.62
40	2.00	1.92	1.84	1.79	1.74	1.69	1.64	1.58	1.51
60	1.92	1.84	1.75	1.70	1.65	1.59	1.53	1.47	1.39
120	1.83	1.75	1.66	1.61	1.55	1.50	1.43	1.35	1.25
∞	1.75	1.67	1.57	1.52	1.46	1.39	1.32	1.22	1.00

APPENDIX

9

Poisson Probability Sums

$$\sum_{x=o}^{r} p(x;\mu)$$

r	μ								
	0.1	0.2	0.3	0.4	0.5	0.6	0.7	0.8	0.9
0	0.9048	0.8187	0.7408	0.6703	0.6065	0.5488	0.4966	0.4493	0.4066
1	0.9953	0.9825	0.9631	0.9384	0.9098	0.8781	0.8442	0.8088	0.7725
2	0.9998	0.9989	0.9964	0.9921	0.9856	0.9769	0.9659	0.9526	0.9371
3	1.0000	0.9999	0.9997	0.9992	0.9982	0.9966	0.9942	0.9909	0.9865
4	1.0000	1.0000	1.0000	0.9999	0.9998	0.9996	0.9992	0.9986	0.9977
5	1.0000	1.0000	1.0000	1.0000	1.0000	1.0000	0.9999	0.9998	0.9997
6	1.0000	1.0000	1.0000	1.0000	1.0000	1.0000	1.0000	1.0000	1.0000

r	μ								
	1.0	1.5	2.0	2.5	3.0	3.5	4.0	4.5	5.0
0	0.3679	0.2231	0.1353	0.0821	0.0498	0.0302	0.0183	0.0111	0.0067
1	0.7358	0.5578	0.4060	0.2873	0.1991	0.1359	0.0916	0.0611	0.0404
2	0.9197	0.8088	0.6767	0.5438	0.4232	0.3208	0.2381	0.1736	0.1247
3	0.9810	0.9344	0.8571	0.7576	0.6472	0.5366	0.4335	0.3423	0.2650
4	0.9963	0.9814	0.9473	0.8912	0.8153	0.7254	0.6288	0.5321	0.4405
5	0.9994	0.9955	0.9834	0.9580	0.9161	0.8576	0.7851	0.7029	0.6160
6	0.9999	0.9991	0.9955	0.9858	0.9665	0.9347	0.8893	0.8311	0.7622
7	1.0000	0.9998	0.9989	0.9958	0.9881	0.9733	0.9489	0.9134	0.8666
8	1.0000	1.0000	0.9998	0.9989	0.9962	0.9901	0.9786	0.9597	0.9319
9	1.0000	1.0000	1.0000	0.9997	0.9989	0.9967	0.9919	0.9829	0.9682
10	1.0000	1.0000	1.0000	0.9999	0.9997	0.9990	0.9972	0.9933	0.9863
11	1.0000	1.0000	1.0000	1.0000	0.9999	0.9997	0.9991	0.9976	0.9945
12	1.0000	1.0000	1.0000	1.0000	1.0000	0.9999	0.9997	0.9992	0.9980
13	1.0000	1.0000	1.0000	1.0000	1.0000	1.0000	0.9999	0.9997	0.9993
14	1.0000	1.0000	1.0000	1.0000	1.0000	1.0000	1.0000	0.9999	0.9998
15	1.0000	1.0000	1.0000	1.0000	1.0000	1.0000	1.0000	1.0000	0.9999
16	1.0000	1.0000	1.0000	1.0000	1.0000	1.0000	1.0000	1.0000	1.0000

Continued on next page . . .

Continued . . .

r	μ								
	5.5	6.0	6.5	7.0	7.5	8.0	8.5	9.0	9.5
0	0.0041	0.0025	0.0015	0.0009	0.0006	0.0003	0.0002	0.0001	0.0001
1	0.0266	0.0174	0.0113	0.0073	0.0047	0.0030	0.0019	0.0012	0.0008
2	0.0884	0.0620	0.0430	0.0296	0.0203	0.0138	0.0093	0.0062	0.0042
3	0.2017	0.1512	0.1118	0.0818	0.0591	0.0424	0.0301	0.0212	0.0149
4	0.3575	0.2851	0.2237	0.1730	0.1321	0.0996	0.0744	0.0550	0.0403
5	0.5289	0.4457	0.3690	0.3007	0.2414	0.1912	0.1496	0.1157	0.0885
6	0.6860	0.6063	0.5265	0.4497	0.3782	0.3134	0.2562	0.2068	0.1649
7	0.8095	0.7440	0.6728	0.5987	0.5246	0.4530	0.3856	0.3239	0.2687
8	0.8944	0.8472	0.7916	0.7291	0.6620	0.5925	0.5231	0.4557	0.3918
9	0.9462	0.9161	0.8774	0.8305	0.7764	0.7166	0.6530	0.5874	0.5218
10	0.9747	0.9574	0.9332	0.9015	0.8622	0.8159	0.7634	0.7060	0.6453
11	0.9890	0.9799	0.9661	0.9467	0.9208	0.8881	0.8487	0.8030	0.7520
12	0.9955	0.9912	0.9840	0.9730	0.9573	0.9362	0.9091	0.8758	0.8364
13	0.9983	0.9964	0.9929	0.9872	0.9784	0.9658	0.9486	0.9261	0.8981
14	0.9994	0.9986	0.9970	0.9943	0.9897	0.9827	0.9726	0.9585	0.9400
15	0.9998	0.9995	0.9988	0.9976	0.9954	0.9918	0.9862	0.9780	0.9665
16	0.9999	0.9998	0.9996	0.9990	0.9980	0.9963	0.9934	0.9889	0.9823
17	1.0000	0.9999	0.9998	0.9996	0.9992	0.9984	0.9970	0.9947	0.9911
18	1.0000	1.0000	0.9999	0.9999	0.9997	0.9993	0.9987	0.9976	0.9957
19	1.0000	1.0000	1.0000	1.0000	0.9999	0.9997	0.9995	0.9989	0.9980
20	1.0000	1.0000	1.0000	1.0000	1.0000	0.9999	0.9998	0.9996	0.9991
21	1.0000	1.0000	1.0000	1.0000	1.0000	1.0000	0.9999	0.9998	0.9996
22	1.0000	1.0000	1.0000	1.0000	1.0000	1.0000	1.0000	0.9999	0.9999
23	1.0000	1.0000	1.0000	1.0000	1.0000	1.0000	1.0000	1.0000	0.9999
24	1.0000	1.0000	1.0000	1.0000	1.0000	1.0000	1.0000	1.0000	1.0000

Continued on next page . . .

Continued . . .

r	μ								
	10.0	11.0	12.0	13.0	14.0	15.0	16.0	17.0	18.0
0	0.0000	0.0000	0.0000	0.0000	0.0000	0.0000	0.0000	0.0000	0.0000
1	0.0005	0.0002	0.0001	0.0000	0.0000	0.0000	0.0000	0.0000	0.0000
2	0.0028	0.0012	0.0005	0.0002	0.0001	0.0000	0.0000	0.0000	0.0000
3	0.0103	0.0049	0.0023	0.0011	0.0005	0.0002	0.0001	0.0000	0.0000
4	0.0293	0.0151	0.0076	0.0037	0.0018	0.0009	0.0004	0.0002	0.0001
5	0.0671	0.0375	0.0203	0.0107	0.0055	0.0028	0.0014	0.0007	0.0003
6	0.1301	0.0786	0.0458	0.0259	0.0142	0.0076	0.0040	0.0021	0.0010
7	0.2202	0.1432	0.0895	0.0540	0.0316	0.0180	0.0100	0.0054	0.0029
8	0.3328	0.2320	0.1550	0.0998	0.0621	0.0374	0.0220	0.0126	0.0071
9	0.4579	0.3405	0.2424	0.1658	0.1094	0.0699	0.0433	0.0261	0.0154
10	0.5830	0.4599	0.3472	0.2517	0.1757	0.1185	0.0774	0.0491	0.0304
11	0.6968	0.5793	0.4616	0.3532	0.2600	0.1848	0.1270	0.0847	0.0549
12	0.7916	0.6887	0.5760	0.4631	0.3585	0.2676	0.1931	0.1350	0.0917
13	0.8645	0.7813	0.6815	0.5730	0.4644	0.3632	0.2745	0.2009	0.1426
14	0.9165	0.8540	0.7720	0.6751	0.5704	0.4657	0.3675	0.2808	0.2081
15	0.9513	0.9074	0.8444	0.7636	0.6694	0.5681	0.4667	0.3715	0.2867
16	0.9730	0.9441	0.8987	0.8355	0.7559	0.6641	0.5660	0.4677	0.3751
17	0.9857	0.9678	0.9370	0.8905	0.8272	0.7489	0.6593	0.5640	0.4686
18	0.9928	0.9823	0.9626	0.9302	0.8826	0.8195	0.7423	0.6550	0.5622
19	0.9965	0.9907	0.9787	0.9573	0.9235	0.8752	0.8122	0.7363	0.6509
20	0.9984	0.9953	0.9884	0.9750	0.9521	0.9170	0.8682	0.8055	0.7307
21	0.9993	0.9977	0.9939	0.9859	0.9712	0.9469	0.9108	0.8615	0.7991
22	0.9997	0.9990	0.9970	0.9924	0.9833	0.9673	0.9418	0.9047	0.8551
23	0.9999	0.9995	0.9985	0.9960	0.9907	0.9805	0.9633	0.9367	0.8989
24	1.0000	0.9998	0.9993	0.9980	0.9950	0.9888	0.9777	0.9594	0.9317
25	1.0000	0.9999	0.9997	0.9990	0.9974	0.9938	0.9869	0.9748	0.9554
26	1.0000	1.0000	0.9999	0.9995	0.9987	0.9967	0.9925	0.9848	0.9718
27	1.0000	1.0000	0.9999	0.9998	0.9994	0.9983	0.9959	0.9912	0.9827
28	1.0000	1.0000	1.0000	0.9999	0.9997	0.9991	0.9978	0.9950	0.9897
29	1.0000	1.0000	1.0000	1.0000	0.9999	0.9996	0.9989	0.9973	0.9941
30	1.0000	1.0000	1.0000	1.0000	0.9999	0.9998	0.9994	0.9986	0.9967
31	1.0000	1.0000	1.0000	1.0000	1.0000	0.9999	0.9997	0.9993	0.9982
32	1.0000	1.0000	1.0000	1.0000	1.0000	1.0000	0.9999	0.9996	0.9990
33	1.0000	1.0000	1.0000	1.0000	1.0000	1.0000	0.9999	0.9998	0.9995
34	1.0000	1.0000	1.0000	1.0000	1.0000	1.0000	1.0000	0.9999	0.9998
35	1.0000	1.0000	1.0000	1.0000	1.0000	1.0000	1.0000	1.0000	0.9999
36	1.0000	1.0000	1.0000	1.0000	1.0000	1.0000	1.0000	1.0000	0.9999
37	1.0000	1.0000	1.0000	1.0000	1.0000	1.0000	1.0000	1.0000	1.0000

APPENDIX

10

Tolerance Interval Factors

Table 10.1a. Values of k for two-sided limits.

n	γ=0.90				γ=0.95				γ=0.99			
	P=0.90	P=0.95	P=0.99	P=0.999	P=0.90	P=0.95	P=0.99	P=0.999	P=0.90	P=0.95	P=0.99	P=0.999
2	15.978	18.800	24.167	30.227	32.019	37.674	48.430	60.573	160.193	188.491	242.300	303.054
3	5.847	6.919	8.974	11.309	8.380	9.916	12.861	16.208	18.930	22.401	29.055	36.616
4	4.166	4.943	6.440	8.149	5.369	6.370	8.299	10.502	9.398	11.150	14.527	18.383
5	3.494	4.152	5.423	6.879	4.275	5.079	6.634	8.415	6.612	7.855	10.260	13.015
6	3.131	3.723	4.870	6.188	3.712	4.414	5.775	7.337	5.337	6.345	8.301	10.548
7	2.902	3.452	4.521	5.750	3.369	4.007	5.248	6.676	4.613	5.488	7.187	9.142
8	2.743	3.264	4.278	5.446	3.316	3.732	4.891	6.226	4.147	4.936	6.468	8.234
9	2.626	3.125	4.098	5.220	2.967	3.532	4.631	5.899	3.822	4.550	5.966	7.600
10	2.535	3.018	3.959	5.046	2.839	3.379	4.433	5.649	3.582	4.265	5.594	7.129
11	2.463	2.933	3.849	4.906	2.737	3.259	4.277	5.452	3.397	4.045	5.308	6.766
12	2.404	2.863	3.758	4.792	2.655	3.162	4.150	5.291	3.250	3.870	5.079	6.477
13	2.355	2.805	3.682	4.697	2.587	3.081	4.044	5.158	3.130	3.727	4.893	6.240
14	2.314	2.756	3.618	4.615	2.529	3.012	3.955	5.045	3.029	3.608	4.737	6.043
15	2.278	2.713	3.562	4.545	2.480	2.954	3.878	4.949	2.945	3.507	4.605	5.876
16	2.246	2.676	3.514	4.484	2.437	2.903	3.812	4.865	2.872	3.421	4.492	5.732
17	2.219	2.643	3.471	4.430	2.400	2.858	3.754	4.791	2.808	3.345	4.393	5.607
18	2.194	2.614	3.433	4.382	2.366	2.819	3.702	4.725	2.753	3.279	4.307	5.497
19	2.172	2.588	3.399	4.339	2.337	2.784	3.656	4.667	2.703	3.221	4.230	5.399
20	2.152	2.564	3.368	4.300	2.310	2.752	3.615	4.614	2.659	3.168	4.161	5.312
21	2.135	2.543	3.340	4.264	2.286	2.723	3.577	4.567	2.620	3.121	4.100	5.234
22	2.118	2.524	3.315	4.232	2.264	2.697	3.543	4.523	2.584	3.078	4.044	5.163
23	2.103	2.506	3.292	4.203	2.244	2.673	3.512	4.484	2.551	3.040	3.993	5.098
24	2.089	2.480	3.270	4.176	2.225	2.651	3.483	4.447	2.522	3.004	3.947	5.039
25	2.077	2.474	3.251	4.151	2.208	2.631	3.457	4.413	2.494	2.972	3.904	4.985

Continued on next page . . .

Table 10.1a—*Continued . . .*

	γ=0.90				γ=0.95				γ=0.99			
n	P=0.90	P=0.95	P=0.99	P=0.999	P=0.90	P=0.95	P=0.99	P=0.999	P=0.90	P=0.95	P=0.99	P=0.999
30	2.025	2.413	3.170	4.049	2.140	2.549	3.350	4.278	2.385	2.841	3.733	4.768
35	1.988	2.368	3.112	3.974	2.090	2.490	3.272	4.179	2.306	2.748	3.611	4.611
40	1.959	2.334	3.066	3.917	2.052	2.445	3.213	4.104	2.247	2.677	3.518	4.493
45	1.935	2.306	3.030	3.871	2.021	2.408	3.165	4.042	2.200	2.621	3.444	4.399
50	1.916	2.284	3.001	3.833	1.996	2.379	3.126	3.993	2.162	2.576	3.385	4.323

Table 10.1b. Values of k for one-sided limits.

	γ=0.90				γ=0.95				γ=0.99			
n	P=0.90	P=0.95	P=0.99	P=0.999	P=0.90	P=0.95	P=0.99	P=0.999	P=0.90	P=0.95	P=0.99	P=0.999
3	4.258	5.310	7.340	9.651	6.158	7.655	10.552	13.857	-	-	-	-
4	3.187	3.957	5.437	7.128	4.163	5.145	7.042	9.215	-	-	-	-
5	2.742	3.400	4.666	6.112	3.407	4.202	5.741	7.501	-	-	-	-
6	2.494	3.091	4.242	5.556	3.006	3.707	50.62	6.612	4.408	5.409	7.334	9.540
7	2.333	2.894	3.972	5.201	2.755	3.399	4.641	6.061	3.856	4.730	6.411	8.348
8	2.219	2.755	3.783	4.955	2.582	3.188	4.353	5.686	3.496	4.287	5.811	7.566
9	2.133	2.649	3.641	4.772	2.454	3.031	4.143	5.414	3.242	3.971	5.389	7.014
10	2.065	2.568	3.532	4.629	2.355	2.911	3.981	5.203	3.048	3.739	5.075	6.603
11	2.012	2.503	3.444	4.515	2.275	2.815	3.852	5.036	2.897	3.557	4.828	6.284
12	1.966	2.448	3.371	4.420	2.210	2.736	3.747	4.900	2.773	3.410	4.633	6.032
13	1.928	2.403	3.310	4.341	2.155	2.670	3.659	4.787	2.677	3.290	4.472	5.826
14	1.895	2.363	3.257	4.274	2.108	2.614	3.585	4.690	2.592	3.189	4.336	5.651
15	1.866	2.329	3.212	4.215	2.068	2.566	3.520	4.607	2.521	3.102	4.224	5.507
16	1.842	2.299	3.172	4.146	2.032	2.523	3.463	4.534	2.458	3.028	4.124	5.374
17	1.820	2.272	3.136	4.118	2.001	2.468	3.415	4.471	2.405	2.962	4.038	5.268
18	1.800	2.249	3.106	4.078	1.974	2.453	3.370	4.415	2.357	2.906	3.961	5.167
19	1.781	2.228	3.078	4.041	1.949	2.423	3.331	4.364	2.315	2.855	3.893	5.078
20	1.765	2.208	3.052	4.009	1.926	2.396	3.295	4.319	2.275	2.807	3.832	5.003
21	1.750	2.190	3.028	3.979	1.905	2.371	3.262	4.276	2.241	2.768	3.776	4.932
22	1.736	2.174	3.007	3.952	1.887	2.350	3.233	4.238	2.208	2.729	3.727	4.866
23	1.724	2.159	2.987	3.927	1.869	2.329	3.206	4.204	2.179	2.693	3.680	4.806
24	1.712	2.145	2.969	3.904	1.853	2.309	3.181	4.171	2.154	2.663	3.638	4.755
25	1.702	2.132	2.952	3.882	1.838	2.292	3.158	4.143	2.129	2.632	3.601	4.706
30	1.657	2.080	2.884	3.794	1.778	2.220	3.064	4.022	2.029	2.516	3.446	4.508
35	1.623	2.041	2.833	3.730	1.732	2.166	2.994	3.934	1.957	2.431	3.334	4.364
40	1.598	2.010	2.793	3.679	1.697	2.126	2.941	3.866	1.902	2.365	3.250	4.255
45	1.577	1.986	2.762	3.638	1.669	2.092	2.897	3.811	1.857	2.313	3.181	4.168
50	1.560	1.965	2.735	3.604	1.646	2.065	2.963	3.766	1.821	2.296	3.124	4.096

Table 10.2. Proportion of population covered with γ% confidence and sample size n.

n	γ=0.90	γ=0.95	γ=0.99	γ=0.995
2	0.052	0.026	0.006	0.003
4	0.321	0.249	0.141	0.111
6	0.490	0.419	0.295	0.254
10	0.664	0.606	0.496	0.456
20	0.820	0.784	0.712	0.683
40	0.907	0.887	0.846	0.829
60	0.937	0.924	0.895	0.883
80	0.953	0.943	0.920	0.911
100	0.962	0.954	0.936	0.929
150	0.975	0.969	0.957	0.952
200	0.981	0.977	0.968	0.961
500	0.993	0.991	0.987	0.986
1000	0.997	0.996	0.994	0.993

Table 10.3. Sample size required to cover (1–α)% of the population with γ% confidence.

α	γ=0.90	γ=0.95	γ=0.99	γ=0.995
0.005	777	947	1325	1483
0.01	388	473	662	740
0.05	77	93	130	146
0.01	38	46	64	72
0.15	25	30	42	47
0.20	18	22	31	34
0.25	15	18	24	27
0.30	12	14	20	22
0.40	6	10	14	16
0.50	7	8	11	12

APPENDIX

11

Durbin-Watson Test Bounds

Table 11.1. Level of significance α=.05

n	p-1=1 d_L	p-1=1 d_U	p-1=2 d_L	p-1=2 d_U	p-1=3 d_L	p-1=3 d_U	p-1=4 d_L	p-1=4 d_U	p-1=5 d_L	p-1=5 d_U
15	1.08	1.36	0.95	1.54	0.82	1.75	0.69	1.97	0.56	2.21
16	1.10	1.37	0.98	1.54	0.86	1.73	0.74	1.93	0.62	2.15
17	1.13	1.38	1.02	1.54	0.90	1.71	0.78	1.90	0.67	2.10
18	1.16	1.39	1.05	1.53	0.93	1.69	0.82	1.87	0.71	2.06
19	1.18	1.40	1.08	1.53	0.97	1.68	0.86	1.85	0.75	2.02
20	1.20	1.41	1.10	1.54	1.00	1.68	0.90	1.83	0.79	1.99
21	1.22	1.42	1.13	1.54	1.03	1.67	0.93	1.81	0.83	1.96
22	1.24	1.43	1.15	1.54	1.05	1.66	0.96	1.80	0.86	1.94
23	1.26	1.44	1.17	1.54	1.08	1.66	0.99	1.79	0.90	1.92
24	1.27	1.45	1.19	1.55	1.10	1.66	1.01	1.78	0.93	1.90
25	1.29	1.45	1.21	1.55	1.12	1.66	1.04	1.77	0.95	1.89
26	1.30	1.46	1.22	1.55	1.14	1.65	1.06	1.76	0.98	1.88
27	1.32	1.47	1.24	1.56	1.16	1.65	1.08	1.76	1.01	1.86
28	1.33	1.48	1.26	1.56	1.18	1.65	1.10	1.75	1.03	1.85
29	1.34	1.48	1.27	1.56	1.20	1.65	1.12	1.74	1.05	1.84
30	1.35	1.49	1.28	1.57	1.21	1.65	1.14	1.74	1.07	1.83
31	1.36	1.50	1.30	1.57	1.23	1.65	1.16	1.74	1.09	1.83
32	1.37	1.50	1.31	1.57	1.24	1.65	1.18	1.73	1.11	1.82
33	1.38	1.51	1.32	1.58	1.26	1.65	1.19	1.73	1.13	1.81
34	1.39	1.51	1.33	1.58	1.27	1.65	1.21	1.73	1.15	1.81

Continued on next page . . .

Table 11.1—*Continued . . .*

n	p-1=1 d_L	p-1=1 d_U	p-1=2 d_L	p-1=2 d_U	p-1=3 d_L	p-1=3 d_U	p-1=4 d_L	p-1=4 d_U	p-1=5 d_L	p-1=5 d_U
35	1.40	1.52	1.34	1.58	1.28	1.65	1.22	1.73	1.16	1.80
36	1.41	1.52	1.35	1.59	1.29	1.65	1.24	1.73	1.18	1.80
37	1.42	1.53	1.36	1.59	1.31	1.66	1.25	1.72	1.19	1.80
38	1.43	1.54	1.37	1.59	1.32	1.66	1.26	1.72	1.21	1.79
39	1.43	1.54	1.38	1.60	1.33	1.66	1.27	1.72	1.22	1.79
40	1.44	1.54	1.39	1.60	1.34	1.66	1.29	1.72	1.23	1.79
45	1.48	1.57	1.43	1.62	1.38	1.67	1.34	1.72	1.29	1.78
50	1.50	1.59	1.46	1.63	1.42	1.67	1.38	1.72	1.34	1.77
55	1.53	1.60	1.49	1.64	1.45	1.68	1.41	1.72	1.38	1.77
60	1.55	1.62	1.51	1.65	1.48	1.69	1.44	1.73	1.41	1.77
65	1.57	1.63	1.54	1.66	1.50	1.70	1.47	1.73	1.44	1.77
70	1.58	1.64	1.55	1.67	1.52	1.70	1.49	1.74	1.46	1.77
75	1.60	1.65	1.57	1.68	1.54	1.71	1.51	1.74	1.49	1.77
80	1.61	1.66	1.59	1.69	1.56	1.72	1.53	1.74	1.51	1.77
85	1.62	1.67	1.60	1.70	1.57	1.72	1.55	1.75	1.52	1.77
90	1.63	1.68	1.61	1.70	1.59	1.73	1.57	1.75	1.54	1.78
95	1.64	1.69	1.62	1.71	1.60	1.73	1.58	1.75	1.56	1.78
100	1.65	1.69	1.63	1.72	1.61	1.74	1.59	1.76	1.57	1.78

Table 11.2. Level of significance α=.01.

	p-1=1		p-1=2		p-1=3		p-1=4		p-1=5	
n	d_L	d_U	d_L	d_U	d_L	d_U	d_L	d_U	d_L	d_U
15	0.81	1.07	0.70	1.25	0.59	1.46	0.49	1.70	0.39	1.96
16	0.84	1.09	0.74	1.25	0.63	1.44	0.53	1.66	0.44	1.90
17	0.87	1.10	0.77	1.25	0.67	1.43	0.57	1.63	0.48	1.85
18	0.90	1.12	0.80	1.26	0.71	1.42	0.61	1.60	0.52	1.80
19	0.93	1.13	0.83	1.26	0.74	1.41	0.65	1.58	0.56	1.77
20	0.95	1.15	0.86	1.27	0.77	1.41	0.68	1.57	0.60	1.74
21	0.97	1.16	0.89	1.27	0.80	1.41	0.72	1.55	0.63	1.71
22	1.00	1.17	0.91	1.28	0.83	1.40	0.75	1.54	0.66	1.69
23	1.02	1.19	0.94	1.29	0.86	1.40	0.77	1.53	0.70	1.67
24	1.04	1.20	0.96	1.30	0.88	1.41	0.80	1.53	0.72	1.66
25	1.05	1.21	0.98	1.30	0.90	1.41	0.83	1.52	0.75	1.65
26	1.07	1.22	1.00	1.31	0.93	1.41	0.85	1.52	0.78	1.64
27	1.09	1.23	1.02	1.32	0.95	1.41	0.88	1.51	0.81	1.63
28	1.10	1.24	1.04	1.32	0.97	1.41	0.90	1.51	0.83	1.62
29	1.12	1.25	1.05	1.33	0.99	1.42	0.92	1.51	0.85	1.61
30	1.13	1.26	1.07	1.34	1.01	1.42	0.94	1.51	0.88	1.61
31	1.15	1.27	1.08	1.34	1.02	1.42	0.96	1.51	0.90	1.60
32	1.16	1.28	1.10	1.35	1.04	1.43	0.98	1.51	0.92	1.60
33	1.17	1.29	1.11	1.36	1.05	1.43	1.00	1.51	0.94	1.59
34	1.18	1.30	1.13	1.36	1.07	1.43	1.01	1.51	0.95	1.59
35	1.19	1.31	1.14	1.37	1.08	1.44	1.03	1.51	0.97	1.59
36	1.21	1.32	1.15	1.38	1.10	1.44	1.04	1.51	0.99	1.59
37	1.22	1.32	1.16	1.38	1.11	1.45	1.06	1.51	1.00	1.59
38	1.23	1.33	1.18	1.39	1.12	1.45	1.07	1.52	1.02	1.58
39	1.24	1.34	1.19	1.39	1.14	1.45	1.09	1.52	1.03	1.58
40	1.25	1.34	1.20	1.40	1.15	1.46	1.10	1.52	1.05	1.58
45	1.29	1.38	1.24	1.42	1.20	1.48	1.16	1.53	1.11	1.58
50	1.32	1.40	1.28	1.45	1.24	1.49	1.20	1.54	1.16	1.59
55	1.36	1.43	1.32	1.47	1.28	1.51	1.25	1.55	1.21	1.59
60	1.38	1.45	1.35	1.48	1.32	1.52	1.28	1.56	1.25	1.60
65	1.41	1.47	1.38	1.50	1.35	1.53	1.31	1.57	1.28	1.61
70	1.43	1.49	1.40	1.52	1.37	1.55	1.34	1.58	1.31	1.61
75	1.45	1.50	1.42	1.53	1.39	1.56	1.37	1.59	1.34	1.62
80	1.47	1.52	1.44	1.54	1.42	1.57	1.39	1.60	1.36	1.62
85	1.48	1.53	1.46	1.55	1.43	1.58	1.41	1.60	1.39	1.63
90	1.50	1.54	1.47	1.56	1.45	1.59	1.43	1.61	1.41	1.64
95	1.51	1.55	1.49	1.57	1.47	1.60	1.45	1.62	1.42	1.64
100	1.52	1.56	1.50	1.58	1.48	1.60	1.46	1.63	1.44	1.65

APPENDIX

12

y Factors for Computing AOQL

c	0	1	2
y	0.368	0.841	1.372
c	3	4	5
y	1.946	2.544	3.172
c	6	7	8
y	3.810	4.465	5.150
c	9	10	11
y	5.836	6.535	7.234

APPENDIX

13

Control Chart Constants

CHART FOR AVERAGES				CHART FOR STANDARD DEVIATIONS					
Observations in Sample, n	Factors for Control Limits			Factors for Central Line		Factors for Control Limits			
	A	A_2	A_3	c_4	$1/c_4$	B_3	B_4	B_5	B_6
2	2.121	1.880	2.659	0.7979	1.2533	0	3.267	0	2.606
3	1.732	1.023	1.954	0.8862	1.1284	0	2.568	0	2.276
4	1.500	0.729	1.628	0.9213	1.0854	0	2.266	0	2.088
5	1.342	0.577	1.427	0.9400	1.0638	0	2.089	0	1.964
6	1.225	0.483	1.287	0.9515	1.0510	0.030	1.970	0.029	1.874
7	1.134	0.419	1.182	0.9594	1.0423	0.118	1.882	0.113	1.806
8	1.061	0.373	1.099	0.9650	1.0363	0.185	1.815	0.179	1.751
9	1.000	0.337	1.032	0.9693	1.0317	0.239	1.761	0.232	1.707
10	0.949	0.308	0.975	0.9727	1.0281	0.284	1.716	0.276	1.669
11	0.905	0.285	0.927	0.9754	1.0252	0.321	1.679	0.313	1.637
12	0.866	0.266	0.886	0.9776	1.0229	0.354	1.646	0.346	1.610
13	0.832	0.249	0.850	0.9794	1.0210	0.382	1.618	0.374	1.585
14	0.802	0.235	0.817	0.9810	1.0194	0.406	1.594	0.399	1.563
15	0.775	0.223	0.789	0.9823	1.0180	0.428	1.572	0.421	1.544
16	0.750	0.212	0.763	0.9835	1.0168	0.448	1.552	0.440	1.526
17	0.728	0.203	0.739	0.9845	1.0157	0.466	1.534	0.458	1.511
18	0.707	0.194	0.718	0.9854	1.0148	0.482	1.518	0.475	1.496
19	0.688	0.187	0.698	0.9862	1.0140	0.497	1.503	0.490	1.483
20	0.671	0.180	0.680	0.9869	1.0133	0.510	1.490	0.504	1.470
21	0.655	0.173	0.663	0.9876	1.0126	0.523	1.477	0.516	1.459
22	0.640	0.167	0.647	0.9882	1.0119	0.534	1.466	0.528	1.448
23	0.626	0.162	0.633	0.9887	1.0114	0.545	1.455	0.539	1.438
24	0.612	0.157	0.619	0.9892	1.0109	0.555	1.445	0.549	1.429
25	0.600	0.153	0.606	0.9896	1.0105	0.565	1.435	0.559	1.420

Continued on next page . . .

Continued . . .

CHART FOR RANGES							
Observations in Sample, n	Factors for Central Line			Factors for Control Limits			
	d_2	$1/d_2$	d_3	D_1	D_2	D_3	D_4
2	1.128	0.8865	0.853	0	3.686	0	3.267
3	1.693	0.5907	0.888	0	4.358	0	2.574
4	2.059	0.4857	0.880	0	4.698	0	2.282
5	2.326	0.4299	0.864	0	4.918	0	2.114
6	2.534	0.3946	0.848	0	5.078	0	2.004
7	2.704	0.3698	0.833	0.204	5.204	0.076	1.924
8	2.847	0.3512	0.820	0.388	5.306	0.136	1.864
9	2.970	0.3367	0.808	0.547	5.393	0.184	1.816
10	3.078	0.3249	0.797	0.687	5.469	0.223	1.777
11	3.173	0.3152	0.787	0.811	5.535	0.256	1.744
12	3.258	0.3069	0.778	0.922	5.594	0.283	1.717
13	3.336	0.2998	0.770	1.025	5.647	0.307	1.693
14	3.407	0.2935	0.763	1.118	5.696	0.328	1.672
15	3.472	0.2880	0.756	1.203	5.741	0.347	1.653
16	3.532	0.2831	0.750	1.282	5.782	0.363	1.637
17	3.588	0.2787	0.744	1.356	5.820	0.378	1.622
18	3.640	0.2747	0.739	1.424	5.856	0.391	1.608
19	3.689	0.2711	0.734	1.487	5.891	0.403	1.597
20	3.735	0.2677	0.729	1.549	5.921	0.415	1.585
21	3.778	0.2647	0.724	1.605	5.951	0.425	1.575
22	3.819	0.2618	0.720	1.659	5.979	0.434	1.566
23	3.858	0.2592	0.716	1.710	6.006	0.443	1.557
24	3.895	0.2567	0.712	1.759	6.031	0.451	1.548
25	3.931	0.2544	0.708	1.806	6.056	0.459	1.541

APPENDIX

14

Control Chart Equations

	np CHART	p CHART
LCL	$LCL = n\bar{p} - 3\sqrt{n\bar{p}\left(1 - \frac{n\bar{p}}{n}\right)}$ or 0 if LCL is negative	$LCL = \bar{p} - 3\sqrt{\frac{\bar{p}(1-\bar{p})}{n}}$ or 0 if LCL is negative
Center Line	$n\bar{p} = \frac{\text{Sum of items with problems}}{\text{Number of subgroups}}$	$\bar{p} = \frac{\text{Sum of items with problems}}{\text{Number of items in all subgroups}}$
UCL	$UCL = n\bar{p} + 3\sqrt{n\bar{p}\left(1 - \frac{n\bar{p}}{n}\right)}$ or n if UCL is greater than n	$UCL = \bar{p} + 3\sqrt{\frac{\bar{p}(1-\bar{p})}{n}}$ or 1 if UCL is greater than 1
	c CHART	**u CHART**
LCL	$LCL = \bar{c} - 3\sqrt{\bar{c}}$ or 0 if LCL is negative	$LCL = \bar{u} - 3\sqrt{\frac{\bar{u}}{n}}$ or 0 if LCL is negative
Center Line	$\bar{c} = \frac{\text{Sum of problems}}{\text{Number of subgroups}}$	$\bar{u} = \frac{\text{Sum of problems}}{\text{Number of units in all subgroups}}$
UCL	$UCL = \bar{c} + 3\sqrt{\bar{c}}$	$UCL = \bar{u} + 3\sqrt{\frac{\bar{u}}{n}}$

Continued on next page . . .

Continued . . .

	X CHART	$\overline{X}$ CHART
LCL	$LCL = \overline{X} - 2.66(\overline{MR})$	$LCL = \overline{\overline{X}} - A_2\overline{R}$
Center Line	$\overline{X} = \frac{\text{Sum of measurements}}{\text{Number of measurements}}$	$\overline{\overline{X}} = \frac{\text{Sum of subgroup averages}}{\text{Number of averages}}$
UCL	$UCL = \overline{X} + 2.66(\overline{MR})$	$UCL = \overline{\overline{X}} + A_2\overline{R}$
	R CHART	
LCL	$LCL = D_3\overline{R}$	
Center Line	$\overline{R} = \frac{\text{Sum of ranges}}{\text{Number of ranges}}$	
UCL	$UCL = D_4\overline{R}$	

♦♦♦
APPENDIX

15

Mil-Std-45662 Military Standard: Calibration System Requirements

CALIBRATION SYSTEM REQUIREMENTS
Mil-Std-45662

1. This standard is approved for use by all departments and agencies of the Department of Defense.
2. Beneficial comments (recommendations, additions, deletions) and any pertinent data which may be of use in improving this document should be addressed to: Commander, US Army Missile Command, ATTN: DRSMI-ED, Redstone Arsenal, AL 35809, by using the self-addressed Standardization Document Improvement Proposal (DD Form 1426) appearing at the end of this document or by letter.

FORWARD

1. This standard contains requirements for the establishment and maintenance of a calibration system used to control the accuracy of measuring and test equipment.
2. Data Item Description (DIDS) applicable to this standard are listed in Section 6.

Paragraph

1. SCOPE

1.1 Scope. This standard provides for the establishment and maintenance of a calibration system to control the accuracy of the measuring and test equipment used to assure that supplies and services presented to the Government for acceptance are in conformance with prescribed technical requirements.

1.2 Applicability. This standard applies to all contracts under which the contractor is required to maintain measuring and test equipment in support of contract requirements.

1.3 Significance. This standard and any procedure or document executed in implementation thereof shall be in addition to and not in derogation of other contract requirements.

2. REFERENCED DOCUMENTS (not applicable)

3. DEFINITIONS

3.1 Calibration. Comparison of a measurement standard or instrument of known accuracy with another standard or instrument to detect, correlate, report, or eliminate by adjustment, any variation in the accuracy of the item being compared.

3.2 Measuring and test equipment. All devices used to measure, gage, test, inspect, or otherwise examine items to determine compliance with specifications.

3.3 Measurement standard (reference). Standards of the highest accuracy order in a calibration system which establish the basic accuracy values for that system.

3.4 Measurement standard (transfer). Designated measuring equipment used in a calibration system as a medium for transferring the basic value of reference standards to lower echelon transfer standards or measuring and test equipment.

3.5 Traceability. The ability to relate individual measurement results to national standards or nationally accepted measurement systems through an unbroken chain of comparisons.

4. GENERAL STATEMENTS OF REQUIREMENTS

4.1 General. The contractor shall establish or adapt and maintain a system for the calibration of all measuring and test equipment used in fulfillment of his contractual requirements. The calibration system shall be coordinated with his inspection or quality control systems and shall be designed to provide adequate accuracy in use of measuring and test equipment. All measuring and test equipment applicable to the contractor, whether used in the contractor's plant or at another source, shall be subject to such control as is necessary to assure conformance of supplies and services to contractual requirements. The calibration system shall provide for the prevention of inaccuracy by ready detection of deficiencies and timely positive action for their correction. The contractor shall make objective evidence of accuracy conformance readily available to the Government representative.

4.2 Quality assurance provisions. All operations performed by the contractor in compliance with this standard will be subject to the Government verification at unscheduled intervals. Verification will include but not to be limited to the following:

a. Surveillance of calibration operation for conformance to the established system.

b. Review of calibration results as necessary to assure accuracy of the system. The contractor's gages, measuring and testing devices shall be made available for use by the Government when required to determine conformance with contract requirements. If conditions warrant, contractor's personnel shall be made available for operation of such devices and for verification of their accuracy and condition.

5. DETAILED STATEMENTS OF REQUIREMENTS

5.1 Calibration system description. The contractor shall provide and maintain a written description of his calibration system covering measuring and test equipment and measurement standards to satisfy each requirement of this standard. The portion dealing with measuring and test equipment shall prescribe calibration intervals and sources and may be maintained on the documents normally used by the contractor to define his inspection operations. The description for calibration of measurement standards shall include a listing of the applicable measurement standards, both reference and transfer, and shall provide nomenclature, identification number, calibration interval and source, and environmental conditions under which the measurement standards will be applied and calibrated. The description of the calibration system and applicable procedures and reports of calibration shall be available to the Government representative.

5.2 Adequacy of standards. Standards established by the contractor for calibrating the measuring and test equipment used in controlling product quality shall have the capabilities for accuracy, stability, range, and resolution required for the intended use.

5.3 Environmental controls. Measuring and test equipment and measurement standards shall be calibrated and utilized in an environment controlled to the extent necessary to assure continued measurements of required accuracy giving due consideration to temperature, humidity, vibration, cleanliness, and other controllable factors affecting precision measurement. When applicable, compensating corrections shall be applied to calibration results obtained in an environment which departs from standard conditions.

5.4 Intervals of calibration. Measuring and test equipment and measurement standards shall be calibrated at periodic intervals established on the basis of stability, purpose, and degree of usage. Intervals shall be shortened as required to assure continued accuracy as evidenced by the results of preceding calibrations and may be lengthened only when the results of previous calibrations provide definite indications that such action will not adversely affect the accuracy of the system. The contractor shall establish a recall system for the mandatory recall of standards and measuring and test equipment within established time limits or interval frequencies.

5.5 Calibration procedures. Written procedures shall be prepared or provided and utilized for calibration of all measuring and test equipment and measurement standards used to assure the accuracy of measurements involved in establishing product conformance. The procedures may be a compilation of published standard practices or manufacturer's written instructions and need not be rewritten to satisfy the requirements of this standard. As a minimum, the procedures shall specify the accuracy of the instrument being calibrated and the accuracy of the standards used. The procedure shall require that calibration be performed by comparison with higher accuracy level standards.

5.6 Out of tolerance evaluators.

5.6.1 Evaluation of suspect product. The contractor shall establish a procedure for the analysis of the impact of out of tolerance measuring and test equipment on product quality. The impact on quality of products examined or tested by equipment found to be out of tolerance during calibration will be determined and appropriate corrective action taken to correct product quality. Records of the result of the analysis and the corrective action taken to maintain the required quality of the product shall be maintained and be available for the Government representative.

5.6.2 Evaluation of calibration system accuracy. The contractor shall establish a procedure to evaluate the adequacy of the calibration system base on out of tolerance data generated from calibrating test and measurement equipment. The procedure shall include but not be limited to adjustment of calibration frequency, adequacy of the measuring or test instrument, calibration procedures and measuring or test procedures. The procedures shall specifically provide for the identification and prevention of use of any equipment which does not perform satisfactorily.

5.7 Calibration sources.

5.7.1 Domestic contracts. Measuring and test equipment shall be calibrated by the contractor or a commercial facility utilizing standards whose calibration is certified as being traceable to the National Standards, has been derived from accepted values of natural physical constants, or has been derived by the ratio type of self-calibration techniques. Standards requiring calibration by a higher level standards laboratory shall be calibrated by a commercial facility capable of providing the required service, a Government laboratory under arrangements made by the contracting officer, or by the National Bureau of Standards. All standards used in the calibration system shall be supported by certificates, reports, or data sheets attesting to the date, accuracy, and environmental or other conditions under which the results furnished were obtained. Statements of certification shall contain as a minimum, the requirements prescribed in paragraph 5.8. All subordinate standards and measuring and test equipment shall be supported by like data when such information is essential to achieving the accuracy control required by this standard. In those cases where no data is required, a suitably annotated calibration level on the item shall be sufficient to satisfy the support data requirements of this paragraph. Certificates or reports from other than the National Bureau of Standards or Government laboratory shall attest to the fact that the standards used in obtaining the results have been compared at planned intervals with the National Standard either directly or through a controlled system utilizing the methods outlined above. The contractor shall be responsible for assuring that the sources providing calibration services, other than the National Bureau of Standards or a Government laboratory, are in fact capable of performing the required service to the satisfaction of this standard. All certificates and reports shall be available for inspection by authorized Government representatives.

5.7.2 Foreign contracts. The provisions in paragraph 5.7.1 shall apply with the exception that the National Standards Laboratories of countries whose standards are compared with International or U.S. National Standards may be utilized.

5.8 Application and records. The application of the above requirements will be supported by records designed to assure that established schedules and procedures are followed to maintain the accuracy of all measuring and test equipment, and supporting standards. The records shall include an individual record of calibration or other means of control for each item of measuring and test equipment and measurement standards, providing description or identification of the item, calibration interval date of last calibration, and calibration results of out of tolerance conditions. In addition, the individual record of any item whose accuracy must be reported via a calibration report or cer-tificate will quote the report or certificate number for ready reference. These records shall be available for review by authorized Government personnel.

5.9 Calibration status. Measuring and test equipment and standards shall be labeled or some other suitable means shall be established for monitoring the equipment to assure adherence to calibration schedules. The system shall indicate date of last calibration, by whom calibrated and when the next calibration is due. The system may be automated or manual. Items which are not calibrated to their full capability or which require a functional check only shall be labeled to indicate the applicable condition.

5.10 Control of subcontractor calibration. The contractor is responsible for assuring that the subcontractor's calibration system conforms to this standard to the degree necessary to assure compliance with contractual requirements.

5.11 Storage and handling. All measuring and test equipment shall be handled, stored, and transported in a manner which shall not adversely affect the calibration or condition of the equipment.

5.12 Amendments and revisions. Whenever this standard is amended or revised subsequent to a contractually effective date, the contractor may follow or authorize his subcontractor to follow the amended or revised military standard provided no increase in price or fee is involved. The contractor shall not be required to follow the amended or revised standard except as a change in the contract. If the contractor elects to follow the amended or revised military standard, he shall notify the contracting officer in writing of this election.

6. MISCELLANEOUS

6.1 Contract data requirements. The following Data Item Description shall be utilized when the contract cites this standard and requires the contractor to develop and deliver data. The approved Data Item Descriptions (DD Form 1664) required in connection with this standard and listed on the DD Form 1423 are as follows:

Data requirements:	Applicable DID	This standard reference para.
Calibration System Description	DI-R-7064	5.1
Equipment Calibration Procedures	DI-R-7065	5.5

Procedures, Array Calibration	UDI-T-23934	5.5
Calibration–Maintenance Test Data	UDI-T-20340A	5.1 and 5.6
Reports; Test Procedures and Results; Calibration of Test Coupons for Propulsion Shafting	UDI-T-23801	5.1

(Copies of Data Item Description required by the contractors in connection with specific procurement functions should be obtained from the procuring activity or as directed by the contracting officer.)

CUSTODIANS:
ARMY-MI
NAVY-OS
AIR FORCE-05

PREPARING ACTIVITY:
ARMY-MI

REVIEW ACTIVITIES:
ARMY-AR, AV
NAVY-AS, EC, SH
DLA-DH

AGENT:
Army-USAMCC

USER ACTIVITIES:
ARMY-ME
DLA-SS

PROJECT NO. QCIC-0003

APPENDIX

16

Table of d_2^* Values

g = # parts x # inspectors	m = repeat readings taken: 2	3	4	5	6	7	8
1	1.41	1.91	2.24	2.48	2.67	2.83	2.96
2	1.28	1.81	2.15	2.40	2.60	2.77	2.91
3	1.23	1.77	2.12	2.38	2.58	2.75	2.89
4	1.21	1.75	2.11	2.37	2.57	2.74	2.88
5	1.19	1.74	2.10	2.36	2.56	2.73	2.87
6	1.18	1.73	2.09	2.35	2.56	2.73	2.87
7	1.17	1.73	2.09	2.35	2.55	2.72	2.87
8	1.17	1.72	2.08	2.35	2.55	2.72	2.87
9	1.16	1.72	2.08	2.34	2.55	5.72	2.86
10	1.16	1.72	2.08	2.34	2.55	2.72	2.86
11	1.16	1.71	2.08	2.34	2.55	2.72	2.86
12	1.15	1.71	2.07	2.34	2.55	2.72	2.85
13	1.15	1.71	2.07	2.34	2.55	2.71	2.85
14	1.15	1.71	2.07	2.34	2.54	2.71	2.85
15	1.15	1.71	2.07	2.34	2.54	2.71	2.85
> 15	1.128	1.693	2.059	2.326	2.534	2.704	2.847

Continued on next page . . .

Continued . . .

		m = repeat readings taken						
		9	10	11	12	13	14	15
g = # parts x # inspectors	1	3.08	3.18	3.27	3.35	3.42	3.49	3.55
	2	3.02	3.13	3.22	3.30	3.38	3.45	3.51
	3	3.01	3.11	3.21	3.29	3.37	3.43	3.50
	4	3.00	3.10	3.20	3.28	3.36	3.43	3.49
	5	2.99	3.10	3.19	3.28	3.35	3.42	3.49
	6	2.99	3.10	3.19	3.27	3.35	3.42	3.49
	7	2.99	3.10	3.19	3.27	3.35	3.42	3.48
	8	2.98	3.09	3.19	3.27	3.35	3.42	3.48
	9	2.98	3.09	3.18	3.27	3.35	3.42	3.48
	10	2.98	3.09	3.18	3.27	3.34	3.42	3.48
	11	2.98	3.09	3.18	3.27	3.34	3.41	3.48
	12	2.98	3.09	3.18	3.27	3.34	3.41	3.48
	13	2.98	3.09	3.18	3.27	3.34	3.41	3.48
	14	2.98	3.08	3.18	3.27	3.34	3.41	3.48
	15	2.98	3.08	3.18	3.26	3.34	3.41	3.48
	> 15	2.970	3.078	3.173	3.258	3.336	3.407	3.472

APPENDIX

17

Power Functions for ANOVA

Table 17.1. v_1=1.

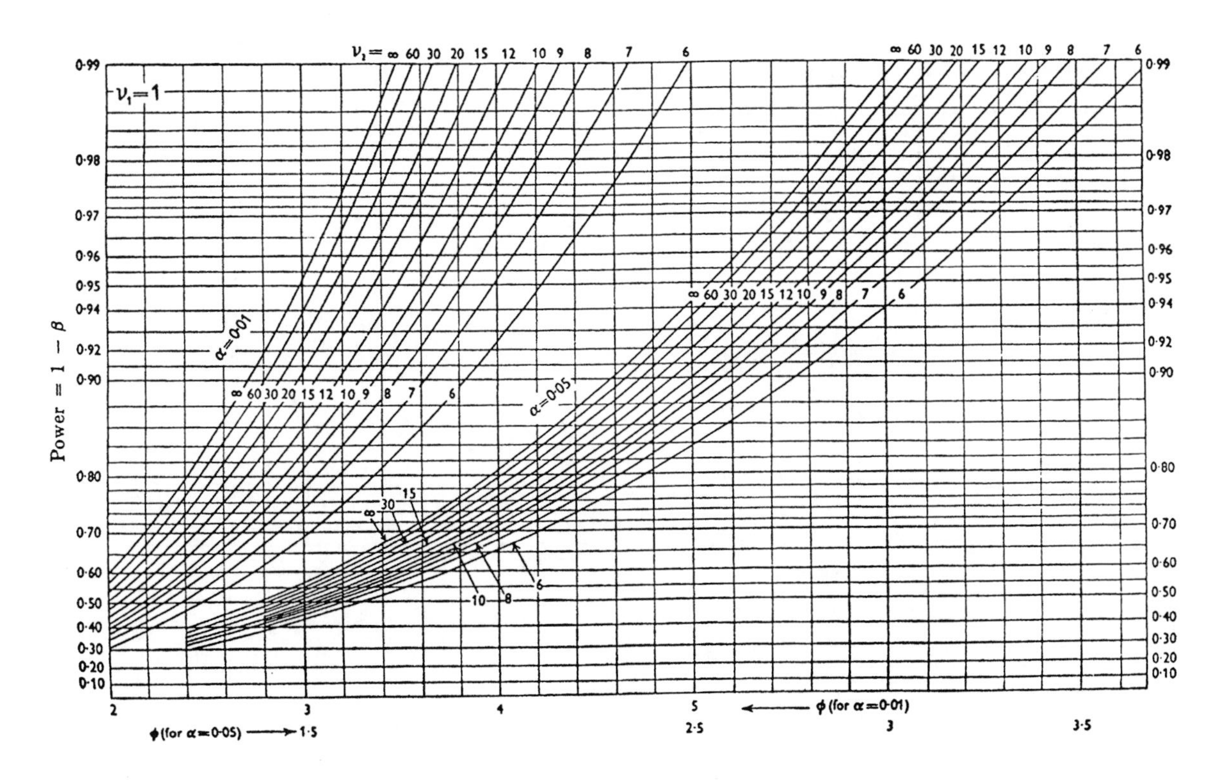

Table 17.2. $v_1=2$.

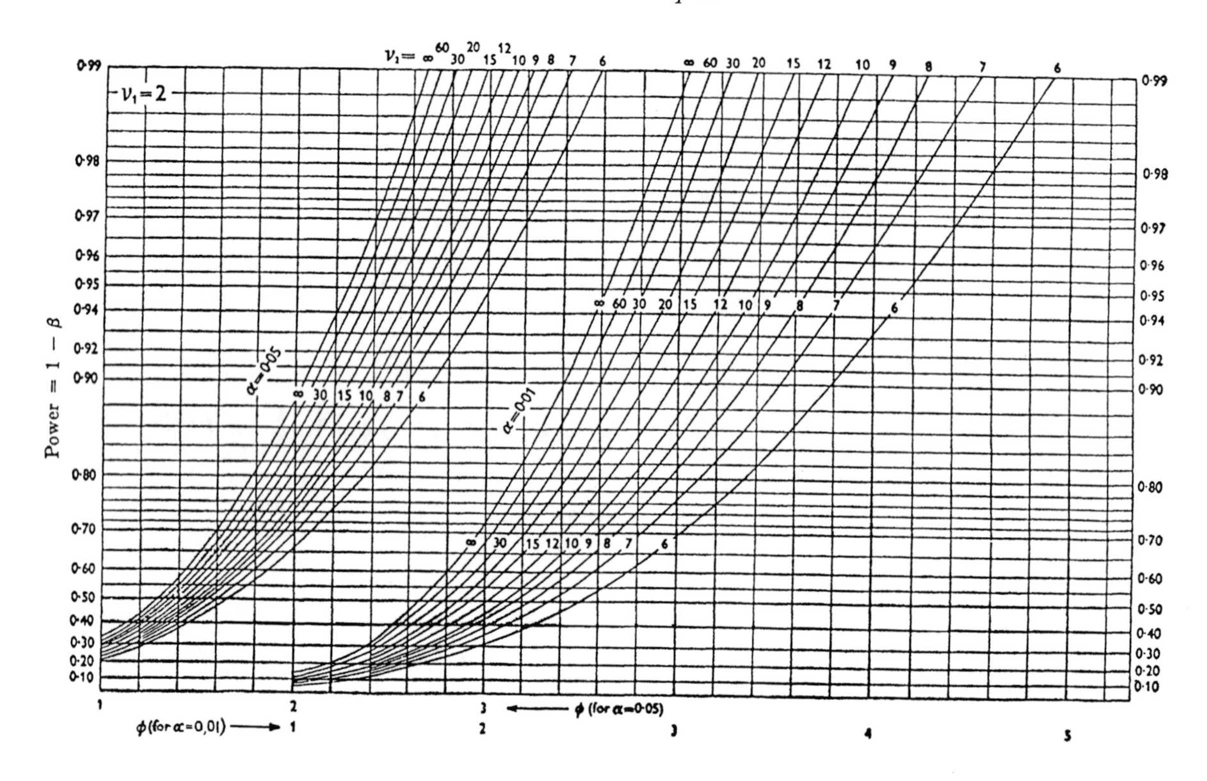

Table 17.3. $v_1=3$.

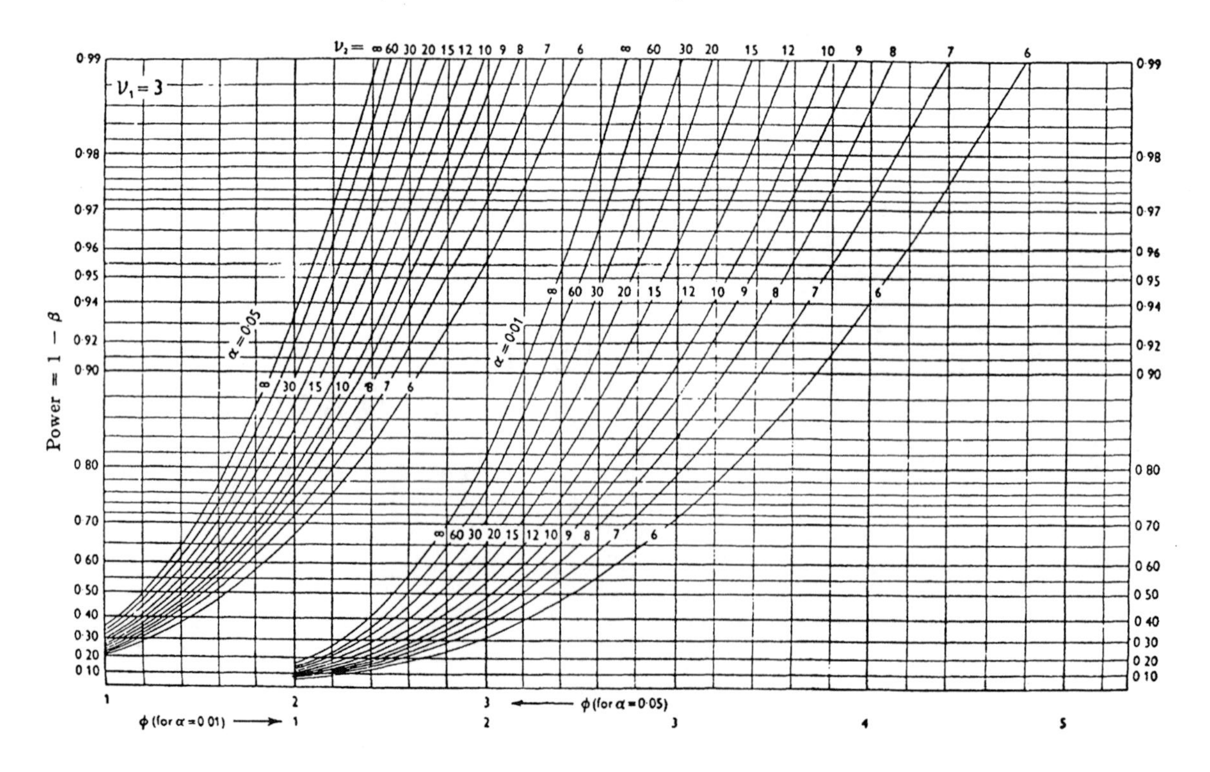

Table 17.4. v_1=4.

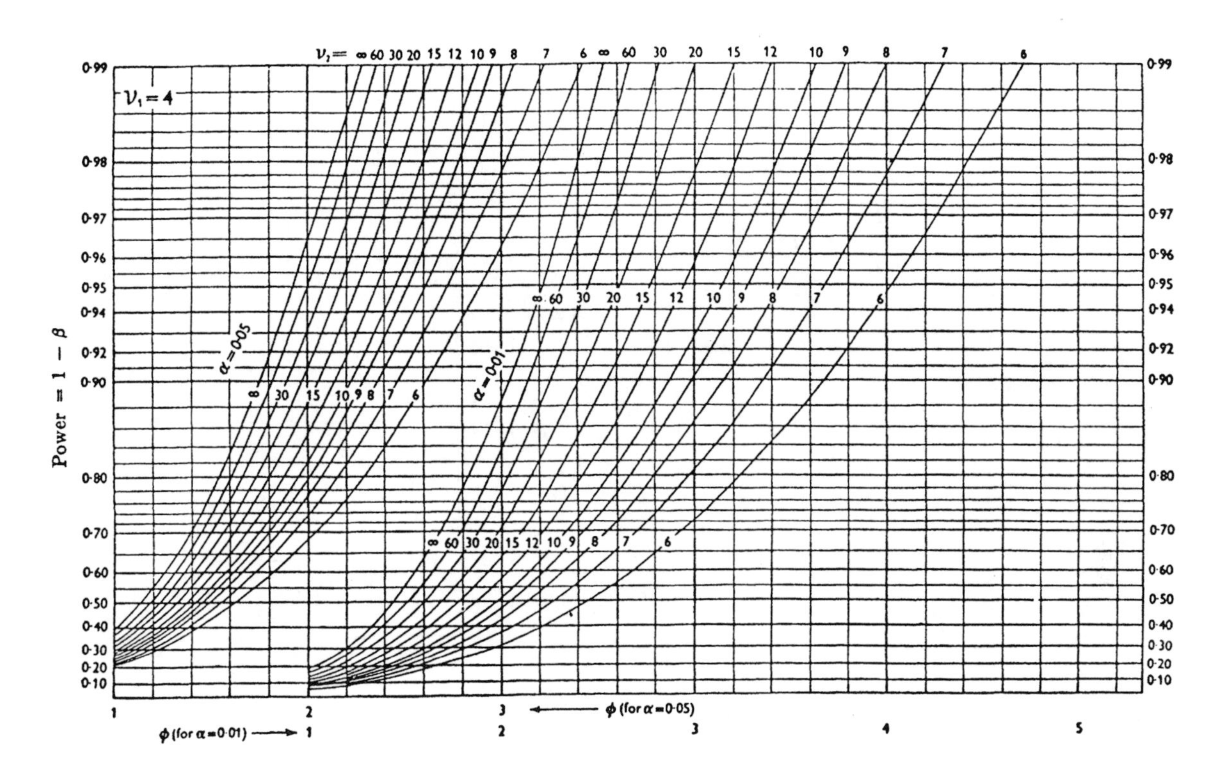

Table 17.5. v_1=5.

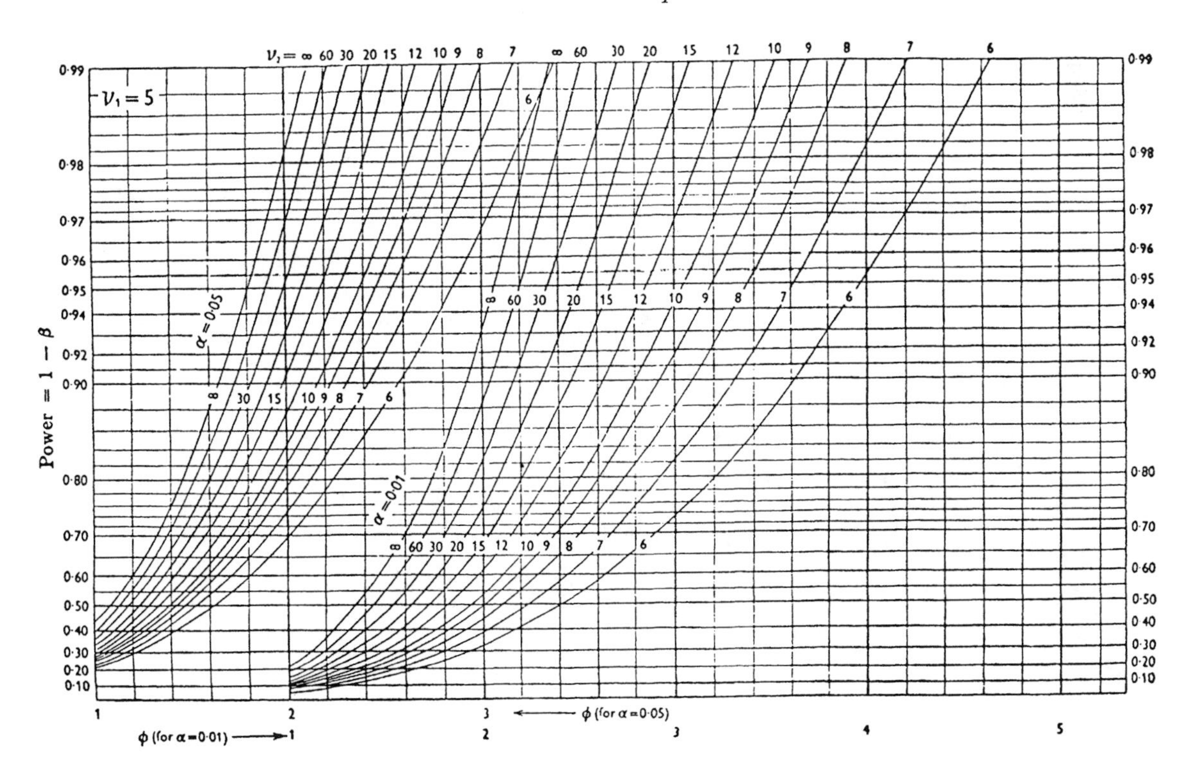

Table 17.6. v_1=6.

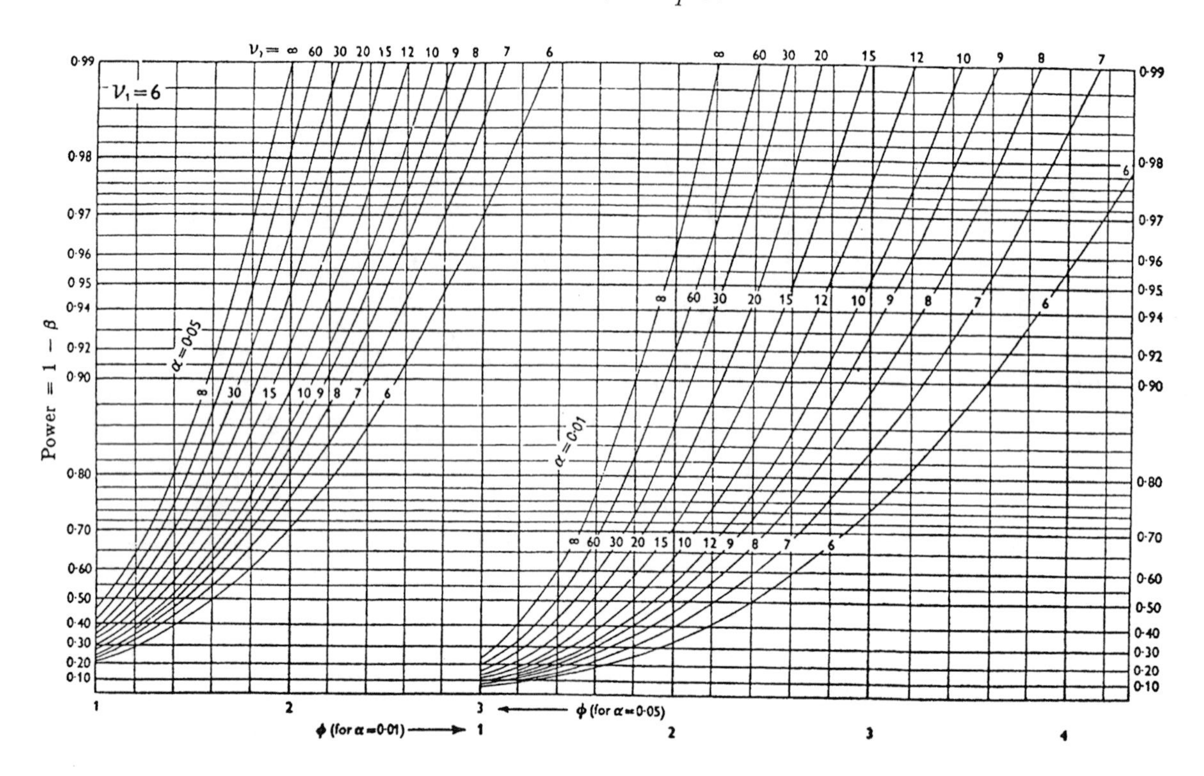

Table 17.7. v_1=7.

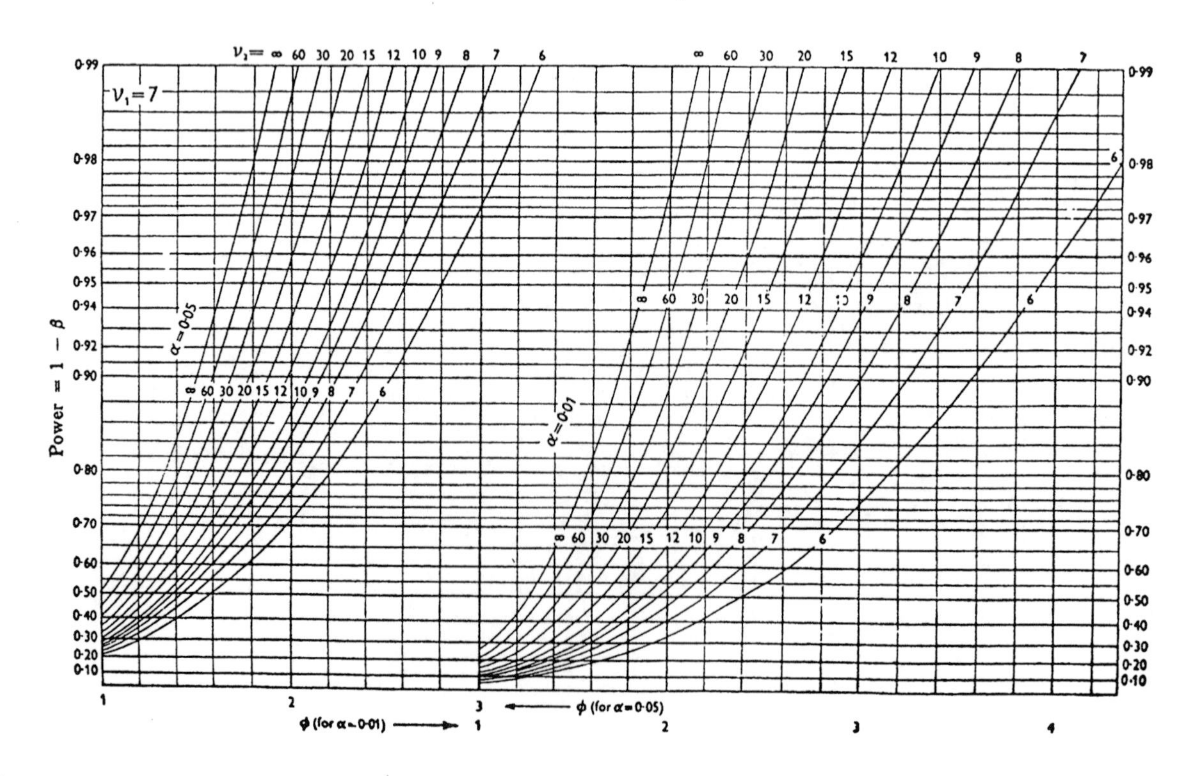

Table 17.8. v_1=8.

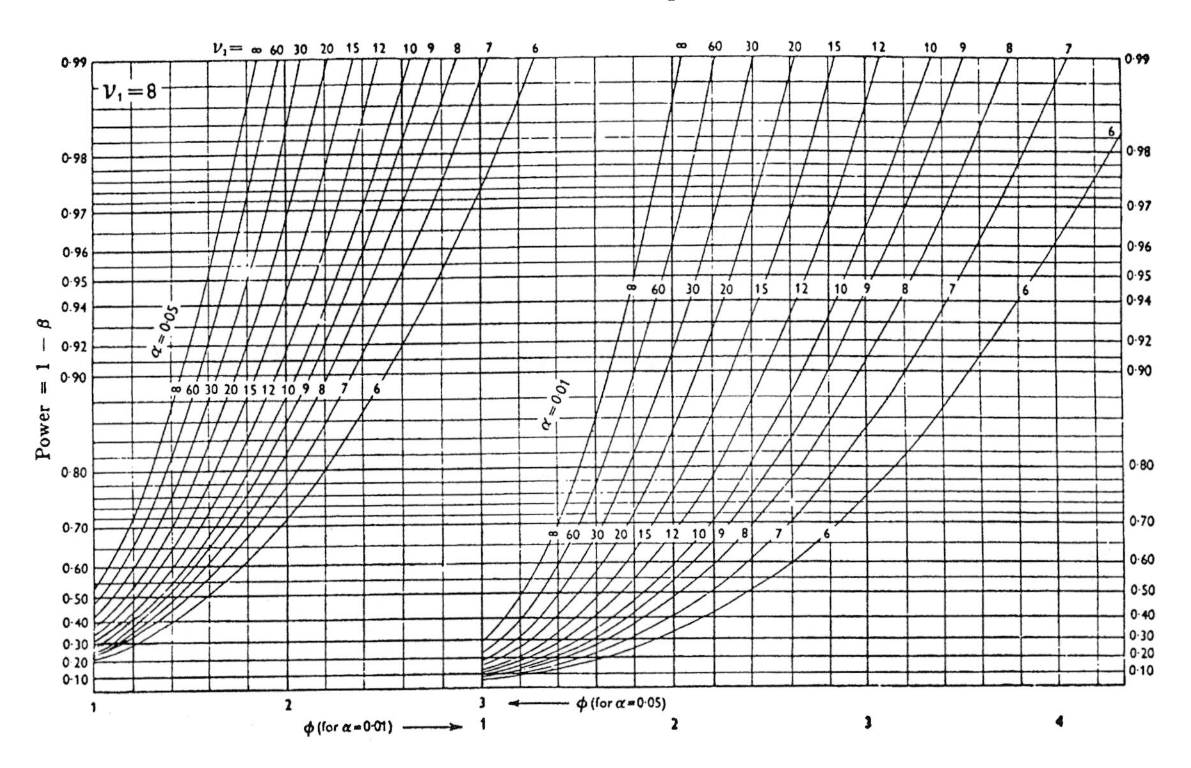

APPENDIX

18

Factors for Short Run Control Charts

g	SUBGROUP SIZE 1 (R based on moving range of 2)				2				3			
	A_{2F}	D_{4F}	A_{2S}	D_{4S}	A_{2F}	D_{4F}	A_{2S}	D_{4S}	A_{2F}	D_{4F}	A_{2S}	D_{4S}
1	NA	NA	236.5	128	NA	NA	167	128	NA	NA	8.21	14
2	12.0	2.0	20.8	16.0	8.49	2.0	15.70	15.6	1.57	1.9	2.72	7.1
3	6.8	2.7	9.6	15.0	4.78	2.7	6.76	14.7	1.35	2.3	1.90	4.5
4	5.1	3.3	6.6	8.1	3.62	3.3	4.68	8.1	1.26	2.4	1.62	3.7
5	4.4	3.3	5.4	6.3	3.12	3.3	3.82	6.3	1.20	2.4	1.47	3.4
6	4.0	3.3	4.7	5.4	2.83	3.3	3.34	5.4	1.17	2.5	1.39	3.3
7	3.7	3.3	4.3	5.0	2.65	3.3	3.06	5.0	1.14	2.5	1.32	3.2
8	3.6	3.3	4.1	4.7	2.53	3.3	2.87	4.7	1.13	2.5	1.28	3.1
9	3.5	3.3	3.9	4.5	2.45	3.3	2.74	4.5	**1.12**	**2.5**	1.25	3.0
10	3.3	3.3	3.7	4.5	2.37	3.3	2.62	4.5	1.10	2.5	1.22	3.0
15	3.1	3.5	3.3	4.1	**2.18**	**3.5**	2.33	4.1	1.08	2.5	1.15	2.9
20	3.0	3.5	3.1	4.0	2.11	3.5	2.21	4.0	1.07	2.6	1.12	2.8
25	**2.9**	**3.5**	3.0	3.8	2.05	3.5	2.14	3.8	1.06	2.6	1.10	2.7

Numbers enclosed in bold boxes represent the recommended minimum number of subgroups for starting a control chart.

Continued on next page . . .

Continued . . .

g	SUBGROUP SIZE 4				SUBGROUP SIZE 5			
	A_{2F}	D_{4F}	A_{2S}	D_{4S}	A_{2F}	D_{4F}	A_{2S}	D_{4S}
1	NA	NA	3.05	13	NA	NA	1.8	5.1
2	0.83	1.9	1.44	3.5	0.58	1.7	1.0	3.2
3	0.81	1.9	1.14	3.2	0.59	1.8	0.83	2.8
4	0.79	2.1	1.01	2.9	0.59	1.9	0.76	2.6
5	0.78	2.1	0.95	2.8	0.59	2.0	0.72	2.5
6	0.77	2.2	0.91	2.7	**0.59**	**2.0**	0.70	2.4
7	**0.76**	**2.2**	0.88	2.6	0.59	2.0	0.68	2.4
8	0.76	2.2	0.86	2.6	0.59	2.0	0.66	2.3
9	0.76	2.2	0.85	2.5	0.59	2.0	0.65	2.3
10	0.75	2.2	0.83	2.5	0.58	2.0	0.65	2.3
15	0.75	2.3	0.80	2.4	0.58	2.1	0.62	2.2
20	0.74	2.3	0.78	2.4	0.58	2.1	0.61	2.2
25	0.74	2.3	0.77	2.4	0.58	2.1	0.60	2.2

Numbers enclosed in bold boxes represent the recommended minimum number of subgroups for starting a control chart.

♦♦♦
APPENDIX

19

Significant Number of Consecutive Highest or Lowest Values from One Stream of a Multiple-Stream Process

On average a run of the length shown would appear no more than 1 time in 100.

# streams, k	2	3	4	5	6	7	8	9	10	11	12
Significant run, r	7	5	5	4	4	4	4	4	3	3	3

APPENDIX

20

Beta Table Values of Reliability

Sample Size	Successes	Confidence Level					
		0.5	**0.75**	**0.90**	**0.95**	**0.99**	**0.995**
1	0	0.293	0.134	0.051	0.025	0.005	0.002
	1	0.707	0.500	0.316	0.224	0.100	0.071
2	0	0.206	0.092	0.034	0.017	0.003	0.002
	1	0.500	0.326	0.196	0.135	0.059	0.041
	2	0.794	0.630	0.464	0.368	0.216	0.171
3	1	0.386	0.243	0.142	0.098	0.042	0.029
	2	0.614	0.456	0.320	0.249	0.141	0.111
	3	0.841	0.707	0.562	0.473	0.316	0.266
4	2	0.500	0.359	0.247	0.189	0.106	0.083
	3	0.686	0.546	0.416	0.343	0.222	0.185
	4	0.871	0.758	0.631	0.549	0.398	0.346
5	3	0.579	0.447	0.794	0.271	0.173	0.144
	4	0.736	0.610	0.333	0.418	0.294	0.254
	5	0.891	0.794	0.681	0.607	0.464	0.414
6	4	0.636	0.514	0.404	0.341	0.236	0.203
	5	0.772	0.659	0.547	0.479	0.357	0.315
	6	0.906	0.820	0.720	0.652	0.518	0.469
7	5	0.679	0.567	0.462	0.400	0.293	0.258
	6	0.799	0.697	0.594	0.529	0.410	0.368
	7	0.917	0.841	0.750	0.688	0.562	0.516
8	6	0.714	0.609	0.510	0.450	0.344	0.307
	7	0.820	0.728	0.632	0.571	0.456	0.415
	8	0.926	0.857	0.774	0.717	0.600	0.555

Continued on next page . . .

Continued . . .

Sample Size	Successes	Confidence Level 0.5	0.75	0.90	0.95	0.99	0.995
9	7	0.742	0.645	0.550	0.493	0.388	0.352
	8	0.838	0.753	0.663	0.606	0.496	0.456
	9	0.933	0.871	0.794	0.741	0.631	0.589
10	7	0.676	0.580	0.489	0.436	0.340	0.307
	8	0.764	0.674	0.585	0.530	0.428	0.392
	9	0.852	0.773	0.690	0.636	0.530	0.492
12	9	0.725	0.638	0.556	0.505	0.412	0.379
	10	0.800	0.720	0.640	0.590	0.494	0.459
	11	0.874	0.806	0.732	0.684	0.587	0.551
	12	0.948	0.899	0.838	0.794	0.702	0.665
15	12	0.775	0.702	0.629	0.583	0.497	0.466
	13	0.836	0.769	0.700	0.656	0.570	0.537
	14	0.897	0.840	0.778	0.736	0.651	0.619
	15	0.958	0.917	0.866	0.829	0.750	0.718
20	17	0.828	0.769	0.709	0.671	0.596	0.568
	18	0.875	0.822	0.766	0.729	0.656	0.628
	19	0.921	0.877	0.827	0.793	0.723	0.696
	20	0.968	0.936	0.896	0.867	0.803	0.777
25	22	0.861	0.811	0.761	0.728	0.663	0.638
	23	0.898	0.855	0.808	0.777	0.714	0.690
	24	0.936	0.900	0.858	0.830	0.771	0.747
	25	0.974	0.948	0.915	0.891	0.838	0.816
30	27	0.883	0.841	0.797	0.768	0.711	0.689
	28	0.915	0.877	0.837	0.810	0.755	0.734
	29	0.946	0.916	0.880	0.856	0.804	0.784
	30	0.978	0.956	0.928	0.908	0.862	0.843
35	31	0.871	0.831	0.791	0.764	0.710	0.690
	32	0.898	0.862	0.824	0.798	0.747	0.727
	33	0.926	0.894	0.859	0.835	0.786	0.767
	34	0.954	0.927	0.896	0.875	0.829	0.811
	35	0.981	0.962	0.938	0.920	0.880	0.863
40	36	0.887	0.851	0.814	0.790	0.742	0.724
	37	0.911	0.879	0.844	0.822	0.775	0.757
	38	0.935	0.907	0.875	0.854	0.810	0.793
	39	0.959	0.936	0.908	0.889	0.849	0.832
	40	0.983	0.967	0.945	0.930	0.894	0.879

Continued on next page . . .

Continued . . .

Sample Size	Successes	Confidence Level					
		0.5	**0.75**	**0.90**	**0.95**	**0.99**	**0.995**
45	41	0.899	0.867	0.834	0.812	0.768	0.751
	42	0.921	0.892	0.860	0.840	0.798	0.781
	43	0.942	0.917	0.888	0.869	0.830	0.814
	44	0.964	0.942	0.918	0.901	0.864	0.849
	45	0.985	0.970	0.951	0.937	0.905	0.891
50	46	0.909	0.880	0.849	0.829	0.789	0.773
	47	0.928	0.902	0.874	0.855	0.816	0.801
	48	0.948	0.925	0.899	0.882	0.845	0.831
	49	0.967	0.948	0.926	0.910	0.877	0.863
	50	0.985	0.973	0.956	0.943	0.914	0.901
75	71	0.939	0.919	0.898	0.884	0.855	0.843
	72	0.952	0.934	0.914	0.901	0.874	0.863
	73	0.965	0.949	0.932	0.920	0.894	0.884
	74	0.978	0.965	0.950	0.939	0.916	0.906
	75	0.991	0.982	0.970	0.961	0.941	0.933
100	95	0.944	0.928	0.910	0.899	0.875	0.866
	96	0.954	0.939	0.922	0.912	0.889	0.880
	97	0.964	0.950	0.935	0.925	0.904	0.896
	98	0.974	0.962	0.948	0.939	0.919	0.911
	99	0.983	0.974	0.962	0.954	0.936	0.929
	100	0.993	0.986	0.978	0.971	0.955	0.949
150	145	0.962	0.951	0.939	0.932	0.916	0.909
	146	0.969	0.959	0.948	0.940	0.925	0.919
	147	0.976	0.966	0.956	0.950	0.935	0.929
	148	0.982	0.974	0.965	0.959	0.946	0.940
	149	0.989	0.982	0.974	0.969	0.957	0.952
	150	0.996	0.991	0.985	0.980	0.970	0.966
200	195	0.972	0.963	0.954	0.948	0.963	0.931
	196	0.977	0.969	0.961	0.955	0.943	0.939
	197	0.982	0.975	0.967	0.962	0.951	0.946
	198	0.987	0.981	0.974	0.969	0.959	0.955
	199	.0992	0.987	0.981	0.977	0.967	0.964
	200	0.996	0.993	0.989	0.985	0.977	0.974

♦ ♦ ♦

References

Akao, Y., ed. (1990). *Quality Function Deployment: Integrating Customer Requirements into Product Design*, Cambridge, MA: Productivity Press.

American National Standards Institute, *ANSI Y14.5M–1994: Dimensioning and Tolerancing.*

American Society for Nondestructive Testing, Columbus, OH, 1–800–222–2768, www.asnt.org.

Ansoff, H.I., (1965). *Corporate Strategy*, New York: John Wiley & Sons.

ASQ Product Safety and Liability Prevention Interest Group, (1999). *Product Recall Planning Guide, 2nd ed.*, Milwaukee, WI: ASQ Quality Press.

ASQ Statistics Division, (1996). *Glossary and Tables for Statistical Quality Control, 3rd ed.*, Milwaukee, WI: ASQ Quality Press.

Aubrey, C.A. and Felkins, P.K., (1988). *Teamwork: Involving People in Quality and Productivity Improvement*, Milwaukee, WI: ASQ Quality Press.

Automotive Industry Action Group, (2002). *Measurement Systems Analysis, 3rd ed.*, Detroit, MI: AIAG.

Baker, R., (1945). "Mensuration" in *Practical Mathematics: Theory and Practice with Applications to Industrial, Business and Military Problems*, R. S. Kimball, ed., New York: National Educational Alliance.

Baysinger, S.M., (2000). *The Complete Guide to the CQA*, Tucson, AZ: QA Publishing, LLC.

Blueprint Reading for Industry, Figure 3-6c. Copyright © 1979 by Goodheart-Willcox Co., Inc.

Brainard, E.H., (1966). "Just how good are vendor surveys?" *Quality Assurance*, August, pp. 22–26.

Brassard, M., (1996). *The Memory Jogger Plus + Featuring the Seven Management and Planning Tools*, Methuen, MA: GOAL/QPC.

Burr, I.W., (1976). *Statistical Quality Control Methods, Statistics: Textbooks and Monographs*, Vol. 16, New York: Marcel-Dekker, Inc.

Burt, D.N. and Doyle, M.F., (1993). *The American Keiretsu: A Strategic Weapon for Global Competitiveness*, Homewood, IL: Times Mirror Books, Business One Irwin.

Buzzell, R.D. and Gale, B.T., (1987). *The PIMS Principles: Linking Strategy to Performance*, New York: The Free Press.

Camp, R.C., (1989). *Benchmarking: The Search for Industry Best Practices That Lead to Superior Performance*, Milwaukee, WI: ASQ Quality Press and White Plains, NY: Quality Resources.

Campanella, J., ed., (1999). *Principles of Quality Costs: Principles, Implementation, and Use, 3rd ed.*, Milwaukee, WI: ASQ Quality Press.

Carder, B. and Clark, J.D., (1992). "The Theory and Practice of Employee Recognition," *Quality Progress*, December.

Chernick, Michael R., (1999). *Bootstrap Methods: A Practitioner's Guide*, New York: John Wiley and Sons.

Continuing Process Control and Process Capability Improvement, p. 4a. Copyright 1983. Ford Motor Company, Dearborn, Michigan.

Cowen, T., and J. Ellig., (1993). "Koch Industries and market-based management," Fairfax, VA: Center for the Study of Market Processes, George Mason University.

Crosby, P.B., (1979). Quality is Free, New York: New American Library.

Daltas, A.J., (1977). "Protecting Service Markets with Consumer Feedback," *Cornell Hotel and Restaurant Administration Quarterly*, May, pp. 73–77.

Deming, W. Edwards, (1975). "On Probability as a Basis for Action," *The American Statistician*, 29(4), pp. 146–152.

Deming, W. Edwards, (1986). *Out of the Crisis*, Cambridge, MA: MIT Press.

Deming, W. Edwards, (2000). *The New Economics for Industry, Government, Education, 2nd ed.*, Cambridge, MA: MIT Press.

DeToro, I., (1995). "The 10 Pitfalls of Benchmarking," *Quality Progress*, January, pp. 61–63.

Dillman, D.A., (1983). "Mail and Other Self-Administered Questionnaires," in *Handbook of Survey Research*, Rossi, P., Wright, J., and Anderson, A., Editors, New York: Academic Press, Inc., pp. 359–377.

Dodson, Bryan and Nolan, Dennis, (2002). *Reliability Engineering Handbook*, Tucson, Arizona: QA Publishing, LLC.

Donovan, R.M., (1995). "Improving manufacturing cycle time," *Quality Digest*, June 1995, pp. 51–53.

Draper, N. and Smith, H., (1981). *Applied Regression Analysis, 2nd ed.*, New York: John Wiley & Sons.

Drucker, P.F., (1993). *Management: Tasks, Responsibilities, Practices*, New York: Harper and Row.

Duncan, A.J., (1986). *Quality Control and Industrial Statistics, 5th ed.*, New York: McGraw-Hill.

Edosomwan, J.A., (1993). *Customer and Market-Driven Quality Management*, Milwaukee, WI: ASQ Quality Press.

Edwards, C.D., (1968). "The meaning of quality," *Quality Progress*, October.

Efron, B., (1982). *The Jackknife, the Bootstrap, and Other Resampling Plans*, Philadelphia, PA: Society for Industrial and Applied Mathematics.

Feigenbaum, A.V., (1956). "Total quality control," *Harvard Business Review*, November–December 1956.

Feigenbaum, A.V., (1961, 1991). *Total Quality Control, 4th ed.*, New York: McGraw-Hill.

Fivars, G., (1980). "The Critical Incident Technique: a Bibliography," Research and Publication Service, American Institutes for Research ED 195 681.

Flanagan, J.C., (1954). "The Critical Incident Technique," *Psychological Bulletin*, 51(4), July, pp. 327–358.

Forsha, H.I., (1992). *The Pursuit of Quality Through Personal Change*, Milwaukee, WI: ASQ Quality Press.

GAO (1986). "Developing and Using Questionnaires—Transfer Paper 7," Washington, D.C.: United States General Accounting Office.

Garvin, D.A., (1988). *Managing Quality*, New York: Free Press.

George, L., (1994). "The bathtub curve doesn't always hold water," *Reliability Review*, 14(3), September, 5–7.

Gilmore, H.L., (1974). "Product conformance cost," *Quality Progress*, June.

Good, Phillip I., (2000). *Permutation Tests: A Practical Guide to Resampling Methods for Testing Hypothese, 2nd ed.*, New York: Springer-Verlag.

Harlow, R.H., Jr., Dotson, C., and Thompson, R., *Fundamentals of Dimensional Metrology, 4th ed.*, Albany, NY: Delmar Publishers, Inc.

Harrington, H.J., (1992). "Probability and Statistics," in *Quality Engineering Handbook*, Pyzdek, T. and R.W. Berger, Editors, Milwaukee, WI: ASQ Quality Press, pp. 513–577.

Hayes, Bob E., (1997). *Measuring Customer Satisfaction: Survey Design, Use, and Statistical Analysis Methods, 2nd ed.*, Milwaukee, WI: ASQ Quality Press.

Hillier, F.S., (1969). "X-bar and R-chart Control Limits Based on a Small Number of Subgroups," *Journal of Quality Technology*, Vol. 1, No. 1, January, pp. 17–26.

Hillier, F.S. and Lieberman, G.J., (2002). *Introduction to Operations Research, 7th ed.*, New York: McGraw Hill.

Hoaglin, D.C., Mosteller, F. and Tukey, J.W., (2000). *Understanding Robust and Exploratory Data Analysis*, New York: John Wiley and Sons.

Hutton, D.W., (1994). *The Change Agent's Handbook: A Survival Guide for Quality Improvement Champions*, Milwaukee, WI: ASQ Quality Press.

Imai, M., (1986). *Kaizen: The Key to Japan's Competitive Success*, New York: McGraw Hill.

Ireson, W.G., Coombs, C.F., Jr., (1995). *Handbook of Reliability Engineering and Management, 2nd ed.*, New York: McGraw-Hill.

Ishikawa, K., (1988). *What is Total Quality Control the Japanese Way?* Englewood Cliffs, NJ: Prentice-Hall, Inc.

Johnson, D.W. and Johnson, F.P., (1999). *Joining Together: Group Theory and Group Skills*, Englewood Cliffs, NJ: Prentice-Hall.

Johnson, R.S., (1993). *TQM: Management Processes for Quality Operations*, Milwaukee, WI: ASQ Quality Press.

Johnson, R.S., (1993a). *TQM: Leadership for the Quality Transformation*, Milwaukee, WI: ASQ Quality Press.

Johnson, R.S., (1993b). *TQM: The Mechanics of Quality Processes*, Milwaukee, WI: ASQ Quality Press.

Joiner, B.L., (1994). *Fourth Generation Management: The New Business Consciousness*, New York: McGraw-Hill, Inc.

Juran, J.M., (1994). "The upcoming century of quality," address delivered at the ASQ Annual Quality Congress, May 24, 1994.

Juran, J.M. and F.M. Gryna, eds., (1951). *Quality Control Handbook*, New York: McGraw-Hill.

Juran, J.M. and F.M. Gryna, eds., (1974). *Quality Control Handbook, 3rd ed.*, New York: McGraw-Hill.

Juran, J.M. and F.M. Gryna, eds., (1988). *Quality Control Handbook, 4th ed.*, New York: McGraw-Hill.

Juran, J.M. and F.M. Gryna, eds., (1999). *Quality Control Handbook, 5th ed.*, New York: McGraw-Hill.

Juran, J.M. and Gryna, F.M., (1993). *Quality Planning and Analysis, 3rd ed.*, New York: McGraw-Hill.

Kano, N., (1993). "A Perspective on Quality Activities in American Firms," *California Management Review*, 35(3), Spring, pp. 12–31.

Keeney, K.A., (1995). *The Audit Kit*, Milwaukee, WI: ASQ Quality Press.

Kimball, Reginald Stevens, ed., "An Introduction to Algebra," in *Practical Mathematics: Theory and Practice with Applications to Industrial, Business and Military Problems*, New York: National Educational Alliance.

Kirkpatrick, D.L. (1996). "Evaluation," in *The ASTD Training and Development Handbook: A Guide to Human Resources Development*, Craig, R.L., Editor-in-chief, New York: McGraw-Hill.

Kohn, A., (1999). *Punished by Rewards: The Trouble with Gold Stars, Incentive Plans, A's, Praise and Other Bribes*, New York: Houghton Mifflin Company.

Kotler, P., (2002). *Marketing Management: Analysis, Planning, Implementation, and Control, 11th ed.*, Englewood Cliffs, NJ: Prentice-Hall.

Kuehn, A. and Day, R., (1962). "Strategy of product quality," *Harvard Business Review*, November–December.

Leffler, K.B., (1982). "Ambiguous changes in product quality," *American Economic Review*, December.

Lewis, C.I., (1929), *Mind and the World Order*, New York: Scribners.

Lofgren, G.Q., (1991). "Quality System Registration," *Quality Progress*, May 1991.

Marquardt, Donald, et al, (1991). "Vision 2000: The Strategy for the ISO 9000 Series Standards in the 1990s", *Quality Progress,* May 1991, pp. 25-31.

Melan, E.H., (1993). *Process Management: Methods for Improving Products and Service,* Milwaukee, WI: ASQ Quality Press.

Minitab Statistical Software. Copyright 2000 by Minitab, Inc., State College, PA.

Mintzberg, H., (1994). *The Rise and Fall of Strategic Planning,* New York: The Free Press.

Mizuno, S., Ed., (1988). *Management for Quality Improvement: The Seven New QC Tools,* Cambridge, MA: Productivity Press.

Montgomery, D.C., (2000). *Design and Analysis of Experiments, 5th ed.*, New York: John Wiley & Sons.

Montgomery, D.C, Peck, E.A., and Vining, G.G., (2001). *Introduction to Linear Regression Analysis, 3rd ed.*, New York: John Wiley and Sons.

Montgomery, D.C., Runger, G.C., and Nairn, A.G., (2002) *Applied Statistics and Probability for Engineers,* New York: John Wiley and Sons.

Morrow, M., (1995). "Defense Department Dumps Mil-Q-9858A," *Quality Digest,* 15(5), May, p. 21.

Moss, M.A., (1985). *Designing for Minimal Maintenance Expense,* New York: John Wiley and Sons.

Moubray, J., (2001). *Reliability-Centered Maintenance,* New York: Industrial Press, Inc.

Natrella, M.G., (1963). *Experimental Statistics: NBS Handbook 91,* Washington, DC: US Government Printing Office.

Nelson. "The Shewhart Control Chart—Tests for Special Causes," *Journal of Quality Technology,* 164, p 238. Copyright © 1986 by Nelson.

Neter, J., Kutner, M.H., Nachtschiem, C.J., Wasserman, W., (1996). *Applied Linear Statistical Models, 4th ed.*, New York, NY: McGraw Hill

Persig, R.M., (1974). *Zen and the Art of Motorcycle Maintenance*, New York: Bantam Books.

Peters, T. and Austin, N. (1985). *A Passion for Excellence: The Leadership Difference*, New York: Random House.

Phillips, J.J. (1996). "Measuring the Results of Training," in *The ASTD Training and Development Handbook: A Guide to Human Resources Development*, Craig, R.L. Editor-in-chief, New York: McGraw-Hill, pp. 313–341.

Proschan, F. and Savage, I.R., (1960). "Starting a Control Chart," *Industrial Quality Control*, Vol. 17, No. 3, September, pp. 12–13.

Provost, L.P., (1988). "Interpretation of Results of Analytic Studies," paper presented at 1989 NYU Deming Seminar for Statisticians, March 13, New York: New York University.

Pyzdek, T., (1990). "Modern Approaches to Software Quality Improvement" paper presented at Australian Organization for Quality Qualcon 1990.

Pyzdek, T., (1992). "Process Capability Analysis using Personal Computers," *Quality Engineering*, 4, No. 3, 432-433.

Pyzdek, T., (1992). *Pyzdek's Guide to SPC Volume Two—Applications and Special Topics*, Tucson, AZ: Quality Publishing, Inc.

Pyzdek, T., (1994). *Pocket Guide to Quality Tools*, Tucson, AZ: Quality Publishing, Inc.

Pyzdek, T., (1998). *Pyzdek's Guide to SPC Volume One—Fundamentals, 2nd ed.*, Tucson, AZ: QA Publishing, LLC.

Pyzdek, T., (2003). *The Six Sigma Handbook*, Tucson, AZ: Quality Publishing, Inc.

Radford, G.S., (1922). *The Control of Quality in Manufacturing*, New York: Ronald Press.

Rand, A., (1964). *The Virtue of Selfishness*, New York: Penguin Books.

Reichheld, F.F., (1996). *The Loyalty Effect: The Hidden Force Behind Growth, Profits, and Lasting Value*, New York: McGraw-Hill.

Rose, K.H., (1995). "A Performance Measurement Model," *Quality Progress*, February, pp. 63–66.

Ruskin, A.M. and Estes, W.E., (1995). *What Every Engineer Should Know About Project Management, 2nd ed.*, New York: Marcel-Dekker, p. 193.

Saaty, T.L., (1988). *Decision Making for Leaders: The Analytic Hierarch Process for Decisions in a Complex World*, Pittsburgh, PA: RWS Publications.

Scholtes, P.R., (1996). *The Team Handbook: How to Use Teams to Improve Quality, 2nd ed.*, Madison, WI: Oriel, Inc.

Schuman, S.P., (1996). "What to Look for in a Group Facilitator," *Quality Progress*, June, pp. 69–72.

Shainin, D. and Shainin, P.D., (1988). "Statistical Process Control," in Juran's *Quality Control Handbook, 3rd ed.*, New York: McGraw-Hill, p. 24.34.

Sheridan, B.M., (1993). *Policy Deployment: The TQM Approach to Long-Range Planning*, Milwaukee, WI: ASQ Quality Press.

Shewhart, (1939, 1986). *Statistical Method from the Viewpoint of Quality Control*, New York: Dover Publications.

Shewhart, W.A., (1931, 1980). *Economic Control of Quality of Manufactured Product*, Milwaukee, WI: ASQ Quality Press.

Simon, J.L., (1992). *Resampling: the New Statistics*, Arlington, VA: Resampling Stats, Inc.

SPC-PC IV Help System. Copyright 2002 by Quality America, Inc., Tucson, AZ.

SPSS for Windows Base System User's Guide, Copyright © 1993. SPSS, Inc., Chicago, IL.

Stapp, Eric H. (2001). *ISO 9000:2000: An Essential Guide to the New Standard*, Tucson, AZ: QA Publishing, LLC.

Stewart, T.A., (1995). "After All You've Done for Your Customers, Why are They Still NOT HAPPY?" *Fortune*, December 11, 1995.

Sullivan, C.D., (1983). *Standards and Standardization: Basic Principles and Applications*, New York: Marcel-Dekker, Inc.

Suminski, L.T., Jr., (1994). "Measuring the Cost of Quality," *Quality Digest*, March, pp. 26–32.

Taguchi, G., (1986). *Introduction to Quality Engineering: Designing Quality into Products and Processes*, White Plains, NY: Quality Resources.

Taha, H., (1976). *Operations Research: An Introduction, 2nd ed.*, New York: Prentice Hall.

Taylor, Frederick W., (1947). *Scientific Management*, New York: Harper & Brothers.

Tuckman, B.W., (1965). "Development Sequence in Small Groups," *Psychological Bulletin*.

Tufte, Edward, R., (2001). *The Visual Display of Quantitative Information, 2nd ed.*, Cheshire, CT: Graphics Press.

Tukey, J.W., (1977). *Exploratory Data Analysis*, Reading, MA: Addison-Wesley.

Vaziri, H.K., (1992). "Using Competitive Benchmarking to Set Goals," *Quality Progress*, November.

Wason, S.K. and Bhalla, S.K., (1994). "Managing the technological process," *Quality Progress*, pp. 81–85.

Weaver, C.N., (1995). *Managing the Four Stages of TQM*, Milwaukee, WI: ASQ Quality Press.

Willborn, W., (1993). *Audit Standards: A Comparative Analysis, 2nd ed.*, Milwaukee, WI: ASQ Quality Press.

Wilson, Paul F., Larry D. Dell, and Gaylord F. Anderson, (1993). *Root Cause Analysis: A Tool for Total Quality Management*, Milwaukee, WI: ASQ Quality Press.

Zaremba, A.J., (1989). *Management in a New Key: Communication in the Modern Organization*, Norcross, GA: The Institute of Industrial Engineers.

Zuckerman, A., (1995). "The future of ISO 9000," *Quality Digest*, 15(5), May, pp. 22–28.

♦ ♦ ♦

Index

—F—

—G—

—L—

—M—

—P—

—Q—